The Adventure of the LARGE HADRON COLLIDER

From the Big Bang to the Higgs Boson

Other Related Titles from World Scientific

A Day at CERN: Guided Tour Through the Heart of Particle Physics
by Gautier Depambour
ISBN: 978-981-122-110-1
ISBN: 978-981-122-064-7 (pbk)

ATLAS: A 25-Year Insider Story of the LHC Experiment
by The ATLAS Collaboration
ISBN: 978-981-3271-79-1

Challenges and Goals for Accelerators in the XXI Century
edited by Oliver Brüning and Stephen Myers
ISBN: 978-981-4436-39-7

The High Luminosity Large Hadron Collider: The New Machine for Illuminating the Mysteries of Universe
edited by Oliver Brüning and Lucio Rossi
ISBN: 978-981-4675-46-8
ISBN: 978-981-4678-14-8 (pbk)

The Adventure of the LARGE HADRON COLLIDER

From the Big Bang to the Higgs Boson

Daniel Denegri
CNRS/IN2P3 and CEA/IRFU, Paris-Saclay University, France

Claude Guyot
CEA/IRFU, Paris-Saclay University, France

Andreas Hoecker
CERN, Switzerland

Lydia Roos
CNRS/IN2P3, France

Published by

World Scientific Publishing Co. Pte. Ltd.
5 Toh Tuck Link, Singapore 596224
USA office: 27 Warren Street, Suite 401-402, Hackensack, NJ 07601
UK office: 57 Shelton Street, Covent Garden, London WC2H 9HE
and
EDP Sciences
17, av. du Hoggar F-91944 Les Ulis, France

Library of Congress Cataloging-in-Publication Data
Names: Denegri, Daniel, author. | Guyot, Claude, author. |
Hoecker, Andreas, author. | Roos, Lydia, author.
Title: The adventure of the Large Hadron Collider : from the Big bang to the Higgs boson /
Daniel Denegri, Claude Guyot, Andreas Hoecker, Lydia Roos.
Other titles: Aventure du Grand Collisionneur LHC. English.
Description: Hackensack, NJ : World Scientific, [2021] | Translation of:
L'aventure du Grand Collisionneur LHC : Du big bang au boson de Higgs. |
Includes bibliographical references and index.
Identifiers: LCCN 2020054673 | ISBN 9789813236080 (hardcover)
Subjects: LCSH: Large Hadron Collider (France and Switzerland) | Standard model
(Nuclear physics) | Particles (Nuclear physics) | Nuclear physics--Experiments.
Classification: LCC QC787.P73 D464 2021 | DDC 539.7/36--dc23
LC record available at https://lccn.loc.gov/2020054673

British Library Cataloguing-in-Publication Data
A catalogue record for this book is available from the British Library.

Based on "L'aventure du Grand Collisionneur LHC: Du big bang au boson de Higgs" published in French by EDP Sciences. © EDP Sciences 2014.

For any available supplementary material, please visit
https://www.worldscientific.com/worldscibooks/10.1142/10881#t=suppl

Desk Editor: Ng Kah Fee

Typeset by Stallion Press
Email: enquiries@stallionpress.com

In memory of Professor Aihud Pevsner

Acknowledgements

First of all, we would like to thank our colleagues, the physicists, engineers, technicians, and administrators from all over the world working on the LHC experiments, in particular those in the ATLAS and CMS collaborations who have contributed to the design, construction and operation of the experiments at the origin of most of the results presented in this book. Our thanks and admiration also go to the physicists, engineers and technicians from CERN and its member states, who have designed and built the LHC and its injectors with contributions from the US, Japan, Russia, Canada and India, and who operate this complex machine with remarkable professionalism and efficiency. Without them none of this adventure would have been possible. We acknowledge the key role played by the CERN directorate and CERN council, who have ensured the necessary human and financial means for the construction and operation of this colossal undertaking, and in first place the successive CERN directors general, Carlo Rubbia, who launched the LHC project and had the kindness to write a foreword for this book, Chris Llewellyn-Smith, Luciano Maiani, Robert Aymar, Rolf Heuer who moved the LHC forward and brought it to successful completion. We must also mention Herwig Schopper who led the construction of the large LEP tunnel which now hosts the LHC, and Fabiola Gianotti who took over the helm at CERN in 2016 and who is about to begin an exceptional second term in office. We extend our appreciation and respect to the worldwide public funding agencies who support the experiments and their computing facilities with a long-term vision and by so doing enable scientific, technological, and cultural progress.

We express our gratitude to our colleagues from the LHC collaborations and to CERN's public relations service, which allowed us to use their results, photos and diagrams in this book. We thank the members of the Particle Data Group, the Planck collaboration, the Supernova Cosmology Project group, for a number of diagrams and plots we used to illustrate this book.

We acknowledge the help of colleagues with whom we had discussions on various topics during the realisation of this work. More specifically, we thank our colleagues from IRFU in Saclay, Vanina Ruhlmann-Kleider and Jim Rich for fruitful discussions on cosmology, and Jean Zinn-Justin for illuminating discussions on theoretical matters. We are also most grateful to our colleagues for useful comments and suggestions on the initial edition in French of this book, specifically to Robert Baratte, Peter Jenni, Carlos Lourenco, Daniel Treille, and for laudatory reviews on the book in specialised journals. Special thanks again to Peter Jenni for his very helpful comments on the present, significantly extended English edition. All remaining mistakes in this book are the entire and sole responsibility of the authors.

Finally, we would like to thank the French Physical Society and our editors, Michèle Leduc and Michel Le Bellac, who proposed us to write this book for the first edition in French. They also put us in contact with World Scientific and in particular Ng Kah-Fee, chief editor of this second, updated and considerably extended edition in English. We thank him for having stoically withstood our many delays over the past years without loosing faith in this project.

Foreword

Carlo Rubbia

This book presents what is beyond doubt the largest purely scientific project ever, the LHC — the Large Hadron Collider at CERN — and associated experiments ATLAS, CMS, LHCb and ALICE, culminating with the discovery of the Higgs boson in summer 2012.

The LHC is the largest scientific instrument ever constructed and detectors ATLAS and CMS are of a complexity and sophistication with no precedent in physics. In the eighties, the physicists discovered at CERN the gauge bosons W and Z, thus confirming the unification of electromagnetic and weak interactions as proposed by Glashow, Salam and Weinberg. What remained to be shown to complete this spectacular scientific advance was to uncover the mechanism through which W and Z boson acquire their masses — as well as the quarks and leptons, whilst leaving the photon massless. The understanding of natural laws, in particular in particle physics, has attained such a level that this became a legitimate scientific question and the development of techniques and technologies was allowing to foresee a successful outcome. The most plausible theoretical scheme proposed to explain particle masses is the Brout–Englert–Higgs mechanism developed in the sixties of the past century. A direct manifestation of this mechanism would be the existence of a particle now usually called the Higgs boson. Discovery of this Higgs boson in 2012, fifty years after it was proposed, is an immense scientific success, certainly the most important progress in particle physics since the discovery of the W and Z almost thirty years ago. Such discoveries, as those of the W, Z and now Higgs are real landmarks in the history of sciences.

The LHC is a purely scientific project whose total cost is about 6 ;billion euros; it has been brought to a successful conclusion by CERN, the European organisation for nuclear research seated in Geneva. It is however a genuinely global project, countries like the United States, Japan, Canada, Russia, India have contributed significantly to the construction of the accelerator-collider itself, and about fifty nations throughout the world have taken part in the construction of detectors, thus this is a truly worldwide undertaking. The contribution to the construction cost of the machine is of the order of 10% for CERN non-member states and of about 25% for the experiments. The total number of scientists, physicists and engineers that have taken part in the realisation of the LHC project is about ten thousand. CERN is a meeting place for scientists the world over and is thus much more than just a laboratory for fundamental scientific research.

This book presents not only the genesis of the LHC project, but also all the main ideas and the theoretical framework subtending and motivating all the research at the LHC. In the recent years, we are witnessing merging of elementary particle physics and of the cosmological Big Bang model describing evolution of matter; this is due to the fact that in initial stages in the evolution of matter in the conditions of extreme temperature and density, nothing but elementary objects, without any structure, could survive, that is elementary particles. Therefore, it is the laws of elementary particle physics that govern initial stages in the evolution of matter. The study of the properties of matter and of its particle content as performed at the LHC can be viewed as a time machine allowing going back to early phases of the Big Bang. Collisions in the LHC in the proton–proton collisions mode allow us to explore the nature of interactions and the particle content at the so-called electroweak transition epoch, from 10^{-15} till about 10^{-12} seconds after the Big Bang, whilst heavy-ion collisions in the LHC allow to study matter at about one microsecond after the Big Bang, during the transition from the quark–gluon plasma phase to the hadronic matter phase. This book thus emphasises the link that exists between the world of elementary particles and the Big Bang, the quest for the origin of the Universe depending as much on the LHC as on the Hubble Telescope or the Planck satellite.

Since 1990 when the LHC project was launched it was evident that this will be a long haul, a long-term effort. Great audacity was required to propose a machine so innovative and complex, pushing technology to its limit. The magnetic system of the LHC with its *two-in-one scheme*, with two beam pipes for two counter-rotating beams in one and the same magnetic and cryogenic enclosure was not only a necessity, due to the lack of space in the LEP tunnel, but allowed also a very significant cost saving, not to talk about the savings represented by the already existing LEP tunnel and the entire CERN infrastructure. The LHC is in fact a prototype for a new type of accelerator-collider. The conceptual, research and development phase for both the accelerator and the experiments, took about ten years. Then the construction and commissioning phase took again almost ten years. In this respect, if there is a delay compared to initial expectations and promises, it is due largely to the fact that the LHC had to be built at a constant budget for CERN.

In the domain of new technologies, the LHC project has generated or stimulated very major advances, in the domain of superconductivity and cryo-magnetism, in the usage of warm superconductors, in the very large-scale usage of vacuum technologies and cryogenics, in the development of new materials, such as scintillating crystals, and in the integrated data acquisition systems of ATLAS and CMS detectors. We cannot forget about the invention at CERN of Internet at the beginning of the nineties to allow physicists to communicate the planet over, and about ten years later, of the WLCG — the world LHC computing grid — to analyse LHC-generated data throughout the world.

Having the LHC as the main instrument in the investigation of the fundamental structure of matter for the next ten to twenty years, we can expect major advances, or at least shed light on a number of key issues in physics. With the question on the origin of particle masses on its way of being solved, we can hope to test the idea of supersymmetry on the path to the possible, if not likely, unification of all fundamental interactions. Discovery of supersymmetry at the LHC could solve the long-standing problem of dark matter of the Universe. The possibility of further spatial dimensions, beyond the three we know of, will also be tested in a more incisive way. The more detailed studies of CP violation in the B-system will allow a deeper understanding of the subtle differences between matter and antimatter and of the emergence, ultimately, of the matter we are ourselves made of. In more general terms, all the foreseeable studies at the LHC will submit the Standard Model, both its electroweak and QCD components, to experimental tests deeper and more incisive than ever before.

These studies at the LHC, as well as those in other very active domains of particle physics research, as is for example the case with neutrino physics, all this should allow in coming years to probe the limits of validity of the Standard Model and should probably indicate the way to go in our search for a new level of understanding of the intimate structure and organisation of matter. All this helps us to understand our place in the Universe and it is precisely this incessant effort and thirst of us humans for understanding that has so much influenced our way of living and allowed the emergence of our present civilisation.

Carlo Rubbia, 7th May 2013

Contents

About the Authors

Daniel Denegri was born in Split, Croatia, from a Croatian father and French mother; today he is research director emeritus at the *Centre national de la recherche scientifique* (CNRS) in France. He studied physics at the University of Zagreb, Croatia, and obtained a PhD in particle physics at the Johns Hopkins University in Baltimore, USA. In 1971 he joined the *Commissariat à l'énergie atomique et aux énergies alternatives* (CEA) in Saclay near Paris. In 1982 and 1983 he took a direct role in the discovery of the W and Z bosons as member of the UA1 experiment. In 1989 and 1990 he participated with Carlo Rubbia in the LHC project launch as one of the coordinators of the physics studies evaluating the LHC's discovery potential. Daniel Denegri is one of the founders of the CMS experiment and was physics coordinator of the collaboration for fourteen years. He took an active part in the discovery of the Higgs boson in 2012 and is now investigating the future physics program of the LHC.

Claude Guyot is a researcher at the *Institut de recherche sur les lois fondamentales de l'univers* (IRFU) at CEA. After a PhD in particle physics in 1984 on neutrino oscillation studies based on results from the CDHSW experiment at CERN, he continued studying neutrino–nucleon interactions investigating the electroweak force and the internal structure of nucleons. While studying the violation of CP and T symmetries with neutral kaons at the CPLEAR experiment at CERN, he joined in 1991 a group of physicists that subsequently founded the ATLAS experiment at the LHC. Claude Guyot was one of the physicists at the origin of the ATLAS muon spectrometer, in particular of its large superconducting toroidal magnet system. He also played a leading role in the preparation of the ATLAS detector for data taking prior to and at the start of the LHC operation. Between 2011 and 2020, he was head of the ATLAS group at CEA/IRFU.

Andreas Hoecker studied physics in Bonn, Germany. He received his PhD in 1997 at the *Laboratoire de l'accélérateur Lineaire* (LAL) in Orsay near Paris on studies of τ-lepton properties and strong interactions using data from the ALEPH experiment at the LEP collider at CERN. After his PhD he became researcher at the French CNRS and joined the BABAR experiment at Stanford, USA, where he spent two years working on a particle identification system and studying the violation of matter–antimatter symmetry. Upon his return to Europe in 2005, Andreas Hoecker changed affiliation to become research physicist at CERN. He joined the ATLAS experiment, where he contributed to various domains of detector operation and data analysis, more specifically to Higgs boson and supersymmetry searches. He was physics coordinator of the ATLAS collaboration in 2014–2015. Between 2017 and 2020, he was deputy spokesperson and in June 2020 he was elected ATLAS spokesperson for a two-year mandate starting in March 2021.

Lydia Roos is a research director at the *Institut national de physique nucléaire et physique des particules* (IN2P3) at CNRS. During her PhD in Marseilles she studied the properties of beauty mesons using data from the ALEPH experiment at LEP. In 1993 she joined a CNRS/IN2P3 laboratory in Grenoble and worked on the construction and operation of a silicon pixel detector for the DELPHI experiment at LEP. Since 1996 she works at the *Laboratoire de physique nucléaire et de hautes énergies* (LPNHE) in Paris. She studied matter–antimatter symmetry with the BABAR experiment. During a four-year stay in Beijing devoted to fostering the Sino-French scientific collaboration,

Lydia Roos led an associated international laboratory for elementary particle physics. At her return to Paris in 2008 she joined the ATLAS experiment, where she was involved in the discovery of the Higgs boson and the measurement of its properties. She also contributed to searches for new particles. In 2018 she became scientific director at IN2P3.

List of Insets and Digressions

Introduction

This book describes one of the most outstanding adventures of modern science: the construction, operation and exploitation of the Large Hadron Collider (LHC) at CERN, the European Laboratory for Particle Physics near Geneva, Switzerland.[1] The LHC is a superconducting proton–proton, proton–ion and ion–ion accelerator and collider that allows to reach unprecedented collision energy and rate. Alongside the LHC, four large experiments, ATLAS, CMS, LHCb and ALICE, have been constructed by international collaborations and they are no less remarkable than the collider itself.

The central theme of this extraordinary scientific project is the study of, in somewhat barbaric terms, *electroweak symmetry breaking*, a mechanism physicists believe to be at the origin of mass of elementary particles, which in conventional theory is forbidden on symmetry grounds. In the 1960s, Robert Brout and François Englert on one hand, and Peter Higgs on the other, have suggested a theoretical solution, further elaborated by others, of the mass problem within the framework of what was to become the most successful theory in the history of physics, the *Standard Model* of elementary particles and their interactions. This solution, now called the *Brout–Englert–Higgs mechanism*, implies the existence of a new particle: the *Higgs boson*. Over the years, alternative theoretical ideas were proposed as possible avenues to give mass to elementary particles. However, the data recorded and analyzed by ATLAS and CMS in 2011 and 2012 pointed

[1]Europeans founded in 1954 the *Conseil Européen pour la Recherche Nucléaire*, subsequently renamed *Organisation*. Today the main activity of this large international laboratory, where more than ten thousand scientists and engineers are active, is fundamental research in particle physics, but the initial acronym remained.

clearly to the existence of a new particle having the properties expected for the Higgs boson. This discovery made the news in the media throughout the globe, culminating in the Nobel Prize for the theoretical idea in 2013. Subsequent studies by ATLAS and CMS confirm so far the Higgs-boson nature of the new particle.

Like a giant microscope, the high-energy collisions of protons in the LHC resolve the smallest structure of space, more than a thousand times smaller than the size of a nucleon. At such short distance, one may wonder whether elementary particles reveal substructure. Protons and neutrons were once also thought to be elementary but then, in the late 1960s, physicist found them to be composed of quarks and gluons, a discovery of fundamental consequences for subatomic physics.

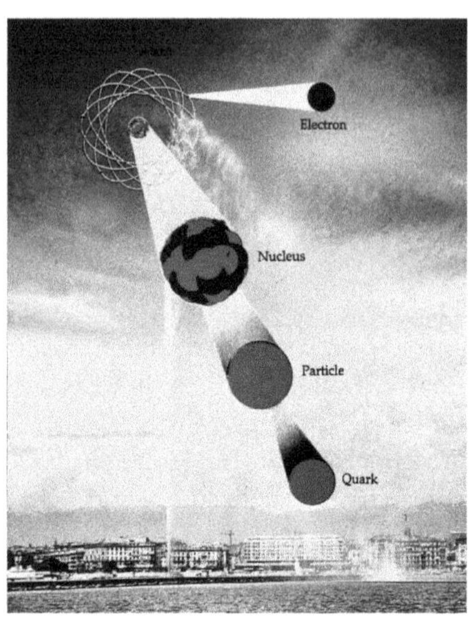

Ancient postcard showing Geneva, its landmark the *jet d'eau* (jetty water fountain), and the structure of matter in reference to CERN.

Because the Universe expands and cools down from its initial heat bath, understanding particles and their interactions at high energy allows physicists to infer the dynamic and thus tell the history of the Universe during the very first decisive moments following the Big Bang. Proton–proton collisions in the LHC, as studied by the ATLAS, CMS and LHCb experiments, empower the experimenters to study the elementary constituents of matter and explore their interactions as they occurred in the first thousandth of a billionth of a second after the Big Bang. The ion–ion collisions, as investigated by the ALICE experiment but also successfully and complementary by ATLAS, CMS and LHCb, shed light on a later moment in the evolution of the Universe, about one microsecond after the Big Bang. Up to that time, matter is believed to have formed a plasma out of quarks and gluons. A transition then occurs to hadronic matter, such as the protons and neutrons making up our present-day world, including ourselves.

The LHC is the most powerful particle accelerator and collider built to this day. Likewise, the multi-purpose ATLAS and CMS detectors surpass

Areal view of the CERN Meyrin site with its many labs and experimental halls. In the background the French Jura mountains. On the right, the wooden "Globe of Science and Innovation". The buildings next to it host the LHC "Interaction Point 1" with the underground ATLAS experiment. Most of the ATLAS and CMS physicists have their offices in the foremost building on the left. The CERN directorate building is at the centre of the second row. *Image: CERN.*

in size and complexity all their predecessors. This exceptional project has been made possible thanks to CERN, its European member states, and the strong support and commitment by the particle physics community and public funding agencies worldwide. In a certain sense, CERN, established in 1954, is the precursor of a unified Europe. To construct the LHC, CERN has moreover been able to mobilise the scientific community globally, bringing together on a common project not only European states, but also the USA, Japan, Russia, China, India, Brazil, Chile, South Africa, Australia, to mention just a few. The ATLAS and CMS collaborations taken together have member institutions in seventy countries.

This book introduces at a qualitative level — so as to be intelligible to a wider audience — the various states of matter and the interactions governing them. For the interested reader with a more complete scientific background, specific inserts provide more detailed digressions, introducing experimental or theoretical concepts where appropriate. Some of these may, admittedly, appear arduous to the general reader without advanced scientific education. However, this knowledge is not required to appreciate the chapters that follow and which address the experimental matters. Sections that may be skipped in a less detailed reading are explicitly indicated. The book also strives to explain what often strikes the non-initiated as a paradox, namely, why are such huge and sophisticated instruments required to explore the infinitesimally small.

In this book, the first three chapters are devoted to the Standard Model, its foundations and experimental confirmation, but we also discuss its deficiencies at the time of the LHC start-up. The fourth chapter provides some ideas on theoretical models that have been proposed to cure the shortcomings of the Standard Model. In the fifth chapter, we discuss the relationship between particle physics as studied at the LHC and the world of the infinitely large as described by the cosmological model of the Big Bang, the other colossal achievement of 20th century physics. Experimental aspects are tackled-with next, starting from the genesis and construction of the LHC in chapter 6, to basic particle detection techniques described in chapter 7. The emphasis is then put, in chapter 8, on the description of the two large general-purpose detectors ATLAS and CMS. Chapter 9 gives an account of the start-up of the LHC and of the first data-taking period, as well as of the LHC restart in 2015 at much higher collision energy. This is followed by an introduction to data analysis techniques in particle physics (chapter 10), a prerequisite for the reader to understand how the main (and most recent) physics results presented in the following chapters have been obtained: the search and discovery of the Higgs boson in chapter 11, the study of other important Standard Model physics phenomena in chapter 12, where we also introduce specific purpose *forward physics* experiments, and in chapter 13 the strenuous and persevering search for new particles indicative of physics beyond the Standard Model. Chapter 14 is essentially devoted to the other two LHC experiments, LHCb and ALICE, whose domains of investigations are more specialised (but no less important). The book concludes with a discussion of options for the future, the approved High-Luminosity LHC project that will allow the LHC to deliver a ten times larger dataset to the experiments than originally planned, and possible new large-scale projects and the physics questions that motivate them.

Since the discovery of the Higgs boson, fundamental particle physics lives in data-driven times. A plethora of theoretical ideas extending the Standard Model exist, but experiment must guide the field to the next stage. The vast and diverse research programme of the LHC, its upgrades and follow-up projects, are key to progress, in the spirit of the scientific procedure initiated by Galileo Galilei some 400 years ago.

Chapter 1

The Standard Model of Elementary Particle Physics

1.1. Particles of matter, interaction fields and Standard Model parameters

Humanity approaches science[1] like a child discovers the world and for whom the bewildering variety of phenomena represents a new challenge every day. Two ways are possible and coexist, the empirical approach providing description and an appropriate answer to every new experimental observation, and categorisation and theorisation attempting to provide an understanding. A child observing cars passing by on the street will, without adult's help, start to recognise the various car brands by understanding the repetition of forms and shapes: the child came to recognise some regularities and symmetries in the surrounding world.

Elementary particle physics really started during the first half of the 20th century. Its beginnings were marked by the observation of totally new and unexpected phenomena, such as the existence of an antiparticle to the electron (the positron), the appearance of new particles with strange

[1]The scientific revolution began with Nicolaus Copernicus in the 16th century, shifting paradigms from the Ptolemaic model of the heavens to the heliocentric one, continued with Galileo Galilei and his *Dialogue Concerning the Two Chief World Systems*, to find its culmination and accomplishment with the 1687 publication of Isaac Newton's *Philosophiæ Naturalis Principia Mathematica*, laying the foundations of modern mathematical physics. The French mathematical physicist Alexis Clairaut wrote *"The method followed by its illustrious author Sir Newton [...] spread the light of mathematics on a science which up to then had remained in the darkness of conjectures and hypotheses."*

properties (thus their name *strange particles*) in cosmic rays,[2] the creation of virtual particles in the vacuum (empty space is thus not quite empty!), the violation of symmetries that up to that time were considered sacrosanct in the world of the infinitesimally small (symmetries between right and left, future and past, between particles and antiparticles).

Particle physics was still in its infancy, but the kid, growing up fast, could be satisfied by just observing natural phenomena. Accelerators were soon constructed allowing the production of these new particles in a laboratory to study their properties in a detailed and controlled way. Discoveries of new particles then followed at such a rate that one could fear that there will not be enough letters in the Latin and Greek alphabets to label them all! Frustration was substantial at that time, in the fifties and sixties of the last century, among the *high-energy nuclear physicists* (as the name *particle physicists* did not yet exist). The famous and humorous Austrian theoretical physicist Wolfgang Pauli is supposed to have said, exasperated by the proliferation of these new objects: *"Had I known things would go this way, I would have become a botanist!"*

Is there some underlying order, or does chaos prevail in this world at the scale of a millionth the size of an atom? Is there possibly a periodic table similar to that Mendeleev established one century earlier for chemical elements? In 1964 George Zweig and Murray Gell-Mann succeeded to put some order in this picture. They (independently) proposed a model whereby hadrons[3] were not elementary particles, but were in fact composed according to well-prescribed rules of either two or three basic particles of various types. These elementary constituents were called *quarks*, a term borrowed by Gell-Mann from James Joyce's Finnegans Wake: *"Three quarks for Muster Mark! And sure he has not got much of a bark, and sure any he has its all beside the mark"*. By careful observation of the road, the child has finally understood that the recurring similarities observed in the passing cars are stemming from the way they are built: it is the beginning of a real understanding.

---

[2]Among the first particles to be discovered in cosmic rays was the electrically charged muon (μ^+, μ^-) few years before World War II, but it was really understood as a heavier relative of the electron only in 1944–1945. Its discovery and properties were so surprising at the time that the well-known physicist Isidor Isaac Rabi supposedly exclaimed *"Who ordered that?!"*

[3]Hadron is the collective name for any particle that is sensitive to strong interactions, we shall discuss them in detail later in this book. The proton and neutron — we are ourselves made of — are the most common among the hadrons.

Indeed, it seems that Nature, despite its prolific variety of manifestations, does follow a path, a principle of great economy. All the phenomena (practically all, see later in this book for exceptions) in the world of the infinitesimally small can be explained by a single theory today called the *Standard Model* of particle physics. It describes three types of fundamental interactions (or fundamental *forces*)[4]: the *electromagnetic interaction*, responsible for electric and magnetic phenomena (and thus of all of chemistry and biology), the *weak interaction*, at the origin of radioactivity and essential to solar energy production and thereby to life on Earth, and the *strong interaction*, responsible for the structure and cohesion of atomic nuclei, this interaction being itself the residue of the more powerful force binding quarks into hadrons. The weak interaction is called so, as its intensity is weak in the present (cold) world, when compared to the electromagnetic or strong forces. It has a very short range of about 10^{-17} metres, about one hundred times smaller than the size of a proton! These three interactions of the Standard Model can all be derived from a geometrical concept, the *principle of gauge invariance*, whose understanding and theoretical elaboration represent the greatest theoretical advance of 20th century physics.

There is a fourth interaction, probably the most celebrated[5] one: gravity. It is responsible for the fall of apples, the motion of planets around the Sun and of the large-scale structures of the Universe, but it is not part of the Standard Model (cf. figure 1.1). This is so as it plays no role at the microscopic level, at least not at the scale that can be studied with present accelerators. It, however, most likely plays a key role — even at the microscopic level — during the very first moments in the existence of our Universe, the Big Bang.

Besides the interactions, the Standard Model also describes the particles, the elementary constituents of matter. These are the six types — also called *flavours* — of quarks, all sensitive to all four types of interactions,

[4]Fundamental interactions cannot be derived from more basic interactions. While this is indeed the case for the three interactions in the Standard Model, many physicists believe that there should be an even more fundamental interaction, a *grand unified force*, realised at very high-energy density that, when the Universe cools down, splits into the known three forces of very different strength. This will be discussed in detail in this book.

[5]It is through the study and gradual understanding of gravity through the works of Kepler, Galileo, Hook, Newton and others that modern science arose in the 16th and 17th centuries. The other big avenue was the study of optical phenomena with Galileo, De Dominis, Leeuwenhoek, Snell, Descartes, Hyugens, Newton, etc.

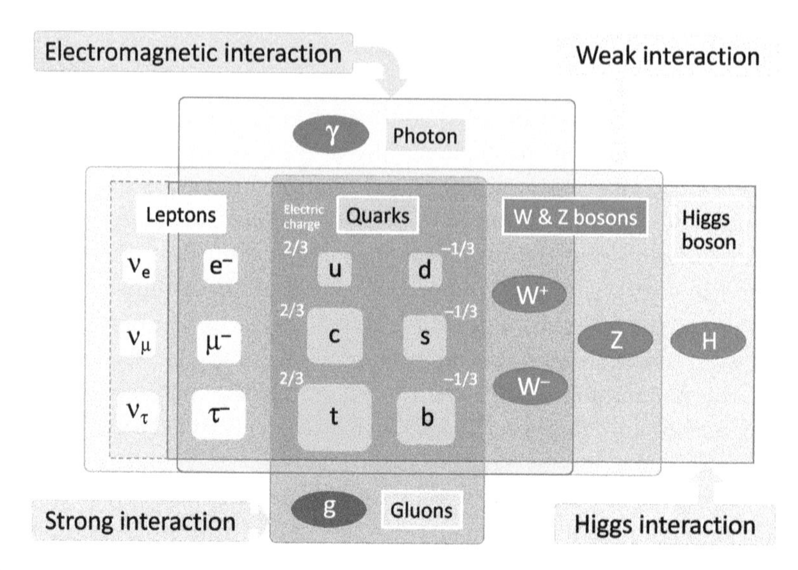

Fig. 1.1. Elementary particles and their interactions within the framework of the Standard Model. Matter particles are leptons and quarks, both groups organised in three families. Each family of leptons is made of a neutral lepton member called a neutrino and an associated charged lepton: the electron (e), the muon (μ) and the τ-lepton. The corresponding neutrinos are called electron-neutrino, muon-neutrino and τ-neutrino. As the electric charge of the electron is taken as the fundamental unit of charge, the charged leptons all have electric charge -1. Quarks however have fractional electric charges, every family having one quark of charge $+2/3$ and one with $-1/3$. The quarks of the first family are called *up* and *down* respectively, of the second *charm* and *strange*, of the third *top* and *bottom* (sometimes also denoted *beauty*). For particles of matter, be it *quarks* or *leptons*, the often used generic term is *fermions*, the exact meaning of this will be explained in the following (see inset 1.1). Each of these particles has an associated antiparticle of opposite electric charge (section 1.7). The interactions among these fermions are realised through the exchange of *bosons* (inset 1.1): the *photon* (γ) for electromagnetic interaction, the W^+, W^- and Z^0 bosons propagating the weak interaction, and the eight *gluons* (g) mediating strong interaction. The Higgs boson H plays a very particular role in this scheme, as it will be explained in detail later in this book. It is neither a particle of matter, nor a boson responsible for gauge interaction. It is a very special kind of a "fifth force".

and the six flavours of *leptons*, all sensitive to weak interaction, but blind to strong interaction. Three among these leptons carry electric charge, whilst the remaining three are electrically neutral and thus do not take part in electromagnetic interaction either. The latter ones are called *neutrinos*. A particle not taking part in any interaction except for gravity is called *sterile*. Such a particle would thus not interact (not even weakly) with the material of any detector and could be seen only indirectly, for example through

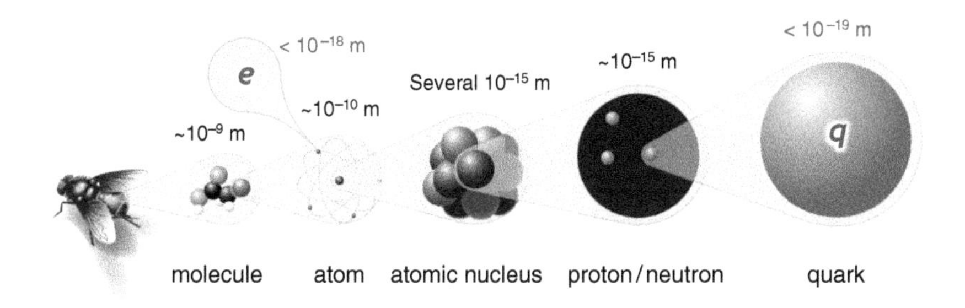

Fig. 1.2. Ordinary matter as we find it on Earth is composed entirely of up and down quarks from the first family, which are the building blocks of protons and neutrons forming the atomic nuclei, and of electrons. The electron-neutrino appears in some radioactive decays; it plays a key role in the nuclear fusion process inside the Sun, whereby four protons fuse into a helium nucleus (made of two protons and two neutrons), releasing energy. According to current knowledge, the electron and quarks are elementary particles. Through measurement at particle colliders one can derive tight upper limits on their size (shown in red).

the disappearance of a known particle that would transform itself into a sterile one. A sterile-type particle is theoretically possible, but has never been observed up to now, although it is recently much speculated about, for example in the context of dark matter (see section 3.1). Neutrinos, on the other hand, subject to weak interactions, are not sterile, but the very weakness of these interactions makes their direct observation rather difficult. The existence of neutrinos (more precisely of the neutrino partner of the electron, ν_e, the first to be discovered) was postulated in 1930 by Wolfgang Pauli, so as to preserve conservation of energy, momentum and angular momentum in radioactive decays. The real observation of a neutrino in a laboratory was achieved only in 1956.

In the Standard Model (figure 1.1) quarks and leptons are organised into three families (also called *generations*) according to their flavour. Each family is made of a doublet of quarks and a doublet of leptons, with one charged lepton and its associated neutrino as members. However, all *stable matter* in the Universe as we know it today is made only of particles from the first family: the up (u) and down (d) quarks and the electron and electron-neutrino (figure 1.2). What is then the role, if any, of the remaining two families, and are there only three families?

With the exception of neutrinos whose mass hierarchy is not yet firmly established, masses of both leptons and quarks increase with family number. Due to the equivalence between energy and mass (for a body of mass m, the energy E is given by $E = mc^2$, where c is the velocity of light

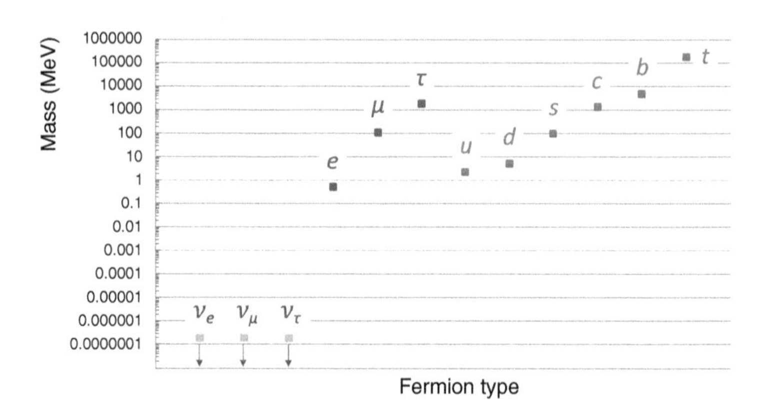

Fig. 1.3. Masses (in MeV) of elementary fermions, quarks (shown in blue) and leptons (orange for neutrinos and red for charged leptons). The logarithmic ordinate scale illustrates the large mass hierarchy between the families. For example, the top quark is almost eighty thousand times more massive than the up quark, which is at least a million times heavier than the electron-neutrino. Only upper limits are yet known for the neutrino masses (whose sum makes up less than 0.17 eV). All fermion masses are free parameters of the Standard Model which have to be determined by experiment. Some speculative grand unification models (see section 4.2) predict relationships among the masses.

in vacuum),[6] production of a more massive particle requires more energy than for a lighter one. With increasing accelerator energies, all the leptons and quarks have gradually been produced and identified (figure 1.3). The heaviest among the leptons, the τ (tau), has been discovered in 1974 at the Stanford Linear Accelerator Centre (SLAC) in California and brought a Nobel Prize to Martin Perl. The discovery of the most massive among the quarks, the *top* or *t-quark*, was announced in 1995 at Fermilab, a large particle physics laboratory near Chicago, USA. Finally, the τ-neutrino (ν_τ), whose existence was indirectly inferred, has been directly observed in 2000 again at Fermilab.

Neglecting for now the tiny masses of neutrinos, known to be at least a billion times lighter than a proton, the Standard Model requires the introduction of eighteen a priori unknown numbers, called *Standard Model*

[6]In this book, we use energy units derived from the electron-volt (eV) to express both energies and masses. An electron-volt is the energy acquired by an electron accelerated by a potential difference of one volt. One mega-electron-volt (MeV) corresponds to 10^6 eV, one giga-electron-volt — approximately the energy required to create a proton — is 10^9 eV, one tera-electron-volt (TeV) corresponds to 10^{12} eV.

parameters.[7] Among these, we have, first, six masses for the quarks and three masses for the charged leptons. This gives nine parameters. What are the remaining ones? Three of them are called *coupling constant*, they determine the strengths of electromagnetic, weak and strong interactions, respectively. We will define later the remaining six parameters.

Quarks take part in all four known interactions, we can thus take the lightest among the quarks, the up quark, to set a scale at which we can compare the relative strength of the interactions. Taking the strong interaction as a reference, the electromagnetic interaction is about a hundred times weaker, and the weak interaction about a million times weaker, whilst gravity is about 10^{38} times weaker still.

How do particles interact among themselves? The modern picture, resulting from a well-defined mathematical framework called *quantum field theory* (see inset 1.3 and also section 1.9) introduces the concepts of sources (or charges) and of field quanta (for more details, see section 1.9). The source of the electromagnetic force (or field) is electric charge. Similarly the charge or source of strong interaction is called *colour* (nothing to do with optical colours, as will be explained later), and for the weak force it is called *weak isospin* (or weak charge). As for the gravitational force, its source is simply mass.

Field quanta propagating a force are elementary particles, existing for a very brief interval of time, that are exchanged between matter particles interacting with each other, as shown in figure 1.4. This can be seen as the charge or source of the corresponding interaction emitting and immediately reabsorbing the field quanta. This creation of a particle from vacuum in principle violates the law of energy conservation. However in 1927 the German physicist Werner Heisenberg showed that it is possible to have a non-conservation of energy in a quantum system during a brief moment in time. His mathematical *uncertainty principle* (see inset 1.2) states that an amount of energy ΔE can be "borrowed", and thus a particle can be created, during a time interval Δt, provided the product $\Delta E \cdot \Delta t$ is of order or smaller than a very small quantity called the *Planck constant* h: $\Delta E \cdot \Delta t \leq \hbar/2 \simeq 3.3 \cdot 10^{-22}$ MeV s, where $\hbar = h/(2\pi)$.

[7]A nineteenth parameter is related to a possible violation of particle–antiparticle symmetry in strong interactions, which is nominally possible but has never been observed experimentally. This parameter must thus have a very small value, possibly zero (see section 3.4).

Inset 1.1: Spin, bosons, fermions

Spin is an intrinsic property of a particle, as is its mass, electric charge or colour charge. At the quantum level spin is connected to the particle's behaviour in rotation-type transformations. It can be thought of as the internal angular momentum of a particle, like for a spinning top, despite the particle being point-like. Spin is characterised by a quantum number that can take only integer or semi-integer values $(0, 1/2, 1, 3/2, 2, \dots)$; particles of integer spin are called *bosons* and those of semi-integer spin *fermions*. These denominations are related to the collective behaviour of these particles: two identical fermions cannot occupy the same quantum state in a quantum system (*Pauli's exclusion principle*), whilst an arbitrary number of bosons can, and even tend to, accumulate in the same quantum state. Fermions obey the *Fermi–Dirac quantum statistics* and bosons the *Bose–Einstein statistics*, the origin of the fermion and boson denominations.

These properties are the reason why in an atom, electrons, which are fermions, fill necessarily progressively higher energy levels and do not accumulate in the lowest energy level. It thus explains Mendeleev's table of chemical elements. The Bose–Einstein statistics on the other hand explains the functioning of a laser beam in which photons (bosons!) are all in the same coherent state, going in the same direction and with same wavelength. This relation between spin and statistics can be proven in the framework of quantum field theory, it is called the *spin-statistics theorem.* As the American theoretical physicist Richard P. Feynman deplores in his famous book *Lectures in Physics*, this theorem, although being one of the most important ones in modern physics, is also among the most difficult ones to explain in a simple way.

In the Standard Model, matter particles (leptons and quarks) are fermions of spin $1/2$. Particles mediating forces (photons, gluons, the W and Z) are bosons of spin 1; a particle of spin 1 is called a *vector particle.* At our present level of understanding, the Higgs boson H seems to be the only fundamental particle of spin zero; such a particle is called a *scalar particle.*

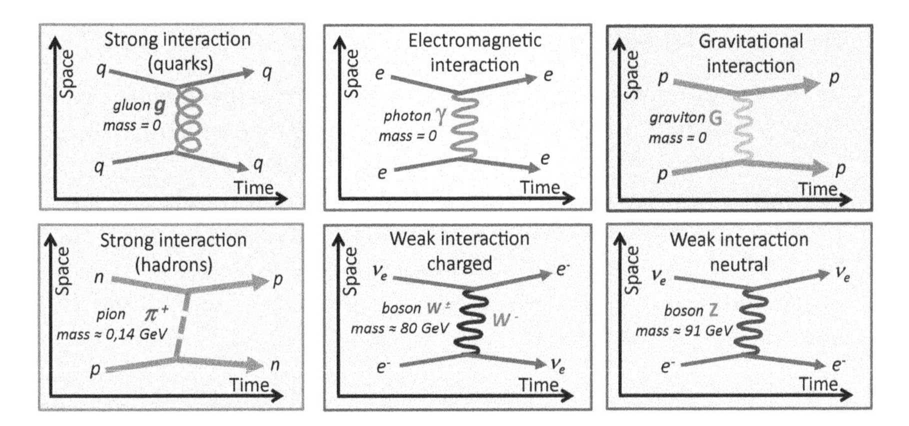

Fig. 1.4. Graphical representations of interactions between particles in quantum field theory propagated by the exchange of *mediating particles* (*bosons*). When the mass of the mediator is zero, the range of the interaction (force) is infinite. For example, in the case of electromagnetic and gravitational forces, the intensity of these forces decreases as $1/R^2$ at large distance R. The range of the weak force is very short, about 10^{-17} metres, due to the large masses of the mediating particles. At still shorter distances quantum effects complicate this simple picture of a single mediating particle exchange (see section 1.9). The nuclear force acting between two nucleons through the exchange of pions (made of a quark–antiquark pair) is just the residual force resulting from quarks, antiquarks and gluons they are made of. This is somewhat analogous to the van der Waals force among molecules. Gravity is not included in the Standard Model of elementary particles. Its strength, as measured by the gravitational force applied by a proton on another proton at a certain distance, is about 10^{38} times weaker than the electric force felt by the protons at the same distance.

A consequence of the uncertainty principle is that the larger the mass m of a particle mediating an interaction, the shorter is its lifetime and thus the interaction range. The range R of an interaction is limited by the maximal distance the exchanged particle can cross at the speed of light during its lifetime[8]: $R \propto 1/m$. It is so that field quanta exchanged in weak interactions (W, Z) are very massive, which explains their short range, whilst the quanta exchanged in electromagnetic and gravitational interactions are of zero mass, giving these forces an infinite range (see figure 1.4).

However, even if gluons, the quanta of the strong interaction fields have zero mass, the *effective range* of strong interactions is nonetheless

[8]In this expression mass is given in units of GeV and distance in GeV^{-1}. This peculiar choice of units is explained in inset 4.3 called the distance–energy relationship in quantum mechanics and the Planck scale (p. 140). In full: $R \leq c \cdot t = c\hbar/E \leq c\hbar/(mc^2) = 1/m$ in so-called *natural units*, where we set $\hbar = 1$ and $c = 1$.

Inset 1.2: Heisenberg's uncertainty relations

At the very base of quantum mechanics, Heisenberg's uncertainty relations, formulated in the twenties of the past century, stem from the wave-like nature of particles. It states that there is a lower limit to the precision with which one can know simultaneously values of two *conjugate* quantities (duals of a Fourier transform) in a coherent quantum system. For example, simultaneous knowledge of position x and momentum p of a particle is necessarily affected by incompressible uncertainties Δx and Δp such that:

$$\Delta x \cdot \Delta p \geq \frac{\hbar}{2}.$$

Here $\hbar$ is the reduced Planck constant, defined from the *quantum of action* h as $\hbar = h/2\pi = 6.6 \times 10^{-22}$ MeV s.

Time t in quantum mechanics is not a dynamical variable (also called an *observable*) as are x or p. There is nonetheless an uncertainty relation connecting it to the energy E of a system:

$$\Delta E \cdot \Delta t \geq \frac{\hbar}{2}.$$

It gives a lower limit on the precision ΔE with which the energy of a system can be known when only a limited time Δt is available to measure it. As the energy of a particle is related in quantum mechanics to the frequency ν of the associated wave ($E = h\nu = hc/\lambda$ for a photon), this relation can intuitively be understood as the precision on the determination of the frequency ν being inversely proportional to the number of observable oscillations during the time interval Δt. Similarly, the above equation implies the possibility to violate the law of energy conservation during a small time interval Δt, allowing the creation and quasi-immediate annihilation of particles.

very short, of the order of 10^{-15} metres, but for a different reason, called *colour confinement.* It is due to the fact that hadrons, otherwise made-up of quarks, are in fact neutral, globally colourless or colour-neutral (remember: colour is the name of the strong charge) with respect to strong interactions. This is somewhat analogous to the effectively short range of electromagnetic forces between atoms or molecules, as these are globally electrically neutral systems. Trying to forcibly separate two coloured quarks interacting strongly requires so much energy, that it is ultimately more economical energy-wise to create new quark–antiquark pairs from the vacuum, and thus new hadrons (as discussed in more detail below). This feature of the strong interaction is called *confinement.* It limits its range to nucleon size of 10^{-15} metres and determines the sizes of hadrons. It also prevents one from observing quarks or gluons as free particles, that are outside hadrons, except at the extremely high energies that existed shortly after the Big Bang, the deconfined quark–gluon plasma phase (see chapter 5 and section 14.2.1).

Let us go back enumerating and defining the parameters of the Standard Model: there are two quanta for the weak force field, an electrically neutral one, the Z, and an electrically charged one, the W, the latter one appearing in two states of opposite electric charge, W^+ and W^-. Both the Z and W are very heavy, although with slightly different masses, as discussed in section 1.6. The mass of the W can be derived from the mass of the Z and the coupling constants of weak and electromagnetic interactions.[9] Therefore, we introduced only one additional parameter here.

We are still left with five Standard Model parameters to be introduced. One is the mass of the Higgs boson, a particle remnant from a mechanism postulated to resolve a serious problem of symmetry breaking that we will address at length later in this chapter. The remaining four parameters govern the mixing of quark flavours. They are connected with transitions among quarks pertaining to different families, transitions induced by the weak interaction. One of these parameters induces the breakdown of matter–antimatter symmetry in the domain of quarks. Such a symmetry breaking is a condition *sine qua non* for the existence of a world as ours, where matter largely predominates over antimatter.

These are thus the eighteen parameters of the Standard Model. They are a priori unknown and must be measured from experiment. With the discovery of the Higgs boson at the LHC in 2012 (as will be discussed in details in chapter 11), these parameters have by now all been observed and measured with greater or lower precision at particle accelerators.

It has been shown not so long ago that neutrino masses, albeit being very small, are not zero.[10] This increases the number of parameters in the theory

[9]This relation from the Glashow–Weinberg–Salam model is explicitly derived in inset 1.10.

[10]Since no symmetry in the Standard Model stipulates that neutrinos should be massless, it is quite natural that they are not. Non-zero neutrino mass, however, requires to extend the minimal form of the Standard Model we discussed so far.

by seven additional units.[11] Three among these are the neutrino masses themselves, and the remaining four are neutrino flavour-mixing parameters, including a parameter breaking matter–antimatter symmetry in the lepton domain similar to the case of quarks. Even if all the neutrino flavour-mixing parameters have by now been measured with a precision better than 20%, not much is really known about neutrinos (and the matter–antimatter symmetry breaking among leptons), what motivates the emergence of large future research projects in neutrino physics.

Another highly interesting topic in particle physics at present is the detailed study of the Higgs boson. Admired by some as the saviour of the Standard Model, the Brout–Englert–Higgs mechanism (section 1.5) is criticised by others as being an *ad hoc* construction generating an interaction not originating from a basic geometrical principle as for the other fundamental interactions. Furthermore, the Higgs boson has the peculiarity of being so far the only fundamental scalar particle, that means of spin zero (see inset 1.1 introducing particle spin). The LHC, surely one of the most audacious scientific projects ever, has been launched largely with the aim of discovering this particle and studying its properties. The challenge was nothing less than the very survival of the Standard Model in its present form, a model whose development took over forty years and the efforts of thousands of particle physicists, experimentalists and theorists.

1.2. The Standard Model in our daily lives

The question of *naturalness* of physics laws and the values of fundamental parameters is deeply inscribed in the minds of researchers. Many physicists believe that the values of the eighteen parameters of the Standard Model (ignoring here again the neutrino sector), which seem rather arbitrary at the level of our present understanding, should ultimately be explained in a natural way by a more general theory encompassing the Standard Model. This theory should explain, not only Standard Model phenomena, but also those at work at very high energies at the time of the Big Bang. Among such phenomena there is the question of the unification of electroweak, strong and, eventually, gravitational interactions, discussed in more detail in chapters 4 and 13.

[11] If the neutrinos are identical to antineutrinos, which is not yet known experimentally, they are said to be *Majorana fermions*. In this case, it would be necessary to introduce two more parameters. We will come back to Majorana neutrinos in chapter 3.

Despite a formulation that is based on abstract mathematical concepts, the theory of particle physics cannot be in contradiction with phenomena we encounter in our daily lives. The concern here is not so much with the rather impenetrable human behaviour, but with the basic structures of the material world that led to the development of stars, planets and — ultimately — life: the forces and particles of matter allowing the creation of complex atoms and stable molecules; a temperature of the quantum vacuum sufficiently low allowing interactions among particles and chemical reactions to depart from thermo-dynamical equilibrium, and leading to a universe made of matter rather than antimatter; the appearance of heavy elements, in particular of carbon essential to life, of stars with lifetimes of several billion years, of planetary systems with habitable zones appropriate for the development of life, even of such sophistication as to be able to reflect upon itself and its own place in the Universe!

Even if we do not have yet an understanding of the values of Standard Model parameters, we can wonder to what extent their values influence our everyday lives. Have these values been tuned as to allow the appearance of life? And if so, how could such a tuning occur? The American physicist Robert Cahn lent himself to a *thought experiment*, imagining a powerful creature able to modify the values of Standard Model parameters. It turns out that consequences of such an experiment can be dramatic. Let us take the mass of the lightest charged lepton, the electron, a particle subject to electromagnetic and weak forces. The electromagnetic force is responsible for chemical reactions and therefore biology, and the weak force for radioactivity. Except for neutrinos, the electron is the lightest, and by far, among all elementary particles of matter. Its mass is about two hundred times smaller than that of its closest cousin, the muon, the charged lepton of the second family. What would happen if the mass of the electron were as large as that of the muon? As the radii of all atoms are determined by the Bohr radius,[12] which is inversely proportional to the mass of the electron, all atoms and all objects of our world would be reduced in size by a factor two hundred. Humans would be about a centimetre tall! Energies emitted in electron transitions between energy levels of such atoms would also be two hundred times bigger. All humans and animals of Lilliputian sizes, would

[12]The Bohr radius a_0, calculable in quantum mechanics, is given by $a_0 = h^2\varepsilon_0/(\pi m_e e^2)$, where h is the Planck constant, ε_0 is the permittivity of vacuum, and m_e and e are the electron mass and charge. The value $a_0 = 0.53 \times 10^{-10}$ m corresponds to the size of the hydrogen atom in its ground state.

be bathed into and surrounded by light of a few nanometre wavelengths,[13] practically X-ray radiation.

Would such a world be liveable? It is not worthwhile contemplating this as other consequences would be much more dramatic. At the quantum level, electron *trajectories* on their atomic levels, the orbitals, and similar proton orbitals within the nuclei overlap each other frequently. Overlapping orbitals allow the electron and proton to encounter, interact, and produce through weak interactions a neutron plus a neutrino. In our world the neutron mass is just a bit larger than the sum of proton and electron masses, and, luckily for us, such a transition is then forbidden (except in some very specific atoms and nuclei, in which case we do have *electron capture*). However, if the electron had the mass of a muon, the situation would be totally different: the sum of masses of an electron and a proton would exceed those of a neutron plus an electron-neutrino and the reaction $p+e^- \to n+\nu_e$ would be allowed. The neutron would become a stable particle, whilst in our actual world the neutron is unstable: it decays through the inverse reaction $n \to p + e^- + \nu_e$ in about 890 seconds on average. In fact a neutron can survive stably only bound within a nucleus, thanks to the nuclear binding energy. Going back to our hypothetical heavy-electron world, sooner or later all electrons would end up captured by protons. For example, the simplest of all atoms, hydrogen, made of just a proton and an electron, would end-up in an unbound system made of a neutron plus a neutrino. The world that would have emerged from the Big Bang would thus be entirely made of neutral particles, as soon as the expansion of the Universe would have lowered its temperature below the energy equivalent of an electron mass. This would be a very strange universe indeed, with no shining stars, the only heavenly bodies would be neutron stars. The Universe would look like a strange billiard game with occasional spectacular, but invisible collisions.[14]

Previously we discussed how lugubrious a world with electrons two hundred times more massive would be. We could now ask ourselves by how much have we escaped such a scenario? What is the possible domain of

[13]Visible light has wavelengths between 400 and 700 nanometres.

[14]Another scenario consists in keeping the electron mass as it is, but inverting the masses of proton and neutron by swapping the masses of up and down quarks. With the proton heavier than the neutron by 1.5 MeV (only 1.5 permil of its current mass), we would have a similar catastrophic world scenario.

masses for the electron that allows the formation of heavy elements, a precondition for the development of life? What would happen if the electron mass were only slightly increased by, say, just 30%? Nitrogen would begin to disappear at the expense of carbon. With a further increase of the electron mass, gradually all elements would disappear and the world would take the unpalatable form described above. When compared to the large differences among the masses of elementary particles of the Standard Model, a 50% increase in the electron mass seems negligible. To avoid believing in fine-tuning of parameters so as to allow the appearance of life (the *anthropic* view of natural laws[15]) and, of course, of physicists asking themselves such questions, scientists have to find a testable (and falsifiable) theory predicting the masses of particles and their interactions.

A possible path is offered by so-called grand unified theories that have the ambition to unify electromagnetic, weak and strong interactions (see section 4.2). Such theories introduce a relationship between quarks and leptons within the same family, and the electron and neutrino masses could be related to the mass difference between the proton and neutron or between u and d quarks. However, the symmetry breaking mechanism that would give rise to observed particle masses is far from understood, and all this still remains at a very speculative level. In the meantime, let us rejoice that nobody can really fiddle with the Standard Model parameters!

In the following two sections we shall introduce fundamental concepts relating elementary particles, their interactions and symmetries of Nature. These aspects may appear somewhat arduous to readers not accustomed with the mathematical tools. The concepts of quantum fields and *Lagrangian* are introduced in inset 1.3 and inset 1.4, respectively. While of fundamental importance for particle physics, it is not necessary to master them to appreciate reading this book.

[15]The anthropic principle according to which we live in a universe that has the characteristics for life to exist, just as we live on the planet Earth and not on Jupiter, may appear esoteric. However, it has scientific grounds in the theoretical possibility of a (very) large number of universes (*multiverse*), which would allow for a statistical rather than a dynamic explanation of the observed parameter values: out of a majority of hostile universes, we live in the one that can accommodate life.

Inset 1.3: A glimpse at the quantisation of free fields

A *field*, for example a scalar field (a real or complex number), or a vector or spinor field (see inset 1.5), is a mathematical entity taking specific values at every point in space–time (x, y, z, t), often shortened to $(\mathbf{x}, t)$. In everyday life we are familiar with classical fields, for instance, temperature fields (scalar) or wind flow speed fields (vector). In fundamental physics, we have for example electric and magnetic fields (vectors) and the gravitational field (vector in its Newtonian version, tensor in Einstein's general relativity).

Quantum fields are the fundamental entities in particle physics. They may get locally excited which gives rise to *field quanta* also called "particles". As shown in this inset, in quantum field theory, the field becomes an *operator*. This generalises a basic concept of quantum theory according to which, for example, a particle's position x or momentum p are operators acting on its wave function.[a]

A field can be seen as a system with an infinite number of degrees of freedom (at least one per space point). Before proceeding to the quantisation of such a complex system, let's start with a simple one, the one-dimensional harmonic oscillator.

In classical mechanics, the displacement $x(t)$ from the equilibrium position of a particle of mass m attached to a spring with strength constant k is the solution of a Newton's equation of motion $m\frac{d^2x}{dt^2} = kx$, which expresses the energy conservation:

$$E = \frac{1}{2}m\left(\frac{dx}{dt}\right)^2 + \frac{1}{2}kx^2. \tag{1.1}$$

This displacement can also, and more appropriately in the present context, be derived from the minimisation of a quantity called *action* (anticipating on inset 1.4), which is an integral over a time interval of the *Lagrangian*

$$\mathcal{L} = \frac{1}{2}\left(\frac{dx}{dt}\right)^2 - \frac{1}{2}m\omega^2x^2, \tag{1.2}$$

where $\omega = \sqrt{\frac{k}{m}}$ is the angular frequency of the oscillator. The solution can be written as (recall that $e^{i\phi} = \cos\phi + i \cdot \sin\phi$ for a phase ϕ)

$$x(t) = A\cos(\omega t + \phi) = \frac{1}{\sqrt{2\omega}}\left(ae^{-i\omega t} + a^* e^{i\omega t}\right), \tag{1.3}$$

where a is complex number (a^* its conjugate) derived from the amplitude A and the phase ϕ of the displacement ($a = \sqrt{\frac{\omega}{2}} A e^{i\omega t}$).

Moving to the quantum level,[b] the dynamic of the system is governed by the Hamiltonian $\hat{H}$ which is the operator version of the energy

$$\hat{H} = \frac{\hat{p}^2}{2m} + \frac{1}{2} m\omega^2 \hat{x}^2, \tag{1.4}$$

where $\hat{p}$ is the momentum operator (given by $\hat{p} = -i\hbar\frac{\partial}{\partial x}$) and $\hat{x}$ is the position operator (given by x). The time independent Schrödinger equation writes $\hat{H}\psi = E\psi$, where the eigenvalue E corresponds to the energy level of the eigenstate ψ. Solving the differential equation leads to eigenstates ψ_n with quantised energy levels $E_n = (n + \frac{1}{2})\hbar\omega$. In this quantisation procedure, the complex numbers a and a^* become respectively annihilation and creation operators $\hat{a}$ and $\hat{a}^\dagger$ of a quantum of energy $\hbar\omega$, which means that they transform respectively a state ψ_n containing n quanta to a state ψ_{n-1} or a state ψ_{n+1} containing $n-1$ or $n+1$ quanta.

Let us move now to the case of a field. For simplicity, we first consider a free (that is, non interacting) relativistic scalar field $\phi(\mathbf{x}, t)$, with one degree of freedom per space point. The equation of motion can be derived from the least action principle with a Lagrangian now expressed as an integral over all space points:

$$\mathcal{L} = \int d^3x \left[\frac{1}{2}\left(\frac{\partial\phi}{\partial t}\right)^2 - \frac{1}{2}(\nabla\phi)^2 - \frac{1}{2}m^2\phi^2\right], \tag{1.5}$$

where $\phi = \phi(\mathbf{x}, t)$ and natural units ($\hbar = c = 1$) have been used to simplify the notations. Note the similarity with the harmonic oscillator

Lagrangian of equation (1.2). The solution is known as the Klein–Gordon equation (see also inset 1.5)

$$\left(\frac{\partial^2}{\partial t^2} - \nabla^2 + m^2\right)\phi(\mathbf{x}, t) = 0\,. \tag{1.6}$$

Through a Fourier transform of this equation, its solutions can be expressed as a sum of plane-waves or normal modes (oscillating systems with a given frequency) as follows:

$$\phi(\mathbf{x}, t) = \int \frac{d^3p}{(2\pi)^3}\frac{1}{\sqrt{2\omega_{\mathbf{p}}}}\left(a_{\mathbf{p}}e^{-i\omega_{\mathbf{p}}t + i\mathbf{p}\cdot\mathbf{x}} + a^*_{\mathbf{p}}e^{i\omega_{\mathbf{p}}t - i\mathbf{p}\cdot\mathbf{x}}\right), \tag{1.7}$$

where $a_{\mathbf{p}}$ is a complex number and $\omega_{\mathbf{p}} = \sqrt{|\mathbf{p}|^2 + m^2}$ is the frequency of the normal mode, with $\mathbf{p} = (p_x, p_y, p_z)$ the three-dimensional momentum. Note again the similarity with the harmonic oscillator displacement $x(t)$ of equation (1.3). Hence, each normal mode corresponding to a value of $\mathbf{p}$ can be seen as a harmonic oscillator with frequency $\omega_{\mathbf{p}}$.

In the process of field quantisation, by the same token as the promotion of a classical harmonic oscillator to a quantum harmonic oscillator, the real scalar field $\phi(\mathbf{x}, t)$, which corresponds to the displacement $x(t)$ in the case of the one-dimensional harmonic oscillator, is promoted to a quantum field operator $\hat{\phi}(\mathbf{x}, t)$. The complex numbers $a_{\mathbf{p}}$ and $a^*_{\mathbf{p}}$ are promoted to or reinterpreted as annihilation and creation operators $\hat{a}_{\mathbf{p}}$ and $\hat{a}^\dagger_{\mathbf{p}}$ of a (free) particle of energy $\hbar\omega_{\mathbf{p}}$ and momentum $\mathbf{p}$. Thus, while the space of states of the one-dimensional quantum harmonic oscillator contains all the discrete energy states of one oscillating particle, the space of states of a quantum field, called *Fock space*, contains the discrete energy levels of an arbitrary number of particles.

As we shall see in the example of the scalar Higgs field, the Lagrangian may contain a self-interaction term, e.g. $\frac{1}{2}\lambda\phi^4$, in addition to the free terms in equation (1.5). In such a case, the ground state (or vacuum) ϕ_0 of the system which corresponds to the minimum of the potential in the Lagrangian, $\frac{1}{2}(m\phi^2 - \lambda\phi^4)$, may be different from zero. The first term of the Taylor expansion of the potential around this minimum with $\phi = \phi_0 + h$ will exhibit a term h^2. The free-field quantisation described above then applies to the field h and the associated particles corresponding to its normal modes are the *excited states.*

A similar procedure can be used to quantise complex scalar fields (charged particles), Dirac (or spinor) fields, and vector fields (e.g. the electromagnetic field). This particular procedure is called *canonical quantisation.* Another quantisation scheme, called *path integral* formulation, has been developed later, in particular by Richard P. Feynman (see inset 1.4). Both approaches have been shown to be equivalent.

So far, we have only considered free fields. The case of interacting fields will be discussed in section 1.9.

Let us finish with a remark to illustrate the difference between the classical and the quantum worlds. In inset 1.1 we introduced the concepts of bosons and fermions, fermions being characterised by single occupancy per quantum state of a system, while bosons can occupy in unlimited numbers the same quantum state. The fact that photons and gravitons, quanta of respectively the electromagnetic and gravitational fields, are bosons and can occupy in innumerable numbers the same quantum states leads to macroscopic manifestations of these quantum fields that can even be treated by methods of classical mechanics and electromagnetism of Maxwell and Newton. The most obvious case of such a macroscopic (quasi) single electromagnetic quantum state is a laser beam. Electrons (like muons, quarks and protons) being fermions, a single electron occupies a particular quantum state. Thus there are no macroscopic manifestations of the quantum electron field.

---

[a]This is why the process of field quantisation is called *second quantisation*, as it promotes the wave function of quantum mechanics, forming the space of states of the system, to the status of an operator acting on the space of states describing the particle content of the system.

[b]See textbooks on basic quantum mechanics.

Inset 1.4: Lagrangian and the principle of least action

Joseph-Louis Lagrange (1736–1813).

Almost all theories aiming at describing the dynamical behaviour of a physical system can be formulated from the minimisation of a mathematical quantity called *action*. The action is expressed as the integral of a function of the variables describing the system's state, the *Lagrangian*. The Lagrangian encapsulates both the dynamical properties of the system, like its kinetic energy, and the properties of the interactions among the components of the system and with the external world. It is named after the Italian-French mathematician and astronomer Joseph-Louis Lagrange (see picture on the right), who established the principles of this variational procedure at the end of the eighteenth century.

For a classical, non-relativistic system described at time t by n generalised coordinates q_i $(i = 1, \ldots, n)$ and their time derivatives $\dot{q}_i = dq_i/dt$, the Lagrangian is given by

$$\mathcal{L}(q_i, \dot{q}_i, t) = T - V,$$

where T and V are, respectively, the kinetic and potential energies of the system. The evolution of the system in the time interval $[t_1, t_2]$ is derived from the minimisation of the action

$$S = \int_{t_1}^{t_2} \mathcal{L}(q_i, \dot{q}_i, t)dt,$$

which leads to differential equations expressing Newton's laws of classical mechanics.

In quantum mechanics, the path of a particle loses its meaning so that the principal of least action has to be reformulated. In the path integral formalism, the probability $P(b, a)$ for a particle to move from state a to state b is given by the square of a complex number, the amplitude function $K(b, a)$: $P(b, a) = |K(b, a)|^2$. The amplitude function is given by the integral

$$K(b,a) \propto \int_a^b D[\mathbf{x}] e^{iS[\mathbf{x}]/\hbar},$$

where $D[\mathbf{x}]$ indicates integration over all possible paths $[\mathbf{x}]$ connecting a to b, and $e^{iS[\mathbf{x}]/\hbar}$ is a phase factor with the action $S[\mathbf{x}]$.

In the classical limit ($\hbar \to 0$ or $S[\mathbf{x}] \gg \hbar$), the path with the minimum action where the variation $\delta(S/\hbar)$ is small dominates, while at large $S/\hbar$ the exponential undergoes large rapid oscillations which cancel each other and average out to zero.

In relativistic quantum field theory, this principle is generalised to the case of an infinite number of degrees of freedom corresponding to the field values and their derivatives at any point in space–time. The action S can then be expressed as an integral of a Lagrangian (also called Lagrange density) over space–time coordinates $S = \int d^4x\, \mathcal{L}(\mathbf{x})$, where now $\mathbf{x} = (t, x, y, z)$. The evaluation of the probability amplitude of a system starting in a certain initial state (for example an electron and a positron colliding with a certain centre-of-mass energy) and ending in a certain final state (for example two W bosons) involves the calculation of an integral over all field configurations of a functional F weighted by a phase factor

$$\langle F \rangle = \frac{1}{Z} \int D[\psi] D[A] F(\psi, A) e^{iS([\psi],[A])/\hbar},$$

where $[\psi]$ and $[A]$ represent a configuration (values at all space–time points) of the fields involved in the process (in the above example, the fermions fields ψ and the bosons fields A). The path integral is performed over all the paths in the space $([\psi], [A])$ connecting the considered initial and final states. Z is a normalisation factor given by

$$Z = \int D[\psi] D[A] e^{iS([\psi],[A])/\hbar}.$$

The concept of path integrals was developed by Richard P. Feynman in 1948 to generalise the Lagrangian approach to quantum mechanics and later to quantum electrodynamics (QED). Like in the classical case, the Lagrangian $\mathcal{L}$ describing the dynamics of the fields is composed of a kinetic term, involving the space–time derivatives (e.g. $\partial_\mu \psi = d\psi/dx_\mu$, $\mu = 0, 1, 2, 3$, where $\mu = 0$ corresponds to the *time* coordinate) and of a potential term, describing the interaction between the fields. As we shall see later, the interaction terms derive naturally from symmetry principles acting on the fields of the theory.

An example of a simple (though non-natural) theory is provided by a scalar field ϕ associated to a particle of mass m and with a self-interaction term ϕ^4. Such a theory is described by the Lagrangian

$$\mathcal{L} = \frac{1}{2}\partial_\mu\phi\,\partial^\mu\phi - \frac{1}{2}m^2\phi^2 + \frac{\lambda}{4!}\phi^4,$$

where λ stands for the coupling strength of the self-interaction. The sum over the space–time indices μ is implied (although not explicit as is customary in relativistic theories, see inset 1.5).

In the calculation of probability amplitudes of the type given above, one often makes use of the fact that coupling strength parameter are small allowing to make a Taylor expansion of the exponential factor in the path integral (more precisely of the part corresponding to the interaction term). This procedure leads to the so-called *perturbative expansion*, each term of the series in λ being associated to a set of Feynman diagrams to simplify the calculation and to provide more intuitive understanding of the involved sub-processes (see section 1.9). The structure of the Feynman diagrams used to calculate the probability amplitude of a process can be directly derived from the interaction term of the Lagrangian.

1.3. External and internal symmetries

Symmetry is a crucial concept in modern fundamental physics. In our everyday lives it is a perfectly clear notion: we have an approximately symmetric face, while the double helix of the DNA molecule is not. A car viewed from the front is not fully symmetric (in France or the USA we drive on the right side), but a billiard ball is symmetric. Is this what physicists mean when they talk about symmetry? To some extent, yes. More quantitatively, a notion of symmetry in physics implies that a certain quantity is not measurable. In mathematical terms, this is equivalent to saying that an equation describing or governing a certain dynamic is invariant. That is, the solution of the equation is not modified by an operation under which it is symmetric: symmetry entails invariance. For example, a transformation consisting in the reflection relative to a plane perpendicular to the axis of a left-handed screw will transform it into a right-handed one: the screw is not invariant under such a transformation. On the other hand, a human face is invariant under a left-right symmetry, at least approximately. Similarly, a billiard ball can be rotated in all directions without this modifying its appearance or properties. The ball is said to be *rotation invariant.* A French (right-hand drive) car reflected in a vertical plane would (approximately) transform it into a British (left-hand drive) car. There is no invariance in this case.

The German mathematician Emmy Noether published in 1918 a theorem of fundamental importance for modern physics (figure 1.5). It states that to every continuous symmetry of a system corresponds a conservation law. Let us take, for example, the invariance of dynamical equations relative to arbitrary spatial translations, $\vec{x} \longmapsto \vec{x} + \vec{s}$, which expresses in mathematical terms the homogeneity of space, i.e. all points in space are equivalent and can be chosen as the origin of a coordinate system. This symmetry guaranties that if a trajectory of a free particle is possible in the reference frame $\vec{x}$, it is also possible in the $\vec{x} + \vec{s}$ frame. This in turn implies the conservation of momentum. Similarly, the invariance of physical laws relative to a rotation of the reference frame around an arbitrary axis, owing to the isotropy of space, implies conservation of angular momentum. And invariance with respect to time relocation — or time translation $t \longmapsto t + T$ — (any moment in time can be taken as the origin of counting time) implies the fundamental law of energy conservation.

Fig. 1.5. Emmy Noether (1882–1935).

Rotations in the three-dimensional physical space belong actually to a more general class of coordinates transformations in the four-dimensional space–time, the Lorentz transformations. According to special relativity, a fundamental theory developed by Albert Einstein in 1905, laws of physics are invariant (i.e. identical) in all inertial frames of reference (i.e. related to each other with a constant velocity). See inset 1.5 for details.

Inset 1.5: Lorentz transformations and relativistic invariance

The principle of special relativity is at the foundation of all particle physics theories.[a] According to it, the laws of physics stay the same (are invariant) for all observers that are moving with respect to each other within an inertial frame (reference frames that are moving at a constant speed with respect to each other). In particular, the speed of light in vacuum c should be the same in any inertial frame — as postulated by Einstein —, a statement which is implicitly contained in the Maxwell equations of electromagnetism. This constraint underlies the Lorentz transformations derived by Henri Poincaré and Hendrik Lorentz before Albert Einstein's formulation of the full theory in 1905.

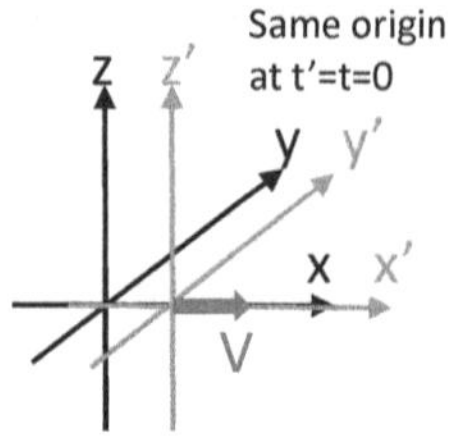

The *Lorentz boost* is the simplest Lorentz transformation. A boost in x-direction with velocity v applied to a space–time interval ΔX is given in its matrix form by

$$\begin{pmatrix} c\Delta t' \\ \Delta x' \\ \Delta y' \\ \Delta z' \end{pmatrix} = \begin{pmatrix} \gamma & -\beta\gamma & 0 & 0 \\ -\beta\gamma & \gamma & 0 & 0 \\ 0 & 0 & 1 & 0 \\ 0 & 0 & 0 & 1 \end{pmatrix} \begin{pmatrix} c\Delta t \\ \Delta x \\ \Delta y \\ \Delta z \end{pmatrix},$$

where $\beta = v/c$, and $\gamma = 1/\sqrt{1-\beta^2}$ is the *Lorentz factor.*[b] Other Lorentz transformations are pure space rotations. A general Lorentz transformation is a product of a pure boost and a pure rotation.

Mathematically, this fundamental symmetry of space–time is expressed by the invariance under coordinate transformations which leave invariant the *Minkowski metric* $ds^2 = cdt^2 - dx^2 - dy^2 - dz^2$ (for an infinitesimal space–time interval). The ensemble of Lorentz transformations, the *Lorentz group*, can be identified with the group $O(3,1)$ of orthogonal transformations acting on a four-dimensional space R^4 endowed with the Minkowski metric tensor $\eta_{\mu\nu} = (1,-1,-1,-1)$.[c]

A physical quantity is said to be Lorentz covariant if it transforms under a given representation of the Lorentz group.[d] These quantities are built out of *scalars*, *four-vectors* (or 4-vectors), *four-tensors*, and *spinors*. Under a Lorentz transformation Λ, a 4-vector X transforms as

$X \longmapsto \Lambda X$ or, with space–time indices ($\mu = 0, 1, 2, 3$), $X^\mu \longmapsto \Lambda^\mu_\nu X^\nu$. Here the explicit sum $\sum_{\nu=0}^{3} \Lambda^\mu_\nu X^\nu$ is omitted to simplify the notation.

Example of 4-vectors are:

- Four-displacement: $\Delta X^\mu = (c\Delta t, \Delta x, \Delta y, \Delta z)$.
- Four-gradient: $\partial_\mu = (\frac{1}{c}\frac{\partial}{\partial t}, -\frac{\partial}{\partial x}, -\frac{\partial}{\partial y}, -\frac{\partial}{\partial z})$.
- Four-velocity: $V^\mu = \gamma(c, \frac{dx}{dt}, \frac{dy}{dt}, \frac{dz}{dt})$.
- Four-momentum: $P^\mu = (\frac{E}{c}, p_x, p_y, p_z) = mV^\mu$, where m is the particle's rest mass.
- Four-current: $J^\mu = (c\rho, j_x, j_y, j_z) = \rho_0 V^\mu$.
- Electromagnetic four-potential: $A^\mu = (\phi/c, A_x, A_y, A_z)$.

Scalar quantities which are invariant under Lorentz transformations can be built out of the norm of four-vectors. For example:

- Space–time interval: $\Delta s^2 = \eta_{\mu\nu}\Delta x^\mu \Delta x^\nu = c\Delta t^2 - \Delta x^2 - \Delta y^2 - \Delta z^2$.
- Mass: $\eta_{\mu\nu}P^\mu P^\nu = E^2/c^2 - p_x^2 - p_y^2 - p_z^2 = m^2c^2$. In the rest frame of the particle, one finds the well-known formula $E = mc^2$.

A four-tensor, or simply tensor, is a quantity with two (or more) indices which transforms like a bilinear form, a matrix (or multilinear form), $T' = {}^t\Lambda T \Lambda$, or written with indices (and implicit summation over λ and ρ): $T'_{\mu\nu} = T_{\lambda\rho}\Lambda^\lambda_\mu \Lambda^\rho_\nu$. An example is given by the electromagnetic field tensor

$$F_{\mu\nu} = \begin{pmatrix} 0 & \frac{1}{c}E_x & \frac{1}{c}E_y & \frac{1}{c}E_z \\ -\frac{1}{c}E_x & 0 & -B_z & B_y \\ -\frac{1}{c}E_y & B_z & 0 & -B_x \\ -\frac{1}{c}E_z & -B_y & B_x & 0 \end{pmatrix}.$$

The metric tensor $\eta_{\mu\nu}$ is by definition invariant under Lorentz transformations: $\eta_{\mu\nu} = \eta_{\alpha\beta}\Lambda^\alpha_\mu \Lambda^\beta_\nu$.

An equation is said to be *Lorentz covariant* if it can be written in terms of Lorentz covariant quantities (also said to be Lorentz invariant or simply *relativistic*). The key property of such equations is that if they hold in one inertial frame they do so in any such frame. For example, Maxwell's equations in vacuum are classically written as (∇ is the gradient operator)

$$\nabla \cdot \mathbf{E} = \frac{\rho}{\epsilon_0}, \quad \nabla \cdot \mathbf{B} = 0 \quad \text{Gauss' laws,}$$

$$\nabla \times \mathbf{B} = \mu_0 \left(J + \epsilon_0 \frac{\partial \mathbf{E}}{\partial t} \right) \quad \text{Ampère's law,}$$

$$\nabla \times \mathbf{E} = -\frac{\partial \mathbf{B}}{\partial t} \quad \text{Maxwell-Faraday's law.}$$

When written with Lorentz covariant quantities, they reduce to

$$\partial_\mu F^{\mu\nu} = \mu_0 J^\nu,$$
$$\partial_\mu \epsilon^{\mu\nu\lambda\rho} F_{\lambda\rho} = 0,$$

where $\epsilon^{\mu\nu\lambda\rho}$ is the completely anti-symmetric tensor.[e]

The Schrödinger equation for a single non-relativistic particle in quantum mechanics is the operator version of the equation in classical mechanics giving its energy: $\frac{\mathbf{p}^2}{2m} + V(r,t) = E$ where V is the potential function. Using the quantum operator ansatz $\mathbf{p} \rightarrow i\hbar\nabla$ and $E \rightarrow i\hbar\frac{d}{dt}$, where ∇ is the gradient operator, it writes:

$$\left[-\frac{\hbar^2}{2m}\nabla^2 + V(r,t) \right] \psi(r,t) = i\hbar \frac{d\psi(r,t)}{dt},$$

where ∇^2 is the Laplace operator. The symbol ψ denotes the scalar complex wave function of the particle.

This equation is clearly not Lorentz covariant. To extend Schrödinger's equation in the relativistic regime, the same operator ansatz can be used for the equation: $E^2 - \mathbf{p}^2c^2 = m^2c^4$. It leads to the *Klein–Gordon* equation:

$$\frac{1}{c^2}\frac{\partial^2\psi}{\partial t^2} - \nabla^2\psi + \frac{m^2c^2}{\hbar^2}\psi = 0,$$

which can be written in a Lorentz covariant form:

$$(\eta^{\mu\nu}\partial_\mu\partial_\nu - m^2)\psi = (\Box - m^2)\psi = 0,$$

where the so-called *natural units* $\hbar = c = 1$ have been used and $\Box$ is the d'Alembert operator.

Whereas this differential equation is supposed to apply to any free particle, it contains only a second order derivative with respect to time, which leads to inconsistencies if we keep the interpretation of the wave function ψ — a Lorentz scalar — as a way to determine the position probability for a single particle.[f]

In an attempt to derive an equation with a first-order derivative with respect to time, the British physicist Paul Adrien Maurice Dirac realised (in 1928) that one has to turn the scalar wave function into a four-component one, ψ_α, called a Dirac spinor (or bispinor). Spinor and Minkowki indices are both related to space–time but in different representation spaces. Lorentz transformations act on spinors in a different (but closely related) way than on 4-vectors. The Dirac equation for a free electron can be written in a compact Lorentz covariant form as the matrix equation in natural units

$$(i\gamma^\mu \partial_\mu - m)\psi = 0$$

with γ^μ being the so-called 4×4 Dirac matrices verifying the anti-commuting relations $\gamma^\mu\gamma^\nu + \gamma^\nu\gamma^\mu = 2\eta^{\mu\nu} I_4$, where I_4 is the four-dimensional identity matrix.[g] Here μ is a Minkowski index, showing that the set of four Dirac matrices behaves as a 4-vector in the Minkowski space–time. The first two components of a Dirac spinor, (ψ_L, ψ_R), can be associated to the fermion with its two helicity (or spin orientation) states and the third and fourth components, $(\bar{\psi}_L, \bar{\psi}_R)$, to its antiparticle.

In the presence of a electromagnetic field described by the four-potential A^μ, the equation of motion of an electron is obtained by replacing the standard derivative ∂_μ by the so-called *covariant derivative* $D_\mu = \partial_\mu - iqA_\mu$ where q is the electron electric charge (see inset 1.6).

[a]For theories attempting to include gravity, the Minkowski metric $\eta_{\mu\nu}$, which corresponds to a flat space–time, must be replaced by a more complex metric tensor $g_{\mu\nu}$ of a curved space–time deduced from the Einstein equation of general relativity.

[b]As a direct consequence of the Lorentz transformation, the time interval in the laboratory frame between the production and decay of a particle moving at speed

v and with a proper lifetime τ (i.e. lifetime in its rest frame) is $\Delta t' = \gamma\tau$. For example, cosmic muons produced at high altitude with a speed close to c can have a long enough lifetime to reach the Earth surface despite their short proper lifetime of 2.2 microseconds.

[c]The Lorentz group can be augmented by translations in space and time, leading to the so-called Poincaré group.

[d]To make it simple, in mathematics a representation of a group is a description of this group as a set of $n \times n$ invertible matrices acting on an n-dimensional vector space. For example, the group of spatial rotations SO(3) and the special unitary group SU(2) are different representations of the same abstract group: the first representation acts on R^3 (rotations in the physical space) and the second on a two-dimensional complex space C^2. In the SU(2) representation the elements of C^2 are called spinors and describe the spin of fermions, namely how they behave internally under a rotation (see inset 1.1).

[e]The Levi-Civita tensor equals to $+1$, -1 or 0 whether the sequence of indices corresponds to an even or an odd permutation or if two indices are identical.

[f]The Klein–Gordon equation recovers its consistency in quantum field theory for the description of scalar spinless particles.

[g]One can check that, by multiplying on the left the left-hand term of the Dirac equation by its hermitian conjugate and making use of the anti-commuting properties of the Dirac matrices, one recovers the Klein–Gordon equation. The Dirac equation can thus be seen as a kind of square root of the Klein–Gordon equation.

All these *external symmetries* and resulting conservation laws are relatively easy to visualise. This is not the case for *internal symmetries* of particles, the understanding of which requires a certain degree of abstraction. Internal symmetries deal with invariance relative to transformations not related to space or time coordinates. They were introduced in the description of the micro-world to explain specific quantum phenomena observed in a laboratory. Take, for example, the wave function $\psi(\vec{x}, t)$, a complex-valued probability amplitude that describes the state of a physical system. Only its modulus $|\psi(\vec{x}, t)|$ is a measurable quantity. The transformation $\psi(\vec{x}, t) \longmapsto e^{i\theta}\psi(\vec{x}, t)$, where $i = \sqrt{-1}$ and θ is an arbitrary constant phase, is an internal symmetry that does not alter the modulus (as $|e^{i\theta}| = 1$). The conservation law related to this symmetry is the one of preserving the system in space. It is called *unitarity* and implies that no object can disappear from the Universe without leaving a trace, i.e. the sum of probabilities of all possible outcomes of any event equals one.

Another example of an internal symmetry is given by the *weak isospin*. We saw earlier that fermions were grouped into three families of quarks (u, d), (c, s), (t, b) and likewise for leptons (ν_e, e), (ν_μ, μ), (ν_τ, τ). Weak isospin distinguishes the u-quark from the d-quark within the first family, and likewise the electron from the neutrino ν_e. If we could neglect the electric charges and masses of the particles and consider only weak interactions Nature would be invariant under global weak isospin transformations.

1.4. Gauge invariance and fundamental interactions

Transformations of the type we introduced so far, a space translation by a vector $\vec{s}$, a time translation by T, a phase shift by θ, are called *global* as $\vec{s}$, T and θ are constant, the same, in every point of space and time. However, $\vec{s}$, T or θ may also depend on their space–time coordinate $(\vec{x}, t)$, which would make the corresponding transformation *local.* At this stage we should introduce the notion of *gauge freedom*, which denotes mathematical redundancy if two different expressions describe the same physical system (so that an observer cannot distinguish which expression is right). The redundant (or *gauge dependent*) degrees of freedom in the expression are non-physical as their values cannot be measured, even in principle.[16] A global phase shift θ of a wave function represents such a redundancy (it corresponds to a *global gauge transformation*), and the concrete value used in a calculation corresponds to a chosen gauge. Accordingly, we can introduce the term *local gauge transformation* and the invariance of an equation or a physical law under such a transformation is called *local gauge invariance.*[17] Is such a local invariance possible at all, and what would be its meaning?

Let us take as an example the translation of a coordinate $\vec{x}$ by a vector $\vec{s}(\vec{x}, t)$, which is a function of x and t. At first sight the transformation $\vec{x} \to \vec{x} + \vec{s}(\vec{x}, t)$ would change in a drastic, if not chaotic, way the trajectory of a particle as a function of its position in space and time (cf. figure 1.6), and it seems difficult to imagine any form of underlying invariance! It is, however, conceivable that there exists a non-uniform field generated by a force that, when applied to the particle, could restore its initial straight trajectory. Thus, invariance under a translation $s(\vec{x}, t)$ does not describe anymore the trajectory of a *free* particle. This observation suggests that gauge invariance, if it exists, must be intimately related to forces.

[16] A familiar case is the electric scalar potential V. Conventionally the static potential is set at zero for the ground, grounding an electrical system, but what is really measurable and relevant is the potential difference ΔV, not the absolute value.

[17] In the following, we will consider only local gauge transformations and thus shall drop the term *local.*

The first example of a gauge theory encountered in physics is the general theory of relativity developed by Albert Einstein during the years 1912–1915. His goal was to describe gravitational phenomena in a manner consistent with the special theory of relativity he formulated about ten years earlier. The theory is based on the *equivalence principle* stating, in its so-called strong version, that, as long as we limit ourselves to a sufficiently small region of space, the laws of physics are identical in a uniformly accelerating reference frame and in a reference frame at rest but under the influence of a gravitational field of force. It has as a consequence the identity of inertial and gravitational masses — as shown to be valid initially by the Eötvös-type experiments,[18] and confirmed by all further experiments until now. More specifically, in a change of reference frame through a translation $\vec{x} + \vec{s}(\vec{x}, t)$ acceleration is induced locally through the space–time dependence. For an observer in the new reference frame everything looks like as if being in a reference frame at rest under the influence of gravitational forces generated by a surrounding distribution of masses. These two situations cannot be distinguished, and it is in this sense that we talk of invariance of physical laws under local coordinate transformations.

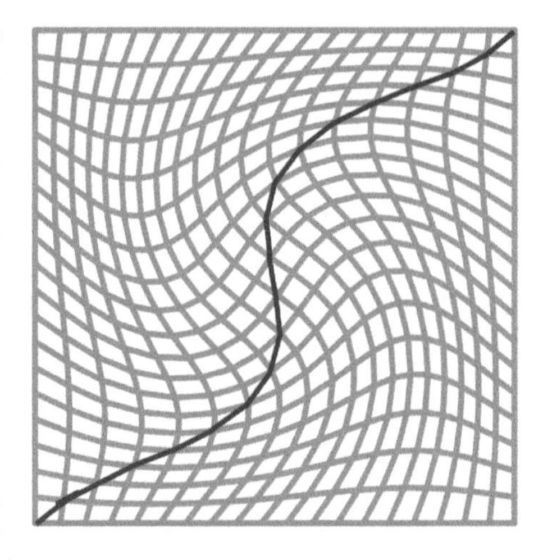

Fig. 1.6. Deformation of the trajectory of a free particle resulting from a local coordinate change $(x, y) \longmapsto (x, y) + \vec{s}(x, y)$ on a two-dimensional square surface.

Requiring invariance under local gauge transformations may seem arbitrary, but it is not. The resulting interactions are physical indeed. The equations describing the classical dynamics of a system of massive particles

[18]This type of experiment, the first one performed around 1885 by the Hungarian physicist Lorand Eötvös, aims at comparing the action of the Earth gravity (associated with the gravitational mass) and the action of the centrifugal force induced by the Earth rotation (associated with the inertial mass) on two different masses located at the two ends of a torsion balance.

that remain invariant under local coordinate transformations[19] are those of Einstein's general relativity. Similarly, requiring the wave function describing an electrically charged electron or positron to be invariant under a local phase transformation $\theta(\vec{x}, t)$ gives rise to a gauge field corresponding to the four-vector potential of electromagnetism (see inset 1.6). In quantum field theory, we call the quantum excitations (quanta) of the electromagnetic field *photons*, usually denoted by the symbol γ. The electromagnetic interaction can thus be viewed as the coupling between a charged particle and a photon. Its strength is proportional to the electromagnetic *coupling constant* we mentioned earlier in this chapter, and thus to the charge of the particle. Going a bit further into the mathematics of gauge transformations we can say in this specific case that electromagnetism stems from the gauge invariance under the group of transformations on a circle, U(1), acting on a one-dimensional complex space.

The extension of gauge invariance to more complex internal particle symmetries than local phase symmetry has led to the so-called Yang–Mills theory (named after the names of the Chinese and American theorists Chen-Ning Yang and Robert Mills who introduced these ideas in the years 1950). Yang–Mills theory is at the origin of the modern description of the weak and strong interactions. The weak interaction results from the invariance under local transformations of the special unitary group SU(2) acting on the complex two-dimensional space of weak isospin. The mathematical structure of SU(2) implies the existence of three gauge fields (and hence three field quanta) rather than a single one as in electromagnetism. In the next section, we shall see how, thanks to the Brout–Englert–Higgs mechanism, one of these fields mixes with the gauge field of U(1) to give two neutral vector bosons, the massive Z boson and the massless photon, whilst the other two fields give rise to the massive charged vector bosons W^+ and W^-. As for the strong interaction, it is associated with the invariance under a group of unitary transformations in three dimensions called SU(3) and acting in a space describing the colour of quarks. In this theory, called *quantum chromodynamics* (see inset 1.13), there are eight gauge fields to which are associated eight massless vector bosons, the *gluons* (usually denoted by the symbol g), which carry the strong interaction.

[19]Coordinate transformations include space–time translations as well as rotations described by the Lorentz group SO(3,1), which encompass spatial rotations and Lorentz transformations in the four-dimensional Minkowski space.

Thus we conclude that all four fundamental interactions known today are related to gauge invariance; they all stem quasi miraculously from a geometrical principle.

Inset 1.6: Electromagnetism as a gauge theory

In quantum mechanics the state of a system, of a particle, is described by a complex wave function $\psi(\vec{x}, t)$ and only its modulus $|\psi(\vec{x}, t)|$ is a measurable quantity. The description of the motion of the particle also requires introducing the modulus of the four-vector gradient $\partial\psi(\vec{x}, t)/\partial x_\mu$, where $\mu = (0, 1, 2, 3)$ are the time and space indices, respectively. The phase transformation $\psi(\vec{x}, t)$ into $e^{i\theta}\psi(\vec{x}, t)$ is an internal symmetry not modifying these two quantities. The invariance under this phase transformation is termed global, as θ is independent of the space–time coordinates $(\vec{x}, t)$. The transformation becomes local if the phase $\theta(\vec{x}, t)$ can be different at every space–time point. In this case we speak of a (local) gauge transformation. The modulus of the wave function still remains invariant, but the modulus of the gradient is not invariant anymore. The transformation of the gradient reads:

$$\frac{\partial\psi(\vec{x}, t)}{\partial x_\mu} \longmapsto \frac{\partial e^{i\theta(\vec{x}, t)}\psi(\vec{x}, t)}{\partial x_\mu} = e^{i\theta(\vec{x}, t)}\left(\frac{\partial\psi(\vec{x}, t)}{\partial x_\mu} + i\frac{\partial\theta(\vec{x}, t)}{\partial x_\mu}\psi(\vec{x}, t)\right)$$

Taking the modulus the phase $e^{i\theta(\vec{x}, t)}$ disappears, but the term with $\partial\theta/\partial x_\mu$ remains. The equation of motion would thus not be gauge invariant. To restore gauge invariance one introduces a four-vector field $A_\mu(\vec{x}, t)$ that transforms as:

$$A_\mu(\vec{x}, t) \longmapsto A_\mu(\vec{x}, t) + \frac{1}{q}\frac{\partial\theta(\vec{x}, t)}{\partial x_\mu},$$

where q is an a-priori arbitrary constant. The next step, the trick, consists in replacing the partial derivative $\partial/\partial x_\mu$ by the so-called covariant derivative D_μ defined by

$$D_\mu\psi(\vec{x}, t) = \left(\frac{\partial}{\partial x_\mu} - iqA_\mu(\vec{x}, t)\right)\psi(\vec{x}, t).$$

A simple calculation shows that this covariant derivative transforms as:

$$D_\mu\psi(\vec{x}, t) \longmapsto e^{i\theta(\vec{x}, t)}D_\mu\psi(\vec{x}, t).$$

The modulus of the gradient in now invariant under a local phase transformation. We just constructed a gauge invariant theory.

What happened? With the introduction of the covariant derivative we generated terms containing the product $qA_\mu(\vec{x}, t)\psi(\vec{x}, t)$, which we interpret as describing the local interaction between the field $A_\mu(\vec{x}, t)$ and the fermion $\psi(\vec{x}, t)$ of charge q at point $(\vec{x}, t)$. We call the field $A_\mu(\vec{x}, t)$ a gauge field. It corresponds to the scalar potential A_0 and the vector one $A_i, i = 1, 2, 3$ from which one derives the electric and magnetic fields of the classical electromagnetic theory of Maxwell. Thus we may say that the electromagnetic interaction follows in a natural way from the requirement of invariance under a local phase transformation.

We have just underlined the *locality* of the interaction between the quantum gauge field $A_\mu(\vec{x}, t)$ and the point-like matter field $\psi(\vec{x}, t)$. This property may be at the root of the difficulty in marrying quantum mechanics with general relativity (a classical, non-quantum theory). The introduction of non-local objects such as strings instead of point-like objects to describe fundamental constituents may be a route to solve that problem.

1.5. When a symmetry breaks down spontaneously: The Brout–Englert–Higgs mechanism

Among the great scientific advances of the 20th century is surely the understanding that all fundamental interactions derive from the principle of gauge invariance. This description does not go without serious difficulties however, as gauge invariance requires that all particles with spin different from zero must be massless! This is in contradiction with experimental facts. We have seen that all quarks and leptons have spin $1/2$, they are not massless and some, on the contrary, are very heavy. The top quark, for example, has as large a mass as an atom of gold! As for the gauge bosons, they have all spin 1 and, although gluons and photon are indeed massless, this is not the case for the W and Z. The large mass of these mediator bosons explains the extremely short range of weak interactions, an indisputable experimental fact.

Fig. 1.7. Robert Brout, François Englert and Peter Higgs, co-inventors of the mechanism carrying their names. Peter Higgs was the first to predict the existence of a boson of spin zero as a consequence of this mechanism. Rather often, when talking about the mechanism and the boson, by an abuse of language only the name of Higgs is mentioned, but this is just for shortness or simplification.

Thus gauge invariance does not seem to be realised in Nature, at least not in an obvious way. How to give masses to elementary particles, to the W, Z bosons and to all fermions, without manifestly breaking gauge invariance that is so desirable? That is the question!

The clever way to circumvent the problem, known today as the *Brout–Englert–Higgs mechanism* (BEH in short), if not the *Brout–Englert–Higgs–Hagen–Guralnik–Kibble* mechanism, has been proposed in 1964 by three teams: Francois Englert and Robert Brout in Brussels, Peter Higgs in Edinburgh (figure 1.7) and few months later Gerald Guralnik, Carl Richard Hagen and Thomas Kibble in London. The idea is inspired by concepts developed in condensed matter physics.[20] It predicts the existence of a quantised manifestation of the field in form of a new spin-zero particle, called the *Higgs boson*, sometimes the *boson of Brout–Englert–Higgs* or just simply the *Higgs*. This boson has been searched for during several decades at accelerators of ever increasing energy, until it was finally discovered at the LHC in 2012. This book is largely devoted to this undertaking.

---

[20]The Lagrangian of inset 1.8 giving the mathematical expression for the BEH mechanism is very similar to the one written in 1951 by the Russian physicists V. Ginzburg and L. Landau to describe superconductivity (see inset 1.9). From the theory point of view, the Higgs boson has a role analogous to the fluctuation of the amplitude of Cooper pairs in superconductivity.

Inset 1.7: Spontaneous symmetry breaking

A simple case

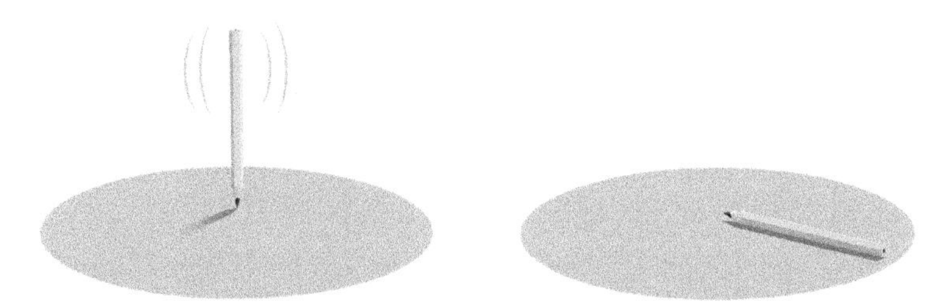

A vertically standing pencil is symmetric with respect to rotations around its axis. Once it has fallen, a unique direction is selected: the symmetry is broken. This simple case illustrates well two important aspects of *spontaneous symmetry breaking*: the first one is that we cannot predict in which direction the pencil will fall (all directions have equal probability), the second being that the description of the system, once the symmetry is broken, does not depend on the chosen direction.

It should not be thought that the breakdown of the initial symmetry here is the result of some minute imperfection on the tip of the pencil. This would then correspond to a system that is not symmetric to begin with, which is not the case we are discussing here.

An example in geometry

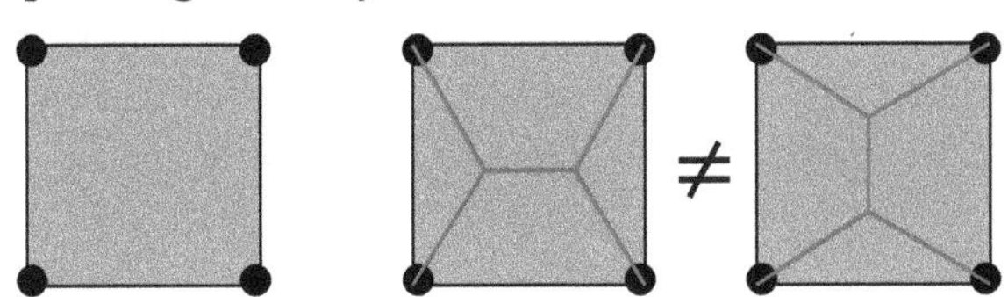

How to connect the four corners of a square, while minimising the length of the segments? The problem is initially perfectly symmetric (the left-hand drawing), whilst the two possible and equally good solutions do not have the symmetry of the initial state (center and right-hand plots).

An example in solid-state physics

Another well-known case of spontaneous symmetry breaking occurs in ferromagnetic materials. When the temperature exceeds a certain critical value called the *Curie temperature*, which depends on the material, the atomic spins (more precisely the directions of the microscopic magnetic domains) of the magnet are randomly oriented, thus not generating any macroscopic magnetic field. The system is invariant under any spatial rotation (figure below on the left).

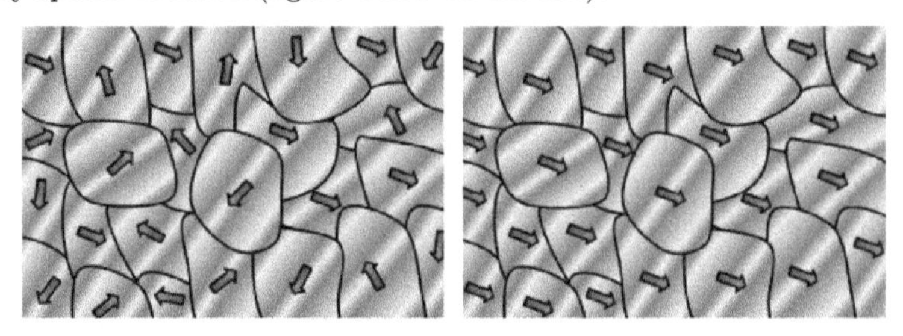

However, when the temperature falls below the critical value, the magnet enters an ordered state in which the magnets are oriented according a privileged direction (figure on the right). There is no rotational symmetry anymore, but the chosen direction is one among an infinity of possible solutions, all obeying the initial symmetry.

This is a case of spontaneous symmetry breakdown. It should not be confused with an explicit symmetry breaking provoked, for example, by the application of an external magnetic field ϕ (in which case we talk about magnetisation). The phase transition of a metal from normal electrical conduction to superconductivity below a critical temperature is yet another case of spontaneous symmetry breaking.

The Brout–Englert–Higgs mechanism rests on two theoretical observations. The first one is that scalar particles (particles of spin zero) can be of non-zero mass without breaking the gauge invariance of the Standard Model equations. The second observation is that it is possible to generate in these equations, in a certain sense spontaneously, terms that correspond to non-zero particle masses, although they were not allowed in the initial equations. This somewhat magic appearance is based on the concept of spontaneous symmetry breaking (insets 1.7 and 1.8) following a phase transition, a concept particle physics theorists borrowed from condensed

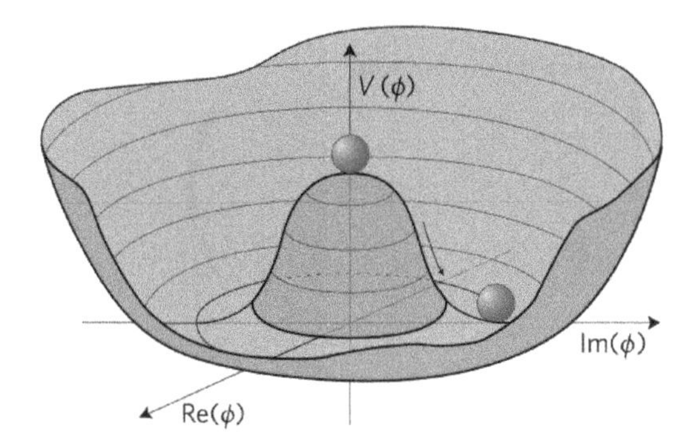

Fig. 1.8. The Higgs potential can be written as a function of the complex Higgs field ϕ as follows: $V(\phi) = \mu^2|\phi|^2 + \lambda|\phi|^4$ where μ^2 and λ are parameters related to the Higgs mass and Higgs self-coupling. When the temperature of the Universe fell under a certain value, μ^2 became negative and the potential took the shape of the Mexican hat shown in this picture. The ground state, i.e. the lowest energy state of the Universe, corresponds then to a non-zero value of the Higgs field, $v = 246$ GeV (see inset 1.8).

matter physics. We talk about spontaneous symmetry breaking when the ground state of a physical system (the lowest energy state of the system) has less symmetry than the equations of motion describing the dynamics of the system. This is also sometimes called a state of *hidden symmetry.* We shall explain a bit further how this looks like. Let us assume there exists a new scalar field which permeates all space. To this field we can associate a potential energy, the Higgs potential, responsible for the self-interaction of the field, as well as for its mass (figure 1.8). At the very beginning of our Universe the shape of this potential is parabolic as a function of the field (which is a complex quantity), with a minimum at zero for a zero field value. At about 10^{-11} seconds after the Big Bang the temperature of the Universe has already fallen by seventeen orders of magnitude, being then only about a million billion kelvin. This corresponds to an average kinetic energy of about 100 GeV for particles present at this epoch. At this stage the Higgs potential transforms and takes the shape of a Mexican hat! The minimum of the potential is now at a non-zero value of the field, more precisely for an infinity of values located on a circle around zero in the complex plane (the Higgs potential is symmetric under rotation in the field space). This behaviour may seem rather astonishing, but it is exactly that of the potential energy of a metallic bar under compression (see figure 1.9). Beyond a certain critical value for the applied force, the symmetry of the bar breaks down, a phenomenon called *buckling* in material sciences.

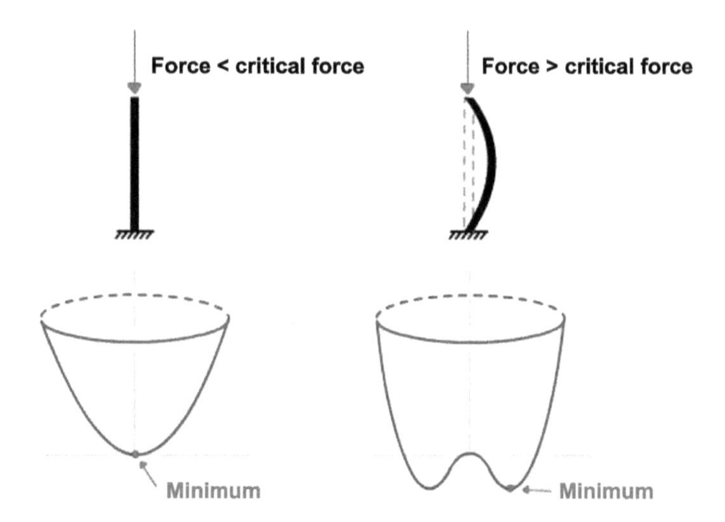

Fig. 1.9. Buckling of an iron bar. As long as the vertical compression force remains smaller than a critical value, the bar remains straight as this corresponds to a minimum of its potential energy. When the force exceeds the critical value, the bar buckles in a certain direction and is not any more symmetric with respect to the rotation axis. However, the shape of the potential energy remains symmetric.

In the present-day world the Higgs field is in this potential minimum, except in some rare cases of high-energy interactions produced either by cosmic rays or by scientists in particle accelerators. Indeed, thanks to the very high localised energy generated in such collisions, it is possible to excite locally the Higgs field in the vicinity of this minimum and to create briefly one or more of these field quanta which we now call Higgs bosons. This is what was achieved at CERN in 2012.

Inset 1.8: Spontaneous symmetry breaking and massive vector fields

Gauge invariance forbids terms of the form $\frac{1}{2}m^2A_\mu^2$ for a vector field $A_\mu = A_\mu(x)$, where $x = (\vec{x}, t)$ is a point in four-dimensional space–time, in the Lagrange density (a function that allows to derive the evolution of particles — or fields — in quantum field theory, see inset 1.4). However, such a mass term is allowed in case of a scalar field $\phi(x)$. We show in this inset how to use the field $\phi(x)$ to generate a mass term for the vector field $A_\mu(x)$ through the mechanism of spontaneous symmetry breaking.

Let us consider a theory with a local gauge invariance under U(1) transformations representing the dynamics of a complex scalar field $\phi(x)$ (see inset 1.6 on electromagnetism as a gauge theory). The Lagrangian of our theory has the following expression:

$$\mathcal{L} = |D_\mu \phi|^2 - (\mu^2|\phi|^2 + \lambda|\phi|^4) - \frac{1}{4}\sum_{\mu\nu} F_{\mu\nu}F^{\mu\nu},$$

where dependences on x are omitted for simplicity. The first term describes the propagation of the scalar field and contains the covariant derivative $D_\mu = \partial_\mu - ieA_\mu$; the second term represents the potential of the scalar field (its mass and self-interaction terms, see figure 1.8); the last term encodes the dynamics of the force field associated to the gauge field A_μ with $F_{\mu\nu} = \partial_\mu A_\nu - \partial_\nu A_\mu$.

The above equation is invariant under a local gauge transformation $\phi \to e^{i\alpha(x)}\phi$. Let us now write this complex field $\phi(x)$ in the following way: $\phi(x) = \rho(x)e^{i\theta(x)}$ where now $\rho(x)$ and $\theta(x)$ are two real-valued functions. Gauge invariance means that the phase $\alpha(x)$ is not a measurable physical parameter, and thus can be used to cancel the contribution from $\theta(x)$ at every space–time point x. The field $\phi(x)$ then becomes a real quantity and can be written as: $\phi(x) = \rho(x) = (v + \chi(x))/\sqrt{2}$ where $v = \sqrt{-\mu^2/\lambda}$. This specific form of the gauge parameter is called a *unitary gauge*.

The first term in our Lagrangian now takes the form

$$\begin{aligned}|D_\mu \phi(x)|^2 &= \frac{1}{2}|(\partial_\mu - ieA_\mu)(v + \chi(x))|^2\\ &= \frac{1}{2}(\partial_\mu \chi(x))^2 + \frac{e^2}{2}A_\mu^2(v^2 + 2v\chi(x) + \chi^2(x)),\end{aligned}$$

which exhibits a mass term for the vector field $A_\mu(x)$ of the form $\frac{1}{2}m^2 A_\mu^2$ if we identify $m = ev$. The spontaneous symmetry breaking has been introduced through a redistribution of fields: one of the two real fields making up the complex field $\phi(x)$ has been removed through a specific choice of gauge, but has reappeared in a new mass term for the vector field A_μ. Overall, the number of degrees of freedom of the system has not changed. This is the Brout–Englert–Higgs mechanism.

Inset 1.9: The Landau–Ginzburg Lagrangian

In 1950, fourteen years before the Brout–Englert–Higgs papers, Vitaly Ginzburg and Lev Landau, two famous Russian theorists, developed a phenomenological theory of metal superconductivity based on Landau's phase-transition theory formulated a few years earlier, which already exhibited most of the features of the BEH mechanism.

Assuming that the transition from normal to superconducting state is a second-order phase transition (denoting a smooth transition that is continuous in the first derivative of the order parameter), Ginsburg and Landau postulated the existence of a complex quantum wave function ψ, which is equivalent to an order parameter characterising the superconducting state: its modulus squared, $|\psi(x)|^2$, is proportional to the density of superconducting particles at position x, identified as the Cooper pairs of the microscopic Bardeen–Cooper–Schrieffer (BCS) theory of superconductivity formulated in 1953. They assumed ψ to be zero above a critical temperature T_c, and a non-zero function $\psi(T)$ for temperatures $T \leq T_c$.

In this solid-state physics case, the physical state, described by the values of the field ψ and of the electromagnetic potential vector A, is derived from the minimisation of *free energy* (analogue to the Lagrangian for field theories), whose density Ginzburg and Landau assumed to be given by

$$F = \frac{1}{2m^*}|(\nabla + ie^*A)\psi|^2 + \alpha|\psi|^2 + \beta|\psi|^4 + \frac{1}{2}(\nabla \times A)^2 + \cdots .$$

Here m^* and e^* are effective mass and charge parameters (m^* and e^* are set to $2m_e$ and $2e$ for electrons in the BCS theory), ∇ is the derivative operator, and α and β are coefficients of a Taylor expansion of the potential at small values of $|\psi(x)|^2$ close to T_c. The first term in the expression is related to a kind of kinetic energy of the scalar field ψ also exhibiting its minimal coupling to the electromagnetic field. To account for the phase transition at T_c, they assumed α to be a linear function of temperature near this critical temperature, $\alpha(T) = \alpha_0(T_c - T)$. Above T_c, the potential is a parabolic function of the ψ

components with a minimum at $\psi = 0$, and for $T < T_c$ it takes the familiar Mexican hat shape (figure 1.8) with degenerate minima at $|\psi| = \psi_0 = \sqrt{\alpha/2\beta}$.

The analogy with the Higgs potential is clear. The consequences are also analogous to those described in the preceeding inset (but which appeared fourteen years earlier). Inserting the value $\psi = \psi_0 e^{i\theta(x)}$ at the minimum in the expression of F, the kinetic term becomes $\psi_0^2(\nabla\theta - e^* A)^2$. The field A can be redefined as $A = A - \nabla\theta/e^*$ (corresponding to a local gauge transformation), which leads to a non-zero effective mass term $(1/2)m_\gamma A^2$ for the photon with $m_\gamma = \psi_0 e^*/\sqrt{m^*}$. This mass corresponds to the inverse (up to some constant) of the short penetration length of the magnetic field $\vec{B} = \vec{\nabla} \times \vec{A}$ inside the superconductor (Meissner effect). As for the BEH mechanism, the massless degree of freedom associated with $\theta(x)$, induced by the spontaneous symmetry breaking at $T = T_c$, has been absorbed to generate an effective mass of the photon. The relativistic extension of this mechanism to particle physics was mainly a technical issue.

1.6. Electroweak unification, the Glashow–Weinberg–Salam model

In 1961 the American physicist Sheldon Glashow working at Harvard made the first attempt to unify electromagnetic and weak interactions based on the gauge group SU(2) × U(1) describing interactions of fermions at high energy. However, he had no satisfactory way to explain the short range of the weak force. In 1967, two physicists, the American Steven Weinberg and the Pakistani Abdus Salam, working at Imperial College London, had independently the idea to complement this gauge theory of Glashow with the Brout–Englert–Higgs mechanism and thus to provide a coherent framework for electroweak unification (see inset 1.10). John Ward at the Johns Hopkins University in Baltimore nearby Washington D.C. was working along these lines, too, in those years. In expressing the excitations of the Higgs field in the Yang–Mills equations (see section 1.4), they found that, on top of the appearance in the theory of a massive scalar boson already suggested by Peter Higgs, new terms appear having exactly the form of mass terms for the gauge bosons W and Z.[21] These terms are proportional to the value v of the Higgs field in the ground state of its potential. This value must thus be different from zero, which means that there is a spontaneous symmetry breaking so as to have these gauge bosons acquire mass. On the contrary, the masses of the photon and the gluons, which do not directly interact with the Higgs field,[22] remain zero.

How can it be that equations after spontaneous symmetry breaking still respect gauge invariance, whilst having gauge bosons of non-zero mass? The initial equations are indeed gauge invariant, adding a potential related to a massive scalar particle, the invariance is preserved. The appearance of a minimum for a non-zero, i.e. non-symmetric, value of the Higgs field is spontaneous in the sense that it is not imposed by an external action, and no direction is favoured. In the final equations the symmetry is thus hidden. In this model neutral gauge bosons of electromagnetic and weak interactions, the photon and the Z boson, are the result of mixing of the two neutral gauge bosons associated to groups $SU(2) \times U(1)$. This is the reason why we talk of *electroweak unification*, to emphasise the common origin of the two interactions. An example how electroweak unification acts experimentally on high-energy electron–proton scattering is given in figure 1.10.[23]

Introducing in the theory the Higgs field whose expectation value in the vacuum is non-zero (or, in other words, which *condensates* in vacuum) thus gives rise to electroweak symmetry breaking with the W and Z bosons acquiring masses. But how to give mass to the fermions? The solution is, at least formally, relatively simple: it is sufficient to add by hand to the Lagrangian terms corresponding to the interaction of each individual (massless) fermionic field with the Higgs field. Each one of these terms

[21]The computation described here in a simplified way is based on perturbation theory allowing to consider the behavior of the scalar field around its minimum and treating this as a small mathematical perturbation (see inset 1.3). The value of the Higgs field is then equal to the sum of the field at the minimum of the potential plus the perturbation. After insertion of this value in the equations of the Standard Model, which initially contained only massless gauge bosons, one obtains terms corresponding to massive W and Z bosons. The initial symmetry is hidden.

[22]We will later see that the Higgs boson can decay to two photons or be produced in the interaction of two gluons through quantum loop effects.

[23]Looking in more detail into figure 1.10, the neutral current cross-sections for e^-p and e^+p are almost identical at small Q^2 but start to diverge as Q^2 grows. This is due to γ–Z interference, which has the opposite effect on the e^-p and e^+p cross-sections. The charged current cross-sections also differ between e^-p and e^+p scattering, with two effects contributing: the helicity structure of the $W^\pm$ exchange (inset 2.4) and the fact that charged current e^-p (e^+p) scattering probes the u-valence (d-valence) quarks.

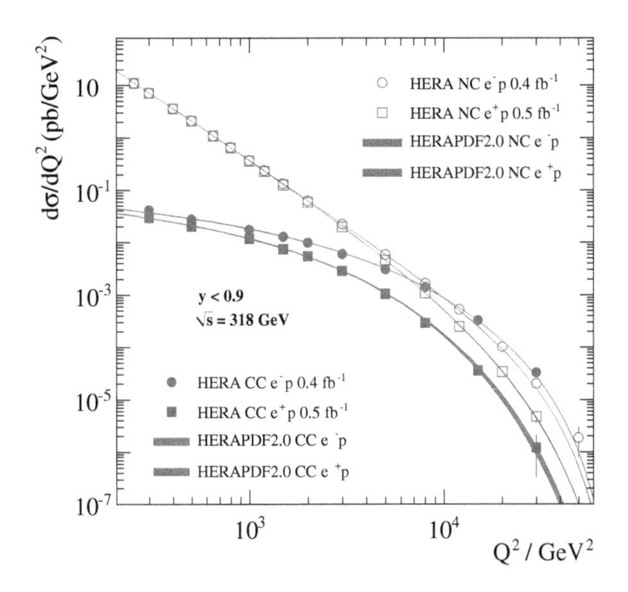

Fig. 1.10. Seminal plot illustrating the unification of the electromagnetic and weak forces as electroweak force. The figure presents data from the electron–proton and positron–proton collider HERA at DESY, taken in the years 1990s and the first decade of 2000. Shown are differential cross-sections versus momentum transfer squared to the collision products, measured for neutral (mainly γ^*/Z, blue) and charged (W boson, red) current scattering processes. At low momentum transfer, neutral current processes with photon exchange producing an electron or a positron in the final state dominate over charged current processes mediated via W bosons. Above 100 GeV ($Q^2 = 10^4$ GeV^2), however, neutral and charged current processes are of similar size: electromagnetic and weak interactions are unified. Electroweak unification relates the electromagnetic and weak coupling strengths to each other (the latter coupling is given at lowest order by the ratio squared of weak gauge boson masses). This relation has been tested experimentally to high precision. *H1 and Zeus Collaborations, Eur. Phys. J. C 75, 12 (2015).*

contains a coupling constant characterising the intensity of the interaction between these two fields. After electroweak symmetry breaking and with an appropriate choice of these constants one can construct the experimentally observed fermion mass spectrum. This procedure is somewhat *ad hoc* and a frequent reason of criticism of the Brout–Englert–Higgs mechanism. The large differences among the masses of the fermions, or, what is equivalent, among these fermionic couplings to the Higgs field, are not explained and remains one of the weak spots of the Standard Model. Only the top quark, whose coupling to the Higgs field is of order one, seems to have a natural mass.

As we have seen already, the Higgs field is not only a constant background field, but it has its own massive quantum, the scalar Higgs boson. Being a boson, we might want to call it a fifth force. However, unlike the other forces, this new force is not a gauge interaction. Its non-universal coupling to masses of fermions and gauge bosons may remind us of classical gravitation, but the BEH force is much stronger than gravity and short-ranged.

In less formal or mathematical terms, one can understand or visualise the effect of the Higgs field (after electroweak phase transition) filling the entire space as a sort of viscosity. In some analogy with an object moving at high speed through a viscous liquid and being thus slowed down through its interaction with the liquid (what we call drag), the bosons and fermions are slowed down by these interactions with the omnipresent condensed Higgs field.[24] A particle moving with a speed of light is a particle of zero mass. If its speed is smaller then it implies it has mass, and the stronger it interacts with the Higgs field the larger is its mass. One could also say that the velocity of the particle is reduced through the multiple scattering with the Higgs field while moving through space–time due to its mass. Another possible analogy is with light moving through a transparent refractive medium, that is, with an index of refraction $n > 1$. Due to its interaction with the atoms of the medium, a photon has a constant speed c/n, smaller than the speed of light c in vacuum: in water it is about $0.75c$, in glass about $0.6c$.

A proof of the mathematical coherence of a theory based on BEH-like broken gauge symmetry was obtained at the beginning of the 1970s by Gerard 't Hooft, Martinus Veltman (Nobel Prizes in 1999), Benjamin Lee and Jean Zinn-Justin. It led to a gradual change of paradigm among theoretical physicists, who then adopted gauge theories as the proper description of Nature at the scale of the infinitesimally small.[25]

[24]This analogy is only of limited value: viscosity is a dissipative phenomenon, provoking a loss of energy of the particle and generating heat, whilst this Higgs friction does not give rise to any deceleration, loss of velocity of the particle.

[25]In the mid-seventies, the up-to-then favourite approach to strong interaction phenomena in terms of S-matrix and Regge-poles theory was gradually abandoned in favour of the QCD gauge theory (see inset 1.13), then in full development.

Inset 1.10: The Glashow–Weinberg–Salam model

At very high energies, in the history of the Universe before the occurrence of a phase transition leading to spontaneous symmetry breaking in the Higgs potential, the electromagnetic and weak interactions are described by a single, unified *electroweak* gauge theory based on the group SU(2) × U(1). The gauge bosons W_1, W_2, W_3 associated to SU(2), and B associated to U(1) are all massless and the range of their interactions is infinite. The strength of the interactions is given by the coupling constants g for the W_i bosons and g' for the B.

In quantum field theory, two particles having same quantum properties (spin, parity, charge) with respect to a given interaction can in certain conditions mix and give rise to two new particles, result of a combination of the initial ones.

In the Brout–Englert–Higgs mechanism, the ground state of the Universe (the vacuum state of the theory) corresponds to the minimum of the Higgs potential in which the Higgs field has an average value $v/\sqrt{2}$ in a chosen direction (cf. figure 1.8). In inset 1.8, we have seen how the symmetry breaking gives rise to masses of the gauge bosons of the theory. With the gauge symmetry group SU(2) × U(1) mass terms can appear only for simple linear combinations of gauge bosons W_i and B. The bosons W_1 and W_2 mix to become the massive charged bosons W^+ and W^-

$$W^+ = \frac{1}{\sqrt{2}}(W_1 - iW_2), \quad W^- = \frac{1}{\sqrt{2}}(W_1 + iW_2).$$

Their (equal) mass is given in terms of the coupling constant g and the parameter v by

$$m_W = \frac{1}{2}gv.$$

It is clear from this expression that spontaneous symmetry breaking, manifesting itself through the non-zero value of the parameter v (*vacuum expectation value*), is indeed responsible for the mass of the physical W bosons.

Similarly, the bosons W_3 and B mix to give the massive Z boson and the photon (γ) remains massless:

$$Z = W_3 \cos\theta_W - B \sin\theta_W, \quad \gamma = W_3 \sin\theta_W + B \cos\theta_W.$$

The mixing is characterised by an angle θ_W often called Weinberg's angle or *weak mixing angle.* It is directly related to the coupling constants before symmetry breaking, g and g': $\tan\theta_W = g'/g$. The mass of the Z boson is given by

$$m_Z = \frac{m_W}{\cos\theta_W}.$$

The strength of the interactions is also modified by the Brout–Englert–Higgs mechanism: the coupling to W bosons remains g whilst the coupling to the photon (which determines the electric charge e of the electron) is related to g by

$$e = g \sin\theta_W = g' \cos\theta_W.$$

The strength of the charged weak interaction, the one mediated by the W, is well measured at low energies and is characterised by the Fermi constant $G_F = 1.166 \times 10^{-5}$ GeV^{-2}. Enrico Fermi introduced this constant in 1934 to describe radioactive β-decays, in particular the decay of the neutron n to $p+e+\bar{\nu}$, many years before the introduction of gauge theories and the W boson. We now know that these decays are mediated by W bosons, and their mass can be related to the Fermi constant and g by

$$G_F = \frac{g^2}{4\sqrt{2}m_W^2},$$

which also explains why G_F has the unit GeV^{-2} as the coupling g is dimensionless, i.e. a pure number. Historically, θ_W has been initially determined in neutrino scattering experiments with a value of $\sin^2\theta_W$ close to 0.23, well before the direct experimental observation of W and Z bosons (we shall come back to this point in section 2.3). Knowing

θ_W it was possible to deduce the value of g and predict the mass of the W:

$$m_W = \sqrt{\frac{g^2}{4\sqrt{2}G_F}} = \sqrt{\frac{\pi\alpha}{G_F\sqrt{2}}} \cdot \frac{1}{\sin\theta_W},$$

where $\alpha = e^2/(4\pi) \approx 1/137$ is the electromagnetic *fine structure constant.*

Numerically, one obtains from these expressions $m_W \approx 80$ GeV and $m_Z \approx 91$ GeV. One can also deduce the value of $v = (\sqrt{2}\,G_F)^{-1/2} = 246$ GeV. This average value of the Higgs field in vacuum is the energy scale at which electroweak symmetry breaking occurs. The only unknown parameter in the theory, before its discovery in 2012, was thus the Higgs boson mass, but this and the very existence of the Higgs boson was a very major unknown motivating and justifying a long and uncertain search (chapter 11).

The Glashow–Weinberg–Salam model described successfully charged-current weak interactions mediated by the exchange of W^+ and W^- bosons, which were known and experimentally studied already for many years. This model, however, was also predicting a *neutral-current weak interaction*, that is a weak interaction mediated by a neutral boson, the Z. The observation of these weak neutral currents at CERN in 1973,[26] with the subsequent discovery through direct observation of the W and Z bosons in 1982 and 1983 (see section 2.3) again at CERN, has made of the Glashow–Weinberg–Salam model a definitive and key component of the Standard Model of elementary particle interactions describing the electroweak sector.[27] The Brout–Englert–Higgs mechanism is its keystone, and the discovery of the Higgs boson at CERN in 2012 crowns its ultimate success.

[26]It took almost a year to convince the community that the observed neutral current signal was not due to an underestimated neutron background.

[27]The other part of the Standard Model is QCD, the gauge theory describing strong interactions (see inset 1.13).

Inset 1.11: The God Particle?

This designation "God particle" was widely used in the media for the Higgs boson being at the origin of all other particles, at least of their mass (remember though: it is not the Higgs boson but the underlying symmetry breaking mechanism that gives mass). What is at the origin of this rather bold and provocative term vehemently criticised by Peter Higgs himself?

In 1993 the American physicist Leon Lederman, Nobel Prize winner for the discovery in 1962 of the muon-neutrino and in 1977 of the bottom quark (the first quark member of the third family of fundamental fermions to be discovered), published together with the American writer Dick Teresi a book entitled *The God Particle: If the Universe Is the Answer, What Is the Question*? Among other things, the book describes the up-to-then unsuccessful hunt for the Higgs boson at CERN and Fermilab, as well as the pharaonic Superconducting Super Collider (SSC) project (cf. section 6.3.1) that was at the time under construction in Texas, USA, and whose main task was to clarify in a definitive way the issues related to electroweak unification and ultimately to find the Higgs boson.

In connection with this so important but also elusive particle, and motivating these huge technical and financial efforts, Lederman said that for that matter the book could as well have been entitled *The Goddamn Particle*. However, this title was not to the liking of his editor who suggested that without the word *damn* the book would sell much better! Lederman also possibly wanted to see in this reference to God an analogy to the story of the Babel Tower (the SSC) built by humans after the episode of the Deluge to reach to Heavens and to God himself and unveil his secrets.

1.7. Dirac theory and antiparticles

We already mentioned antiparticles a number of times: the antielectron also called positron, antineutrinos, antiquarks, etc. What are those about? The idea of antiparticles does not come originally from experimental observation, but rather from mathematical reasoning.

Fig. 1.11. On the left, Paul Dirac (1902–1984), co-laureat with Erwin Schrödinger of the Nobel Prize in physics in 1933 for their fundamental contributions to the development of quantum mechanics. On the right, Carl Anderson (1905–1991), Nobel Prize in physics in 1936 for discovering the positron. *Images: public domain.*

In 1928 the British physicist Paul Dirac (figure 1.11) derived an equation describing the space–time evolution of the wave function of the electron. The linear first-order differential equation he found was for the first time consistent with the principles of both quantum mechanics and special relativity. Dirac's equation (see also inset 1.5) had to respect Einstein's relation between energy[28] and matter, $E = \sqrt{m^2 + p^2}$, where m is the electron rest mass, p its momentum and E its total relativistic energy.[29] Dirac understood that for this constraint to be satisfied, the equation had to be a matrix equation of dimension 4. The wave function $\psi(\vec{x}, t)$ of the electron should thus be a (complex) four-component object called a *spinor*. The use of matrix algebra introduces particle spin in the theory in a natural way. Furthermore, when the particle is non-relativistic, i.e. with small velocity compared to the speed of light, that is when the ratio p/m is small, Dirac's equation reduces to the known non-relativistic equation describing a spin-$1/2$ particle: the Pauli equation established in 1927 (which itself is a generalisation of the earlier Schrödinger equation). Dirac's equation also explains why spin-$1/2$ particles behave as small magnetic dipoles, and it predicted correctly the value of the quantum magnetic dipole moment of the electron. The net result is that Dirac's equation explains the *fine structure* (splitting of lines) of the optical spectrum of the hydrogen atom.

---

[28] From Einstein's relation, we can understand that the energy content in a material body is enormous. The total annihilation of one kilogram of matter would produce an energy of 9×10^{16} g m^2 s^{-2} = 9×10^{16} joules: this is about equal to the entire solar energy falling onto the Earth in one second, or to a quarter of the overall electricity consumption of Norway in 2008.

[29] As customary in particle physics, we use here the convention in which the speed of light c is equal to 1. Mass and energy can then be expressed in same units, the electron-volt (eV), or in units derived from its multiples or fractions.

This was exactly what Dirac was looking for. What was not expected is that the equation also yielded solutions describing particles with negative energy! The interpretation of these solutions was a serious problem and generated much discussion and controversy among the physicists at the time. It is only in 1931 that Dirac found the solution, the correct interpretation: he postulated the existence of an antielectron, reinterpreting the negative-energy solutions as particles having positive energy, same mass and spin as an electron, but opposite electric charge.[30] The following year the positron was discovered by the Swedish-American physicist Carl Anderson (figure 1.11 right) studying cosmic rays. Dirac received the Nobel Prize in 1933 and Anderson in 1936.

In subsequent years, the antiparticle concept was extended to all particles, including those composed of several fundamental particles, as is the case for hadrons or even nuclei. A particle and its antiparticle have opposite internal (additive) quantum numbers (electric charge, leptonic or baryonic number,[31] strangeness, etc.), but have same mass, spin and lifetime. Some electrically neutral particles are identical to their antiparticles. This is the case for the Z, the photon, the Higgs boson or, for example, the π^0, a light hadron composed of a symmetric combination of quarks and antiquarks from the first family: $(u\bar{u} + d\bar{d})/\sqrt{2}$, and possibly for neutrinos as discussed in section 3.5. On the contrary, this is not the case for the neutron, as it is made of a combination of quarks that is not symmetric under the exchange of quarks with antiquarks: the quark content of a neutron being udd, that of the antineutron is $\bar{u}\bar{d}\bar{d}$. Neutron and antineutron are thus different particle states.

After the positron, Carl Anderson and Seth Neddermeyer discovered the antimuon in 1936. The antiproton — a composite particle although

[30] Of the four components of the Dirac spinor describing an electron, two components are related to the spin state of the particle, the other two components to the particle–antiparticle nature. The subsequent development of quantum field theory, based on group theory applied on the Poincaré group (combining the Lorentz transformation of relativity and translation in space–time), gives a mathematically rigorous framework in which antiparticles can be introduced.

[31] The baryonic number B of a system is defined as one third of the difference in the number of quarks and antiquarks composing the system. Thus $B = 1$ for a baryon (a system composed of three quarks, as is the case for a proton or neutron) and $B = -1$ for an antibaryon (made of the corresponding antiquarks). Similarly, the leptonic number L of a system is defined as the difference in the number of leptons and antileptons it contains. In the Standard Model, these two quantum numbers are conserved in the interactions, which means in the transitions between initial and final states of a scattering process. In theories beyond the Standard Model, like grand unified ones, B and L are not conserved.

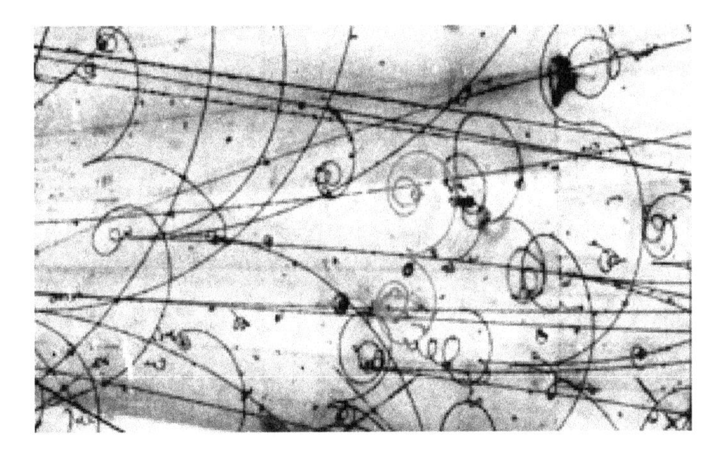

Fig. 1.12. Creation of an electron–positron pair (shown in purple) along the trajectory of an energetic charged particle — most likely a muon in this case — as visible in a bubble chamber picture. This process can occur when a charged particle passes nearby an atomic nucleus: thanks to the electromagnetic field of the nucleus, the pair can be created, as part of the incident momentum is transferred to the nucleus. In vacuum such a pair-production would not be possible because of energy and momentum conservation.

not known as such at the time of its discovery — was discovered at Berkeley in the USA in 1955, and the antineutron the following year at the same place. The first antinucleus to be discovered was the antideuteron in 1965, produced simultaneously at CERN and at the Brookhaven National Laboratory (BNL) in the USA. Finally the first antiatom, the antihydrogen, was produced at CERN in 1995. Since then millions of antihydrogen atoms are being produced at CERN in a controlled way for detailed studies of their properties.[32]

What happens when a particle meets its antiparticle? Just as in science fiction novels, they mutually annihilate producing energy, for example in the form of photons. This is a typical case of transformation of matter into energy according to Einstein's celebrated relation $E = mc^2$. Vice versa, photons of sufficiently high energy and in presence of a strong electromagnetic field can convert into particle–antiparticle pairs (figure 1.12). These types of metamorphosis form the basis of the operation

[32]A fundamental theorem of quantum field theory, the *CPT theorem* (see inset 1.12), predicts that the spectroscopic properties of hydrogen and antihydrogen must be identical. It also also asserts that the mass, width and lifetime of a particle and its antiparticle are identical. A question that is also at present experimentally addressed is whether these antimatter systems fall (exactly like matter does) or rise in the gravitational field.

of particle accelerators and colliders. Let us take as an example an electron–positron collider: electrons and positrons are accelerated, applying radio-frequency electric fields, to velocities close to the speed of light and are brought into collision. The sum of their kinetic energies in their common centre-of-mass is transformed into mass, allowing the production of particles much heavier than is the rest-mass of the initial electron and positron. At the Large Electron–Positron (LEP) collider at CERN, which operated at the end of the past century and about which we are going to talk in section 2.4, such collisions made possible the creation of millions of Z bosons whose mass is 180 000 times more than that of an electron!

1.8. Violation of matter–antimatter symmetry

One of the basic conditions necessary for the creation of a universe made of matter (and not antimatter) as is ours today, is the existence of interactions not respecting the symmetry between particles and antiparticles,[33] a symmetry that was present at the earliest moments of the Big Bang. The violation of this symmetry was observed experimentally for the first time in 1964 in the USA, in the decay of neutral kaons governed by weak interactions. Kaons are hadrons, particles made of quarks d (or antiquarks $\bar{d}$) from the first family and antiquarks carrying strangeness number $\bar{s}$ (or strange quarks s) from the second family of fundamental fermions.

An elegant explanation of this phenomenon was given in 1973 by two Japanese physicists, Makoto Kobayashi and Toshihide Maskawa. In the 1960-ties, the Italian physicist Nicola Cabibbo had shown how quarks from the first and second family can mix through weak interactions. What Kobayashi and Maskawa realised is that, if — and only if — there existed a third family of quarks (there was no experimental evidence for such a third family at the time), then the quark mixing could be different for quarks and antiquarks. In other words, a Standard Model with at least three families of quarks would incorporate in a natural way an asymmetry between matter and antimatter. This hypothesis was strengthened by the discovery in 1975 of the τ-lepton and then in 1977 by the discovery of the first quark from the third family, the *beauty* or *bottom* quark (b-quark).

[33]More exactly, this interaction must violate the C and CP symmetries described in inset 1.12.

The Cabibbo–Kobayashi–Maskawa model also predicted matter–antimatter symmetry violation in processes involving neutral B mesons, which are made of an anti-b-quark and a d-quark (or of a b-quark and an anti-d-quark). The phenomenon was indeed observed in 2001, thanks to dedicated experiments at electron–positron colliders at SLAC in California, USA, and at KEK, the Japanese laboratory for particle physics near Tsukuba. The very detailed studies made in the following years at these two colliders have shown the correctness of this model; we shall come back to this point in the next chapter.

Mixing through weak interactions among quarks of the three families is described by a 3×3 matrix called Cabibbo–Kobayashi–Maskawa (CKM) matrix. Owing to probability conservation (unitarity) in the quark transitions, the matrix' original 18 unknown parameters (each entry of the matrix is a complex number) reduce to only four, of which one is the phase of a complex number (see section 1.1). The CKM model is thus highly predictive with the single phase explaining all the observations made up to now in laboratory measurements on the matter–antimatter symmetry violation. The experimental verification led to the attribution of the Nobel Prize in Physics to Kobayashi and Maskawa in 2008.

In spite of this success it was found by means of difficult calculations that the amount of matter–antimatter symmetry violation present in the Standard Model through the CKM matrix is badly insufficient to explain the excess of matter over antimatter observed in the present-day Universe.

The recent discovery that neutrinos possess mass, although a very small one, opens fascinating perspectives. Firstly, with three families of massive neutrinos one may expect an additional source of violation of matter–antimatter symmetry in the leptonic sector similar to the one observed for quarks. Secondly, it is difficult to imagine that the same Brout–Englert–Higgs mechanism is at the origin of neutrino masses that are so much smaller than those of their charged counterparts (see section 3.5). A more satisfactory theoretical mechanism assumes the existence of additional more massive neutrinos. These might be so heavy, up to 10^{14} times heavier than a proton, that they cannot be produced in foreseeable accelerators on Earth. The matter–antimatter symmetry-violating decays of these hypothetical particles produced in the very early epochs of our Universe, in conjunction with an expected instability of the electroweak vacuum during that epoch, might be at the origin of the leptonic and baryonic matter excess we all descend from.

Inset 1.12: P, C, CP, and CPT symmetries

For a long time physicists have thought that natural laws are the same for two processes which are spatial reflections, mirror images, of each other. This symmetry is called *parity* and the mathematical symbol (*operator*) performing the transformation is denoted P. In 1957, two Chinese-American theoretical physicists, Chen-Ning Yang and Tsung-Dao Lee, suggested that to explain some puzzling features in kaon decays, weak interactions might not respect P symmetry and that this could be verified experimentally. The same year Chien-Shiung Wu and her collaborators at Columbia University proved P symmetry violation through the study of the weak-interaction-mediated radioactive β-decays at low temperature of polarised cobalt (^{60}Co) nuclei into nickel (^{60}Ni) accompanied by an electron and an antineutrino.

To explain parity violation in weak interactions it is useful to introduce the notions of *helicity* and *chirality*.[a] Helicity is a quantum property of a particle possessing spin, and it is defined as the magnitude and sign of the projection of the particle's spin onto its momentum (its direction of motion). Spin is a vector quantity (more precisely, an axial vector) related to the intrinsic rotational properties of a particle. Physicists use the same word, spin, to designate both the vector, the physical observable, and the quantum number related to the modulus of this vector. Chirality is an intrinsic property of a particle related to its invariance (or non-invariance) in a mirror transformation. For a particle of (quasi) zero mass, chirality is (quasi) identical to helicity and here we are not going to make a difference between the two.

Chirality is related to handedness: a negative chirality is called *left-handed* and a positive one is called *right-handed*. For example, photons in left (respectively right) circularly-polarised light are left-handed (respectively right-handed). From the theoretical point of view, parity violation (P violation) stems from the fact that fermions of left as compared to right chirality have different couplings to the gauge vector bosons W and Z mediating the weak interactions. For example, only left-handed fermions couple to the charged bosons W^+ and W^-, whilst the right-handed fermions do not couple to these at all, and thus have no charged weak interaction! The results of the experiment of Wu and others that followed immediately provoked a revolution in particle physics, with Lee and Yang being awarded the Nobel Prize immediately the same year.

Let us now combine the parity transformation P (permutation between left and right) with the charge conjugation transformation C, corresponding to the interchange of a particle with its antiparticle. The new symmetry operation CP amounts to reversing the spin of a particle and transforming it into its antiparticle with opposite electric charge, as all other additive quantum numbers, S, L, B, etc. The figure below illustrates an electron with spin up that is transformed into a positron with spin down. CP symmetry seems to be preserved: antiparticles of right chirality (respectively left chirality) behave the same way as particles of left chirality (respectively right chirality). At least this was the expectation after 1957.

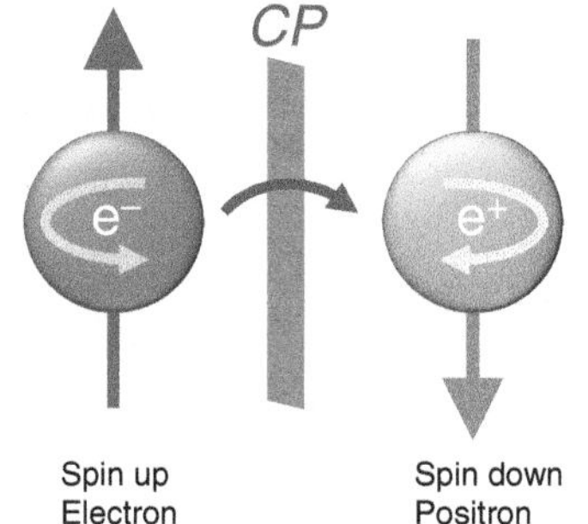

However, again, a small violation of CP symmetry was discovered in the decays of neutral kaons in an experiment at BNL in 1964 (James Cronin and Val Fitch, Nobel Prizes in 1980). In 2001 CP violation has also been detected in transitions involving neutral B mesons. Since then, CP violation has been studied in great detail in both kaon and neutral and charged B-meson decays and this domain of research remains a very active one at the LHC as we will see later.

Full invariance is, however, obtained for the CPT symmetry combining charge conjugation C, parity P, and the time-reversal transformation T. The absolute validity of this invariance property of natural laws can be shown theoretically on general grounds such as *causality* and the validity of special relativity. It implies that an antiparticle behaves exactly as a particle seen in a mirror and going backwards in time. A consequence of this is that a particle and its antiparticle have exactly the same mass, spin and lifetime. Invariance under CPT means that, at the microscopic level, a world cannot be distinguished from its anti-world, seen in a mirror and with a reversed flow of time.

---

[a]The term chirality was introduced end of the 19th century by the British scientist Lord Kelvin into chemistry, adopting it from the Greek $\chi\varepsilon\rho\iota$ (chéri), or *hand*.

1.9. Quantum field theory and virtual corrections

We already mentioned that in order to describe fundamental phenomena characterised by the appearance of new particles in particle collisions, or the decays of unstable ones, it was necessary to introduce the concept of a quantum field. *Quantum field theory* (see also inset 1.3) was developed since the 1930s by a number of renowned physicists whose names remain for posterity: Paul Dirac, Werner Heisenberg, Max Born, Pascual Jordan, Wolfgang Pauli, Lev Landau, Vladimir Fock, Richard Feynman, Sin-Itiro Tomonaga, Julian Schwinger and a number of others. A quantum field is a function of space–time coordinates ($\vec{x}$, t) and its mathematical structure depends on the nature of the particle it is associated with, in particular of its spin. The fields are the fundamental entities of the theory and the particles then appear just as quanta of the fields. A particle can then be interpreted as a materialisation of a local excitation of the quantum field in the vicinity of its value in the vacuum defined through the minimum of the overall potential energy. All the quanta of a given field are identical, indistinguishable, particles.

This holds both for particles mediating interactions and associated to bosonic gauge fields, as for matter particles associated to fermionic fields. The interaction between a gauge boson and a fermion is described in the equations by a term containing the product of the two corresponding fields, multiplied by a real number called the *coupling constant* (see insets 1.6 and 1.13). The coupling constant is a measure of the intensity of the interaction and depends on its nature (weak, electromagnetic or strong) and the charge (weak hypercharge, electric charge or strong charge, i.e. colour) carried by the interacting particles. In the case of weak and strong interactions, the bosonic quanta of the force field carry themselves the charge of their own field of force and can thus interact among themselves. For example, in the case of the four quanta of the unified electroweak interactions (γ, W^+, W^-, Z), one can have interactions among W^+W^-Z or $W^+W^-\gamma$.[34] Similarly, gluons can interact with each other.

In inset 1.2 we introduced Heisenberg's uncertainty relations limiting the precision with which the energy (or mass) of a system can be measured versus the time available to measure it. A consequence of this relation, the essence of quantum mechanics, is that a quantum system can "borrow"

[34]However, interactions among three neutral bosons such as γZZ, ZZZ, etc. are strongly suppressed on symmetry grounds in the Standard Model.

energy ΔE from the quantum vacuum for some limited time Δt such that $\Delta E \cdot \Delta t$ remains smaller than $\hbar/2$ (the larger the borrowed energy, the shorter it may last).

Let us take as an example the weak-interaction-mediated scattering of a neutrino on a d-quark through the exchange of a W boson: $\nu_e + d \to e^- + u$ shown in figure 1.13. This can be visualised as the decay $\nu_e \to e^- + W^+$, i.e. the production of a very massive boson from a quasi zero-mass neutrino, immediately followed, within the allowed time Δt, by the recombination $W^+ + d \to u$. It can also be thought of as the emission of a W by the neutrino whilst transforming itself into an electron, with the W immediately reabsorbed by the d-quark and giving a u-quark. Note that during the entire process the electric charge is conserved at every step, while overall the energy is preserved between the initial state $\nu_e + d$ and the final state $e^- + u$. The energy is only non-conserved within the small time interval of the W transition, owing to Heisenberg's relations.

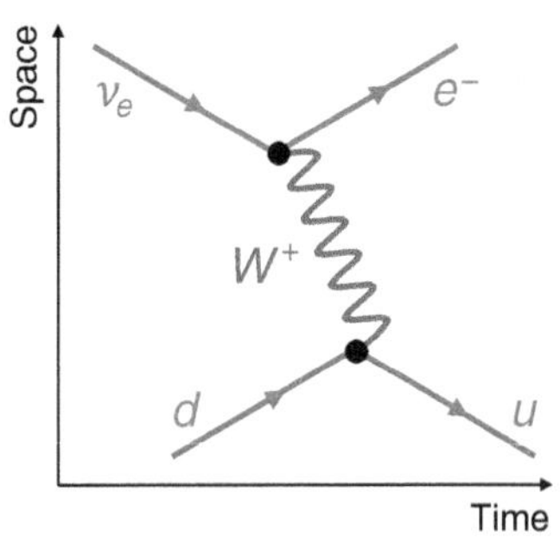

Fig. 1.13. Feynman diagram for the scattering of a neutrino on a d-quark through the exchange of a W boson. The time flow in the diagram is from left to right. This is the simplest process contributing to the scattering amplitude.

The visionary American physicist Richard Feynman invented in the late 1940s a diagrammatic system as the one shown in figure 1.13 (and also figure 1.4) to illustrate particle physics processes. Actually Feynman diagrams (as they are called today) are more than just graphical representations; they are symbolic codes of mathematical rules to calculate the probability of a specific final state to occur from a given initial state (see also inset 1.4). To each element of the diagram (a line, external or internal, a junction between lines called a vertex) corresponds a well-defined mathematical term for the calculation of the *probability amplitude* describing the process. This amplitude is in general a complex number and the square of its modulus gives the probability of occurrence of the process. If several diagrams (or amplitudes) can lead from the same initial state to the same final state, the overall probability of the process is equal to the square of the modulus of the sum of the amplitudes related to each diagram. This is the principle of *superposition of amplitudes*, which allows the individual diagrams to interfere, a fundamental feature reflecting the wave nature of quantum phenomena.

The W exchanged in the diagram in figure 1.13 on an internal line is said to be *virtual*, as distinguished from the particles in the initial and final states which are *real*. The process shown in figure 1.13 can take place even if the energies of the particles in the initial state are low. The mass of the exchanged W can then not be its rest mass, as this one is about 80 GeV. We say that this W is *off its mass shell*, or short *off-shell*, and this characterises virtual particles. However, even if virtual, such a particle can make its presence felt!

A consequence of the non-conservation of energy during a very short time interval is that the quantum vacuum, contrary to the classical one, is not empty. It is instead permanently seething, agitated, with particle pairs emerging from the vacuum and instantly returning into it. These pairs can be e^+e^- or $u\bar{u}$, but also, if only for a much shorter time, $t\bar{t}$ pairs, despite the large top-quark mass.

The diagram of figure 1.13 features a single virtual particle exchange. However, in quantum field theory numerous elementary processes allowed by the uncertainty relation and relating the same initial and final states can occur, and will unavoidably contribute to the overall scattering amplitude. An example is shown in figure 1.14. The larger is the number of vertices, internal loops and virtual particles away from their mass shell a diagram contains, the smaller is its contribution to the overall amplitude.

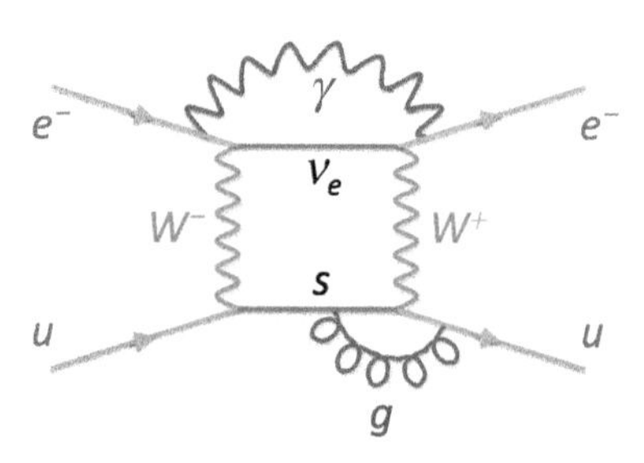

Fig. 1.14. Feynman diagram containing not less than eight vertices and contributing, albeit very weakly, to the scattering amplitude of an electron on a u-quark.

The so-called *perturbative development* of scattering amplitudes, represented by a series of more and more complicated Feynman diagrams, is valid, i.e. convergent only in so far as we are in a kinematical regime where coupling constants are much smaller than order one. Such an approach is, for example, not valid to represent the quarks interacting through gluon exchanges and confined inside hadrons. In fact the values of the couplings of all Standard Model interactions vary with the energy scale of the phenomena investigated (see section 1.10). The processes leading to quarks being confined inside hadrons are low-energy processes, and in this kinematical

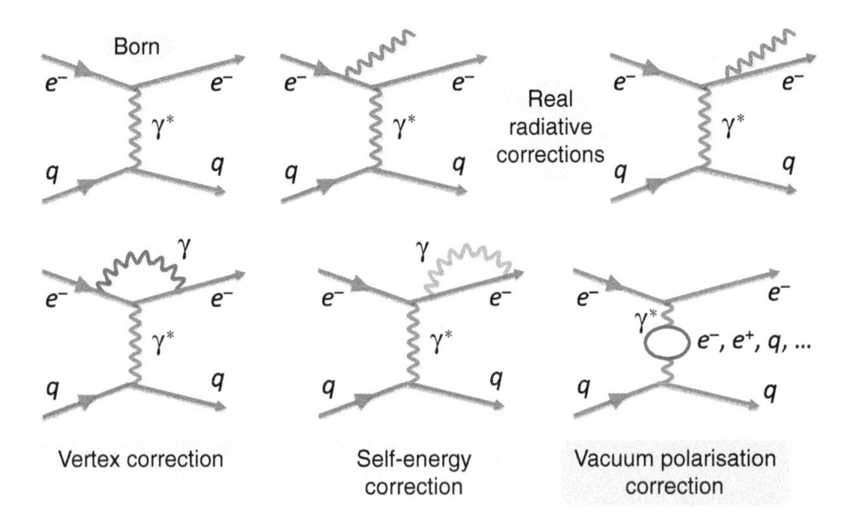

Fig. 1.15. Feynman diagrams for the scattering of an electron on a quark. The diagram on the top left corresponds to the simplest sub-process contributing to the scattering amplitude, the Born approximation. The other diagrams show examples for radiative corrections, omitting the ones with corrections attached to quark lines. The virtual processes in the lower row shown can be arranged according to three categories: the *vertex correction* corresponds to the emission of a photon by the incident electron followed by its absorption by the electron after its interaction with the quark (scattered electron); similarly, in the *self-energy correction* case, a photon is emitted and immediately absorbed by the scattered (or incident) electron; the *vacuum polarisation* diagram correspond to a closed loop containing a fermion–antifermion pair created from the vacuum.

regime the strong coupling constant is large, too large for the perturbative approach to be applicable (see also inset 1.13).[35]

As a typical example of a process that can be described as a perturbative series with Feynman diagrams let us consider the scattering of an electron on a quark (figure 1.15). The basic diagram is the one on top left corresponding to the exchange of a single virtual photon. It gives the largest contribution to the overall amplitude. When only this contribution is taken into account, we say that we work in the *Born approximation*. However, other and more complex diagrams can also contribute. Only some of these are shown here. These additional contributions correspond to what

[35]Specific QCD calculation techniques have been developed to address these problems, called *Lattice QCD*. This is a highly developed field, but is outside the scope of this book.

is called *radiative corrections.* There are two types of corrections. The *real* radiative corrections change the number of particles in the final state. For example, additional photons can be radiated by the incident or outgoing final-state particles. The *virtual* radiative corrections do not change the number of final state particles. The exchanged virtual particles essentially do not modify the kinematic properties of the reaction, but their contribution is taken into account in the calculation of the full scattering amplitude and thus affects the probability of a given process to occur.

The computation of the radiative corrections can reach incredible precision. An astonishing example is the calculation of the magnetic dipole moment (the gyromagnetic factor) of the electron or muon. Dirac's equation predicts a value of exactly 2 for this quantity, but radiative corrections modify this value at the level of one permil. Today it is possible to calculate these tiny deviations up to the tenth decimal place for the muon and the thirteenth for the electron magnetic moment! The experimental data are at the same level of precision, allowing thus to test the theory in a most incisive way. In the case of the muon this level of precision was required to detect a small discrepancy between experimental measurement and theoretical calculation whose origin is not yet understood.

Radiative corrections do not only play a role in calculations of reaction or decay rates, they also affect particle masses, like the W, Z, Higgs boson, or top-quark masses for example. They can involve virtual particles that have not yet been discovered because they are too heavy to be produced in collider experiments, but which nonetheless can make their presence felt. Precise measurement of predictable quantities is thus a way to indirectly detect virtual particles and thereby a possible signal of new physics. That way constraints on the masses of the top quark, and subsequently of the Higgs boson, were obtained before their actual experimental discovery. We shall come back to these points in the following chapters.

1.10. Running coupling constants

In 1924 the French physicist Louis de Broglie surmised that to each particle of momentum[36] p it is possible to associate a wave with wavelength $\lambda = h/p$ (h is the Planck constant). This relation, expressing the wave

[36]Momentum is the term used by physicists to designate the quantity $p = mv$ (v is velocity) in the non-relativistic regime and $p = mv/\sqrt{1 - v^2/c^2}$ in the relativistic regime, where the velocity approaches the speed of light c.

nature of particles and manifesting the particle–wave duality, is at the origin of Heisenberg's uncertainty relations that were formulated the following year (inset 1.2). It shows that a particle's wavelength diminishes (or its frequency increases) with increasing momentum.

As a consequence the higher is the collision energy of accelerated particles, the smaller are the scales λ that can be probed. This is the reason why particle physicists aspire towards higher and higher energy accelerators.

Let us now come back to the creation of virtual particles in an interaction, and let us consider the scattering of two electrons. An example for a relevant diagram is shown in figure 1.16. Each of the two electrons can appear for a brief moment as containing an additional electron–positron pair. The electron from this additional pair is repelled whilst the positron is attracted by the parent negatively-charged electron. The net result is the screening, i.e. reduction, of the net charge of the parent electron as seen by the one it is scattering with. This phenomenon is called *vacuum polarisation.* It is analogous to the polarisation in dielectric materials. At high energies, however, the probing distance of the incident beam, λ, can become smaller than the size of the electron–positron cloud surrounding the electron. The screening effect is then reduced and the effective charge of the electron as seen by the other one is larger. In other words, the electromagnetic coupling constant increases.

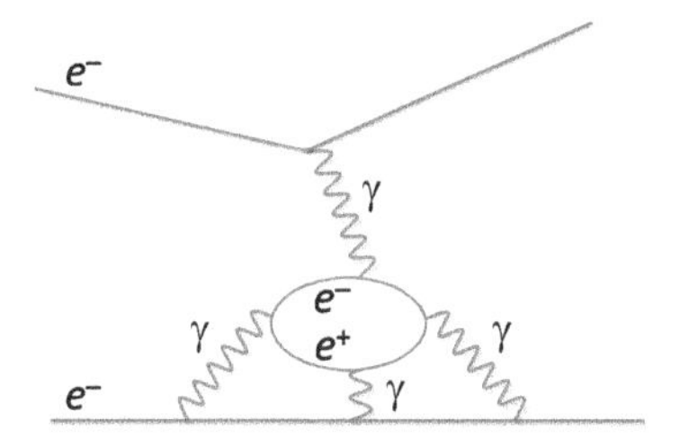

Fig. 1.16. Feynman diagram for the scattering of two electrons involving the creation of a positron–electron pair. In quantum electrodynamics (QED, the quantum version of Maxwell electromagnetic theory) the interaction between the two incoming electrons proceeds through the exchange of a photon, which can couple to one of the virtual particles resulting in a partial screening of the initial electron charge.

Charge screening, that is the dependence of the effective charge on the energy at which it is probed, has been experimentally tested to great precision. For example, the so-called *fine structure constant* α, which is proportional to the square of the electron charge ($\alpha = e^2/4\pi$) and is a measure of the intensity of the electromagnetic coupling, at low energy amounts to $1/137$. This is the value used to describe all atomic physics phenomena. But at the energy equal to the Z mass, about 90 GeV, as tested at the LEP collider, α increases to $1/128$.

The evolution ("running") of the coupling constants with energy can be calculated in quantum field theory for the three fundamental interactions of

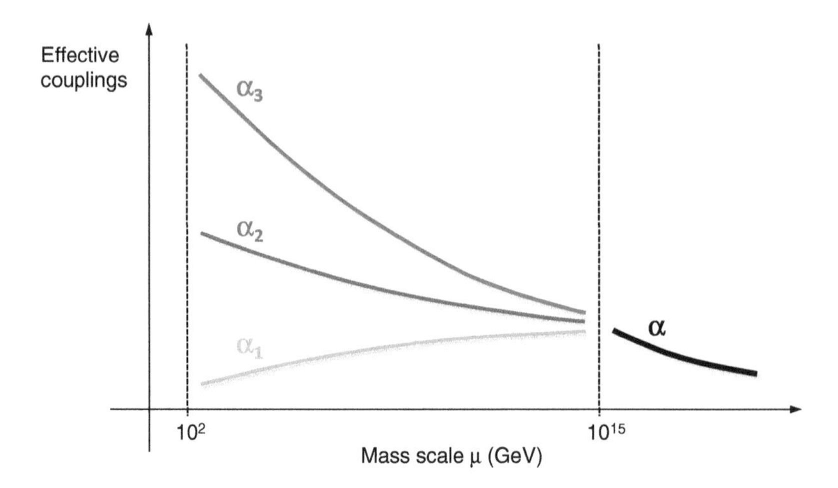

Fig. 1.17. Schematic evolution of the Standard Model coupling constants with energy, shown here on a logarithmic scale. α_1 and α_2 are the electroweak coupling constants associated to the gauge groups U(1) and SU(2) (see inset 1.10) and α_3 is the coupling constant for the strong interactions described by the gauge group SU(3). The right part of the plot illustrates the expected evolution of couplings in a grand unified scenario where there would be a single coupling α beyond about 10^{15} GeV; we discuss this conjecture in chapter 4.

the Standard Model, provided all interacting particles are known, at least in the energy domain under study. The outcome of such a computation is sketched in figure 1.17. In the case of strong interactions (coupling constant α_3), the coupling strength is largest at low energy. This is because gluons also carry colour charge and interact among themselves, in contrast to the photon that couples only to electrically charged particles, but does not carry charge itself. In strong interactions, instead of a screening of the electric charge at large distance (low energy) by electron–positron pairs, the increasing number of gluons and thus the amount of colour charge produces an *anti-screening* effect. At smaller distance (higher energy) the effective colour charge diminishes: we say that quarks and gluons are *asymptotically free*. The development of quantum chromodynamics (see inset 1.13) and the discovery of this dramatic feature of strong interactions earned the American physicists David Gross, Franck Wilczek and David Politzer the Nobel Prize in 2004. Looking again at figure 1.17, it seems as if the two unified electroweak couplings and the strong coupling may converge to a common value at an energy of the order of 10^{15} GeV. This suggests a possible unification of these two forces into a grand unified one, which we shall discuss in chapter 4.

Inset 1.13: Quantum chromodynamics: A theory for strong interactions

Shortly after the introduction of quarks in 1964 to explain the spectroscopy of hadrons (see section 2.1 and inset 2.2), the American physicist Wally Greenberg introduced a new degree of freedom for quarks to explain how three u quarks in the same spin state $+1/2$ could coexist in the baryon Δ^{++} without violating the Pauli exclusion principle (inset 1.1 and inset 2.2). At the quantum level, this degree of freedom appears as a new charge, which can take values along three dimensions. Similarly to the electric charge, which is associated to a global invariance under U(1) transformations acting in a one dimension space (see inset 1.4), these new charges are associated to invariance under the special unitary group SU(3) acting in three dimensions.

As the observed hadrons do not appear to carry such a charge, it has been postulated that only assemblies of quarks which are neutral under this new charge (or *singlets* in the terminology of group theory) can be observed. In analogy with the theory of colours in optics, this new quark charge can take three values called *colours*: red, green and blue. An antiquark can take one of three anticolours: antired, antigreen, and antiblue (represented as cyan, magenta, and yellow, respectively). Only *white* hadrons, with an overall colour charge equal to zero, can exist in Nature. The following figure shows assemblies of quarks and antiquarks with zero colour charge.

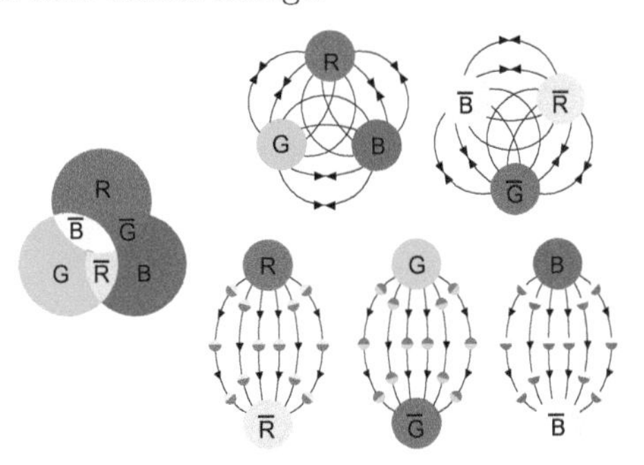

For instance, baryons (and antibaryons) are composed of three quarks (or antiquarks) with three different colours (left and upper parts of the figure) and mesons are made of a quark of some colour and an antiquark of the corresponding anticolour (lower part). Colourless configurations of four or five quarks and antiquarks (tetraquark and pentaquarks) have also been observed (cf. section 14.1.5).

In the seventies of the last century, after the success of the electroweak theory based on the local invariance under the gauge group SU(2) × U(1) associated to the electric and weak charges (after symmetry breaking, see section 1.6 and inset 1.8 on the GWS model), it became very tempting to promote the global SU(3) colour group invariance to a local one. The corresponding gauge theory, called quantum chromodynamics (QCD), predicts the existence of 8 gauge bosons, called gluons, carrying all possible combinations (not colour neutral) of a colour and an anticolour (exchanges of gluons within a meson are sketched on the above figure). Similarly to quantum electrodynamics (QED) and in contrast to the GSM model, the gauge symmetry stays unbroken and the gluons are massless vector bosons. However, in contrast to QED where the gauge boson (the photon) is electrically neutral and cannot interact with itself, the coloured gluons exhibit self-interaction.[a] The upper part of the following figure shows the Feynman diagrams corresponding to the fundamental vertices of QCD. The lower part shows examples of possible colour flow along these vertices. The coupling strength, labelled α_S (the subscript stands for "strong"), is a parameter of the theory.

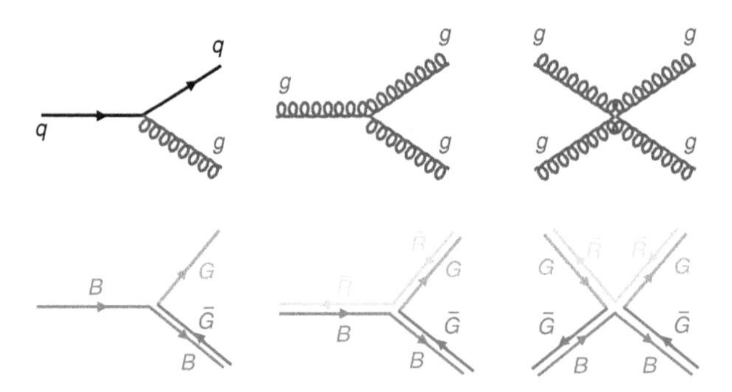

The gluon self-coupling leads to dramatic differences with respect to QED for the quark–quark interaction. The upper part of the following figure shows the electric field lines for QED (left-hand side) and colour field lines for QCD (right-hand side) for a static situation with two charges (electric or coloured) at rest.

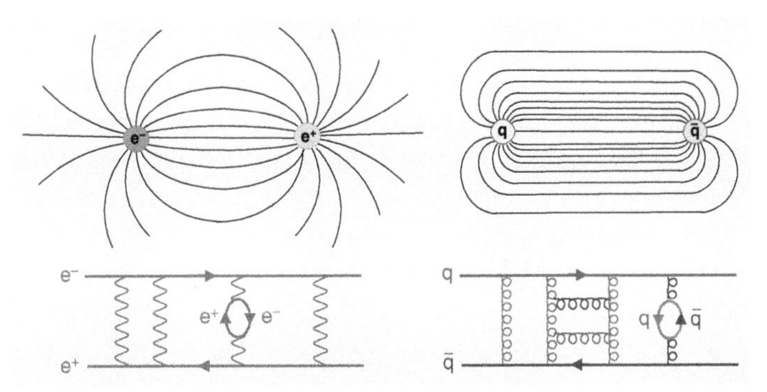

The gluon–gluon interaction constrains colour fields to be confined in "flux tubes", which exert a constant force when stretched. Beyond a certain distance between the quark and the antiquark, of the order of the nucleon size ($\sim 10^{-15}$ metres), the energy of the flux tube increases linearly. At a large enough distance, it becomes energetically more favourable to create a quark–antiquark pair out of the vacuum rather than continuing to increase the length of the flux tube. This property leads to what is called quark or colour *confinement*: the quarks and gluons are never observed as free particles outside hadrons. Colour confinement has been experimentally verified by (negative) free quark searches (searches of fractional charges). The confinement property is also at the origin of the formation of jets (figure 2.4) in the final states of high-energy interactions involving quarks and gluons.

The lower part of the above figure illustrates in form of Feynman diagrams the interaction between incoming electron and positron through virtual exchange of photons (left-hand side) and incoming quark and antiquark through gluons (right-hand side), respectively. Although classical QCD is a scale invariant theory,[b] at the quantum level the treatment of virtual processes (through loop effects in the Feynman diagrams) leads to the introduction of a scale parameter Λ through the renormalisation procedure required to cure the divergences. As a consequence, the renormalised coupling constant varies with the energy scale of the process (see section 1.10). At the one loop level, the QCD coupling constant α_S (which is indeed not a constant) depends on the energy scale μ according to:

$$\alpha_S(\mu) = \frac{4\pi}{\beta_0 \cdot \ln(\mu^2/\Lambda^2)}$$

with $\beta_0 = 11 - \frac{2}{3}n_f$, and n_f being the number of quarks flavours involved at the energy scale of the process. Λ is a free parameter which characterises the scale of strong interactions. It has to be determined experimentally (for instance in deep-inelastic scattering experiments or in multijet production in e^+e^- colliders, see next chapter). For QCD, Λ is of the order 200 MeV. As shown in figure 1.17, whereas the QED coupling constant increases with the energy scale, the strong coupling constant decreases rather rapidly. At very high energy, interactions between hadrons can be described by individual interactions between almost free quarks and gluons.[c] The experimental verification, in the late 1970s, of this property, called *asymptotic freedom*, has been a crucial step towards establishing QCD as the theory describing the strong interaction between hadron constituents.

At low energy, for instance in the interaction between nucleons inside a nucleus, the coupling constant is so large that the effective interaction between nucleons can be described by the exchange of quark–antiquark bound states such as π^+, π^- or π^0 mesons (figure 1.4).

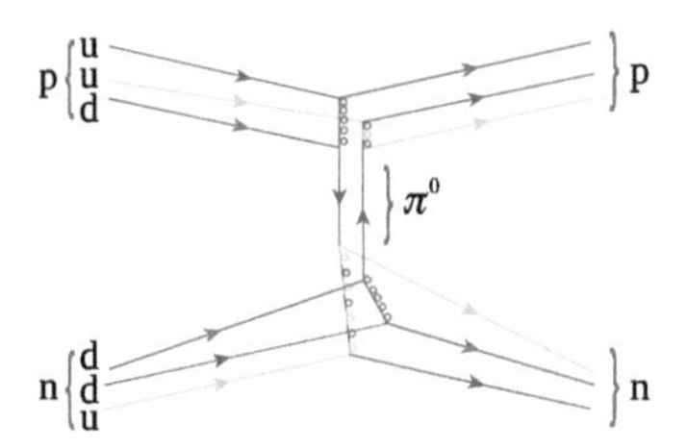

The figure above illustrates how this kind of nuclear interaction between a proton and a neutron can be described at the quark–gluon level as a residual of the more fundamental QCD processes.

[a]This feature is linked to the non-Abelian nature of the SU(3) group, whereas the QED gauge group U(1) is Abelian (also called commutative).

[b]The pure QCD Lagrangian with zero quark masses does not contain any dimensionful parameter as the only free parameter, the coupling constant α_S, is a dimensionless real number.

[c]This is also true for interactions between leptons and nucleons when the invariant mass of the lepton-nucleon system exceeds a few GeV.

Inset 1.14: The Holy Grail of physics

The almost entire present knowledge on elementary particles and their interactions is contained in a very compact mathematical expression called the Standard Model Lagrangian (cf. inset 1.4). From this expression, so short that it can fit onto a mug (see picture above), every physical process can be computed, at least in principle.

In its compactified form, the mathematical formula shown here extends and complements the one given in inset 1.8. The first line describes the dynamics of the force fields $F^{\mu\nu}$, associated with the gauge fields responsible for the electromagnetic, weak and strong forces. The second line represents the dynamics of the matter fields (quarks and leptons) ψ. As a consequence of gauge invariance, the partial derivatives with respect to space and time are here replaced with covariant derivatives denoted with a slashed D and incorporating the gauge fields and the coupling constants. Thus this second line describes the interactions between matter fields ψ and force fields (see also inset 1.6 on *Electromagnetism as a gauge theory*). The third line specifies the way matter fields are coupled to the Higgs field, here denoted as ϕ, a coupling which is at the origin of matter particle (fermion) masses.

The last line contains the dynamics of the Higgs field. The first term, called kinetic term, describes the propagation of the scalar field and contains also covariant derivatives. Similarly as for the matter-fields, this generates an interaction between the Higgs field and the gauge boson fields, giving rise to masses of the W and Z. The second term of this last line describes the peculiar Higgs potential in the shape of a Mexican hat (figure 1.8) that is responsible for electroweak symmetry breaking. Finally, the terms *h.c.* (hermitian conjugate) are here to indicate that antiparticles must be included too.

Owing to its simplicity, this compact mathematical expression incorporating the most fundamental laws of physics is of great beauty to physicists, albeit to the non-initiated it may look like a collection of cabalistic signs. The realisation that so much of Nature's behaviour is contained in so few symbols is mind-boggling. It is of profound significance and a monument for human inventiveness and rationality.

Nevertheless, once fully written down (see equation below), with all field components, including the internal degrees of freedom and auxiliary non-physical fields required for the consistency of the perturbative gauge theory (so-called *ghost fields*, where the scalar ghost fields are denoted $\phi^{-,0,+}$, $X^{-,0,+}$, Y, G in the notation by Martin Veltman), the expression of the complete Standard Model Lagrangian takes about a full page, which is still pretty compact given the amount and diversity of phenomena it describes. About 50 years of experimental and theoretical work were required to establish its present form.

In practice, however, even with the help of the most powerful computers, physicists are today only able to calculate and resolve the equations that can be derived from this Lagrangian in the simplest cases and with some approximations (as for example with developments in a series of Feynman diagrams). Moreover, despite its beauty, the theory embodied in the Lagrangian contains a number of imperfections at the fundamental level and it is certainly incomplete as will be discussed in the following chapters. This suggests that the present Standard Model is just an effective theory, an approximation of the ultimate theory, valid only up to a certain energy scale that is still to be determined.

The Standard Model Lagrangian in its full and most frightening form:

$$
\begin{gathered}
\boxed{\mathcal{L}_{SM} =} -\tfrac{1}{2}\partial_\nu g^a_\mu \partial_\nu g^a_\mu - g_s f^{abc}\partial_\mu g^a_\nu g^b_\mu g^c_\nu - \tfrac{1}{4} g^2_s f^{abc} f^{ade} g^b_\mu g^c_\nu g^d_\mu g^e_\nu - \partial_\nu W^+_\mu \partial_\nu W^-_\mu \\
- M^2 W^+_\mu W^-_\mu - \tfrac{1}{2}\partial_\nu Z^0_\mu \partial_\nu Z^0_\mu - \tfrac{1}{2c^2_w} M^2 Z^0_\mu Z^0_\mu - \tfrac{1}{2}\partial_\mu A_\nu \partial_\mu A_\nu - igc_w(\partial_\nu Z^0_\mu (W^+_\mu W^-_\nu \\
- W^+_\nu W^-_\mu) - Z^0_\nu (W^+_\mu \partial_\nu W^-_\mu - W^-_\mu \partial_\nu W^+_\mu) + Z^0_\mu (W^+_\nu \partial_\nu W^-_\mu - W^-_\nu \partial_\nu W^+_\mu)) \\
- igs_w(\partial_\nu A_\mu (W^+_\mu W^-_\nu - W^+_\nu W^-_\mu) - A_\nu (W^+_\mu \partial_\nu W^-_\mu - W^-_\mu \partial_\nu W^+_\mu) + A_\mu (W^+_\nu \partial_\nu W^-_\mu \\
- W^-_\nu \partial_\nu W^+_\mu)) - \tfrac{1}{2} g^2 W^+_\mu W^-_\mu W^+_\nu W^-_\nu + \tfrac{1}{2} g^2 W^+_\mu W^-_\nu W^+_\mu W^-_\nu + g^2 c^2_w (Z^0_\mu W^+_\mu Z^0_\nu W^-_\nu \\
- Z^0_\mu Z^0_\mu W^+_\nu W^-_\nu) + g^2 s^2_w (A_\mu W^+_\mu A_\nu W^-_\nu - A_\mu A_\mu W^+_\nu W^-_\nu) + g^2 s_w c_w (A_\mu Z^0_\nu (W^+_\mu W^-_\nu \\
- W^+_\nu W^-_\mu) - 2A_\mu Z^0_\mu W^+_\nu W^-_\nu) - \tfrac{1}{2}\partial_\mu H \partial_\mu H - 2M^2 \alpha_h H^2 - \partial_\mu \phi^+ \partial_\mu \phi^- - \tfrac{1}{2}\partial_\mu \phi^0 \partial_\mu \phi^0 \\
- \beta_h \left(\tfrac{2M^2}{g^2} + \tfrac{2M}{g} H + \tfrac{1}{2}(H^2 + \phi^0\phi^0 + 2\phi^+\phi^-) \right) + \tfrac{2M^4}{g^2}\alpha_h \\
- g\alpha_h M \left(H^3 + H\phi^0\phi^0 + 2H\phi^+\phi^-\right) \\
- \tfrac{1}{8} g^2 \alpha_h \left(H^4 + (\phi^0)^4 + 4(\phi^+\phi^-)^2 + 4(\phi^0)^2\phi^+\phi^- + 4H^2\phi^+\phi^- + 2(\phi^0)^2 H^2\right) \\
- gMW^+_\mu W^-_\mu H - \tfrac{1}{2} g \tfrac{M}{c^2_w} Z^0_\mu Z^0_\mu H \\
- \tfrac{1}{2} ig \left(W^+_\mu (\phi^0 \partial_\mu \phi^- - \phi^- \partial_\mu \phi^0) - W^-_\mu (\phi^0 \partial_\mu \phi^+ - \phi^+ \partial_\mu \phi^0)\right) \\
+ \tfrac{1}{2} g \left(W^+_\mu (H\partial_\mu \phi^- - \phi^- \partial_\mu H) + W^-_\mu (H\partial_\mu \phi^+ - \phi^+ \partial_\mu H)\right) + \tfrac{1}{2} g \tfrac{1}{c_w} (Z^0_\mu (H\partial_\mu \phi^0 - \phi^0 \partial_\mu H) \\
+ M \left(\tfrac{1}{c_w} Z^0_\mu \partial_\mu \phi^0 + W^+_\mu \partial_\mu \phi^- + W^-_\mu \partial_\mu \phi^+\right) - ig \tfrac{s^2_w}{c_w} M Z^0_\mu (W^+_\mu \phi^- - W^-_\mu \phi^+) + igs_w M A_\mu (W^+_\mu \phi^- \\
- W^-_\mu \phi^+) - ig \tfrac{1-2c^2_w}{2c_w} Z^0_\mu (\phi^+ \partial_\mu \phi^- - \phi^- \partial_\mu \phi^+) + igs_w A_\mu (\phi^+ \partial_\mu \phi^- - \phi^- \partial_\mu \phi^+) \\
- \tfrac{1}{4} g^2 W^+_\mu W^-_\mu \left(H^2 + (\phi^0)^2 + 2\phi^+\phi^-\right) - \tfrac{1}{8} g^2 \tfrac{1}{c^2_w} Z^0_\mu Z^0_\mu \left(H^2 + (\phi^0)^2 + 2(2s^2_w - 1)^2 \phi^+\phi^-\right) \\
- \tfrac{1}{2} g^2 \tfrac{s^2_w}{c_w} Z^0_\mu \phi^0 (W^+_\mu \phi^- + W^-_\mu \phi^+) - \tfrac{1}{2} ig^2 \tfrac{s^2_w}{c_w} Z^0_\mu H (W^+_\mu \phi^- - W^-_\mu \phi^+) + \tfrac{1}{2} g^2 s_w A_\mu \phi^0 (W^+_\mu \phi^- \\
+ W^-_\mu \phi^+) + \tfrac{1}{2} ig^2 s_w A_\mu H (W^+_\mu \phi^- - W^-_\mu \phi^+) - g^2 \tfrac{s_w}{c_w} (2c^2_w - 1) Z^0_\mu A_\mu \phi^+ \phi^- \\
- g^2 s^2_w A_\mu A_\mu \phi^+ \phi^- + \tfrac{1}{2} ig_s \lambda^a_{ij} (\bar{q}^\sigma_i \gamma^\mu q^\sigma_j) g^a_\mu - \bar{e}^\lambda (\gamma\partial + m^\lambda_e) e^\lambda - \bar{\nu}^\lambda (\gamma\partial + m^\lambda_\nu) \nu^\lambda - \bar{u}^\lambda_j (\gamma\partial \\
+ m^\lambda_u) u^\lambda_j - \bar{d}^\lambda_j (\gamma\partial + m^\lambda_d) d^\lambda_j + igs_w A_\mu \left(-(\bar{e}^\lambda \gamma^\mu e^\lambda) + \tfrac{2}{3}(\bar{u}^\lambda_j \gamma^\mu u^\lambda_j) - \tfrac{1}{3}(\bar{d}^\lambda_j \gamma^\mu d^\lambda_j)\right) \\
+ \tfrac{ig}{4c_w} Z^0_\mu \{(\bar{\nu}^\lambda \gamma^\mu (1+\gamma^5) \nu^\lambda) + (\bar{e}^\lambda \gamma^\mu (4s^2_w - 1 - \gamma^5) e^\lambda) + (\bar{d}^\lambda_j \gamma^\mu (\tfrac{4}{3} s^2_w - 1 - \gamma^5) d^\lambda_j) \\
+ (\bar{u}^\lambda_j \gamma^\mu (1 - \tfrac{8}{3} s^2_w + \gamma^5) u^\lambda_j)\} + \tfrac{ig}{2\sqrt{2}} W^+_\mu \left((\bar{\nu}^\lambda \gamma^\mu (1+\gamma^5) U^{lep}{}_{\lambda\kappa} e^\kappa) + (\bar{u}^\lambda_j \gamma^\mu (1+\gamma^5) C_{\lambda\kappa} d^\kappa_j)\right) \\
+ \tfrac{ig}{2\sqrt{2}} W^-_\mu \left((\bar{e}^\kappa U^{lep\dagger}{}_{\kappa\lambda} \gamma^\mu (1+\gamma^5) \nu^\lambda) + (\bar{d}^\kappa_j C^\dagger_{\kappa\lambda} \gamma^\mu (1+\gamma^5) u^\lambda_j)\right) \\
+ \tfrac{ig}{2M\sqrt{2}} \phi^+ \left(-m^\kappa_e (\bar{\nu}^\lambda U^{lep}{}_{\lambda\kappa} (1-\gamma^5) e^\kappa) + m^\lambda_\nu (\bar{\nu}^\lambda U^{lep}{}_{\lambda\kappa} (1+\gamma^5) e^\kappa\right) \\
+ \tfrac{ig}{2M\sqrt{2}} \phi^- \left(m^\lambda_e (\bar{e}^\lambda U^{lep\dagger}{}_{\lambda\kappa} (1+\gamma^5) \nu^\kappa) - m^\kappa_\nu (\bar{e}^\lambda U^{lep\dagger}{}_{\lambda\kappa} (1-\gamma^5) \nu^\kappa\right) - \tfrac{g}{2} \tfrac{m^\lambda_\nu}{M} H (\bar{\nu}^\lambda \nu^\lambda) \\
\underline{- \tfrac{g}{2} \tfrac{m^\lambda_e}{M} H (\bar{e}^\lambda e^\lambda) + \tfrac{ig}{2} \tfrac{m^\lambda_\nu}{M}} \phi^0 (\bar{\nu}^\lambda \gamma^5 \nu^\lambda) - \tfrac{ig}{2} \tfrac{m^\lambda_e}{M} \phi^0 (\bar{e}^\lambda \gamma^5 e^\lambda) - \tfrac{1}{4} \bar{\nu}_\lambda M^R_{\lambda\kappa} (1-\gamma_5) \hat{\nu}_\kappa \\
- \tfrac{1}{4} \bar{\nu}_\lambda M^R_{\lambda\kappa} (1-\gamma_5) \hat{\nu}_\kappa + \tfrac{ig}{2M\sqrt{2}} \phi^+ \left(-m^\kappa_d (\bar{u}^\lambda_j C_{\lambda\kappa} (1-\gamma^5) d^\kappa_j) + m^\lambda_u (\bar{u}^\lambda_j C_{\lambda\kappa} (1+\gamma^5) d^\kappa_j\right) \\
+ \tfrac{ig}{2M\sqrt{2}} \phi^- \left(m^\lambda_d (\bar{d}^\lambda_j C^\dagger_{\lambda\kappa} (1+\gamma^5) u^\kappa_j) - m^\kappa_u (\bar{d}^\lambda_j C^\dagger_{\lambda\kappa} (1-\gamma^5) u^\kappa_j\right) - \tfrac{g}{2} \tfrac{m^\lambda_u}{M} H (\bar{u}^\lambda_j u^\lambda_j) \\
- \tfrac{g}{2} \tfrac{m^\lambda_d}{M} H (\bar{d}^\lambda_j d^\lambda_j) + \tfrac{ig}{2} \tfrac{m^\lambda_u}{M} \phi^0 (\bar{u}^\lambda_j \gamma^5 u^\lambda_j) - \tfrac{ig}{2} \tfrac{m^\lambda_d}{M} \phi^0 (\bar{d}^\lambda_j \gamma^5 d^\lambda_j) + \bar{G}^a \partial^2 G^a + g_s f^{abc} \partial_\mu \bar{G}^a G^b g^c_\mu \\
+ \bar{X}^+ (\partial^2 - M^2) X^+ + \bar{X}^- (\partial^2 - M^2) X^- + \bar{X}^0 (\partial^2 - \tfrac{M^2}{c^2_w}) X^0 + \bar{Y} \partial^2 Y + igc_w W^+_\mu (\partial_\mu \bar{X}^0 X^- \\
- \partial_\mu \bar{X}^+ X^0) + igs_w W^+_\mu (\partial_\mu \bar{Y} X^- - \partial_\mu \bar{X}^+ Y) + igc_w W^-_\mu (\partial_\mu \bar{X}^- X^0 \\
- \partial_\mu \bar{X}^0 X^+) + igs_w W^-_\mu (\partial_\mu \bar{X}^- Y - \partial_\mu \bar{Y} X^+) + igc_w Z^0_\mu (\partial_\mu \bar{X}^+ X^+ \\
- \partial_\mu \bar{X}^- X^-) + igs_w A_\mu (\partial_\mu \bar{X}^+ X^+ \\
- \partial_\mu \bar{X}^- X^-) - \tfrac{1}{2} gM \left(\bar{X}^+ X^+ H + \bar{X}^- X^- H + \tfrac{1}{c^2_w} \bar{X}^0 X^0 H\right) + \tfrac{1-2c^2_w}{2c_w} igM \left(\bar{X}^+ X^0 \phi^+ - \bar{X}^- X^0 \phi^-\right) \\
+ \tfrac{1}{2c_w} igM \left(\bar{X}^0 X^- \phi^+ - \bar{X}^0 X^+ \phi^-\right) + igMs_w \left(\bar{X}^0 X^- \phi^+ - \bar{X}^0 X^+ \phi^-\right) \\
+ \tfrac{1}{2} igM \left(\bar{X}^+ X^+ \phi^0 - \bar{X}^- X^- \phi^0\right) .
\end{gathered}
$$

Legend: highlighted in blue are terms involving gluons including gluon ghost fields, gauge-boson interaction terms are highlighted in yellow, pure Higgs and Higgs ghost interactions are in light green, gauge boson to Higgs (and ghost Higgs) interactions are in dark green, highlighted in light-red are terms of weak-boson to matter (fermions) interactions, in purple the interactions of Higgs ghost fields with fermions, and finally in grey further ghost terms, so-called Faddeev–Popov ghosts, which cancel redundancies in weak interactions. *Source: SymmetryMagazine.org, 2016.*

Chapter 2

Key Experiments Establishing the Standard Model

2.1. The beginnings of experimental particle physics

In the 1930s, the knowledge about elementary constituents of matter was limited to the proton (p), the neutron (n), the electron (e^-) and its antiparticle the positron (e^+). To explain features of β-radioactivity Pauli postulated the existence of a very light and electrically neutral particle, the neutrino (ν).[1] About 20 years later this particle was finally observed for the first time in an experiment carried out by Clyde Cowan and Frederick Reines at a nuclear reactor in the US.

However, it soon became evident that Nature is much more complex. In studies of cosmic ray collisions with nuclei in the Earth's atmosphere or in early-type particle detectors such as cloud chambers or photographic emulsions, numerous new particles were found. Even if unstable, these particles seemed not less elementary and fundamental than the previous ones.

The first new particle to be found was the *muon* (μ) of mass 105 MeV. After an initial incorrect interpretation, the muon was finally properly understood, thanks to an experiment carried out in Rome in 1944–1945 by Marcello Conversi, Ettore Pancini and Oreste Picioni. In fact, after its initial discovery by Carl Anderson and Seth Neddermeyer (see section 1.7) before World War II, the muon was thought to be the *Yukawa particle*, a hypothetical particle postulated by the Japanese physicist Hideki Yukawa

[1] Pauli gave his postulated particle the name *neutron*. Upon the discovery of the more massive neutron by James Chadwick two years later in 1932, Enrico Fermi in 1933 introduced the term *neutrino* ("little neutron"), which was then also adopted by Pauli.

as the mediator of strong interactions among nucleons within the nucleus. In 1947, however, Cesar Lattes, Hugh Muirhead, Giuseppe Occhialini and Cecil Powel observed, in photographic emulsions exposed to cosmic rays in high altitude laboratories and in balloon flights, *pions* or π mesons (π^+, π^0, π^-) with masses around 140 MeV. It was rapidly realised that these pions, little heavier than the muon, did really behave as expected for mediators of strong nuclear forces, which was not the case with the muon that appeared, in its interactions with nuclei, to be a weakly-interacting particle.

After the muon and pions followed the discovery of *kaons*, K^+, K^- and K^0 mesons, with masses around 500 MeV, heavier than the pions, followed by the baryon Λ, called a *hyperon*. With its mass of 1125 MeV the Λ-hyperon was the first particle discovered with a mass larger than the proton (940 MeV).[2]

These discoveries were made in a relatively artisanal fashion by small teams of only few physicists. The situation changed in the 1950s, when physicists started to apply technologies developed during World War II, for example electronic coincidence and anti-coincidence circuits, electronic multichannel counting systems and analysers, klystrons or wave-guides used in radar antennas, etc. This led to the development of new detector types and especially of much more powerful new accelerators, in first place synchocyclotrons allowing acceleration of protons up to energies ranging from 200 to 600 MeV. This was a considerable advancement compared to the 25 MeV that could be obtained in cyclotrons at that time (see chapter 6). This period saw the first change of scale in accelerators, detectors and the sizes of teams, but also in the quantity of data collected by the experiments.

Thanks to new higher energy accelerators,[3] physicists started to bombard targets with protons and to produce in the process pions at an industrial scale. These could, in turn, be used to bombard proton targets. It was the set-up with which Enrico Fermi discovered in 1951 the first baryonic *resonance*,[4] the $\Delta^{++}(1232)$. This was the first in a long series of particle discoveries that followed between 1955 and about 1975. These findings

[2]The Λ was in fact discovered in cloud chamber exposures to cosmic rays through its decay into a proton and a pion: $\Lambda^0 \to p\pi^-$.

[3]We may quote here the English-born American physicist and mathematician, Freeman Dyson (1923–2020), who wrote on the importance of tools: "*New directions in science are launched by new tools much more often than by new concepts. The effect of a concept-driven revolution is to explain old things in new ways. The effect of a tool-driven revolution is to discover new things that have to be explained.*"

[4]The term *resonance* is related to the fact that this new particle has a sufficiently large width that it can be directly measured (see inset 2.1). The large width, and thus short lifetime (6×10^{-24} seconds), shows directly that the decay $\Delta^{++}(1232) \to p\pi^+$

showed that the proton and neutron are the lightest members of a large family of particles subject to strong interaction and just as elementary. These particles are collectively called *hadrons*. The family of hadrons is subdivided in two categories, the *mesons* such as pions, kaons etc, and the *baryons* such as the nucleons, Λ-hyperon, $\Delta^{++}(1232)$, etc. The proton and neutron occupy a rather modest place among the baryons (see inset 2.2), but they do have the remarkable property to be the only stable[5] hadrons and we are ourselves made up of them! The variety and organisation of the hadronic states observed was indicative of an internal structure. The *hadron spectroscopy* of both mesons and baryons is one of two pillars on which the quark model of matter stands, as we will discuss in insets 2.1 and 2.2.

Inset 2.1: Width of an unstable particle

Another important consequence of Heisenberg's uncertainty relation is the finite width of a particle, if the particle has very short lifetime. With a single measurement, even of perfect precision, one can only determine the energy — and thus the mass — of an unstable particle with lifetime τ to a precision ΔE such that $\Delta E \cdot \tau \geq \hbar/2$. Indeed, when one performs a large number of mass measurements, its distribution follows a bell-shaped form (called a *Breit–Wigner resonance shape*) with a peak defined to be the particle's mass and a width, Γ, defined as the particle's width. The width is inversely proportional to the particle's lifetime according to: $\Gamma = \hbar/\tau$. It is a property of all unstable quantum systems as the excited atomic or nuclear states.

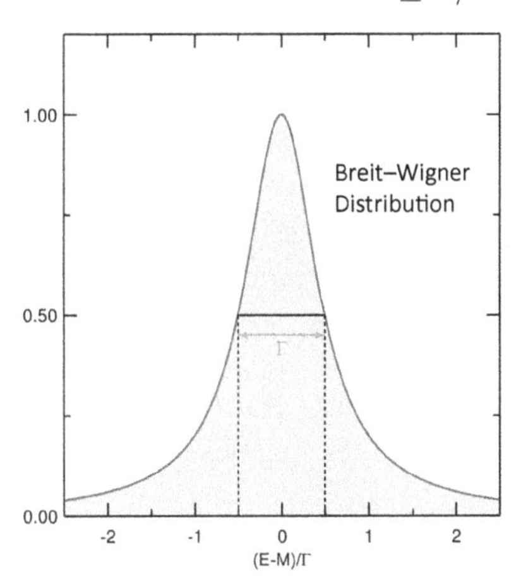

proceeds through strong interactions. The number in parenthesis following the symbol for a hadronic resonance is its mass, usually expressed in MeV.

[5]The neutron will in fact decay when free with a lifetime of about 15 minutes. It is the longest-living unstable particle known (protons are assumed to be unstable, but their decay has not yet been observed, giving them a very large lifetime, see section 4.2). Fortunately for our world, the neutron is stable when trapped in an atomic nucleus.

The width of a particle decaying through strong interactions (a *resonance*) ranges from few MeV to about 100 MeV, as the lifetime is of order 10^{-23} seconds. Particles decaying through weak interactions, however, have much longer lifetimes, of the order of 10^{-8}–10^{-12} seconds. In this latter case it is technically impossible to measure directly the width of the particle, but its lifetime can be measured, for example, through the decay length measurement at production in a suitable detector.

If a particle can decay in several ways (as is usually the case), each decay channel contributes to shortening the particle's lifetime and thus increases its width. The total width is then the sum of the *partial widths* Γ_i for each decay mode i: $\Gamma_{\text{tot}} = \sum_i \Gamma_i$. The *branching fraction* BR_i for each decay mode is given by the ratio of the partial width to the total width: $\text{BR}_i = \Gamma_i/\Gamma_{\text{tot}}$. However, in each particular decay mode we measure the same width, the total width Γ_{tot} , and the same lifetime τ of the unstable particle.

All these particles were initially considered elementary, that is, without internal structure. The physicists Murray Gell-Mann and George Zweig worked in the 1960s on a classification of the observed hadrons, which led to the *quark model* (chapter 1) according to which the baryons were composed of three quarks,[6] $qq'q''$, whilst mesons were made of a quark–antiquark pair, $q\bar{q}'$ (see inset 2.2). The quarks, when introduced, having a fractional electric charge compared to an electron (2/3 and −1/3) — something never seen before — were considered initially as purely mathematical objects, providing a sort of accounting procedure, but having no physical reality. However, in 1969 Jerome Friedman, Henri Kendall and Richard Taylor showed results of an experiment performed at SLAC on the scattering of high-energy electrons on protons. The experiment revealed the presence of hard point-like scattering targets inside protons, very much similar to Rutherford's experiment that uncovered the presence of nuclei at the centre of atoms at the beginning of 20th century. Since that time,

[6]Throughout this book it is implicit that whatever is being said for particles is also valid for antiparticles. Thus, there exist also anti-baryons made up of three antiquarks. When there are specific differences between particles and antiparticles, we will point them out explicitly.

physicists accepted the existence of quarks as real physical entities residing inside hadrons. Friedman, Kendall and Taylor received the Nobel Prize in physics for this discovery in 1990.

Inset 2.2: Quarks and hadrons: where are the elementary particles?

In the Standard Model, quarks are the elementary fermions undergoing strong interactions described by quantum chromodynamics (QCD, see inset 1.13). They carry a strong-interaction charge (*colour*), which can take three values conventionally called *red, blue and green*, although it has nothing to do with optical colours, these are just labels. In the same way, antiquarks carry anti-charges, anti-red, anti-blue and anti-green. QCD predicts that only colour-neutral (or *white* in analogy with the optical terminology) systems of quarks can exist as free states. This property, called quark *confinement*, explains why a free, fractionally charged quark, has never been seen (see also figure 2.4).

Observable particles made of quarks are thus always composite objects, and these are the hadrons. To respect the rule of colour neutrality, they can be made either of a quark–antiquark pair (e.g. a blue quark and an anti-blue antiquark), or of a triplet of quarks each carrying a different colour (q_r, q'_b, q''_g) so that the system is globally white (q may or may not be equal to q' or q''). The first species are the mesons; they are bosons of integer spin and can be produced individually in particle reactions. The second species are baryons; they are fermions with half-integer spin and are always produced in baryon–antibaryon pairs. The figure below shows on the left the quark–antiquark combinations (only of u, d, s, $\bar{u}$, $\bar{d}$, $\bar{s}$ quarks) making up the lightest spin-zero mesons, organised according to their strangeness quantum number (number of s-quarks incorporated) and electric charge. The middle picture shows the combinations of three quarks making up spin-1/2 baryons. In both cases we show here hadrons organised in octets, but baryons can also appear in decuplets; both types of configurations result from Gell-Mann's *eight-fold way* scheme from the early days of the quark model, with only three flavours, or types, of quarks, the u, d, s known. There are also mesons with spin 1, spin 2, even spin 3, and baryons with spin 3/2, among these the Δ^{++} with an electric charge of 2 (made of a combination uuu), and even higher spins.

The meson and baryon families (multiplets) grow when adding also c and b flavoured quarks (the top quark decaying too fast to allow top-hadrons to form). They are then represented not by two-dimensional constructs, but rather polyhedrons in three or four dimensions.

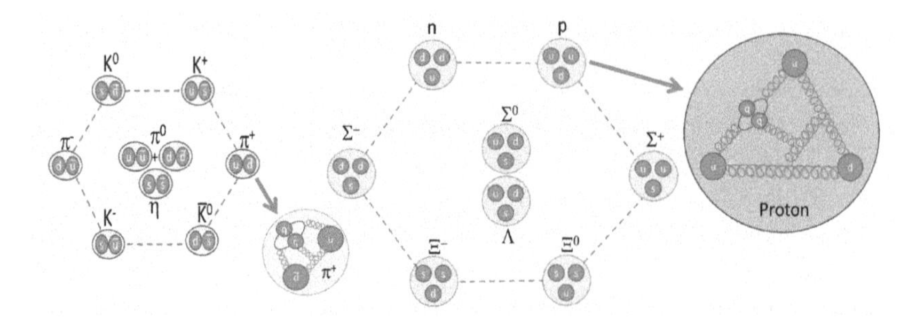

Colour was initially introduced in the theory to resolve the apparent contradiction between the existence of the Δ^{++} and the *Pauli exclusion principle* according to which it is impossible to have two identical fermions in the same quantum state in a physical system. In the case of the Δ^{++} we do have three u-quarks, but since they carry different colours (u_r, u_g, u_b) in fact they are not identical particles.

The dynamics of quarks and gluons (gluons are drawn as little springs in the figure) inside hadrons is complex. When one probes with a high-energy projectile, say a lepton, the structure of a hadron, for example a proton, the interaction and subsequent scattering can either occur on one of the *valence quarks*, or on a quark or an antiquark of the virtual quark–antiquark pair emerging from a gluon for a brief moment, as shown in the picture on the right (such emerging quarks are denoted *sea quarks*).

The spectroscopy of hadrons is at present well understood with the help of a powerful non-perturbative computational method called *lattice QCD*. Perturbative calculations cannot be applied in the quark confinement regime where the strong coupling constant is large. The LHC significantly contributes to increasing the knowledge of the hadron spectrum with the discovery of several heavy hadronic bound states.

2.2. First successes

During all this period of discoveries and studies of hadronic resonances the present-day Standard Model was in the making. The quark model was resting on two bases: first, deep inelastic scattering experiments on nucleons showing presence of point-like scattering centres, and second, the meson and baryon spectroscopy, exhibiting clear evidence of internal structure of hadrons organised in multiplets with well-defined regularities in spin and parity. This became even more evident after the discovery in 1974 of $c\bar{c}$ and in 1977 of $b\bar{b}$ mesonic bound states. The unified electroweak theory took shape between 1964 and 1971, while quantum chromodynamics, the quantum field theory of strongly interacting quarks and gluons, developed between 1971 and 1975. These theoretical advances were accompanied by major experimental discoveries: parity violation in 1956 (section 1.8), the electron-neutrino in 1956, followed by the muon-neutrino in 1962 at BNL, USA by Leon Lederman, Melvin Schwartz and Jack Steinberger (Nobel Prize in 1988); CP violation in the K^0 system again at BNL in 1964 (section 1.8); the proof of existence of quarks as real physical objects in 1969 (section 2.1), discovery of the J/ψ ($c\bar{c}$ bound state) in 1974 independently at BNL and SLAC (Samuel Ting and Burt Richter, Nobel Prize in 1976), the τ-lepton in 1975 at SLAC (section 1.1), the b-quark and $b\bar{b}$ bound state, the Υ, in 1977 at Fermilab, Chicago.

All these major discoveries took place in the United States. At CERN, the proton synchrotron, the PS, as well as the bubble chambers and electronic detectors were functioning very well. These devices often allowed measurements of higher precision, but European physicists most of the time were just able to confirm and improve on discoveries previously made in the US. One of the first beautiful European results at the time was the measurement of the anomalous magnetic dipole moment of the muon in dedicated muon storage rings at CERN by Emilio Picasso, Francis Farley, John Bailey and collaborators, allowing to probe quantum electrodynamics at short distances to unprecedented precision. Then, at the beginning of the 1970s, came the first really major discovery at CERN, the proof of existence of weak neutral currents with neutrino interactions in the heavy liquid bubble chamber Gargamelle (figure 2.1).

By the end of the 1970s several high quality and very successful electronic detector experiments had been designed and used at CERN to investigate deep inelastic scattering of leptons on nucleons, either with neutrinos — such as the experiment CDHS (CERN, Dortmund, Heidelberg,

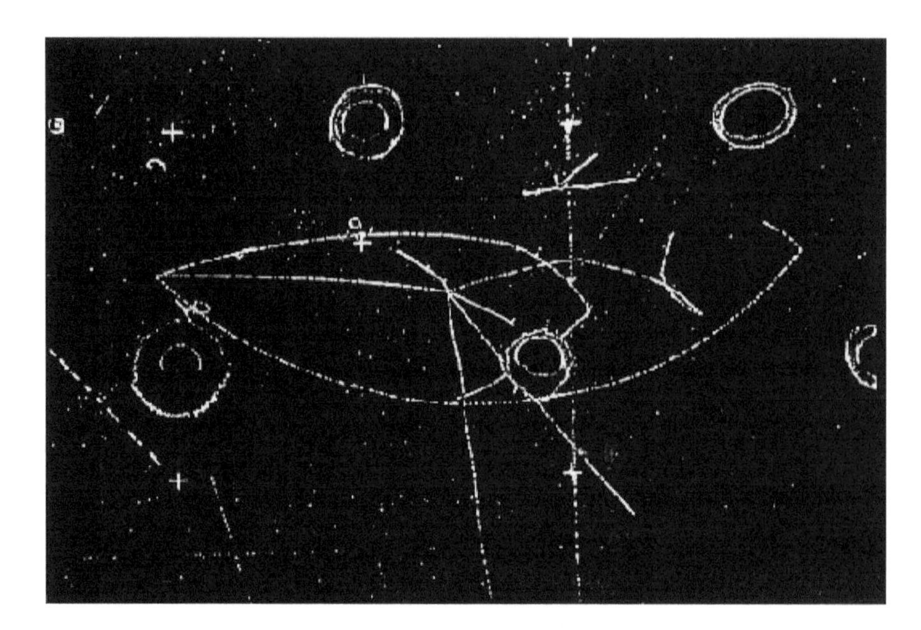

Fig. 2.1. Picture from the Gargamelle bubble chamber with a neutral-current interaction event taken in 1972. The neutrino beam, produced with a beam from the proton synchrotron (PS), is coming from the left. The interaction taking place is $\nu + \text{nucleus} \rightarrow \nu + h + h + h + \text{invisible nuclear residue}$, with the production of three charged hadrons, all interacting with nuclei in the chamber liquid, thereby identifying themselves as hadrons. If a muon were created at the interaction point, as is the case with charged-current weak interactions, it would have left a track from the vertex traversing the chamber without interaction, thus identifying itself unequivocally as a muon.

Saclay), or with muons (figure 2.2) — BCDMS (Bologna, CERN, Dubna, Munich, Saclay) and EMC (European muon collaboration). These experiments allowed detailed studies of the internal structure of nucleons as a function of the *momentum transfer*[7] between the incident lepton and a *parton*, a generic name for hadron constituents, quarks and gluons. These experiments verified that the variations observed for the momentum distributions of partons inside nucleons as a function of the momentum transfer behave as expected according to QCD. This was one of the key step in the confirmation of QCD as the theory of strong interaction.

[7]The momentum transfer (usually denoted p or q) is a measure of the resolution (λ) with which the structure of the nucleon is probed. The larger it is, the better is the resolution and the deeper is the insight into the structure — see discussion beginning of section 1.10.

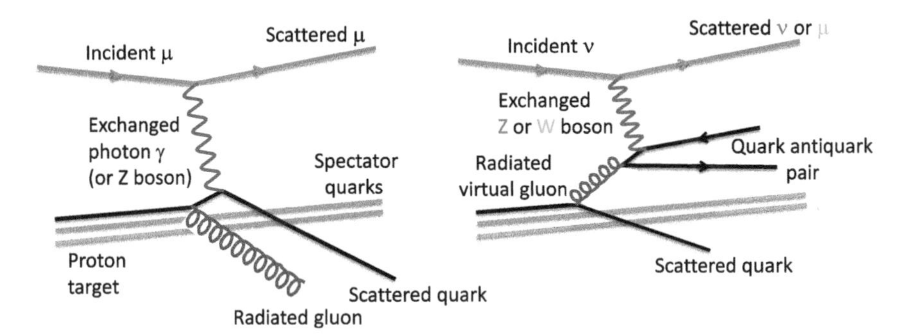

Fig. 2.2. Schematic representations of high-energy lepton scattering on a proton, on the left for an incident muon and on the right for an incident neutrino. When the wavelength associated to the exchanged boson is sufficiently small, or equivalently its momentum (momentum transfer) sufficiently large, the scattering takes place on an individual proton constituent (quark or gluon). In the drawing on the right-hand side, for example, the scattering occurs on an antiquark resulting from quark-pair creation from a virtual gluon radiated by a target-proton quark. The exchange of a W (electrically charged) corresponds to a so-called weak charged-current interaction, and the exchange of a Z (electrically neutral) to a weak neutral-current interaction. In the latter case, the incident neutrino is not changed into a muon in the final state.

Among the main results of these years of experimental confirmation of QCD, we should highlight the observation in 1979 of events[8] with three *jets* (collimated sprays of hadrons, see figure 2.3 for a display of a three-jet event and figure 2.4 for an explanation of the term jets) at the e^+e^- collider PETRA in Hamburg, Germany. This was the first direct evidence for gluons through the reaction $e^+e^- \to q\bar{q}g$.

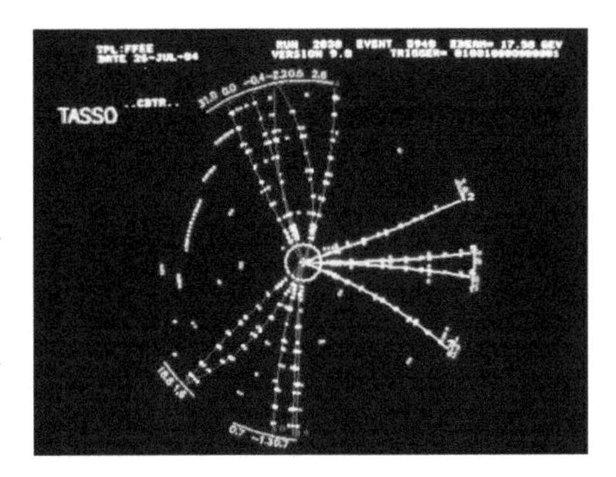

Fig. 2.3. Production of a quark–antiquark pair with a gluon in an e^+e^- collision, as reconstructed in a particle detector. Three *jets* are emerging from the collision point (see figure 2.4 for an explanation of jets and the Feynman diagram explaining this collision).

As for the electroweak segment of the Standard Model,

[8]Throughout this book we use the term *event* (the usual terminology among particle physicists) to designate the result of a specific particle physics collision in a detector. Often we also talk about events XY to talk about all collisions leading to the production of particles X and Y in the final state.

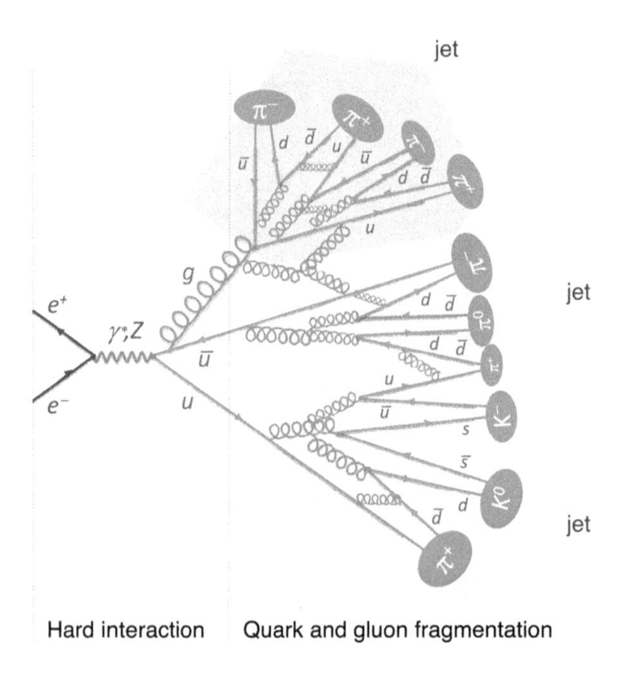

Fig. 2.4. Final state with three jets resulting from the production of a quark–antiquark pair plus a gluon in an electron–positron collision. Due to the property called confinement, gluons and quarks carrying colour charge cannot exist as free particles. Their production is immediately followed, within 10^{-25} seconds, by a process called *fragmentation* during which quarks and gluons *dress-up* to make particles without observable overall colour (in the sense of QCD), thanks to quark–antiquark pairs created out of the vacuum using the energy from the collision. The quark and antiquarks form hadrons (they *hadronise*), mostly pions and kaons. These hadrons keep memory of the parent gluon or quark they originate from, of its direction of motion and momentum, thus producing a collimated spray of particles — a *jet* — materialising the parent quark or gluon (figure 2.3).

the crucial outstanding test was the direct observation of W and Z bosons as real physical particles, not only as virtual ones via the manifestation of weak charged and neutral currents. This being said, the observation of weak neutral currents in 1973 with the Gargamelle bubble chamber was already a magnificent success of the theory. In 1974 Marie-Anne and Claude Bouchiat from the École Normale Supérieure in Paris succeeded in detecting the tiny effect of the interference between photon and Z boson in an atomic physics experiment by studying the symmetry properties of light absorption in caesium atoms. The amount of parity violation they measured agreed with calculations they had performed beforehand. In both these phenomena the Z plays a role only as a virtual particle, not as a real physical particle

Fig. 2.5. Sheldon Glashow, Abdus Salam, Steven Weinberg, 1979 Physics Nobel Prize laureates for their contributions to the construction of the electroweak unification scheme.

appearing in the final state of a particle reaction. Nonetheless, the Nobel Prize for physics was given to Glashow, Salam and Weinberg for the unified electroweak theory already in 1979, before the actual discovery of the W and Z! This shows the level of trust in the theory even before the decisive experimental proof was brought. The discovery of the W and Z bosons at CERN in 1983 was the sign of return on centre-stage of European particle physics research.

2.3. Discovery of the *W* and *Z* bosons

Thanks to measurements of the weak mixing angle $\sin\theta_W$ (see inset 1.10) in neutrino scattering experiments, it was expected that, provided the W and Z really exist, these should have masses in the 80–100 GeV range (section 1.6). With accelerators available at the time, and with the up-to-then standard method of bombarding a fixed target with a beam from an accelerator, it was impossible to produce so heavy particles, as accessible centre-of-mass energies were maximally of 25–30 GeV.[9]

Here came the suggestion in 1976 by David Cline, Peter McIntyre and Carlo Rubbia to transform a proton accelerator (the CERN Super Proton Synchrotron, SPS, for example) into a proton–antiproton collider. This represented a game changer. We shall come back to beam and accelerator

[9]The Intersecting Storage Rings (ISR) at CERN (cf. section 6.2.1) were the world's first hadron collider. The ISR operated between 1971 and 1984 and reached a maximum centre-of-mass energy of 62 GeV, which was, however, insufficient to create W or Z bosons.

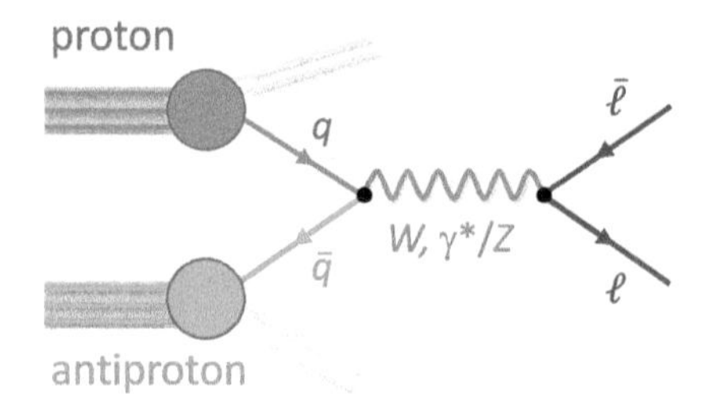

Fig. 2.6. Diagram illustrating the production of a W or Z (or virtual photon, γ^*) through the fusion of a quark and an antiquark, followed by the decay into a lepton and an antilepton, for example: $u\bar{d} \to W^+ \to e^+\nu_e$ or $u\bar{u} \to Z \to e^+e^-$.

aspects of this revolutionary project in more detail in chapter 6. Let us discuss here the physics aspects first. The proton and antiproton beams were brought to collision at two crossing points in the accelerator-collider where were located two large experiments, UA1 and UA2 (Underground Area 1, 2), ready to look for the production and subsequent decay of W and Z bosons (see the Feynman graphs of the processes in figure 2.6). The key idea was to exploit the leptonic decay modes $W^\pm \to \ell^\pm\nu$ and $Z \to \ell^+\ell^-$ ($\ell = e$ or μ) allowing to detect the rare weak-interaction-mediated production of W and Z among the much more abundant production of hadrons and jets through strong interactions.

The search for the W was based on the observation of events with large momentum (or energy) imbalance as measured in the plane transverse to incident beams (see inset 7.5). Namely, the decay neutrinos from the W do not interact with the material of the surrounding detector, thus the energy they carry leaves the detector undetected.[10] The W candidate events should also contain a relatively isolated high-momentum track, with a large momentum component in the plane transverse[11] to the beam lines and signalling the presence of an electron or a muon. Between December 1982 and January 1983 five such events were found in UA1 with backgrounds small enough to unambiguously prove the existence of the W boson (figure 2.8). With these few candidates, the mass of W boson was estimated at $m_W = 81 \pm 5$ GeV, where the second number gives a measure of the one-standard-deviation uncertainty on the measurement (figures 2.7–2.9). The UA2 experiment presented on January 22nd, 1983 the observation of four W boson candidates. The official CERN announcement of the W boson discovery followed shortly after on January 25th (cf. figure 2.10).

[10]The longitudinal momentum imbalance cannot be well measured in hadron colliders, as too many debris from the collision are lost along the beam pipe, or are at too small forward angles to be measurable (see inset 7.5).
[11]In what follows, the projection in the plane transverse to the beam direction of the momentum or energy of a particle is just called *transverse*.

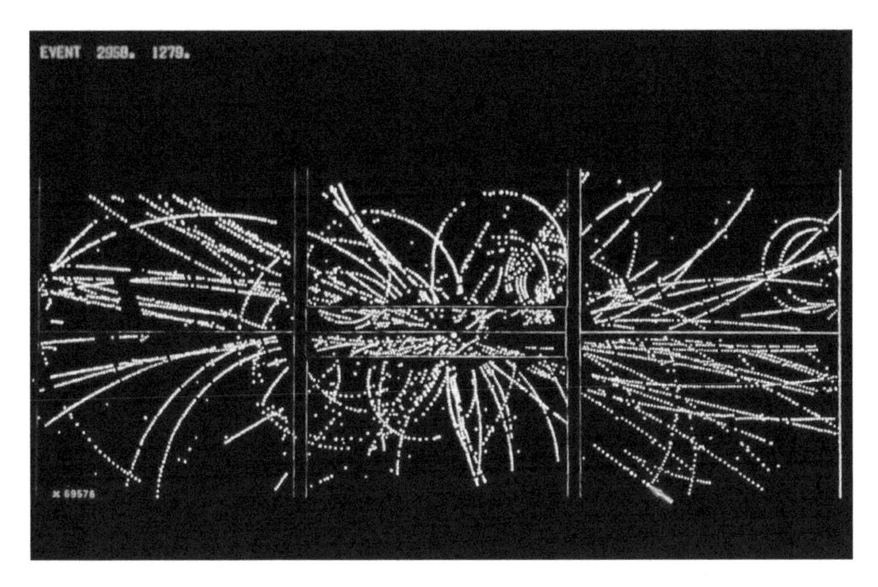

Fig. 2.7. Reconstruction of a W candidate event decaying into an electron and a neutrino, as seen in the tracking detector of UA1. The tracks in the central detector shown in yellow are debris of the incident proton–antiproton collision and consist mostly of pions and kaons. The electron, shown in white and indicated by the red arrow, shows up as an isolated high-momentum track, as visible for the near absence of track curvature in the magnetic field.

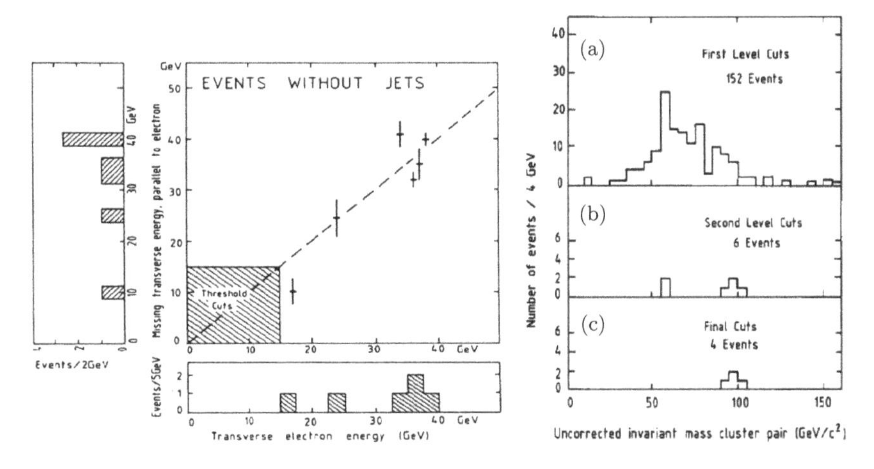

Fig. 2.8. On the left: distribution of the missing transverse energy versus the electron or positron transverse momentum for the first five $W \to e\nu$ candidate events in UA1. On the right: histograms of the measured electron–positron invariant mass (see inset 2.3) with subsequent tightening of selection requirements from top to bottom. The bottom panel shows the first four $Z \to e^+e^-$ events detected by UA1 close to 100 GeV mass.

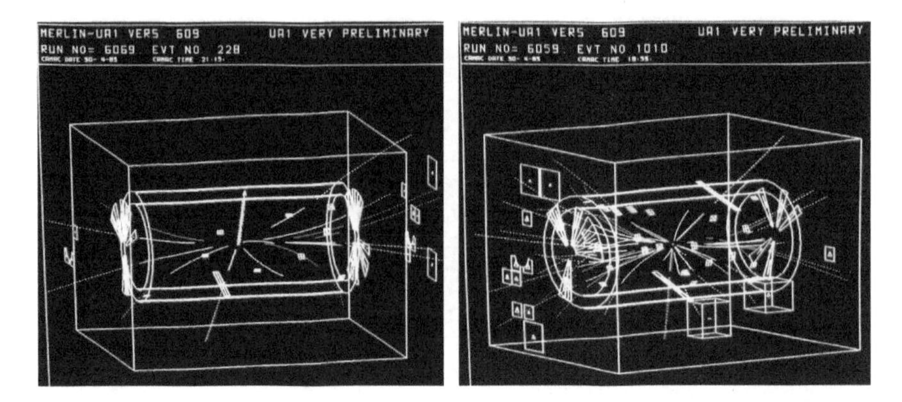

Fig. 2.9. Graphical displays of very first $W \to e\nu$ (on the left) and $Z \to e^+e^-$ (on the right) events measured in the UA1 detector. The colours correspond to coding for particle momentum values, red for small and blue for large values. The rectangular boxes are proportional to energy depositions in cells of calorimeters surrounding the tracking detector. The heavy yellow line on the left-hand display indicates the direction and magnitude of the missing transverse energy, i.e., of the escaping neutrino.

Fig. 2.10. Press conference in 1983 announcing the discovery of the W boson. From left to right: Carlo Rubbia (spokesperson of UA1), Simon van der Meer, Herwig Schopper (Director General of CERN), Erwin Gabathuler (research director at CERN) and Pierre Darriulat (spokesperson of UA2).

In March 1983, the collision rate was increased by an order of magnitude allowing the search for $Z \to e^+e^-$ or $\mu^+\mu^-$ events.[12] In April 1983, UA1 and UA2 found their first $Z \to e^+e^-$ candidates (figures 2.8 and 2.9). UA1's Z mass determination using the 1982–1983 dataset gave $m_Z = 95.2 \pm 3.9$ GeV, while UA2 measured $m_Z = 91.9 \pm 1.9$ GeV, benefitting from a superior calorimeter energy calibration.

In the meantime the selection of W events continued and rapidly the samples were large enough to study angular distributions of electrons or positrons in the $W \to e\nu$ decay. These studies in UA1 revealed that parity was maximally violated, as expected in weak charged-current interactions[13] (see inset 2.4). This was the final confirmation that the discovered particle was indeed the expected mediator of the charged weak force.

In October 1984, the Nobel Prize in physics was awarded to Carlo Rubbia and Simon van der Meer for their contributions to the large projects, which led to the discovery of the W and Z gauge bosons. Among these contributions, a key innovation is the development of the antiproton accumulation technique.

In the years that followed, thousands of W and Z events were produced at the CERN Sp$\bar{\text{p}}$S collider up to its closure in 1990, when the successful operation of a new and more powerful proton–antiproton collider, the Tevatron, with 1.8 TeV collision energy began at Fermilab, thus rendering the CERN Sp$\bar{\text{p}}$S collider obsolete. The discovery of the W and Z gauge bosons, carriers of the weak force, was one of the grand moments of particle physics. It represented the result of a, at the time, tremendous scientific and technical undertaking with the proton–antiproton collider and huge, novel types of particle detectors, altogether almost a gamble. It was also a period of extreme and unprecedented activity, tensions and passions, an unforgettable experience for all those who had the privilege to take part in that adventure. More profoundly, this discovery showed undeniably that gauge invariance is the proper symmetry principle governing the world of elementary particles at this scale.

[12]The product of Z production rate and its probability to decay to two charged leptons is about one-tenth of that of W production and decay to a lepton plus a neutrino.

[13]This means that only left-handed fermions (and right-handed antifermions) couple to Ws, whilst right-handed fermions (and left-handed antifermions) being "blind" to Ws do not participate to charged weak interactions.

Inset 2.3: Invariant mass of a system of particles

A particle with energy E and vector momentum $\vec{p}\,(p_x, p_y, p_z)$ in special relativity is described by a *four-vector* $p = (E, \vec{p})$, often called the *four-momentum.* The modulus squared of the four-momentum, $|p|^2 = E^2 - |\vec{p}|^2 = m^2$, gives the mass of the particle. It is a Lorentz-invariant quantity, that is, it has the same value in whatever inertial frame E and $\vec{p}$ are expressed or measured.[a] The *invariant (or effective) mass* m_{inv} of a system of N particles with individual masses m_i, energies E_i and momenta $\vec{p_i}$, where $i = 1, N$, is defined as the modulus of the sum of the four-vectors $\vec{p_i}$:

$$m_{\text{inv}} = \sqrt{\left|\sum_{i=1}^{N} p_i\right|^2} = \sqrt{\left(\sum E_i\right)^2 - \left|\sum \vec{p_i}\right|^2},$$

and is itself a Lorentz-invariant quantity. Let us take as an example the decay of an unstable parent particle with unknown four-momentum $P = (E, \vec{P})$ into two stable daughter particles with known masses m_1, m_2, measured momenta $\vec{p_1}, \vec{p_2}$, and thus energies $E_i = \sqrt{|\vec{p_i}|^2 + m_i^2}$, $i = 1, 2$:

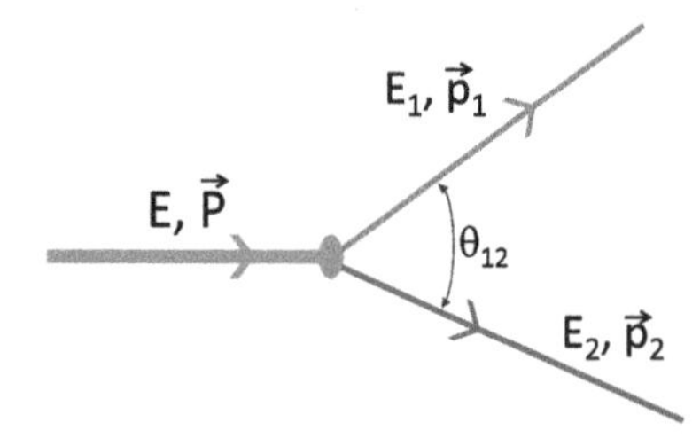

By energy and momentum conservation, the sum of the four-momenta of the final state particles must be equal to the four-momentum of the initial state particle: $P = p_1 + p_2 = (E_1 + E_2, \vec{p_1} + \vec{p_2})$, so that we can compute the mass of the parent particle from the invariant mass of the daughter pair: $M^2 = P^2 = m_1^2 + m_2^2 + 2E_1E_2 - 2|\vec{p_1}||\vec{p_2}|\cos\theta_{12}$, where θ_{12} is the angle between the directions of the daughter particles.

This method is used to search for new unstable particles with very short lifetimes: one draws a histogram with the invariant mass of a large number of measured two-candidate-particle events and watches out for a *peak* in the distribution that could be indicative of a new particle (see for example figure 9.8).

When one of the decay particle is an undetected neutrino (for instance in the $W \to \mu\nu$ decay), a so-called *transverse mass* m_T is used as an approximation of the invariant mass. It makes use of the so-called *missing transverse energy* (see inset 7.5 on page 213) as of measurement of the neutrino p_T. Then only the p_T of the charged lepton and the neutrino are used to calculate the invariant mass m_T (see figure 2.11).

[a] In special relativity, Lorentz transformations are coordinate transformations between coordinate frames that move relative to each other at constant velocity, see inset 1.5.

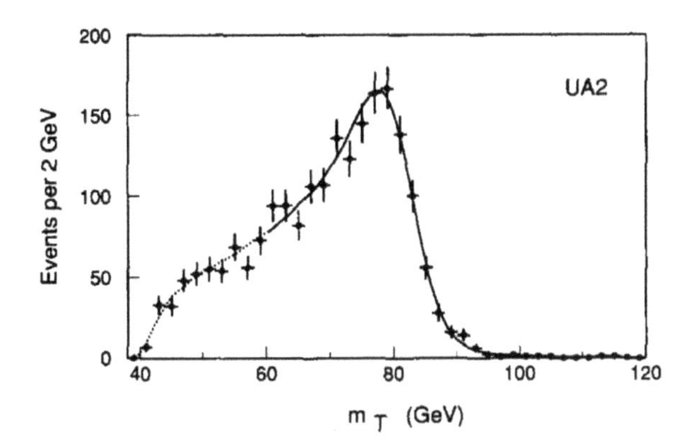

Fig. 2.11. Transverse mass distribution of 2065 $W \to e\nu_e$ candidate events measured by UA2 (mass of the electron/positron and the neutrino, approximating the neutrino momentum with the missing transverse energy — see inset 2.3). The curve is the best fit to determine the W mass. *Image: UA2 Collaboration, Phys. Lett. B 276, 354 (1992).*

In the final years of the proton–antiproton CERN collider (1987–1990), while UA1 was engaged in an (abortive) upgrade of its electromagnetic calorimeter, the upgraded UA2 detector (mainly with the addition of two large calorimeters in the forward and backward positions to improve the missing E_T measurement) pursued very successfully the study of W and Z bosons. The final *Jacobian peak* distribution for Ws from UA2 is shown in figure 2.11. The final samples of about 2100 $W \to e\nu$ and 250 $Z \to e^+e^-$ decays allowed UA2 to determine for the W and Z masses values of

$m_W = 80.84 \pm 0.22$ GeV (statistical) and $m_Z = 91.74 \pm 0.28$ GeV (statistical error). As the ratio m_W/m_Z was determined with higher precision (cancellation of most of systematic errors) and the LEP data were just producing the first much more precise Z mass measurements, UA2 final result combining both was: $m_W = 80.35 \pm 0.33 \pm 0.17$ GeV (statistical and systematical uncertainties). From this precise (about 0.5%) W mass measurement and the theoretical calculations radiative corrections to m_W, depending quadratically on the top-quark mass (see inset 2.6), it was possible to determine indirectly $m_{\text{top}} = 160^{+50}_{-60}$ GeV, years before its observation. Another important result from UA2 was, although the signal to background ratio was rather poor, the observation of W, Z decays into a quark–antiquark, i.e. a jet–jet pair. This possibility of doing "jet spectroscopy" has played a major role in the design of the ATLAS detector for the LHC (see chapter 8). To conclude, the proton–antiproton CERN collider allowed the historical observation of about 2700 W and 300 Z bosons altogether. By 1990 it was time for the much more powerful machines to take central stage, the Tevatron and the LEP.

Inset 2.4: Test of (maximal) parity violation in *W* decays: handedness of the charged weak interaction

The proton–antiproton centre-of-mass collision energy of the Sp$\bar{\text{p}}$S collider of about 600 GeV was suitable to test a key (but most unusual) feature of the charged weak interaction: the handedness of weak couplings, namely that only left-handed fermions (and right-handed antifermions) couple to the W bosons, whilst the opposite helicity states do not participate at all in that interaction! This maximal violation of parity (see inset 1.12) in W-mediated processes stems from what is historically called a $V - A$ (vector minus axial-vector) interactions, the two couplings being of equal strength.

To understand this let us first consider the production of a W^+: due to the handedness of the weak force and angular momentum conservation, the W^+, a spin-1 particle, is produced by the fusion of a left-handed u-quark with a right-handed anti-d quark. At this energy, the u-quark is predominantly coming from the incident proton and the anti-d quark from the incident antiproton. The W is produced almost collinear with the beam axis and with the direction of alignment of its spin, the handedness, as indicated by the red arrows sketched on the figure below. The W^- is produced correspondingly: the anti-u quark (which is right-handed) is coming from the antiproton and the left-handed d-quark from the proton. As a result of the handedness of the interaction, the spins of the produced W's are polarised preferentially along the antiproton direction of incidence as indicated in the figure.

Let us now consider the W decays, which proceed through the same charged weak interaction. As the W^+ is decaying to a positron and a neutrino, the e^+ (an antifermion!) is right-handed and the neutrino (a fermion) is left-handed, as indicated in the figure. Angular momentum conservation in the decay process favours emission of the decay positron in the direction of the spin of the W^+ with the neutrino just opposite (in the decay frame where the W^+ is at rest). The symmetric process occurs for the W^- decay. Angular momentum conservation, expressed in the rest frame of the decaying W bosons, results in a decay angular distribution for the positrons or electrons of the type $(1 + q \cdot \cos\theta^*)^2$, where θ^* is the decay angle relative to the antiprotons incident direction, and q is the electric charge +1 or −1 of, respectively, the positron or electron originating from the W decays.

In the proton-antiproton i.e. laboratory frame:

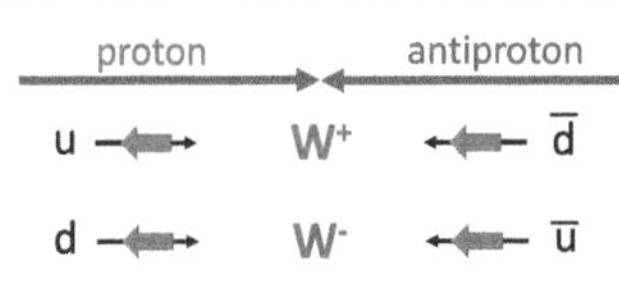

In the W rest frame:

e^+ — W^+ — ν_e

$\bar{\nu}_e$ — W^- — e^-

Electron (positron) angular distribution:

$$\frac{dn}{d\cos\theta^*} \propto (1 + q\cos\theta^*)^2$$

q=+1 for positrons; q=-1 for electrons

θ^* = 0 along antiproton direction

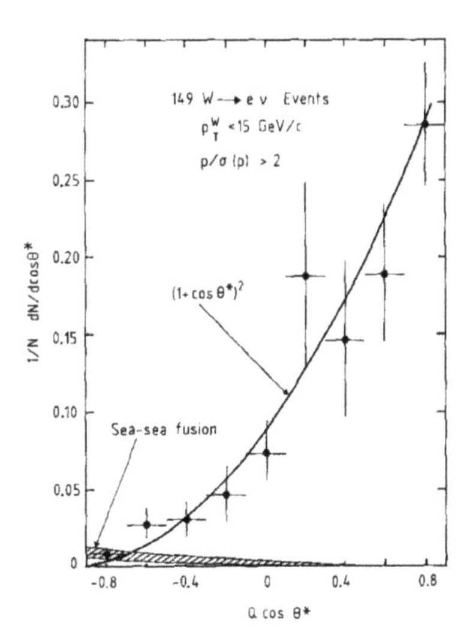

The differentiation between electrons and positrons in the final state was based on the magnetic field in the UA1 inner tracker volume (figure 2.7). The measurement of this angular decay asymmetry, already in 1983, when a sample of only 53 W-boson candidates was accumulated, showed beyond all doubts that the particle observed at a mass of about 80 GeV is indeed the mediator of the charged weak current (spin 1 and maximal parity violation). The right-hand figure above[a] shows data up to 1984 (150 W candidate events) and the shaded area shows the expected background from wrong polarisation states due to sea-quark–sea-antiquark fusion. At higher collider energies, as became available at the Tevatron, the correlation between the quark (resp. antiquark) origin and the parent proton (resp. antiproton) and its direction of motion is diluted, as more sea quarks in the antiproton and sea antiquarks in the proton have sufficient energy to fuse into W bosons.

The asymmetry between the behaviour of left- and right-handed fermions in the charged weak interaction has led some theorists to postulate the existence of another W boson (often labelled W_R) that would only couple to right-handed fermions. Such an extension would render the Standard Model left–right symmetric. The new boson would propagate a new form of charged weak interaction. Its non-observation until now at the LHC (see section 13.2) implies that the W_R must be much heavier, with mass of several TeV at least, than the Standard Model one (the W_L). In left–right symmetric models, the symmetry would thus only be restored at very high-energy scale, much beyond the W_R mass.

[a]Reference: UA1 Collaboration, Z. Phys. C — Particles and Fields 44, 15 (1989).

2.4. LEP and the consolidation of the Standard Model

By the end of the 1970s, even if neither the W nor the Z were yet found, there was so much confidence in the Standard Model that CERN started to plan the construction of an e^+e^- collider, the Large Electron–Positron collider (LEP), with around 100 GeV collision energy to study in detail electroweak unification.

For that purpose, a 27-kilometre long circular tunnel with internal diameter between 3.8 and 5.5 metres at a depth between 50 and 175 metres underground was built, the same tunnel where, twenty years later, the

LHC should be located.[14] With the discovery of the W and Z bosons with masses as predicted by theory, the LEP construction was accelerated[15] and brought to successful completion by 1989. At the four collision points around the ring four large detectors were constructed: ALEPH, DELPHI, L3 and OPAL. Each of these experiments was run by collaborations strong of about 400 physicists, about four times as many as in UA1 and UA2.

The main idea behind LEP was to study the Z, its decay modes, their angular distribution and polarisation, to probe in detail Z couplings to leptons and quarks, improve the precision of the Standard Model predictions and challenge them. The search for the Higgs boson in the debris of the collisions was also among the plans. The LEP collider was thus designed as a Z *factory*, what it indeed became during the first phase (LEP-1) of data taking between 1989 and 1995 (figure 2.12). The beam energy was tuned around 45.6 GeV so that the e^+e^- collision energy was exactly equal to the Z mass which maximised the Z production rate. Whilst during the six years of functioning of the Sp$\bar{\text{p}}$S collider, the UA1 and UA2 detectors observed less than one thousand Z bosons, the LEP experiments measured each about 17 millions of these!

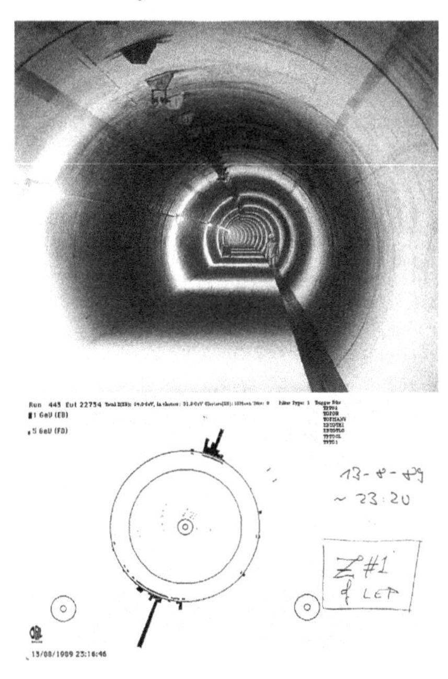

Fig. 2.12. Top: LEP tunnel before installation of the bending magnets. Bottom: OPAL logbook entry for the first Z-boson candidate recorded late on 13 August 1989. *Images: CERN.*

[14]The excavation of the LEP tunnel using three tunnel-boring machines started in 1985 and was completed three years later. It was Europe's largest civil-engineering project before the Channel Tunnel.

[15]In fact, LEP at start-up was in a neck-and-neck race for the first observation of the Z with the e^+e^- linear collider SLC at SLAC in the United States, an adaptation of the SLAC linear accelerator. The SLC could not match LEP in intensity (luminosity), but successfully achieved polarised beams allowing sensitive measurements, complementary to LEP, of the weak mixing angle.

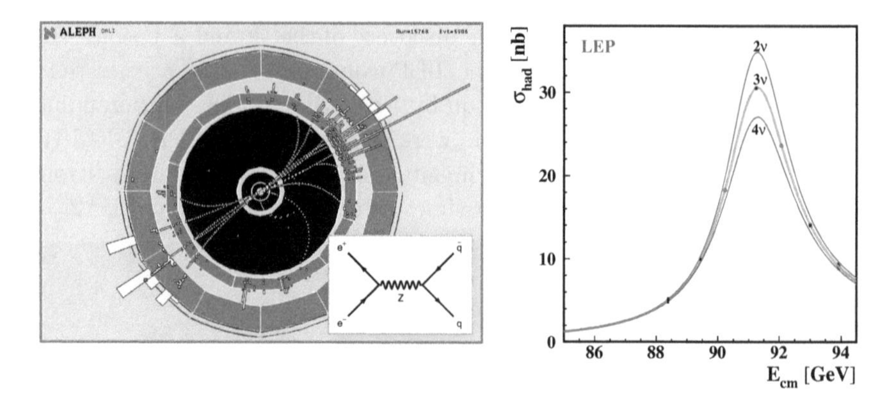

Fig. 2.13. On the left-hand plot, decay of a Z boson into a quark–antiquark pair, both members hadronising into a jet of particles, as detected in the ALEPH experiment. On the right, measured cross-sections for production of hadrons for different e^+e^- collision energies around the Z-boson mass peak. The curves give the theoretically expected cross-sections as a function of the number (2, 3 or 4) of light neutrino species (with mass less than $m_Z/2$ so the Z boson can decay to a pair of them). The experimental data coincide with the three-neutrino curve.

The first important LEP result was the determination of the number of light neutrino species N_ν. In the Standard Model the Z boson can decay into any flavour i of neutrino ν_i, $Z \to \nu_i \bar{\nu}_i$, provided its mass m_{ν_i} is smaller than $m_Z/2$. At that time three types of neutrinos, ν_e, ν_μ, and ν_τ corresponding to the three fermion generations, were already known.[16] The question was whether there may exist a fourth lepton generation, and thus another neutrino species, or even more.

Very soon after the start of the new collider in summer 1989, the experiments measured the Z spectral line-shape by counting the number of produced Z bosons at different collision energies around the Z mass as shown in figure 2.13. This measurement allowed to infer the total Z width Γ^Z_{tot} (see inset 2.1), which is the sum of the partial widths of Z decaying to fermion–antifermion pairs: neutrino pairs, charged-lepton pairs, and quark–antiquark pairs, except for the top quark which is too heavy (and whose existence or mass were not yet known). All these partial widths are predicted by the Standard Model, and they can be individually measured, except for the neutrino modes as these escape the detectors unseen. It is, however, enough to subtract from the total measured width the sum of all

[16]The τ neutrino had not yet been directly observed but its existence was evident from kinematic considerations of τ-lepton decays.

measured partial widths to obtain the contribution from the N_ν families of neutrinos. The LEP experiments obtained this way already in October 1989 the combined result $N_\nu = 3.12 \pm 0.19$, consistent with and only with three light neutrino species. Within a few days difference, SLC produced a comparable although less precise result.[17]

The number of light neutrino species plays a major role in cosmology. For example, it influences the abundance of helium nuclei (^{4}He) relative to hydrogen (^{1}H) during the primordial nucleosynthesis period in the first 3 to 17 minutes following the Big Bang (see section 5.3). Well before LEP times, this ratio was determined by astrophysical methods to be 24%, a value consistent with just three neutrino species.

During the second phase of LEP, denoted LEP-2, from end of 1995 until 2000, the energy of the e^+e^- collider was gradually raised up to 209 GeV in the centre-of-mass. The main goal was then to exceed the W-pair production threshold, $e^+e^- \to W^+W^-$, so as to be able to measure in detail the

Inset 2.5: Production cross-section of a process

In elementary particle physics, as in nuclear physics, the cross-section is a measure of the probability for a given process to occur. For example, when we talk of the cross-section for production of two W bosons in an electron–positron collision, we talk about the probability that a final state of such a collision contains two W bosons.

The unit of cross-section is a measure for surface, the barn (1 barn = 10^{-24} cm^2). It corresponds to the surface of a square with sides of 10 fermi (equal to 10 fm = 10^{-12} cm), the approximate cross-sectional size of a nucleus. By analogy with collisions in classical mechanics, we can say that the cross-section is the fictitious surface a target particle should have to reproduce the observed reaction probability, assuming a point-like projectile particle and impenetrable material bodies. For the rather rare phenomena we are dealing with in this book, most often sub-units are used as, for example, the picobarn (1 pb = 10^{-12} barn = 10^{-36} cm^2) or the femtobarn (1 fb = 10^{-15} barn = 10^{-39} cm^2). For the calculation of cross-sections, see inset 10.2.

[17]The combined result, taking into account all LEP and SLC data, is $N_\nu = 2.9840 \pm 0.0082$, which is extremely precise but exhibits a small tension of about two standard deviations with the expectation of $N_\nu = 3$.

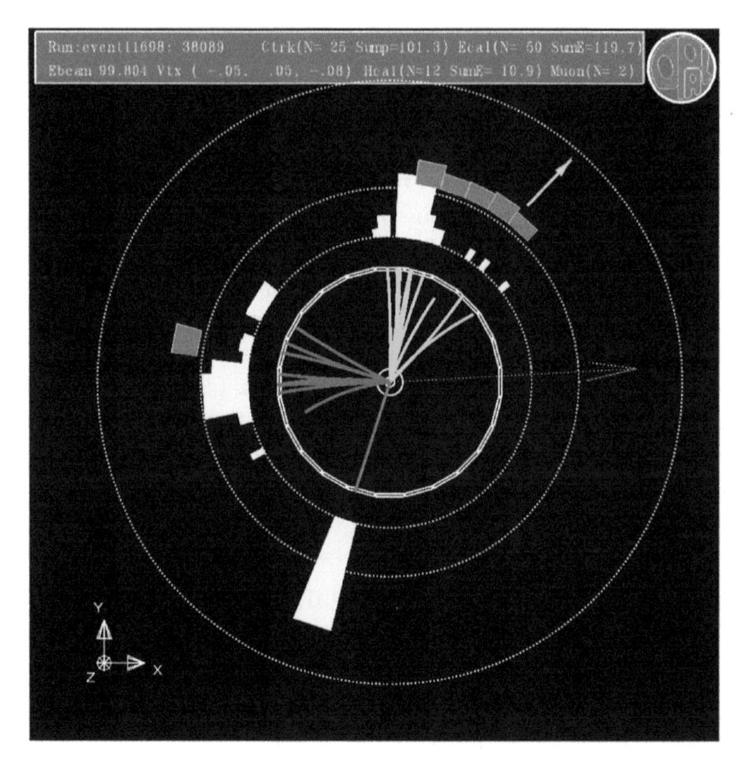

Fig. 2.14. *WW* production event, as seen in the OPAL experiment at LEP. One *W* decays into a quark–antiquark pair (jets in blue and green) and the other into an electron (in red) and an antineutrino, which is not directly detectable but whose momentum can nonetheless be measured (dashed arrow) by virtue of the constraint from overall energy conservation at the collision point (see inset 7.5 on page 213).

properties of this boson, too, in first place its mass. Among the very first priorities of LEP-2 was also, and naturally, the search for direct production of the Higgs boson already initiated at LEP-1, but which could now be extended to higher Higgs boson masses at LEP-2. We shall discuss these searches in chapter 11.

A total of about 40 000 W-pair events were collected and studied by the four LEP experiments. Among these studies, the measurement of the cross-section for W^+W^- production as a function of the e^+e^- collision energy was one of the most beautiful ones. The theoretical calculation invokes three Feynman diagrams (three quantum amplitudes — see section 1.9) corresponding to the three processes leading to the same final

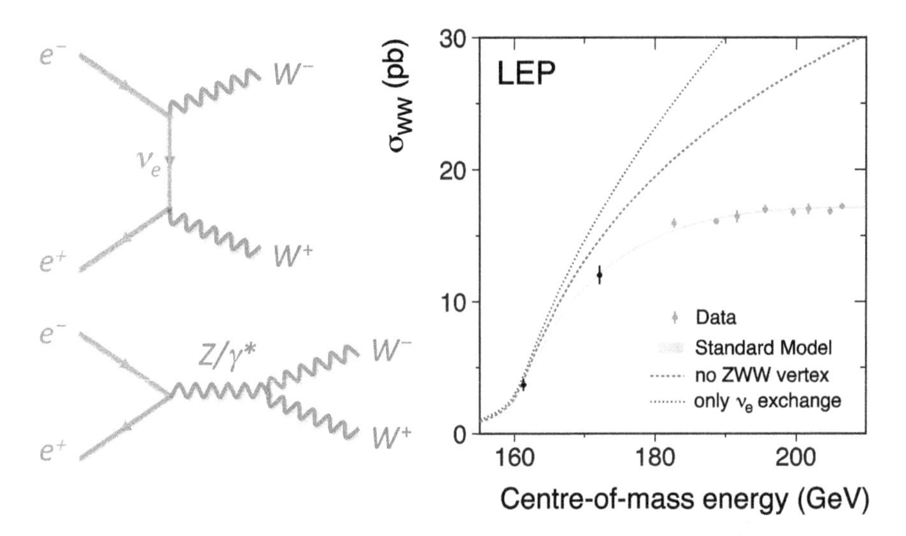

Fig. 2.15. On the left-hand side: diagrams leading to the production of a W^+W^- pair in e^+e^- collisions. Three processes are distinguished: neutrino exchange, annihilation via a photon (the asterisk indicates that the photon is virtual, i.e. off its mass shell) and annihilation via a Z. Each of the W's subsequently decays (in 10^{-24} seconds) into a pair of fermions: either a quark–antiquark pair materialising as two jets of hadrons (see figure 2.4), or as a charged lepton plus a neutrino (see figure 2.14). The figure on the right shows the cross-section for W pair production measured at LEP-2 as a function of collision energy. The points in black and green represent experimental measurements — in excellent agreement with Standard Model prediction (green band whose thickness indicates the theoretical uncertainty). The dotted curves correspond to theoretical predictions without including the amplitude via Z annihilation (red) or via Z and photon annihilation (blue). These predictions rise strongly with collision energy and eventually become non-physical.

state (figure 2.15). If the diagram with the Z in the intermediate state were not included in the calculation, the production cross-section would increase without bounds with increasing collision energy and would ultimately cross a limit where the result would make no physical sense (the so-called *unitarity limit*). In the Standard Model, the three amplitudes corresponding to the three diagrams compensate each other exactly, through appropriate couplings, so as to limit the rise of the cross-section, which was beautifully experimentally verified at LEP (figure 2.15).

To compare experimental results obtained at LEP with Standard Model predictions it was necessary to take into account in the calculations the effects of radiative corrections (mentioned in section 1.9) affecting

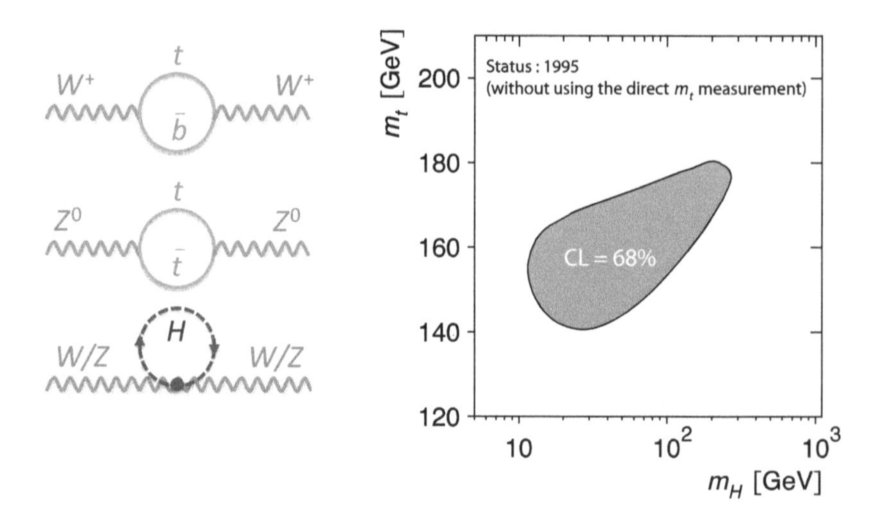

Fig. 2.16. On the left, Feynman diagrams with loops corresponding to quantum corrections to W and Z masses. Precise measurements of electroweak parameters of the Standard Model allowed LEP physicists in collaboration with theorists to constrain masses of the top quark, m_{top}, and the Higgs boson, m_H, thanks to their presence as virtual particles in the loops. The picture on the right shows the domain in the (m_H, m_t) plane favoured by constraints resulting from LEP measurements. This result was available before the discovery of the top quark at Fermilab's proton–antiproton collider Tevatron in 1995.

experimental observables.[18] As an example, the top quark, too massive to be directly produced at LEP, is nonetheless signalling its presence as a virtual particle through quantum corrections (figure 2.16) with a strength depending on the top-quark mass. The same is true for the Higgs boson, although its effect is much smaller than that of the top quark. Among the observables sensitive to the presence of top quark and Higgs boson is the W boson mass (see inset 2.6). A precise measurement of the W mass can thus provide numerical constraints on the, at the time unknown, top-quark and

[18]To get a feeling about the importance of quantum corrections, let us compare the prediction for the weak mixing angle (section 1.6) as given by the electroweak theory at Born level where all radiative corrections are ignored, $\sin^2\theta_W = 1-(m_W/m_Z)^2$, with its experimentally measured value. Using the most precise W and Z mass measurements available, $m_W = 80.385 \pm 0.015$ GeV and $m_Z = 91.1876 \pm 0.0021$ GeV, one finds $\sin^2\theta_W = 0.22290 \pm 0.00029$. This prediction differs by 26 standard deviations from the experimental value $\sin^2\theta_W = 0.23153 \pm 0.00016$. The comparison of the two values shows that the effect of radiative corrections is of order 4%. This example also emphasises the importance of measurement precision to be able to exploit the information contained in the radiative corrections (see inset 2.6).

Higgs-boson masses. A number of such measurements sensitive to quantum corrections were performed at LEP with great accuracy. Fitting the theoretical predictions with freely varying top-quark and Higgs-boson masses to these measurements allowed the LEP physicists to obtain significant constraints well before these particles were directly observed.

The LEP collider dominated the experimental particle-physics scene up the year 2000 when it stopped operating. The main accelerators in the world then became the B factories (section 2.5) and the Tevatron.[19] The Tevatron, if renowned firstly for the discovery of the top quark, which we will discuss in detail in section 12.3, contributed also magnificently to the confirmation of the Standard Model, even in the domain of B physics. A beautiful Tevatron result is the most precise W mass measurement to date, whose importance we just mentioned above.

Measurements of partial widths for decays of the Z into leptons and quarks, of decay angular and polarisation distributions, of couplings to gauge bosons, and of the W width at LEP and Tevatron represented the most complete and accurate tests of Standard Model parameters. Never before has any physics theory been subjected to so thorough and precise tests. No statistically significant deviation between experimental data and theoretical prediction were observed, and the Standard Model emerged rock-solid from this era. Its only missing piece before the start-up of the LHC was the Higgs boson, and its discovery was naturally the main motivation for the LHC project.

Inset 2.6: Radiative corrections to the *W* mass

In the Standard Model, neglecting quantum correction effects due to loop diagrams, the W mass is given by

$$m_W^{(0)} = \sqrt{\frac{\pi\alpha}{\sqrt{2}G_F}} \cdot \frac{1}{\sin\theta_W}.$$

Here α is the fine structure constant, G_F the Fermi constant of weak interactions ($G_F = 1.166 \cdot 10^{-5}\ \mathrm{GeV}^{-2}$) and θ_W the weak mixing angle,

[19] We should mention other experiments too, as for example the one installed on the SLC (Stanford Linear Collider), the prototype of a new concept of linear e^+e^- collider (not a circular one as LEP), where about 500 000 Z's were produced between 1992 and 1998.

precisely measured at LEP and SLC. When including radiative quantum corrections, due to the higher order diagrams such as those drawn in figure 2.16, this expression gets modified as follows ($m_H \gg m_W$):

$$m_W^2 = (m_W^{(0)})^2 \cdot (1 + \Delta r_{\text{top}} + \Delta r_{\text{Higgs}} + \cdots),$$

where Δr_{top} is the correction due to the virtual top quark with mass m_t, and Δr_{Higgs} the correction for the virtual Higgs boson with mass m_H:

$$\Delta r_{\text{top}} \simeq \frac{3G_F m_t^2}{8\sqrt{2}\pi^2} \cdot \frac{1}{\tan^2\theta_W},$$

$$\Delta r_{\text{Higgs}} \simeq \frac{3G_F m_W^2}{8\sqrt{2}\pi^2} \cdot \ln\left(\frac{m_H^2}{m_W^2}\right),$$

where the dependence of Δr_{top} on m_t is quadratic, while that of Δr_{Higgs} on m_H is logarithmic, hence the stronger sensitivity of radiative corrections to the top quark than the Higgs boson. More precisely, whilst a variation by 10 GeV of the top mass modifies the W mass by 61 MeV, the same modification of m_H changes m_W by only 5 MeV.

During the first phase of LEP, neither m_t nor m_H were known. However, the Z mass, m_Z, and the weak mixing angle were measured with great precision. Both taken together allowed to constrain the top mass. Once the top quark was discovered and its mass measured, as well as that of the W boson, it became possible to constrain the mass of the Higgs boson. The fit revealed the preference for a light Higgs boson, not too far away from the masses of the Z and W bosons.

2.5. *CP* violation

We have seen previously that CP violation in weak interactions, which means breaking up the symmetry between weak-interaction processes involving particles as compared to antiparticles, can be introduced in the Standard Model through a non-zero phase of a complex number in the Cabbibo–Kobayashi–Maskawa (CKM) matrix.

Most of the aspects of CP violation can be graphically shown through what is called the *unitarity triangle*[20] (figure 2.17). A flat, squashed, triangle is indicative of a vanishing CKM phase, which would mean weak interactions preserve CP symmetry. The aim of the experiments is thus to precisely measure the shape of the triangle. It is thereby important to verify whether the variety of phenomena of CP violation in rather different, but theoretically related processes, all correspond to the same triangle with the same apex position.

Indeed, a large number of such measurements has been performed both with K mesons and, especially, with B mesons. Two dedicated electron–positron colliders, one at SLAC in the USA and the other at KEK in Japan, were built in the late 1990s with exactly this goal. Their centre-of-mass collision energy was optimised for 10.6 GeV, just above the threshold for production of pairs of beauty mesons: $B_d^0\bar{B}_d^0$ (bound states of $\bar{b}d$ and $b\bar{d}$, respectively) and B^+B^- ($\bar{b}u$ and $b\bar{u}$). In the years from 1999 until 2009, more than a billion B pairs were recorded and analysed, justifying the name of B factories for these colliders. The CDF and DØ experiments at the Tevatron also contributed to this domain of research, more specifically by adding to the long list of measurements those concerning the B_s^0 and $\bar{B}_s^0$ (bound states of $\bar{b}s$ and $b\bar{s}$), too heavy to be produced with high enough rate at the B factories.

Figure 2.17 shows the status of these CP studies as of summer 2019. It is remarkable that all measurements are consistent with a unique position for the apex of the unitarity triangle.[21] The CKM phase, given by the angle γ of the unitarity triangle, has a value $\gamma = \arctan(\bar{\eta}/\bar{\rho}) = 65.8 \pm 1.5°$, thus clearly establishing CP violation. This impressive success of the Cabibbo–Kobayashi–Maskawa paradigm shows that the unique phase of the CKM matrix of the Standard Model is the main, if not the only, source of CP violation in the quark system at the electroweak scale.

The domain of B physics studies, in particular of CP-violation aspects, is one of the most active ones in particle physics. In chapter 14, we discuss

[20]This denomination is related to the one we met in section 1.3 in connection with scattering amplitudes. Here it results from a relation among transition amplitudes involving CKM matrix elements and can be related to the sides of the triangle and insuring that the triangle is closed. When considering also B_s^0 type states other unitarity triangles enter the game.

[21]The exact position of the vertex is given by $\bar{\rho} = 0.16 \pm 0.01$ and $\bar{\eta} = 0.35 \pm 0.01$, with $\bar{\rho}$ and $\bar{\eta}$ as defined in the legend of figure 2.17.

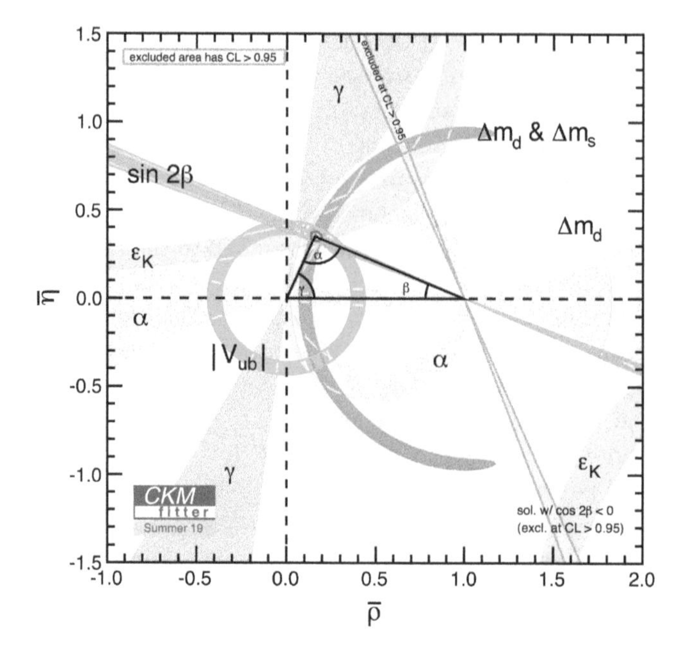

Fig. 2.17. Unitarity triangle of the CKM matrix in the $(\bar{\rho}, \bar{\eta})$ reference frame, where $\bar{\rho}$ is the real part and $\bar{\eta}$ the imaginary part of the unique phase of the CKM matrix. *CP* violation requires a non-zero value for $\bar{\eta}$ (i.e. the triangle is not squashed, not reduced to a line). Constraints stemming from a variety of measurements sensitive to this phase are shown in colour. Constraints on angles of the triangle all depend on differences among interfering processes involving particles and antiparticles. These differences can be small, as is the case in the K, D or B_s meson systems, or large, as for B_d or $B^{\pm}$ mesons (but involving rare transitions thus requiring large amounts of data to be measurable). Experiments determining the lengths of the sides of the triangle, except the normalisation segment [0, 1] that is fixed along the abscissa, are largely those measuring the B-meson decay rates involving transitions between quark families, or oscillations among B mesons. The tiny elliptical region in orange is the result of an overall fit to all available measurements. It is consistent with a closed triangle, that is, it matches every individual measurement, indicating that a single parameter, the CKM matrix phase, provides an appropriate description for all *CP* violation effects observed in transitions among quark flavours. Yet another superb success of the Standard Model!

the achievements and successes of the LHC in this domain. In 2019, a much upgraded e^+e^- B factory collider at KEK entered operation at 10.6 GeV collision energy and is expected to achieve a two orders of magnitude higher luminosity than in its initial version.

Chapter 3

What the Standard Model Cannot Explain

Experimental searches and studies of the past thirty to forty years have gradually uncovered and measured all the ingredients of the Standard Model, including the Higgs boson that was still missing from the particle list just a few years ago. Despite all its successes, beauty, and accurateness of its predictions, the model cannot be the ultimate theory of fundamental physics. It has problems, deficiencies, even major ones, and the LHC in the coming years could bring clues for some of them. These problems may be put into three categories.

- The Standard Model has eighteen free parameters,[1] unconstrained by the theory, whose values must be determined experimentally to make the theory predictive. Among these parameters are the masses of the elementary fermions (quarks and leptons). Particularly troubling is the large mass hierarchy observed between the members of the three families, which is the result of widely varying coupling strengths to the Higgs field. The most massive among the quarks, the t-quark, is about 80 000 times heavier than the lightest quark, the u-quark. Similarly, the heaviest charged lepton, the τ, is 3500 times as heavy as the electron, the lightest among the charged leptons. In what regards the neutral leptons, the neutrinos, their mass is at least 500 000 times smaller than that of the electron.

[1] Ignoring neutrino masses, their mixing and CP-violation parameters, which are not included in the original version of the Standard Model as neutrinos were initially assumed to be massless. We also ignored one parameter describing CP violation in the strong-interaction sector, which empirically turns out to be very small, or zero (see section 3.4).

- The Standard Model cannot describe a certain number of observations in astrophysics and particle physics as we are going to discuss below.
- The Standard Model describes well the state in which the Universe was about 10^{-13} seconds after the Big Bang, corresponding to an ambient energy of the order of a TeV where the electroweak phase transition occurs, as well as the subsequent evolution. However, it becomes problematic to extrapolate the model towards much earlier epochs, with larger ambient energies, closer to the very first instants in the history of the Universe.

The first problem, although fundamental, seems more of an aesthetic nature and has no confirmed solution at present.[2] Often science has to limit itself to being descriptive rather than explanatory, or to being explanatory only on basis of some measured parameters or some hypothetical principles. The remaining two problems, however, show beyond doubt that the Standard Model is only an effective theory, and even an incomplete one.

Inset 3.1: Baryons, the Higgs mechanism, and the mass of the observable Universe

Before exploring concepts such as dark matter and dark energy in sections 3.1 and 3.2 it is useful to discuss the origin of mass in the observable Universe, which is mainly made of baryons (protons and neutrons), and its relation to the Higgs mechanism.

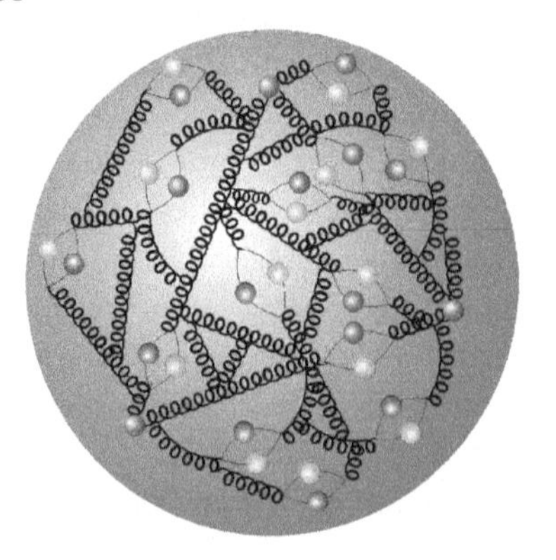

The figure is a schematic representation of the structure of a proton, with the three valence quarks and a multitude of virtual quark–antiquark pairs stemming from the energy content of the gluon field, the whole being contained in a small sphere with a radius of the order of a femtometre (10^{-15} metres). One notices that one can barely identify

[2]Some extensions of the Standard Model, such as models with additional spatial dimensions, propose solutions termed natural to explain the observed hierarchy among fermionic masses, but these explanations are still highly speculative as no experimental evidence of such extra dimensions has yet been found.

the three valence quarks that define the proton properties such as its charge (see also inset 2.2).

While the Higgs boson is indeed thought to be at the origin of the quark masses (amounting to a few MeV for u and d quarks), most of the proton mass of 938 MeV is generated by the energy of motion of quarks and antiquarks and the interaction (*binding*) energy content in the gluon field (the gluons themselves are massless), through the relation $m = E/c^2$. The proton and neutron masses are thus essentially of dynamical origin. Were the u and d quarks massless, the masses of protons and neutrons would not be much different (though this would have other important implications, such as vanishing CP violation in the electroweak quark sector). However, even if the Higgs field is not directly responsible for most of the mass of the observable Universe, its contribution is nonetheless fundamental. For example, with massless electrons there would be no atoms, as the average distance between an electron and a proton in a hydrogen atom (called *Bohr radius*) would be infinite, and with massless gauge bosons W and Z, the weak force would have an infinite range, etc.

3.1. Invisible matter: dark matter

A number of astrophysical and cosmological observations, some of which decades old, show consistently that there appears to be more matter present in the large structures of the Universe (galaxies and clusters of galaxies) than what we see as visible matter (stars, clouds of gas or dust, luminous or dark ones) with detection instruments. Among the first observations of this phenomenon is the one by Dutch astronomer Jan Oort which dates back to the early 1930s. He studied the motion of stars on the periphery of galaxies using the Doppler shift and, to explain the observed motion, he concluded that the matter present must exceed by a factor of at least three the amount of visible matter. In 1933 the Swiss-American astrophysicist Fritz Zwicky, who studied the motion of galaxies in a cluster of galaxies (Coma cluster), showed that about 400 times more matter than what is visible would be needed to gravitationally bind the fast moving galaxies within the cluster. Although this value turned out to be an over-estimation, Zwicky's hypothesis of missing matter seems correct from today's point of view: about 90% of the Coma cluster's matter remains invisible.

The most compelling observation suggesting the presence of invisible matter, or *dark matter*, is related to rotation curves of stars within galaxies. Kepler's laws teach us that the orbital velocity of a planet far away from the Sun is smaller than for a nearby one. For example, the Earth is circling around the Sun at a velocity of around 110 000 kilometres per hour, whilst Saturn, about ten times more distant, has a three times smaller orbital velocity. The same law is valid for galaxies, many of which, due to the gravitational force, have a spiral structure with a much larger matter density close to the centre. One would thus expect that the revolution velocity v of stars or gas clouds at the periphery of a galaxy diminishes compared to the one for stars closer to the centre according to the law $v \propto 1/\sqrt{r}$, where r is the distance of a star from the centre of the galaxy. In the years 1960–1970, Louise Volders (The Netherlands) and Vera Rubin (USA) studied this problem systematically and found that the velocity remains approximately constant or may even grow with increasing distance r (figure 3.1).

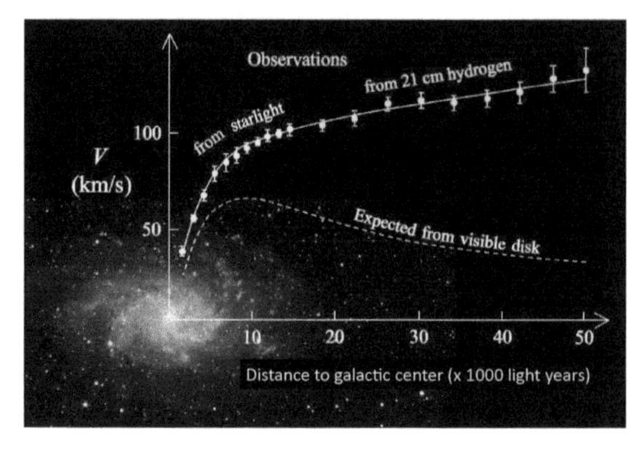

Fig. 3.1. Revolution velocity vs. distance from the centre of celestial bodies — stars or gas clouds — in the Triangle galaxy M33.

This measurement was made using a large number of spiral galaxies. A way to explain the observed orbital velocity vs. distance, while respecting Newton's law of gravity, is to invoke a large *halo* of dark matter of approximately spherical shape embedding the entire galaxy. Indeed, given the relation $v = \sqrt{GM/r}$, where G is the gravitational constant, and M all visible and invisible mass between the galactic centre and the probed star, a constant or increasing velocity with r can be obtained if M increases proportionally with r. The hypothesis that this halo might be made up of dead stars, such as neutron stars or black holes, was found to be inconsistent (or hardly consistent) with observational data. A more appealing explanation is dark matter in form of *weakly interacting massive elementary particles* (WIMP).

The first thought was that this halo could be made of cosmological neutrinos (see section 5.5) at a temperature of about two kelvin (thus with a kinetic energy of about 10^{-4} eV), and with masses of the order of ten

Fig. 3.2. Temperature fluctuations in the fossil microwave background radiation created about 380 000 years after the Big Bang, as measured by the Planck satellite (data of March 2013). The magnitude of these fluctuations is of the order of 0.2 milli-kelvin, about 10^{-4} in relative magnitude compared to the average background radiation temperature of 2.7 kelvin. These fluctuations are seeds from which galaxies, galaxy clusters and large structure will evolve and develop billions of years later. The statistical analysis of their distribution on the sky in terms of size and intensity, which is in deep connection with the expansion of the Universe, allows to determine a number of cosmological parameters, among these the amount of dark matter and of dark energy (section 5.4).

electron-volt. Almost at rest, such neutrinos could have been attracted gravitationally and captured by galaxies. But this hypothesis is in contradiction with the currently accepted models to explain the formation of large structures[3] in the Universe. To allow formation of medium and small size galaxies these models require dark matter being cold, that is made of particles whose velocity is small compared to the velocity of light. But at seeding and formation time of galaxies these neutrinos were relativistic and thus could not have congregated into such small structures. It was thus necessary to look for a source of dark matter outside, beyond the Standard Model. At present, the most precise determination of the amount of dark matter in the Universe is obtained from the statistical analysis of the small temperature fluctuations present in the microwave background radiation (section 5.2). These fluctuations have been measured precisely since the years 2000 by the WMAP satellite experiment. In 2013 the Planck satellite has made even more precise measurements (figure 3.2). The outcome

[3]The upper limit on the ν_e mass from direct experimental measurements is less than one electron-volt, however the upper limits from direct measurements on ν_μ and ν_τ masses are much looser.

of these measurements (figure 3.3) shows that at most about 5% of the matter-energy content of the Universe is made of ordinary baryonic matter, in the form of protons or neutrons in atomic nuclei. WMAP could thus be called the first *baryometer* in human history. The results also show that about 27% of the matter-energy content is dark matter, this is about five times as much as for ordinary visible matter. At this stage we are still missing about 68% of the matter-energy of the Universe! This is what is called *dark energy*, which we will briefly discuss in the next section.

It should, however, be mentioned that the existence of dark matter, albeit plausible, is not the only explanation that has been put forward to explain the obscure gravitational phenomena discussed above. A small fraction of physicists proposed an explanation in terms of a possible breakdown of Newton's law of gravity at large distances, on the scale of galaxies or the Universe, the so-called MOND (*modified Newtonian dynamics*) hypothesis. This hypothesis is confronted with observational difficulties of its own.

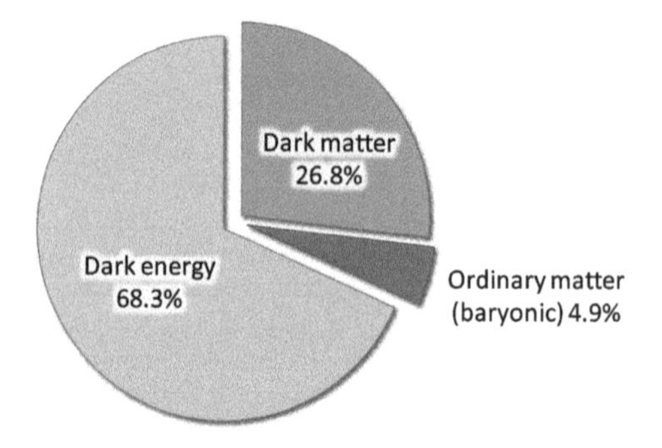

Fig. 3.3. Following initial results of the WMAP satellite, high-precision data from the Planck satellite show that ordinary, visible, baryonic matter makes only a small fraction of energy-matter in the Universe. Most of it is of still unknown nature in the form of dark matter and dark energy.

Another indication in favour of the presence of dark matter comes from observations of collisions among clusters of galaxies. Thanks to a *gravitational lensing effect* on the light from background galaxy clusters, behind the colliding ones, it is possible to measure the overall mass distribution of two colliding clusters and compare it to the distribution of visible matter in the form of gas and stars in these clusters. A famous study of this phenomenon, made in 2006 on the Bullet cluster in the Carina constellation 3.7 billion light-years away from the Earth in the southern sky (figure 3.4), came to the conclusion that dark matter was necessary to explain the magnitude of the phenomenon observed. Dark matter interacts with ordinary visible matter only gravitationally and through weak interactions (which could be of a sort even weaker than those of the Standard Model). The dark matter halos of the two clusters thus pass through each other with little mutual interaction and start separating, whilst the ordinary-matter

Fig. 3.4. Bullet cluster, formed from the collision of two clusters of galaxies, as captured by NASA's Hubble Space Telescope and Chandra X-ray Observatory (visible and X-ray images are superimposed). The right side picture illustrates the time evolution of the collision. Most of the ordinary, visible matter in form of hot interstellar gas, shown in pink, which mutually interacts through electromagnetic forces (which makes it visible), has been distorted into a bullet shape by the collision. The density of dark matter, shown together with the non-interacting galaxies and stars in blue, can be estimated by looking at the deformation of the background galaxies (behind the scene) via the so-called *gravitational lensing* effect, which originates from the curvature of light rays traversing in-homogeneous matter distributions. The clusters of dark matter, which interact mutually and with ordinary matter only gravitationally and, possibly, through weak interactions, appear to have passed straight through each other. *Image: X-ray: NASA/CXC/CfA/M.Markevitch; Optical and lensing map: NASA/STScI, Magellan/U.Arizona/D.Clowe; Lensing map: ESO WFI.*

gaseous content of the two clusters, mutually interacting through electromagnetic forces and thus subject to drag, stay behind intermingled in the collision region. There are several similar cases of galaxy clusters in collisions displaying these same features. This type of observation is difficult to explain within the MOND hypothesis.

The first observation of gravitational waves in September 2015 resulting from the merger of two black holes with masses in the ranges of 20–30 times that of the Sun, and several similar subsequent events indicate that such massive black holes are more common than initially thought and might be constituents of galactic halos. Previous experiments looking for massive compact halo objects, denoted MACHO, were able to exclude these as possible source of the observed dark matter for black holes reaching a few

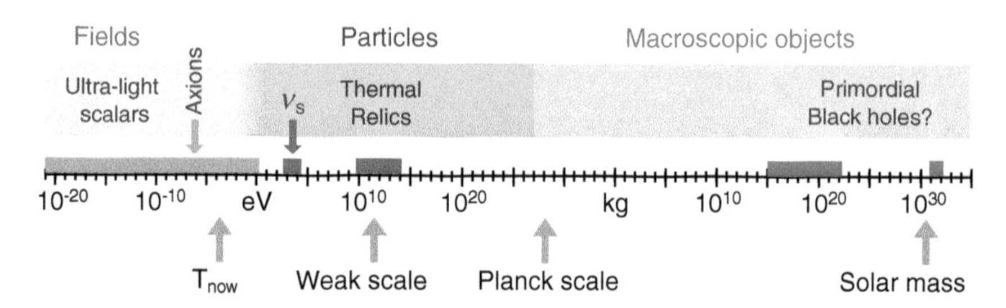

Fig. 3.5. Allowed range for the mass of particles (in units of electron-volt) or macroscopic objects (in kilogram) forming the observed dark matter in the Universe. The favoured windows for each region are indicated by thick coloured lines. Possible candidates are also indicated. Today's temperature of the Universe is $T_{\text{now}} = 2.7\,\text{K} = 2.3 \times 10^{-4}$ eV. *Image: M. Cirelli, lecture notes on dark matter phenomenology, July 2018; illustration modified by the authors.*

solar masses. The viability of heavier black holes as dark matter candidates is still under study.

Despite decades of intense experimental and theoretical work, the mass of dark matter particles or macroscopic objects, M, is still largely unconstrained. The viable range spans more than ninety orders of magnitude, 10^{-21} eV $< M < 10^{35}$ kg, as shown in figure 3.5 (note the change in units from electron-volt for particles to kilogram for macroscopic objects).

This range contains three qualitatively different regions. Cosmological observations indicate that dark matter should be cold, that is, non-relativistic with $v_{\text{DM}} \ll c$. This requirement would normally exclude very light particles after reaching thermal equilibrium, unless the cold dark matter particles never thermalised with the rest of the Universe due to small interactions, which would allow them to be very light. Such very weakly interacting sub-eV particles (WISP) behave more like classical fields for our detectors[4] (green region in figure 3.5). A particularly interesting candidate is the *axion* (cf. section 3.4).

If dark matter particles are in thermal equilibrium, their mass should exceed about 1 keV, which excludes the Standard Model neutrinos, but would allow heavier sterile neutrinos proposed in some extensions of the Standard Model to be viable dark matter candidates. Encompassing the

[4]Very light non-relativistic particles can have a large de Broglie wavelength $\lambda = h/Mv$, above ten centimetres for $M < 1$ meV and about two kiloparsecs (the size of a small galaxy) for $M = 10^{-21}$ eV and $v = 10$ km/s.

weak mass scale, in the range from 1 GeV to 100 TeV, WIMPs are attractive candidates (sections 4.1 and 5.5), which are actively searched for in underground, in space, and with particle accelerators such as the LHC.

Among macroscopic objects, primordial black holes are arguably the most interesting candidates. They are hypothetical objects, which could have been produced as a consequence of very large primordial density fluctuations in a class of multi-stage inflation scenarios (section 5.5), or from the collapse of such peculiar objects as cosmic strings or cosmic walls. There are no predictions for their mass, which are nevertheless severely constrained by theoretical and observational studies. As of 2020, data from supernovae microlensing, stability of stellar clusters, and cosmic microwave background measurements, almost exclude the possibility that MACHOs or primordial black holes could constitute the entire dark matter in the observable Universe, except maybe for a window around 10^{-13} solar masses. If one assumes that primordial black holes constitute only a fraction of dark matter, the window between 10 and 50 solar masses is compatible with a fraction of less than 30% of the dark matter relic density, which is consistent with the hypothesis of merging primordial black holes as the source of observed gravitational waves.

3.2. A repulsive energy: dark energy

In 1998, two experiments investigating the redshift (see section 5.1) of distant supernovae showed that the *Hubble constant*, describing the expansion rate of the Universe, is increasing with time (so it is not actually a constant, see figure 5.7). The Universe therefore expands with increasing velocity, and does so since approximately five billion years! This surprising result on the accelerating expansion of the Universe brought the 2011 Nobel Prize in Physics to Saul Perlmutter, Adam Riess, both from the USA, and Brian Schmidt from Australia. The result was surprising as, due to the effect of gravitational attraction among the galaxies, a gradual slow-down of the expansion was expected with, possibly — depending on the overall matter-energy density, reversal of the motion leading to contraction of the Universe and ultimately to a *Big Crunch*. We shall come back to the observations leading to this stunning result in more detail in chapter 5, and shall only mention here the currently most widely accepted description of the phenomenon.

A repulsive energy, residing in space itself and called *dark energy*, is suspected of being the cause of the accelerating expansion. There are several models attempting to explain it. The presently most common one is based on the cosmological constant (Λ) introduced by Albert Einstein a long time ago, in 1916, well before any experimental confirmation, to complement his theory of General Relativity. The history behind it is quite interesting. The solutions of Einstein's initial equations applied to the Universe as a whole were showing that the Universe is unstable, that it must either expand or contract, in flagrant disagreement with the general thinking of the time, a preconceived idea also shared by Einstein himself, that the Universe is static, unchanging, eternal. Thus Einstein's aim by introducing the cosmological constant in his equations was to counterbalance the term generating the expansion (or contraction) of the Universe.

The cosmological constant represents an energy density filling uniformly the entire space. It could be called the energy of the vacuum, of empty space. Indeed, we know from quantum mechanics that empty space is never really empty: it is a place of ceaseless quantum fluctuations. These fluctuations result from the brief creation of particle–antiparticle pairs, as allowed by Heisenberg's uncertainty relation, followed by their immediate annihilation. They give rise to a non-zero vacuum energy that can be calculated within the Standard Model. The value obtained is about 10^{120} (you read correctly: 120 orders of magnitude) times larger than the cosmological constant, the measured dark energy![5] The inclusion of new physics such as supersymmetry could help lowering the factor of disagreement to possibly 10^{60}, but not cure the problem. This *cosmological constant problem* is among the most serious ones in all modern physics. The Standard Model prediction of the vacuum energy was judged in a textbook by Hobson, Efstathiou and Lasenby (2006) to be the *"probably worst theoretical prediction in the history of physics"*. Clearly, vacuum energy as calculated with the present version of quantum field theory does not gravitate, and something very essential must be escaping our present understanding. A solution may be in sight if, and when, a satisfactory theory of quantum gravity is found (section 3.6).

[5]According to many experts this factor of 10^{120} is questionable, as the energy of the vacuum in the Standard Model is calculable only up to an arbitrary constant. The difference between the vacuum energies before and after spontaneous electroweak symmetry breaking due to the Higgs potential can be unambiguously calculated, but the result for this difference is still 10^{50} times too large.

There are other models trying to explain dark energy. For example, in the *quintessence model*[6] a spin zero (scalar) field permeating space is introduced, but with an energy density that can vary through space and time. This is of course still very speculative, but the discovery of the Higgs boson, also a scalar field and the first fundamental field of this type ever observed,[7] provoked renewed interest in this approach.

3.3. Asymmetry between baryonic matter and antimatter

Despite the fact that we can produce antimatter in our laboratories and that we observe it in cosmic ray showers, anti-galaxies, anti-planets, not even anti-dust, seem to exist, at least not in the parts of the Universe we can observe. Numerous experiments[8] have been undertaken, both in the Earth's atmosphere and in outer space, looking for signs of static antimatter which differs from that produced dynamically in high-energy collisions of cosmic rays. Up to now no such signal has been spotted, for example, in the search for anti-helium nuclei in cosmic rays or soft X-ray emission signals expected to be produced in the contact zones between matter and antimatter regions in the Universe. These negative results led the majority of physicists suspect that antimatter disappeared soon after the Big Bang by annihilation with matter. As matter and antimatter are thought to be produced initially in equal amounts in the Big Bang (figure 3.6), the question is how matter — the one we are

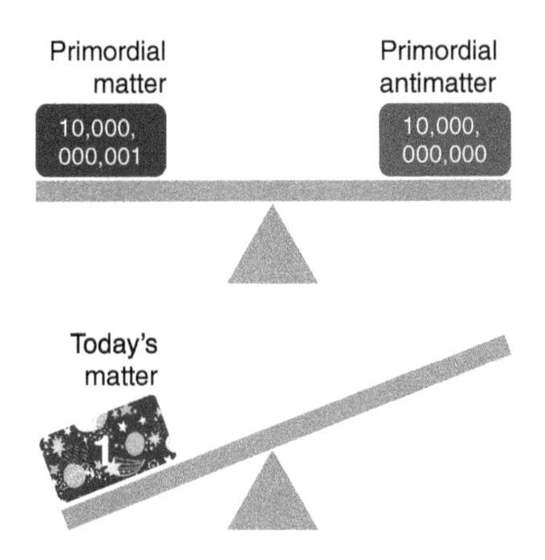

Fig. 3.6. From the (almost) symmetry between matter and antimatter in the primordial Universe to the present observed baryonic asymmetry.

[6]This quintessence field has nothing to do with the fifth and quintessential Platonic polyhedron, the dodecahedron, but rather with the fifth element of the Aristotelian world which the non-terrestrial, supralunar, world is made of.

[7]The question of whether the Higgs boson is really a fundamental scalar or possibly a composite object is actually still an open issue, discussed in chapter 15.

[8]For example the ongoing Alpha Magnetic Spectrometer (AMS) experiment, a particle physics detector installed on the International Space Station.

made of — survived the cataclysm of post-Big-Bang matter–antimatter annihilation?

The present-day measurements, based on primordial nucleosynthesis (BBN, the production of nuclei heavier than hydrogen after the Big Bang, see section 5.3) and the study of the fossil cosmic microwave background radiation (CMB, section 5.3 and figure 3.2), show that the number of surviving baryons is about a billion times smaller than the number of photons produced during the primeval matter–antimatter annihilation phase. The corresponding primordial matter–antimatter asymmetry must thus have been very small. Probably very simple to explain in the Standard Model with the Kobayashi–Maskawa mechanism of CP violation, would one be tempted to assume. Unfortunately, this mechanism invoking only charged weak interactions among quarks predicts a far too small asymmetry. Elaborate calculations have shown that it can generate an asymmetry at the level of only 10^{-19} at the scale of the Universe, which is ten billion times less than required to explain the observed matter–antimatter asymmetry.

New sources of asymmetry are therefore needed, beyond that known in the Standard Model. Even if no theoretical model has the support of the entire particle physics community, the *leptogenesis* scenario, favoured by numerous physicists since neutrinos are known to have mass, would provide an elegant solution to the problem. In this model, CP violation in the lepton sector[9] could give rise to an asymmetry between leptons and antileptons in the early Universe well before electroweak symmetry breaking occurs. A non-perturbative lepton and baryon number violating phenomenon predicted by the Standard Model, which involves the rather exotic concept of *sphalerons*, allows the transformation of the lepton asymmetry into a baryonic one. During the evolution of the Universe in the pre-spontaneous symmetry-breaking phase, sphaleron transitions among baryons and leptons are possible. For example baryons could be converted to antileptons and antibaryons to leptons, allowing the transformation of the initial leptonic asymmetry into a baryonic one, which is *baryogenesis*. Only a tiny CP violation in the leptonic sector is needed to generate the required baryon asymmetry.

[9]Matter–antimatter symmetry violation in the leptonic sector has, as of yet, not been unambiguously observed. Present and future very large-scale experiments in Japan and the USA, however, are expected to lift this secret during the next decade.

3.4. Matter–antimatter symmetry in strong interactions

The breakdown of matter–antimatter symmetry in weak interactions, described in such a simple and natural way in the electroweak theory (through the CKM matrix), has been probed and tested with great precision in a large variety of experiments. An analogous asymmetry in strong interaction processes seems to be infinitesimal or even totally absent.[10] But quantum chromodynamics (QCD) does contain a term that can and should give rise to such an asymmetry. It is parameterised by a variable traditionally called θ (one more θ!), which — as experiment shows — must be very small, or even zero. Rare are indeed the cases where Nature chooses to suppress a phenomenon allowed by equations, unless there is a symmetry that forbids it. The physicist Richard Feynman was known to have said *"Whatever is not explicitly forbidden in physics is allowed, and even mandatory"*. For example the parameter describing symmetry breaking between matter and antimatter in weak interactions of quarks, the phase of the CKM matrix we discussed before, is not constrained by any symmetry and has a "natural" value of 68 degrees (see section 2.5). In contrast, the only known particles with zero mass, the photon and the gluons, are precisely so due to intrinsic symmetry properties of electromagnetic and strong interactions, respectively.[11]

A solution to this problem could thus be brought by a new symmetry that would insure the smallness of θ. In 1977, Roberto Peccei and Helen Quinn proposed such a new symmetry whose spontaneous breaking, similar to the BEH mechanism, as Frank Wilczek and Steven Weinberg independently found, would give rise to yet another spin-zero particle, called the *axion*.[12] The axion would have non-zero mass, although most likely a very small one (of order 10^{-5}–10^{-3} eV). The parameter θ would then not be arbitrary anymore, but become a quantum field associated to the axion, which pervades all of space. The Peccei–Quinn mechanism naturally leads to a vanishingly small ("relaxed") θ. The axion is also a potentially interesting

[10]A non-zero value for the electric dipole moment of the neutron, for example, would be a proof of CP violation in strong interactions. This has, however, not yet been observed. The current experimental upper limit is 3×10^{-26} $e \cdot \text{fm}$, implying $\theta < O(10^{-10})$.

[11]Similarly, chiral symmetry (the symmetry under rotation of the left-handed and the right-handed components of a quark field) naturally leads after spontaneous breaking to a small pion mass.

[12]Wilczek came to the idea for the name from a brand of laundry detergent, associating it to the new particle that "cleaned up" a problem with an "axial" current. Weinberg had proposed the name "Higglet" instead, which he gave up in favour of the axion.

candidate particle to explain cold dark matter and has thus been actively searched for over many years, with increasing recent emphasis. Numerous experiments and searches, be it in the laboratory or through astrophysical observations, have given no positive result until now.[13]

3.5. Neutrino masses

In the Standard Model, neutrinos and antineutrinos are particles without mass.[14] Neutrinos are thus left-handed, antineutrinos right-handed (see inset 2.4 on page 88). It could even be that neutrinos and antineutrinos are identical particles differing only by their helicity state, in which case they are called *Majorana fermions*, by the name of the Italian physicist Ettore Majorana who developed this theoretical approach in the years 1930. Otherwise they are *Dirac fermions* as the charged leptons and quarks.

Neutrinos appear in three flavours, ν_e, ν_μ and ν_τ, in weak interaction processes, the only ones they are taking part in. However, if their mass differs from zero, even by a very small amount, these states of definite leptonic flavour could be quantum mixtures of states of definite mass ν_1, ν_2 and ν_3. As a consequence, a neutrino produced in a particular process with a specific flavour could subsequently appear, at a certain distance from the source, as a neutrino with a different flavour. This is called *neutrino oscillations*, an idea first proposed by the Italian physicist Bruno Pontecorvo in 1957.[15]

An experiment dedicated to the detection of solar neutrinos led by Ray Davis at the end of the 1960s in the Homestake gold mine in South Dakota,[16] USA observed a deficit in the number of measured electron-neutrinos when compared to the expected flux from the Sun[17] on basis of

---

[13]Another, simpler, solution to the strong CP problem would be possible if at least one Standard Model quark were massless. In this case, the parameter θ would disappear from the strong interaction equations. However, there is convincing empirical evidence that all quarks are massive.

[14]This initial choice of massless neutrinos in the Standard Model was chosen for simplicity reasons, as there is no fundamental principle requiring it.

[15]This property is analogous to the mixing of quarks introduced by the Cabbibo–Kobayashi–Maskawa mechanism and responsible for CP violation.

[16]Because of the rarity of signals from neutrinos interacting weakly with active detector material, neutrino experiments are usually buried under several hundred or even thousand meters of rock to reduce background from cosmic rays.

[17]Reduced to essentials, the dominant exothermic fusion process inside the Sun fuelling its energy output is: 4 protons $\rightarrow {}^2_4\mathrm{He} + 2e^+ + 2\nu_e$, with the protons originating from ionised hydrogen atoms.

its energy production. Numerous experiments at various places in the world have been performed since the pioneering Homestake measurement to study in detail neutrino oscillations and to measure the neutrino mixing parameters. Some experiments are based on the detection of neutrinos produced in the Earth's atmosphere by impinging cosmic rays and where ν_μ and $\bar{\nu}_\mu$ are the more abundantly produced ones (ν_μ:$\nu_e \approx 2$:1), other experiments are located near nuclear reactors producing essentially only anti-ν_e. Still others again measure the flux of ν_μ and $\bar{\nu}_\mu$ produced by an accelerator located hundreds of kilometres away from the detectors. In this context, mention must be made of the Super-Kamiokande experiment in Japan, a 50 000 tons reservoir of ultra-pure water surrounded and constantly scrutinised by thousands of photo-multipliers. In 1998 this experiment, led by Masatoshi Koshiba, has shown that muonic neutrinos produced in the atmosphere by cosmic rays disappear in favour of electronic neutrinos exactly as expected if the neutrinos were oscillating. This result contributed to the 2002 Nobel Prize for physics to Koshiba, sharing it with Davis. Observation of neutrino oscillations implies that at least two out of the three neutrino species must have non-zero mass. The Super-Kamiokande detector is being used since 2013 by the T2K experiment for the detailed study of muonic-neutrino oscillations produced by a proton accelerator located at a distance of three hundred kilometres from the detector. The main task is to try to find evidence for a possible CP violation in the lepton sector; results obtained by 2021 are indicative, although not yet entirely conclusive. The study of neutrinos has allowed to determine the differences of masses squared among the mass states ν_j and ν_i:

$$\Delta m_{21}^2 \simeq 7.6 \cdot 10^{-5}\,\mathrm{eV}^2,$$

$$\left|\Delta m_{32}^2\right| \simeq 2.4 \cdot 10^{-3}\,\mathrm{eV}^2.$$

Attempts to determine directly the mass of electron-neutrinos, through the study of the endpoint in the energy spectrum of electrons emitted in the radioactive β-decay of tritium ($^3\mathrm{H} \to {}^3\mathrm{He} + e^- + \bar{\nu}_e$), allowed establishing an upper limit on the ν_e neutrino mass of about one electron-volt. More indirectly, constraints from cosmology provided by the measurements of WMAP and Planck satellite experiments indicate that the upper limit on the mass of the heaviest among the neutrinos should be less than 0.23 eV. Combining cosmological and oscillation experiments results, the most recent results as of summer 2019 yield an upper bound on the lightest neutrino mass species of 0.09 eV.

Neutrinos having a mass, albeit a very small one, it is necessary to extend the Standard Model to accommodate the neutrino of right helicity and the antineutrino of left helicity (remember that in the Standard Model, only left-handed neutrinos and right-handed antineutrinos take part in weak interactions). Thus some symmetry in the treatment of neutrinos and quarks has been restored, with, however, the notable difference that neutrinos could still be Majorana fermions. The mass of neutrinos is exceedingly small compared to quark masses: there is a difference of at least six orders of magnitude. It is very difficult to imagine that it could be due to the same mass-generating Brout–Englert–Higgs mechanism operative for quarks and charged leptons. The observation of non-zero neutrino masses can thus be taken as a manifestation of physics phenomena beyond the Standard Model. Some models, based either on the symmetrisation between left and right handed weak interactions, or, more fundamentally, on the unification of all forces at an energy scale of 10^{15}–10^{16} GeV, provide a possible, even plausible, interpretation.

3.6. Gravity

All the points we discussed up to now could be thought of as mere omissions in the Standard Model. They could be corrected once the particles making up dark matter would be found, or until new sources of matter–antimatter asymmetry were discovered. Let us now consider the third category of problems, and these could be of a more fundamental nature.

Up to now we did not talk about the fourth fundamental interaction, gravity. The gravitational force plays a negligible role in particle physics at energy scales now accessible in our accelerators: the gravitational attraction between two protons at a distance of one Fermi from each other is 10^{36} times weaker than the electromagnetic force between them (the Coulomb force, a repulsive one in this case). The gravitational force, however, becomes significant when you consider very compact objects: it is comparable to the other interactions for point-like objects with masses of the order of the so-called *Planck mass*: $M_P \sim 10^{19}$ GeV (see inset 3.2).

At the beginning of twentieth century Einstein brought a definitive and relativistic reformulation of the gravitational force, in a form of a theory essentially geometrical in nature, connecting the geometry of space–time with the distribution of energy-matter. You may say that in Einstein's General Relativity the distribution of energy-matter shapes space–time, which

Inset 3.2: Planck mass and Planck scale

For any object of mass m one can define its Schwarzschild radius:

$$R_S = 2G_N \frac{m}{c^2} ,$$

where G_N is Newton's constant ($G_N = 6.7 \times 10^{-11}$ J cm kg^{-2}) and c is the speed of light. This is the radius (or *event horizon*) that would make the object become a black hole, if all its mass were concentrated within it. For example, the Schwarzschild radius of the Sun is three kilometres, whilst for the Earth it is about one centimetre. The Planck mass is defined as

$$M_P = \sqrt{\frac{\hbar c}{G_N}} ,$$

where $\hbar$ is the "reduced" Planck constant, $\hbar = h/(2\pi) = 6.6 \times 10^{-22}$ MeV s. The Planck mass is about 10^{19} GeV and corresponds approximately to the mass for which Schwarzschild radius and Compton wavelength, $\lambda_C = h/(M_P c)$, are of the same order. This latter quantity can be considered as the fundamental lower limit with which one can localise a particle taking into account both quantum theory and special relativity.

The Planck mass thus sets the energy scale at which one expects quantum gravity effects to be important. It is also the mass scale at which the gravitational force is of the same order of magnitude as the electrostatic force (within a factor of the fine structure constant α)

$$\frac{1}{4\pi\varepsilon_0} \frac{e^2}{r^2} = \alpha \, G_N \frac{M_P^2}{r^2} .$$

Here r is the distance over which the force acts, ε_0 is the *permittivity* of vacuum (i.e. the resistance encountered when forming an electric field) and e is the elementary electron charge. The Planck mass represents thus the natural scale at which the still hypothetical quantum theory of gravity might unify with the three interactions of the Standard Model.

in turn determines the motion of matter through it. In this theory space–time is not any more a passive stage on which our particles and fields play their roles, but is now an active dynamical participant. This theory was explaining quantitatively the motion of the perihelion of Mercury (an excess of 43″ per century), observed before, but left unexplained by Newtonian mechanics, despite all attempts and efforts deployed. It was also predicting a bending of the trajectory of light, if passing nearby a large mass, as Sun's one. The observation of this deviation at the expected level by Sir Arthur Eddington in 1919 played a major role in the acceptance of General Relativity. A number of additional observations were made since then, all agreeing with the predictions of the theory. For example, the energy loss in the binary pulsar PSR B1913+16 is perfectly explained as a loss due to *gravitational radiation*. The observation of this double pulsar and its gravitational wave radiation interpretation goes back to the 1990s, before gravitational waves were ever detected. These have been finally observed in September 2015 when the gravitational-radiation detector LIGO started operating in the USA, a detector with sufficient sensitivity to detect the minute ripples produced in the fabric of space–time by the collapse and merger of two stellar black holes of thirty solar masses each at an estimated distance of 1.3 billion light-years from us. Virgo, a detector of comparable sensitivity and located near Pisa in Italy, began operation in 2017 and contributed jointly with LIGO to subsequent observations. By the end of 2020, more than 50 cases of black holes or neutron stars mergers have been observed through the detection of produced gravitational waves. Another example is gravitational lensing, it is due to bending of light produced by a distant galaxy by intervening matter on the line-of-sight, a galaxy or a cluster of galaxies. Over the past decade it has grown into a full-fledged observational technique in cosmology (see, for example, figure 3.4).

Whilst the other fundamental interactions of the Standard Model, the electromagnetic, weak and strong ones, are all successfully described by a relativistic (special relativity) quantum field theory, up to this day it has not yet been possible to formulate a theory encompassing both quantum theory and General Relativity, which is a classical theory, in a mathematically coherent way. The hypothetical quantum of the gravitational field is called the *graviton* and should have spin 2. There exist, however, some specific models taking into account quantum effects in the presence of strong gravitational fields, helpful and appropriate for the understanding of specific situations in the Universe. In particular, this is the case with the works of Stephen Hawking in connection with neutron stars and stellar or cosmological black holes.

3.7. The Higgs boson is light

The new particle with a mass of 125 GeV observed at CERN in 2012 has all the properties to be the long-sought Higgs boson (see chapter 11).

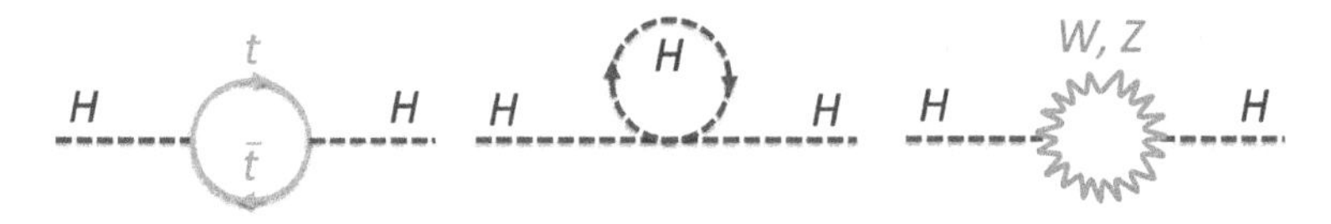

Fig. 3.7. One-loop Feynman diagrams contributing to radiative corrections to the Higgs boson mass.

From a theoretical point of view the bare mass of a particle is defined as the mass it would have had it no interaction with the rest of the world. The mass we measure in an experiment corresponds to the bare mass modified by quantum corrections, just as we saw for the W boson in section 2.4. These corrections are, however, much more pernicious in the case of a scalar particle as the Higgs boson.

In figure 3.7, we show Feynman graphs of some of the leading quantum corrections contributing to the Higgs boson mass. For example, a Higgs boson transforms itself through the creation of a fermion–antifermion pair, followed immediately by the annihilation of the pair within the time allowed by Heisenberg's uncertainty relations. The fermions taking part are preferentially the heavy ones, a top–antitop pair, as the Higgs coupling to fermions is larger the heavier is the particle. Other virtual processes contribute too, such as those involving electroweak gauge bosons or the Higgs boson itself (via the Higgs-boson self interaction).

The problem is that these virtual particles in the loops can have very large energies, up to infinity, provided uncertainty relations are respected. From the mathematical point of view, the quantum corrections to the square of the Higgs mass m_H^2 due to such diagrams are proportional to the square of the maximal energy allowed Λ for the virtual particles in the loops. The corrections thus become infinite when Λ goes to infinity. This is particularly annoying as the physical, that is experimental mass of the Higgs is obviously finite.

Problems of divergence in quantum field theories can be circumvented through a procedure called *renormalisation*. It consists of cancelling infinite quantities through a redefinition of parameters of the theory (such as masses or couplings) valid at a certain energy scale in terms of physics quantities directly measured by experiment. When such a procedure is possible, the theory is said to be renormalisable, in which case it is possible to make definite and finite predictions that can be observed experimentally. This is the case with the Standard Model and was the cause of its success and acceptance already in the early 1970s (see section 1.6). If there were no new physics beyond the Standard Model, no particle heavier than the top quark, the renormalisation procedure allows preserving a finite mass for the Higgs boson by redefining the bare mass such as to compensate for the diverging corrections (see inset 3.3). Indeed, the Standard Model is a self-consistent quantum field theory up to very high energies, exceeding the Planck mass by many orders of magnitude, where it eventually breaks down due to the presence of a pole increasing beyond limits the weak hypercharge.

However, for the various reasons already discussed in this chapter, it is likely and even essential that new interactions and new elementary particles coupling to the Higgs boson appear at an energy scale Λ well beyond the TeV. Then the quantum corrections to the Higgs boson mass squared would become proportional to the square of this new physics scale

$$\delta m_H^2 \sim \frac{\lambda^2}{8\pi^2}\Lambda^2.$$

The constant λ is related to the strength of the couplings of the Higgs boson to the particles present in the loops of the diagrams in figure 3.7. Its value is dominated by the coupling to the top quark for which $|\lambda| \sim 1$.

But what is that energy scale Λ? It could be the scale of the grand unification of fundamental interactions (see section 4.2), which could be $\Lambda \sim 10^{15}$ GeV, or even the Planck mass $\Lambda \sim 10^{19}$ GeV, where gravity becomes important at the quantum level. In any case, these virtual corrections δm_H^2 are huge compared to the square of the physical Higgs mass

$$m_H^2 = m_0^2 + \delta m_H^2.$$

The small physical Higgs boson mass requires the square of the bare mass, m_0^2, to be precisely adjusted and of opposite sign to cancel δm_H^2 in the sum, a fine-tuning of two a priori independent numbers over at least 26 orders of magnitude! This looks quite artificial.

The huge difference between the electroweak scale determined by the W, Z, and Higgs boson masses, and the very large scale at which new physics is expected to occur is called the *hierarchy problem* of particle physics, leading to a problem of *naturalness* of the Standard Model theory. It reflects the dependence of the physics of one scale on the physics of a very different scale, which should factorise in naturally behaving systems. It would thus have been more natural for the Higgs boson to have a mass of the order of the grand unification or even Planck scale, rather than the much smaller one we observe. Vice versa, as the Planck scale is determined by the gravitational constant (see inset 3.2), it is the extreme weakness of the gravitational force in comparison with the other fundamental interactions in the Standard Model that is the root of the hierarchy problem.

Inset 3.3: Is the Standard Model an effective theory?

The renormalisation procedure allows physicists to make quantitative predictions despite the occurrence of diverging amplitudes related to Feynman diagrams with loops. These predictions can reach spectacular precision as is the case with some atomic physics processes described by quantum electrodynamics (the quantum version of electromagnetism). However, for some mathematical physicists, this procedure really consists in *ultima linea* in hiding infinities under the rug, and it is thus not fully satisfactory. More specifically, renormalisation amounts to considering the bare masses and couplings in the Lagrangian as infinite quantities. Thus, only the perturbative approach in terms of renormalised Feynman diagrams would have a fundamental meaning, not the Lagrangian itself. It would then limit this approach to weak couplings only when the perturbative series can converge.

The appearance of divergences in the theory and the need to resort to a procedure such as renormalisation to provide meaningful results may be a sign that *local* quantum field theories with *point-like particles* are not the ultimate theory. The Standard Model might just be an *effective* theory,[a] an approximation valid at low energies of a more fundamental theory, finite, without problems of divergences, that is still to be discovered. To avoid divergences at very high energies, and thus very small distances, it is likely that this theory will have to introduce *non-point-like fundamental objects* mutually interacting in a *non-local way*. The theory of superstrings is a possible candidate for such a theory, among others, as is *loop quantum gravity*, or non-commutative geometries.

[a]Effective theory has to be understood here with a different meaning than the *effective field theories* discussed in inset 11.3. The latter is a parametrisation of some unknown new physics and is valid only at energies much below the scale of this new physics, whereas the former is renormalisable and in principle mathematically valid up to the highest energies.

Some extensions of the Standard Model (supersymmetry, additional spatial dimensions) discussed in the next chapter propose a solution to this problem. Another family of models, called *composite models*, avoid introducing scalar particles as fundamental particles of the theory, and interpret the Higgs boson as a bound state of a new class of fundamental fermions whose presence could manifest itself at energies in the TeV range. Instead of depending quadratically on Λ, the virtual quantum corrections would, then, depend only logarithmically on it (as is the case in supersymmetry), which greatly moderates the problem. As explained in inset 3.3, the physics at an energy scale beyond Λ will have to be described by a theory unifying quantum mechanics and gravity, it would not be plagued by mathematical divergences, but such a theory is still to be found.

3.8. The origin of the Higgs potential

Among the problems of the third category we must also mention the somewhat mysterious origin of the Higgs potential. Since the very first publications in the 1960s, the *ad hoc* character of the BEH mechanism was questioned. Neither the potential in the shape of a Mexican sombrero (figure 1.8), nor the interactions of fermions with the Higgs field, stem from a geometrical principle, as is the case with all other interactions in the Standard Model, that result from the local gauge invariance principle. The recent discovery and ongoing measurements at CERN seem, nonetheless, to confirm the validity of this mechanism considered as simplistic by many. An important subject for future theoretical and experimental studies will be to explain how this potential and the particle interactions with this field emerged during the evolution of the Universe.

Coupling constants and particle masses, far from being constant, depend on energy through radiative quantum corrections (section 1.9). Their values thus change as a function of time elapsed since the Big Bang, the expansion and cooling down of the Universe, and the decrease in the kinetic energy contributing to the interactions among particles (see chapter 5). The same happens to the parameters μ^2 and λ of the Higgs potential (figure 1.8). Some extensions of the Standard Model, such as supersymmetry (section 4.1), explain in this manner the evolution of the shape of the Higgs potential from the symmetric phase with a minimum at zero energy to the *Mexican hat* shape in the broken phase.[18]

[18] See also inset 12.2 on page 395 on the stability of the electroweak vacuum.

Chapter 4

How Could New Physics Look Like?

To overcome the hierarchy problem in a "natural" way some new physical phenomena at an energy scale much below that of (the putative) grand unification or the Planck mass should occur. If this were the case at the TeV level, the quantum corrections to the bare Higgs mass, dominated by the top quark, would be only around 20%, becoming 50% at 4 TeV. The task of the LHC is precisely to explore this energy range.

What form could a solution to the hierarchy problem take? An elegant one is certainly supersymmetry. While supersymmetry allows to moderate the quantum corrections contributing to the Higgs boson mass, it does not solve the problem of the large disparity between the intensities of gravity and the other interactions. Another type of model of new physics addresses precisely this problem and proposes a solution for the apparent weakness of gravity: it introduces one or more additional space dimensions, small and compact enough to have escaped detection up to now. Finally, an important class of models directly addresses the Higgs mass problem by assuming that this boson is not a fundamental particle.

4.1. Supersymmetry

Supersymmetry theory, often called by its diminutive name SUSY, was born in the 1970s initially in connection with the development and application of string theory to strong interactions. The pioneering ideas came from Pierre Ramond, André Neveu, John Schwartz, Joel Scherk, and others, with transposition to field theory with point-like particles as an extension of our present-day Standard Model by Julius Wess and Bruno Zumino.

Supersymmetry connects bosons to fermions through a new symmetry, unifying in a way particles of matter with particles propagating force. It predicts the existence of a partner for every known elementary particle, having identical properties but differing by 1/2 unit of spin. Thus a supersymmetric operation would transform an integer spin boson into a fermion of exactly same mass and internal symmetry properties (charge, isospin, flavour, colour, etc.) and, similarly, a fermion of half integer spin would acquire a partner boson of integer spin. To characterise these new particles a new quantum number is introduced, R-parity, sometimes called *matter parity*, and denoted as R_P. Its value is $R_P = 1$ for Standard Model particles, including the Higgs boson, and $R_P = -1$ for the supersymmetric partners. The world would then be invariant under a global supersymmetry transformation.

At the beginning of the 1980s, in good part under the influence of, among others, Pierre Fayet in Paris, Ricardo Barbieri from Pisa and John Ellis from CERN, the application of supersymmetry to the particle world led to development of the MSSM, *Minimal Supersymmetric Standard Model* offering a rich phenomenology (figure 4.1).[1] It is a minimal supersymmetric extension in the sense that its doublets {boson, fermion} all contain one Standard Model particle. The supersymmetric partner particles are often called *sparticles* (see figure 4.2).

Fig. 4.1. "One day, all of these will be supersymmetric phenomenology papers." *Credit: Cluff.*

Supersymmetry requires a more complex Higgs sector than in the Standard Model. The MSSM introduces two doublets of Higgs fields (compared to a single doublet in the Standard Model) leading to five Higgs bosons of spin zero and $R_P = 1$ as well as their fermionic partners, the higgsinos, with $R_P = -1$. At breakdown of electroweak symmetry by a *super-Higgs* mechanism, particles carrying the same quantum numbers *mix* in a quantum

[1]This tremendous wealth of publications was spectacular during the years 2000–2015, but has decreased significantly since then for reasons which will be developed in section 13.3.

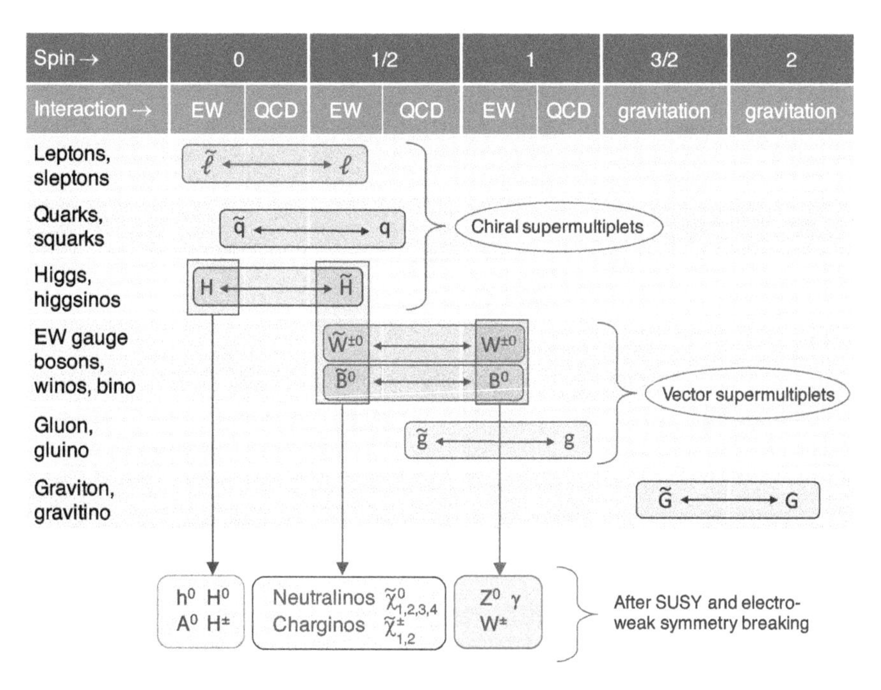

Fig. 4.2. Table of particles in the minimal supersymmetric extension of the Standard Model (MSSM). Particles are organised according to their spin and type of interactions they take part in: electroweak (EW), strong (QCD) or gravitational (mentioned here explicitly only for particles of spin 3/2 and spin 2, although all of them take part as soon as they have mass). A horizontal rectangular box connects particles related by a supersymmetry transformation (boson ↔ fermion). Supersymmetric particles resulting from the application of a supersymmetry operator on a Standard Model particle are denoted by a tilda. These *supermultiplets* are called chiral or vector, depending on whether the highest spin particle is a fermion of spin 1/2 or a boson of spin 1. The vertical rectangular boxes connect particles that mix quantum mechanically after breakdown of both supersymmetry and electroweak symmetry.

mechanical sense, a behaviour we already met several times in previous chapters. Thus the fermionic supersymmetric partners of the Z (the *zino*), of the photon (*photino*), and the two neutral *higgsinos*, mix giving rise to four *neutralinos*, labelled $\tilde{\chi}^0_{1,2,3,4}$ (figure 4.2). In the same way, the fermionic partners of the charged W and charged Higgs, the *winos* and charged *higgsinos*, mix to two *charginos* ($\tilde{\chi}^{\pm}_{1,2}$).

As two particles of same mass but spins differing by 1/2 unit have never been observed, supersymmetry — if it exists — must be broken at a scale larger than that of electroweak symmetry breaking. The mechanism responsible for breakdown of supersymmetry must give a much larger mass

Fig. 4.3. Loop diagrams contributing to radiative corrections to the Higgs boson mass with, on the left, a top quark, and its superpartner, the *stop* (for *scalar top*), on the right.

to the supersymmetric partner in the doublet {boson, fermion} thereby explaining why it has not yet been observed in experiments until now, although it could still be observed at the LHC. An unpleasant feature of the MSSM is not only that the number of fundamental particles more than doubles, but also that the number of free parameters of the theory increases a lot, by more than one hundred, when compared to the Standard Model.

Despite these unpleasant aspects, supersymmetry has such appealing features that over the past forty years it has been thoroughly studied resulting in a plethora of experimental and phenomenological articles (figure 4.1). First among its appeals is that it solves the hierarchy problem. Particles differing by half a unit of spin contribute with opposite signs in the sum of amplitudes making up the radiative corrections to the Higgs boson mass. Therefore quantum corrections due to creation and subsequent annihilation of Standard Model particles are compensated, or almost so, by those of the superpartners that have the same couplings, provided their masses are not exceedingly different, i.e. that the differences stay below about a TeV (figure 4.3).

Another virtue has to do with grand unified theories which we discuss in section 4.2. Introducing supersymmetry in the theory allows the three coupling constants to converge to a same and common value at an energy scale of about 10^{16} GeV (figure 4.4). The expected lifetime of the proton (figure 4.5) is then sufficiently lengthened not to be any more in contradiction with the non-observation of proton decay in experiments up to now.

Furthermore, supersymmetry could also bring a solution to the dark matter problem. Most supersymmetric models and scenarios assume R-parity conservation. This has as the consequence that it is impossible to produce single supersymmetric particles in the collision of Standard Model

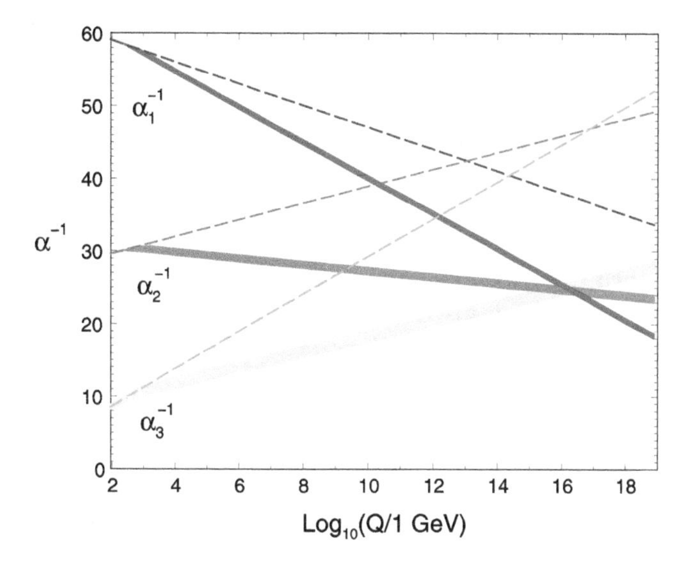

Fig. 4.4. Evolution of the Standard Model coupling constants with energy scale Q. The ordinate displays the inverse of the coupling constants: α_1 and α_2 correspond to the electroweak couplings related to the groups U(1) and SU(2), at an energy scale exceeding the electroweak breakdown (at about 100 GeV) and α_3 corresponds to the strong interaction described by the group SU(3) of QCD. The dashed lines are the predicted evolution within the Standard Model, whilst the full lines depict possible scenarios within its minimal supersymmetric extension. The thickness of the lines indicates the theoretical uncertainty due to the lack of knowledge of the sparticle spectrum in the model. Introducing supersymmetry in the theory (with sparticles masses in the 1–10 TeV range) greatly improves the convergence of the couplings to a single value, and increases the unification energy scale so that the proton lifetime extends beyond the current experimental limits.

ones.[2] Sparticles must be produced in pairs and they can only decay to another sparticle (plus additional Standard Model particles). The lightest among the sparticles, denoted LSP (for *lightest supersymmetric particle*), is then necessarily stable.[3]

The LSPs produced in the first moments after the Big Bang, in the period between 10^{-15} and 10^{-12} seconds, must still be present in large

[2]In the initial state of the collision $R_P = 1$, whilst in the final state it would be $R_P = -1$, which would violate the hypothesis of R-parity conservation. It should be mentioned that there are also theoretical supersymmetry scenarios without R_P conservation as there is no fundamental symmetry requiring R_P to be conserved. These are somewhat less popular, but this might just be a prejudice.

[3]In many supersymmetric scenarios the LSP is the lightest among the four neutralinos, the $\tilde{\chi}_1^0$, but could also be a sneutrino or the gravitino, for example.

numbers as fossil particles from that period. To fulfil the cosmological requirements for dark matter the LSP must be subject to only weak and gravitational forces, which would also explain why they have not been detected yet. In general terms such particles are called WIMPs (for *weakly interacting massive particles*) and they are candidates to contribute to the cold dark matter in the Universe (see section 3.1). It is a remarkable coincidence that the expected mass range of these WIMPs of several hundred GeV, and their expected fossil abundance driven by their production and self-annihilation rates, leads to the correct order of magnitude to match present-day dark matter observations. This is what is called the *WIMP miracle.*

A great unknown is, of course, the supersymmetry breaking mechanism, which grants sparticles their masses and defines their experimental characteristics. Among the many existing theoretical schemes, let us mention *mSUGRA* (for *minimal supergravity*), which was very popular for a long time. This model is based on a minimal version of supergravity in which spontaneous symmetry breaking is due to the presence of a set of new particles interacting only gravitationally with the MSSM particles. This set is the so-called *hidden sector.* The mSUGRA scheme requires only five new free parameters, which greatly simplifies experimental investigations of features of the theory (this is the reason why it was so popular, especially among experimentalists designing ATLAS and CMS!) and the interpretation of data if supersymmetry were to be discovered. However, this model is at present in difficulty in view of the experimental results of WMAP and Planck, as well as due to unsuccessful direct searches for supersymmetry up to now at LEP, the Tevatron, and at the LHC (section 13.3). More recent searches for supersymmetry employ agnostic, bottom-up *simplified models* that serve as benchmarks to interpret the reach of a given search, without the pretension of describing a full supersymmetric phenomenology.

Despite the exclusively negative results so far which constrain more and more the MSSM parameter values, there is enough flexibility in the theory to adapt and survive these constraints. Moreover, supersymmetry could have a more complicated form than the MSSM. It could also manifest itself at much higher energy, possibly beyond the reach of the LHC, than that expected if it were to solve the hierarchy and dark matter problems.

4.2. Unification of strong and electroweak forces

On several occasions we already mentioned grand unified theories (GUT). What is this all about? One of the yardsticks in the development of modern

Inset 4.1: Supersymmetry: an extension of space–time symmetries

Every Standard Model particle with a definite mass and spin is described by a quantum field (section 1.1). These characteristics stem from invariance properties of Nature under space–time transformations of the so-called Poincaré group, containing translations, rotations and the Lorentz transformations of special relativity (see inset 1.5 on page 24).

The spin-statistics theorem in quantum field theory (see inset 1.1 on page 8) is related to the fields' commutation properties: boson fields commute ($\phi_1\phi_2 = \phi_2\phi_1$), while fermion fields anticommute ($\psi_1\psi_2 = -\psi_2\psi_1$). In the 1970s, a natural extension of Poincaré invariance was proposed: to the usual four space–time dimensions where the positions are described by four commuting real numbers, four new dimensions are added where positions are now given by mathematical entities that anticommute (in a way analogous to spinors describing spin-1/2 fermions). The components of the field along these new dimensions can be viewed as a quantification of the degree of *bosonic* or *fermionic state* of the particle. This way supersymmetry, transforming a boson into a fermion and vice versa, corresponds to a translation in this *superspace*.

Requiring invariance of physical laws under the group of translations–rotations in this superspace results in every fundamental particle having a bosonic and a fermionic state, with same mass and same internal symmetry properties such as gauge or flavour symmetries. It is both remarkable and bewildering that when appropriately combining two supersymmetry transformations one gets as a result a translation in space–time. This is why it is sometimes said that a supersymmetry transformation is equivalent to the square-root of an ordinary space–time translation. If, on top, one requires invariance of physical laws under local (gauge) supersymmetric transformations, one finds automatically a theory that is invariant under local space–time transformations, characteristic of general relativity describing the gravitational force (section 3.6). Such a type of theory is called *supergravity*.

physics is the wish for simplicity of the fundamental theoretical principles. In this sense, being able to explain Nature's phenomena through and by a single overarching interaction would be a tremendous step towards ultimate understanding and the reductionist ideal pervading fundamental physics.

Gravity was the first fundamental interaction identified by Galileo Galilei towards the end of the sixteenth century.[4] It also provides the first obvious unification, as Newton's law of universal gravity (around 1690) allowed to explain both the fall of apples or the ballistics of objects on Earth, as the motion of celestial bodies. The second major unification was the one of electricity and magnetism achieved by James Clerk Maxwell in the 1860s and embodied in his famous equations. These equations also brought optics within the realm of electromagnetic theory. With the development of quantum mechanics at the beginning of the twentieth century and thereby the interpretation of atomic and molecular structures in conjunction with the use of electromagnetic theory, the only phenomena not encompassed within this theoretical framework were those related to natural radioactivity discovered by Henri Becquerel, and Pierre and Marie Curie around 1900. The first direct experimental manifestation of strong interactions was α-radioactivity, and β-radioactivity for weak interactions. As described in section 1.6, the Glashow–Weinberg–Salam model subsequently successfully unified weak and electromagnetic interactions. Thus today, at the beginning of 21st century, physics rests on three fundamental forces: gravity, electroweak interaction and strong interaction.[5]

The next episode could be unification of electroweak and strong interactions (figures 4.4 and 4.6). The coupling constants of these interactions seem to converge towards a common value at a colossal energy of 10^{15}–10^{16} GeV, well beyond the reach of present-day particle accelerators,

[4]Galilei dropped two objects of different mass from the leaning tower of Pisa to show that the speed of their descent was independent of mass. Actually, most historians believe that it was rather a *thought experiment*. The real measurements were done on inclined planes.

[5]Let us emphasise that the unification of terrestrial (Galileo) and celestial (Newton) gravity, as well as of electricity and magnetism, is not of the same nature as electroweak unification or grand unification. In the first two cases it was the realisation that two apparently different physical phenomena were in fact manifestations of one same interaction. In the latter cases two distinct interactions emerge from a common one as a consequence of a symmetry-breaking phenomenon due to the cooling of the Universe since the Big Bang.

or even conceivable ones on Earth (section 1.10).[6] It suggests that the two forces (electroweak and strong) we experience as distinct ones could be manifestations of a single unified interaction, expression of a still higher symmetry, which is broken at the scale of the present-day cold Universe and the experiments physicists performed so far.

The central idea of grand unified theories is to choose a gauge symmetry group, such as SU(3) for QCD, in a way to have a single coupling constant.[7] This group must be larger than the Standard Model one, SU(3) $\times$ SU(2) $\times$ U(1), so as to encompass all its symmetries. The consequence will be that *new interactions* and *new gauge bosons* should appear. To understand this aspect, it is worth recalling that it is the unification of electromagnetic and weak interactions (at the time when only weak charged currents were known) that led to the prediction of the existence of neutral weak currents (i.e. a new form of weak interaction) and the Z boson.

The simplest group satisfying these criteria is SU(5), which was proposed by Howard Georgi and Sheldon Glashow in 1974 (see inset 4.2). The GUT based on this group predicts that the proton can decay into a π^0 and a positron (figure 4.5). In the Standard Model such a process is forbidden as it violates conservation[8] of both lepton number L (0 in the initial state, -1 in the final) and baryon number B (1 in the initial state and 0 in the final state). Baryon and lepton numbers are additive quantum numbers; their quasi-conservation is an observational fact, it is not required by any known fundamental principle such as gauge symmetry.

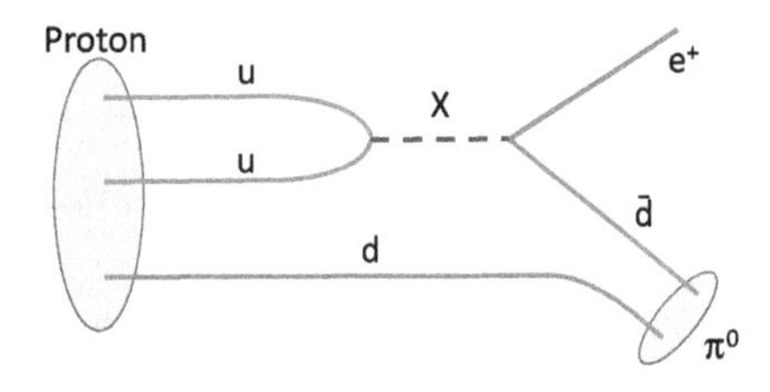

Fig. 4.5. Proton decay mediated by the hypothetical particle X appearing in a model of grand unification of all forces of the Standard Model and possessing quantum numbers of both quark and lepton (a *leptoquark*).

[6]An accelerator circling the Earth and equipped with 20 tesla magnets would allow to accelerate protons to only 2.5×10^7 GeV, a factor 3,500 times the maximum beam energy of the LHC.

[7]This is not the case with the electroweak theory based on SU(2) $\times$ U(1) and thus requiring two distinct coupling constants, g and g' (see inset 1.10 on the Glashow–Weinberg–Salam model, page 45).

[8]Transitions changing (i.e. not conserving) the baryon number (but preserving $B - L$) are in fact possible in the Standard Model by a so-called sphaleron process that is expected to occur at very high-energy density of above 10 TeV (section 3.3). Despite the LHC's proton collision energy of 14 TeV, it does not allow to initiate sphaleron transitions as the energy is shared amongst the proton's constituents.

Inset 4.2: The SU(5) grand unified theory

In chapter 1, we introduced the grouping of quarks and leptons in doublets. These doublets are constructs, vector spaces of dimension 2 onto which the SU(2) group of weak interaction acts. For example, the doublets (ν_e, e) and (u, d) correspond to the first family of quarks and leptons. Similar doublets exist for antiparticles. The exchange, emission or absorption, of a weak gauge boson W allows transition from one member of the doublet to the other member of the same doublet. Thus the transition $d \to u + W^-$ (where the W boson is virtual), followed immediately by the decay $W^- \to e^- + \bar{\nu}_e$, corresponds to the neutron to proton β-decay $n \to p + e^- + \bar{\nu}_e$, which, as seen at the level of quarks, reduces to the transition $d \to u + e^- + \bar{\nu}_e$. In the case of quantum chromodynamics, each quark type is put in a multiplet of dimension 3, corresponding to the three colour states, for example (u_r, u_g, u_b), and it is on these triplets that the group SU(3) of strong interaction acts.

Accordingly, in the SU(5) grand unification scheme, quarks and leptons of one same family are grouped into two multiplets, one of dimension 5 (thus with five members), the other of dimension 10. For example, for the first family, the multiplet **5** is made of the three antiquarks $\bar{d}$ in the three colour states, the electron and the electron neutrino. The ten-dimensional multiplet contains the positron, the u and d quarks, and the $\bar{u}$ antiquark, each in three colour states. This theoretical scheme with quarks and leptons in the same multiplets also contains 24 gauge bosons: the eight gluons of SU(3), the W^+, W^-, Z and γ of SU(2) × U(1) from the Standard Model, but also 12 additional gauge bosons! These new gauge bosons have fractional charges, of $-4/3$ and $-1/3$ and allow direct transitions between quarks and leptons within the same multiplet. Therefore the proton decay, $p \to e^+ + \pi^0$ (figure 4.5), which at quark level occurs as the (virtual) transition $d \to e^+ + X^{-4/3}$ followed by the annihilation $X^{-4/3} + u \to \bar{u}$, with $X^{-4/3}$ being one of the new gauge bosons of charge $-4/3$, becomes possible. Through a process similar to the Brout–Englert–Higgs mechanism with very massive scalar fields, spontaneous SU(5) symmetry breaking will generate very large masses for these bosons, of the order of 10^{15} GeV. Due to these large masses, the transition rates through virtual $X^{-4/3}$ bosons are heavily suppressed, which explains the large proton lifetime expected and observed!

The framework of such a grand unified theory allows us to understand why quarks have fractional electric charges: fundamental group theoretical arguments require the sum of electric charges in each SU(5) multiplet to be zero. For example in the **5**-plet the $\bar{d}$ antiquarks appear in three colour states, their charge must thus necessarily be fractional, equal to 1/3. Furthermore, it can be shown that the internal coherence of the Standard Model requires the sum of electric charges of quarks (with three colours) and leptons to also be zero, and this in turn implies that the charge of u quarks be $+2/3$, if one assume that it is $-1/3$ for d quarks. Grand unification would thus also explain why the charge of the proton is exactly equal and opposite to the electron charge.

A direct consequence of SU(5) unification would be that the proton has a finite lifetime of about 10^{31} years. This means that in an ensemble of 10^{31} protons, which corresponds to about 70 tons of pure water, one would have on average one proton decay per year. In the 1980s appeared the first experiments attempting to detect these decays, with however no positive result up to now, in 2021. The largest among these experiments is the Super-Kamiokande detector, already mentioned in section 3.5 in connection with the discovery of neutrino oscillations. The lack of evidence for proton decay allows to set a stringent lower limit on the lifetime of the proton at about 10^{34} years. This clearly is in contradiction with the prediction from the SU(5) model, at least in its simplest form. The theoretical prediction of the lifetime is sensitive to the value of the grand unification scale which is still very uncertain. An increase of its value (see inset 4.2) by a factor of ten, from 10^{15} to 10^{16} GeV, could recover compatibility of the model with the experimental limit.[9] A value of about 10^{16} GeV is precisely what supersymmetric models suggest, as discussed earlier in section 4.1.

The SU(5) grand unified theory also predicts for the weak mixing angle θ_W a value such that $\sin^2\theta_W = 0.21$. This is an encouraging result as it is rather close to the value measured at LEP: $\sin^2\theta_W = 0.23149 \pm 0.00017$.

[9] In supersymmetric GUT models the dominant proton decay mode may shift from $p \to e^+\pi^0$ to $p \to K^+\bar{\nu}$, which is more difficult to detect.

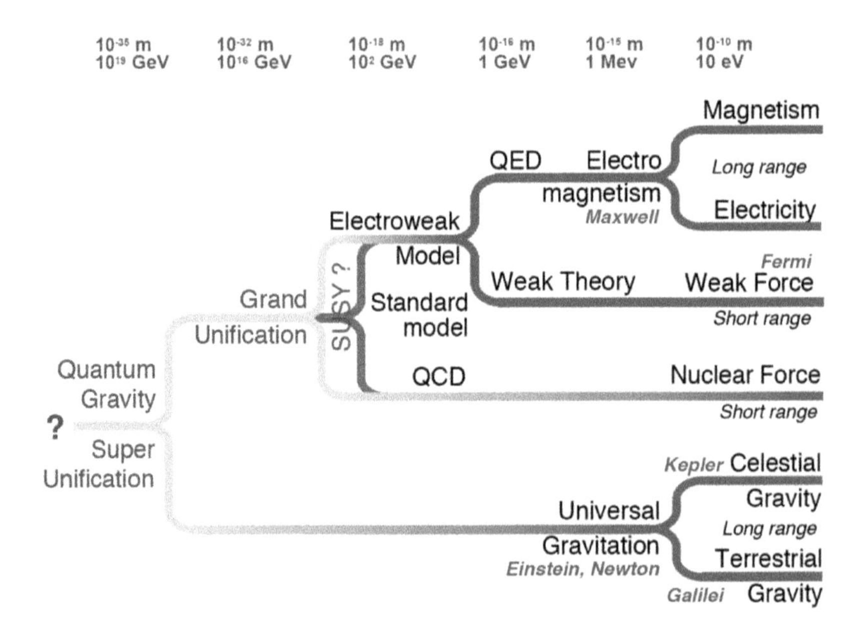

Fig. 4.6. Unification of fundamental forces in the Universe, from Newton, who understood that forces binding material objects to Earth and governing the motion of celestial objects are the same, gravity, through Maxwell and the unification of electricity and magnetism, to the Standard Model with electroweak unification, up to present-day attempts to achieve grand unification and ultimately also incorporate gravity.

Nonetheless, in view of the great precision of this measurement, these two values are incompatible. However, in the supersymmetric version of SU(5) the predicted value is $\sin^2\theta_W = 0.23$, now in very good agreement with experiment. This together with the improved convergence of the three couplings of the Standard Model (see figure 4.4) are arguments in favour of supersymmetry playing a relevant role in grand unification models.

A major problem of SU(5)-type models is that its multiplets are too small to accommodate the neutrino of right helicity whose existence is now unavoidable after the discovery of a small, but finite, non-zero neutrino mass. Theorists are therefore turning their attention to models based on larger groups, such as the SO(10) group (the group of real orthogonal matrices of dimension 10) or to more exotic groups as E_6 (exceptional Lie group of order 6) inspired by string theory models.

As we just saw for the proton decay, grand unified theories allow transitions between quarks and leptons, where neither baryon number nor lepton number are conserved. This is one of the requirements for *baryogenesis* to occur, that is creation of a permanent asymmetry between baryons and antibaryons. Assuming that there exists in a grand unified scenario a mechanism breaking matter–antimatter symmetry with larger intensity[10] than the CKM mechanism of the Standard Model, in an evolving Universe (not in thermal equilibrium) the antibaryons could disappear more easily than the baryons. The three conditions for baryogenesis: proton decay, CP violation and non-equilibrium conditions, were formulated in 1967, well before the introduction of grand unified theories, by the Soviet physicist Andrei Sakharov to explain the observed matter–antimatter asymmetry in Universe, with its residue of baryons we are made of.

4.3. Hidden dimensions in the Universe

In the framework discussed in the preceding sections, the unification of all interactions including gravity could be envisioned at an energy scale of the order of the Planck mass, as sketched in figure 4.6. An alternative to this scheme was proposed at the end of the 1990s in terms of theories assuming the existence of additional spatial dimensions of almost macroscopic size, theories inspired by string and superstring theories.

The idea that the Universe might have more dimensions than just the Minkowski space M_4 with three spatial dimensions plus one time dimension, goes back to the years 1920. The German physicist Theodor Kaluza and the Swede Oskar Klein proposed in those years a model with five dimensions (four spatial ones and one temporal) in an attempt to unify electromagnetism and the general theory of relativity developed just a few years earlier by Albert Einstein to explain the gravitational force.

The key idea is that the fourth spatial dimension is compactified, described by a circle C of fixed radius R attached to each point of M_4. This radius, related in the original Kaluza–Klein model to the electron charge, is sufficiently small that, seen from a distance (energy) scale much larger (smaller) than R $(1/R)$, the Universe appears as having only the usual three space dimensions. This is what is called *compactification* (see inset 4.3).

[10] What would be needed is, for example, that the rate for $\bar{X} \to \bar{\ell} + \bar{q}$ differs from that of $X \to \ell + q$ at the level of 10^{-9}.

Inset 4.3: Energy–distance relation and the Planck scale

In quantum mechanics energy and distance are related: to every particle of momentum p is associated a *de Broglie wavelength* $\lambda = h/p$ where h is the Planck constant. An incident beam particle interacting with a target nucleus, for example, can probe substructure at length scales larger than or at best of order λ. Directly related, a particle of mass m (rest energy mc^2) in quantum mechanics cannot be localised to a distance smaller than its so-called Compton wavelength $\lambda_C = \hbar/mc$, where $\hbar$ is the reduced Planck constant $\hbar = h/2\pi$.

Accordingly, to an energy scale E one can associate a wavelength $\lambda = \hbar c/E$. For the phenomena studied in particle physics suitable units are MeV for energies and fermi (fm) for distances (one fm is the approximate size of a nucleon). In these units, the above relation can be written as

$$\lambda \approx \frac{200\ \mathrm{MeV\,fm}}{E}.$$

If we choose, as it is often done in particle physics, a system of units where $\hbar = c = 1$, a distance can be expressed in MeV^{-1}, the equivalence relation to remember being approximately 1 fm $\sim$ 200 MeV^{-1}. The distance scale associated to the Planck energy scale ($M_P = 10^{19}$ GeV) is then 10^{-20} fermi or 10^{-35} metres, often called Planck's length. At this scale quantum mechanics must affect the structure of space–time. At the LHC, the energy scale of the interacting partons within the proton beams is about a TeV allowing to probe distances of the order of 10^{-4} fermi or 10^{-19} metres.

Figure 4.7 shows in a simplified way the Kaluza–Klein geometrical construct. There may be several additional compactified dimensions, and these could be visualised by introducing at every point of an M_4 space for example a two-dimensional space in the shape of a sphere S_2 or a torus.[11]

[11] As the original Kaluza–Klein model, $M_4 \times C$, connecting electromagnetism (associated to the gauge group U(1)) with general relativity, the $M_4 \times S_2$ model developed in the years 1930 by Wolfgang Pauli gives rise to interactions described by the SU(2) $\times$ U(1) group. At Pauli's time gauge theories were unknown and spontaneous symmetry breaking mechanisms for mass generation even less, so this type of theory was abandoned. These examples nonetheless show how the geometry of space–time can give rise to interactions that are a priori based on internal symmetry properties of particles.

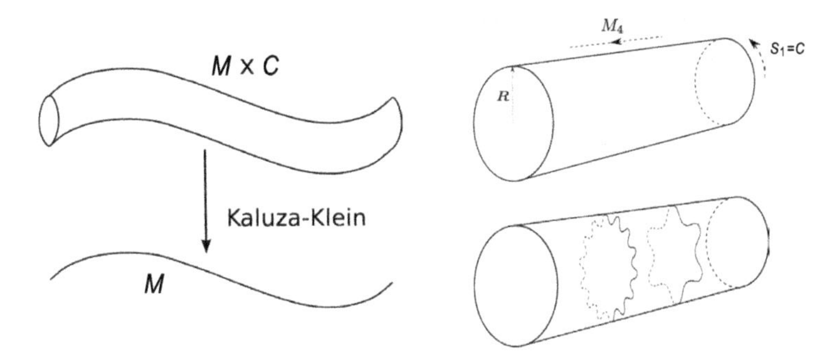

Fig. 4.7. Geometry of the Universe in the Kaluza–Klein model. The three-space-dimensional Universe of ours is represented by the one-dimensional wavy line M. An additional space dimension is rolled up on a circle C of radius R in such a way that at large distances the overall space $M \times C$ appears just like M. The figure on the right shows the propagation of excited field modes in a $M_4 \times C$ universe, where M_4 is the relativistic four-dimensional Minkowski universe (three space dimensions plus one temporal dimension). The larger the number of oscillations of the field along the additional dimension (along the circle) the more massive is the effective field propagating in M_4 space.

One of the main consequences of additional spatial dimensions would be the appearance of excited propagation modes for an oscillating field associated to a zero mass particle in the space of overall dimensions $4 + n$. Indeed, the propagation in M_4 space of a mode k corresponding to k oscillations on the circle of the space component in the additional dimension, would effectively look in M_4 as if the field has acquired a mass $m_k = k/R$. One thus introduced on top of the fundamental (massless) ground state a *Kaluza–Klein tower* of massive excited modes.[12]

Kaluza–Klein type theories became fashionable again during the 1990s in connection with developments related to string and superstring theories. To be mathematically coherent and provide a framework within which all interactions including gravitation could be unified, these theories must be formulated in more than four space–time dimensions. In superstring theories nine (if not ten) spatial dimensions and one temporal dimension are required. The six (or seven) additional spatial dimensions are compactified

[12]It is possible to make a parallel with the theory of wave-guides. An electromagnetic wave in free space moves with the speed of light c and the photon is massless. In a wave-guide, however, only certain discrete modes k can propagate and this with a speed smaller than c, so that the associated field quantum effectively acquires a non-zero mass.

in the Kaluza–Klein manner with tiny proportions of the order of the Planck scale (10^{-33} cm). The corresponding excited Kaluza–Klein modes have mass of the order of the Planck scale and are thus experimentally inaccessible.

However in 1998, Nima Arkani Hamed, Gia Dvali and Savas Dimopoulos (ADD) proposed a model with extra spatial dimensions whose size could reach almost a millimetre without being in contradiction with experiment! In this model, the Standard Model particles remain confined in the non-compactified three-dimensional space, which in the jargon of this part of theoretical physics is called a *brane*, a terminology borrowed from string theory. Only gravity is assumed to propagate through all space dimensions. The intensity of gravity turns out much larger when considering the entire space and it is diluted in the three-dimensional brane making up the particle world. This could explain why gravity looks so much weaker than the other interactions among particles at length scales much larger than R (see inset 4.4).

The ADD model opened the prospect that gravity might become strong, comparable to other interactions, at energies accessible with the LHC (probing length scales of order 10^{-19} metres) and that phenomena related to *quantum gravity*, in which case the production of gravitons (zero modes or excited Kaluza–Klein modes) or even *microscopic quantum black holes* could be observed in the ATLAS and CMS detectors (see section 13.5).

If there were only two additional compactified dimensions, for example on a torus of radius 0.1 mm, the fundamental energy scale of gravity would be brought down to a value of about 10 TeV (rather than $M_P = 10^{19}$ GeV $= 10^{16}$ TeV)! The value $R = 0.1$ mm is not a haphazard choice, it corresponds to the minimal distance between two masses for which the validity of Newton's law of gravity has been verified experimentally. The ADD model predicts that the gravitational potential has its usual, classical form with a $1/r$ variation at distances large compared to the size of extra dimensions, while a deviation with a behaviour of the form $1/r^{1+n}$, where n is the number of extra dimensions, should be observed at smaller distances. For $n \geq 3$ the distance R, for which the energy scale of gravity would be of order of few TeV, is of atomic size and this is out of reach of any presently foreseeable experiments testing gravity directly.

Inset 4.4: The gravitational force and the size of extra dimensions

In its usual form in four-dimensional Minkowski space the gravitational potential between two masses m_1 and m_2 is given by:

$$V(r) = -G_N^{(3)} \frac{m_1 m_2}{r} = -\frac{m_1 m_2}{M_P^2 \cdot r},$$

where $G_N^{(3)} = G_N = \hbar c / M_P^2$ (=$1/M_P^2$ in usual quantum units, see inset 4.3) is Newton's gravitational constant for three space dimensions, M_P is the Planck mass and r the distance between the masses.

For $3 + n$ spatial dimensions and $r \ll R$, where R is the size of additional compactified space dimensions, we can use Gauss' theorem and, taking into account that $V(r)$ must have a dimension of energy or mass or the inverse of length, the potential takes the form

$$V(r) = -G_N^{(3+n)} \frac{m_1 m_2}{r^{1+n}} = \frac{m_1 m_2}{M_f^{2+n} \cdot r^{1+n}},$$

where M_f is the new fundamental mass scale of gravity in $3 + n$ space dimensions, analogous to Planck's mass in three dimensions. The masses M_P and M_f are related by

$$M_P^2 = M_f^{2+n} V_n,$$

where V_n is the volume spanned by the extra dimensions ($V_n \sim (2\pi R)^n$).

In the Arkani Hamed, Dvali, Dimopoulos (ADD) scenario (see text) M_f is the energy scale where gravity becomes of comparable strength to other interactions. From the above equations one finds

$$M_f = \frac{1}{2\pi R} (2\pi R \cdot M_P)^{\frac{2}{2+n}}.$$

For $n = 2$ and $R = 0.1$ mm, M_f is about 10 TeV. The striking outcome is that additional dimensions dilute the strength of gravity apparent in the three non-compactified dimensions. In this scheme the small value of the gravitational constant $G_N^{(3)}$ is just a residue of the real and much larger constant $G_N^{(3+n)}$.

4.4. The Higgs boson as a composite particle

As discussed in the previous chapter, the problem of the Higgs boson mass stabilisation at the electroweak scale stems from the fact that the Higgs is a fundamental scalar field. Supersymmetry appears to be the only known theoretical framework that can accommodate a fundamental scalar. In case we find no experimental evidence for supersymmetry (see section 13.3), an alternative path to alleviate this fine-tuning or "naturalness" problem of the Standard Model is to consider that the Higgs boson is not a fundamental field but rather a bound state of a new strong interaction sector which may show up at a much higher energy scale, typically a few TeV. It would be a bit like the pion for QCD, which is a composite state of a quark and an antiquark bound by gluons.

Actually, the first and simplest theory working along these lines to be proposed, called *Technicolour*, just postulates the existence of a strong sector analogous to QCD, but transposed at a scale of a few TeV (instead of a few hundred MeV for QCD), also based on a SU(3) gauge group, with confined *techniquarks* and *technigluons* carrying a charge called *technicolor*. Such a theory contains composite pseudo-scalar *technipions* which can play the role of the Higgs boson to break electroweak symmetry. Unfortunately, theories simply based on this type of hypothesis can hardly explain the smallness of the Higgs boson mass, compared to the fundamental scale of the theory (remember that the pion mass is of the same order as the QCD scale which is about 200 MeV). Reducing this fundamental Technicolor scale would induce too strong experimental deviations from the Standard Model predictions. In addition, in the absence of a simple scheme like the Yukawa couplings between technipions and Standard Model fermions, it does not explain in a natural way the origin of the fermions masses.

A more promising class of theories proposes that the Higgs is a *pseudo-Nambu–Goldstone boson* of a spontaneous symmetry breaking occurring at higher energy scale (a few TeV). What does that mean? There is a general theorem in field theory, called *Nambu–Goldstone theorem*,[13] which states that the spontaneous breaking of a continuous global symmetry of the Lagrangian generates massless scalars, called Nambu–Goldstone bosons.

[13]Named after the Japanese-American physicist Yoichiro Nambu, who received the Nobel prize in 2008 for his contributions to the studies of spontaneous symmetry breaking in sub-atomic physics, and the British physicist Jeffrey Goldstone.

Inset 4.5: The Randall–Sundrum model

Subsequent to the publication of the ADD idea a number of theoretical models based on extra spatial dimensions with effects on TeV-scale physics were developed. In the various models proposed all or part of the force fields, those of the Standard Model or of the grand unified fields, or of the matter fields, could propagate through one or more extra dimensions, and these could even be of different sizes.

In a model proposed by the American physicists Lisa Randall and Raman Sundrum in 1999, a fifth spatial dimension separates two branes of (3 + 1) spatial dimensions. One of the branes, called the TeV brane, corresponds to our Standard Model world, whilst in the other one, called Planck's brane, gravity is strong, that is of similar strength as the other interactions.

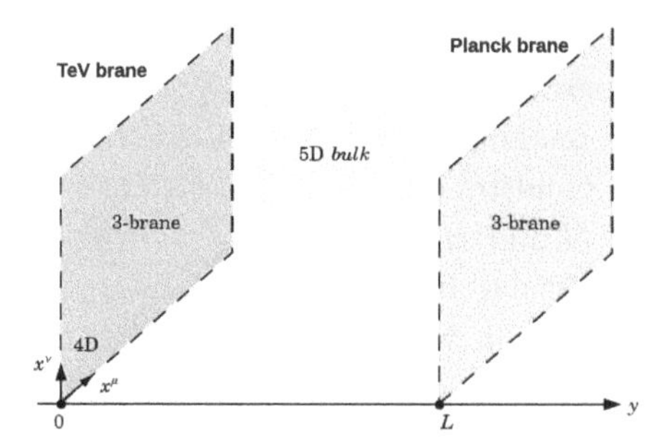

The metric that defines distance and energy scales differs by about 16 orders of magnitude between the two branes. In the intervening five-dimensional space (the bulk) connecting the two branes, the metric is highly deformed with an exponential variation as a function of the fifth coordinate y. There is a dilatation of distances and time as one approaches the TeV brane, whilst mass and energy are contracted. This has as the consequence that gravity, transmitted by the graviton which can propagate through the bulk, is felt with much reduced intensity in the Standard Model brane.

If the symmetry is only approximate, the scalar fields are not exactly massless and are called "pseudo-Nambu–Goldstone" bosons.

We have already encountered this situation with the electroweak symmetry breaking in section 1.5. The "Mexican hat" shape of the Higgs potential exhibits a rotation symmetry which is spontaneously broken when a vacuum is chosen as the ground state at a given point of the circular minimum (see inset 1.8 on page 38). Excitations around this vacuum (do not forget that particles correspond to field excitations around the ground state) correspond, on one hand to the massive Higgs particle for excitations along the direction perpendicular to the circle of minima (the parabolic shape of the local potential corresponding to its mass) and, on the other hand, to a massless particle for excitations along the tangent to the circle, which is the Nambu–Goldstone boson of this spontaneous symmetry breaking. In the special context of a gauge theory, this massless particle disappears in the BEH mechanism to give rise to the masses of the W and Z bosons.

In the class of composite Higgs models, as for Technicolour theories, one assumes the existence of a new sector of particles and forces at a higher mass scale. The corresponding Lagrangian is invariant under a global symmetry described by a group G. This symmetry is partially spontaneously broken, leaving a subgroup H, containing the Standard Model gauge group $SU(2) \times U(1)$, as the remaining symmetry of this G/H theory. In this process, several Nambu–Goldstone bosons are generated. With an appropriate choice of the groups G and H, it is possible to generate the equivalent of the doublet of Higgs scalar fields. If the symmetry G is not exact, due to terms in the Lagrangian which explicitly break it, the corresponding pseudo-Nambu–Goldstone fields are not massless. These terms are at the origin of the Higgs potential leading to a Higgs mass much smaller than the typical mass of composite particles of this theory. If the G/H symmetry breaking occurs at a mass scale f_G, then deviations from the Standard Model (like for instance the values of the Higgs coupling to the W and Z bosons) are of the order v^2/f_G^2, where $v = 246$ GeV is the electroweak vacuum expectation value. For $f_G \sim 1$ TeV, effects of a few % are expected.

Another ingredient of most composite Higgs models is the existence of fermionic composite states with the same quantum numbers as the fundamental fermions of the Standard Model. A consequence is that they can mix, leading to *partial compositeness*: each observed fermion would be a quantum mixture of a fundamental Standard Model fermion and of a composite fermion of the G/H theory. The amount of composite fermion in the mixture would be large for the heaviest fermion (the top quark could even be almost entirely composite) and very tiny for the other much lighter fermions. Actually the fermions would acquire their masses through this mixing and the coupling of the composite fermion with the composite Higgs. Another consequence is the existence of composite partners which differ from the Standard Model fermions by their electric charge and their mass: for instance top/antitop partners with charge $5/3$ or $-4/3$ would be expected at higher mass scale ($\sim$TeV).

Chapter 5

Back to the Big Bang

With the formulation of the General Theory of Relativity in 1915 by Albert Einstein, physicists had for the first time at their disposal a mathematical framework that would allow them to address questions in relation with the Universe as a whole, its past evolution and possible future. In the years 1920, the Russian physicist Alexander Friedmann (1922) and the Belgian priest-physicist Georges Lemaître (1928) found and studied a solution of Einstein's equations describing an expanding Universe evolving from an initial state of very high density. Einstein himself a few years earlier discovered the possibility for such solutions. However, he was so much convinced of an unchanging, eternal Universe that he introduced in his equations an additional term, the *cosmological constant*, to avoid solutions where the Universe could be in expansion or collapse back onto itself (see section 3.2). When about a decade later the *universal expansion* was discovered by Hubble and coworkers, Einstein said that the introduction of the cosmological constant was the biggest mistake he ever made. Thus in the 1930s the cosmological constant disappeared from physics. It reappeared only in the years 2000 in an amazing reversal of fortune to explain the unexpected discovery of the *accelerated expansion* of the Universe.

This idea of an expanding Universe that would have evolved from a state of extremely high density and temperature[1] has been formalised in the Cosmological Big Bang Model. The model imposed itself as a result of three basic observations: the *recession of galaxies*, the existence of *fossil microwave background radiation*, and the *primordial nucleosynthesis* of

[1]The idea of a *hot* and not only dense initial phase of matter in the Big Bang was subsequently formulated by George Gamow in the mid-forties of the last century.

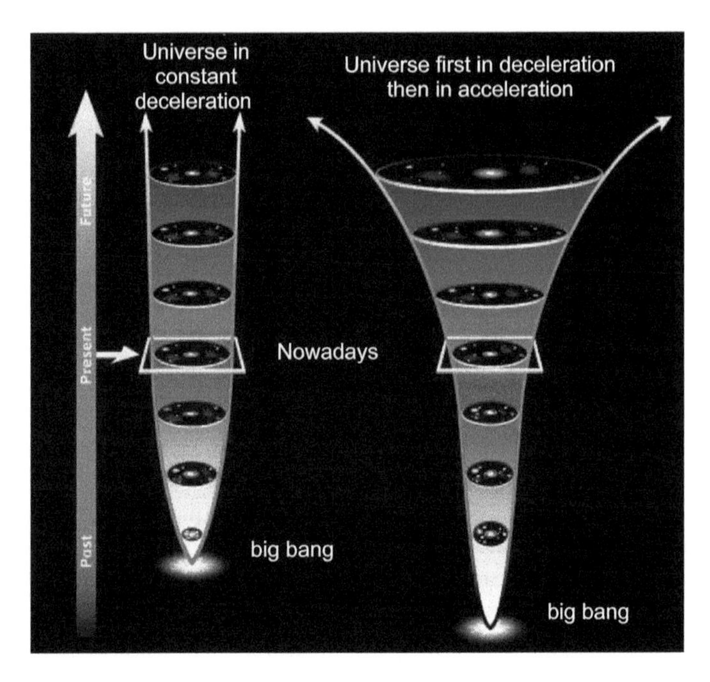

Fig. 5.1. Schematic picture of the time evolution of the Universe (represented by a disc) since the Big Bang. The scheme on the left corresponds to the picture prevailing before the years 2000 with a decelerating expansion. The right-hand picture, with an expansion reaccelerating over the last five billion years due to the presence of dark energy, corresponds to the presently accepted model. *Image: public domain.*

light elements. This model gradually developed over most of the twentieth century and is generally accepted today. If some aspects still require clarification and confirmation, in particular those related to *inflation* during the initial phases of the expansion, the overall picture will remain in the future. Nevertheless, rather recently in 1998, the model has gone through a major modification. Up to then it was thought that due to the gravitational pull on matter the expansion of the Universe and the flight of galaxies were slowing down with time. The observation and detailed study of distant supernovae over the past two decades revealed, however, that after an initial deceleration phase as expected over the last five billion years the expansion is accelerating (figures 5.1, 5.3 and 5.7)!

At the birth of the Universe, matter was so dense and the temperature so high that no structure could survive in the heat bath. Only elementary, structure-less objects could exist. Thus, the beginning of the history of the Universe is directly governed by elementary particle physics. We shall see at the end of this chapter how the LHC can help elucidating some questions related to primordial epochs in cosmology.

5.1. The flight of galaxies

In the 1920s with the commissioning of the large telescope at Mount Wilson in California and subsequently of a still larger one at Mount Palomar the American astronomers Edwin Hubble, Vesto Slipher and Milton L. Humason were able to show that, with the exception of very few galaxies in the immediate neighbourhood of ours, all galaxies seem to recede, fly away from us. Moreover, their velocity increases proportionally to their distance from us. Hubble, Slipher and Humason measured the *recession velocity* of galaxies using the Doppler–Fizeau effect, a relative velocity-dependent redshift, i.e. an increase in wavelength induced when a source and observer move apart from each other. They identified in the optical spectra well-known spectral lines and noticed that they are always shifted towards the red end of the spectrum. Essentially the same method is still employed today although at cosmological distance scales this redshift is slightly reinterpreted as being due to the expansion of the underlying space.[2] The most remote astronomical objects that can be observed today are at distances of about 13 billion light years corresponding to a redshift $z \sim 11$. Such a large redshift is a direct manifestation of the expansion of the Universe.[3]

The recession velocity of galaxies v is given by Hubble's law $v = H_0 d$, where d is the distance relative to our galaxy and H_0 is *Hubble's constant.* According to the most recent indirect measurement obtained with

[2]The spectral shift of a luminous object is defined by the relation $z = (\lambda_{\text{obs}} - \lambda_0)/\lambda_0$. The quantity λ_0 is the wavelength of the electromagnetic radiation emitted in a well-defined atomic transition as measured in the rest frame of the emitter. When emitted by a distant astronomical source it is being observed on Earth with a wavelength λ_{obs} owing to the expansion of the Universe. It can be shown that $1+z = a_0/a(t)$, where $a(t)$ is the scale factor of the Universe at the time t the object emitted the electromagnetic radiation, and a_0 the present-day scale factor. The scale factor thus depends on the locus in the Universe we are studying.

[3]The currently most distant galaxy named GN-z11 was detected in Hubble space telescope images in 2016. Although it is extremely faint, the galaxy is unusually luminous. Its redshift was determined to be 11.1, which corresponds to light emission just 400 million years after the Big Bang, and between 200 and 300 million years after the first stars were formed. Although GN-z11 is about 25 times smaller than the Milky Way, it has a much faster star formation rate, which explains its brightness.

the Planck satellite,[4] its value[5] is 67.3 ± 0.6 km s^{-1} Mpc^{-1} (obtained under the assumption of a flat Universe — figure 5.7). If one were to unroll backwards the film with galaxies fleeing apart, the recession we observe today implies the existence in the past of a very dense state of matter into which all of the Universe's matter would be concentrated. With Hubble's law one can get as a first approximation the age of the Universe as $1/H_0$, which amounts to about 10 to 15 billion years. A more accurate evaluation must take into account that Hubble's constant is not exactly constant with time (H_0 is the present-day value), as well as the uncertainties concerning our knowledge of the energy density of the Universe, and its exact geometry at large scales.[6] The present best estimates give for the age of the Universe about 13.8 billion years.

During the first years and even decades after its formulation the hypothesis of an initial state of extreme density and temperature from which the Universe would have emerged in a sort of primordial explosion was called derisively "Big Bang" by its detractors, in particular the astrophysicist-cosmologists Fred Hoyle[7] (England) and Jayant Narlikar (India). They were proponents of competing theories, among which the concept of a *stationary state* Universe which was the generally accepted cosmological model until 1963. Of course, there is abuse of language when we talk about an explosion, as there has been no explosion with expansion of matter in a pre-existing space and time, but rather the creation of space, time and matter

---

[4]There is presently a tension (dubbed as "crisis" by some physicists) of several standard deviations between the H_0 measurements involving local and distant observations, respectively. It could indicate unaccounted systematic effects in some of the measurements or that the standard model of cosmology needs to be changed in some way using new physics to resolve the discrepancy.

[5]The megaparsec (Mpc) is the adequate unit of length for measurements at galactic and extragalactic distances. One Mpc corresponds to a distance of about 3.26 million light years, or about 3.1×10^{19} kilometres. The large spiral galaxy closest to ours, the Andromeda galaxy (M31), is about two million light years, or 0.7 Mpc, away. For comparison, the diameter of our Milky Way galaxy is about 100 000 light years or 0.03 Mpc.

[6]The present measurements of the average matter-energy density give a value close to the *critical density* ρ_c corresponding to a flat Universe: $\rho_c = 3H_0^2/(8\pi G_N) \sim 5 \times 10^{-6}$ GeV cm^{-3} where G_N is Newton's gravitational constant. This critical density corresponds to about five proton masses per cubic metre, whilst the real baryonic density, the one of visible matter, is only about 0.2 protons per cubic meter (figure 3.3).

[7]Hoyle became famous for formulating the theory of stellar nucleosynthesis in the 1940s and 1950s, see section 5.3 below.

altogether from that initial (quantum) singularity, not meaning a singularity in the mathematical sense but rather physical, a state of extreme density and temperature.

5.2. Cosmic microwave background radiation

In 1963 two physicists from the Bell Laboratories in the USA, Arno Penzias and Robert Wilson (Nobel Prize in 1978) discovered accidentally a radio background radiation in the domain of microwaves. The radiation was coming uniformly from directions all over the sky, and the spectrum of radiation corresponded to a perfect *black-body radiation spectrum*[8] at a temperature of 2.73 kelvin (−270.42°C). This reminded people of a prediction made in the late 1940s by George Gamow, a brilliant US physicist of Russian origin, that such a fossil radiation permeating the whole Universe should still exist as a trace of the hot past of the Universe: a redshifted remnant of the radiation emitted at the time of formation of hydrogen atoms from the primordial plasma of protons and electrons. The predicted temperature was close to that observed by Penzias and Wilson. This radio background is now usually called *cosmic microwave background*, CMB, sometimes also fossil microwave radiation.

As the Universe expanded, both the plasma and the radiation filling it grew cooler. When the temperature decreased sufficiently so that the average kinetic energy, k_BT (where $k_B = 1.38 \times 10^{-23}\ \mathrm{J\,K^{-1}} = 8.62 \times 10^{-5}\ \mathrm{eV\,K^{-1}}$, is Boltzmann's constant[9]), reaches the electron-volt level, a typical binding energy for an electron in an atom, the electrons that up to then were roaming freely in the plasma were captured by hydrogen and helium nuclei with traces of other light elements, the only ones present at this epoch (see next section), to end-up for the first time stably

[8]A black body is an ideal body absorbing without reflection or diffusion all incident electromagnetic radiation. It re-emits all absorbed energy in the form of electromagnetic radiation. In conditions of thermodynamic equilibrium between absorbed and re-emitted radiation, the frequency spectrum of emitted radiation is given by a universal law (Planck's radiation law), independent of the material and depending only on the temperature of the black body. It is with the formulation of this black body radiation law in 1900 by Max Planck that quantum mechanics entered physics.

[9]Boltzmann's constant k_B can be interpreted as the proportionality factor between the temperature of a system and its thermal energy, that is the average kinetic energy of its elementary constituents. This is the reason why very often the temperature of a multiparticle system, or of a plasma, is expressed in energy units, for example the electron-volt.

bound into atoms. The Universe which was opaque to light up to that moment in time (the photons being constantly captured, absorbed and re-emitted by available free electron charges) suddenly became transparent to light, as if the fog lifted from the universal landscape, and the last photons emitted at that time could propagate freely and reach us today, carrying information from this epoch. This phase in the evolution of matter, called recombination or the separation of light from matter, occurred about 380 000 years after the Big Bang.[10] While the evolution of the Universe was up to then dominated by radiation,[11] it will now be dominated by matter — until about 5 billion years ago, when dark energy took over as the dominant energy form accelerating the expansion of the Universe (see following section). The separation of matter and radiation happens when the temperature of the Universe is a few thousands of degrees ($T \sim 0.35\ \mathrm{eV}/k_B \sim 4000\ \mathrm{K}$). The cosmological microwave background we observe today at 2.73 K, which corresponds to a number density of about 400 photons per cubic centimetre, is the final and cold glimmer of this light emission at atomic capture, cooled down by a factor of 1100, the expansion factor (redshift) of the Universe since then. In other words, with the dilatation of space by a factor of 1100 the emitted wavelengths increased by the same factor. The photon radiation changed from the visible spectrum domain with 0.2–0.6 micrometres at emission, to the millimetre wavelength domain in which we measure it today. The discovery of CMB, testimony of a hot past of the Universe, sounded the death bell for the stationary state cosmological model, and triggered the general acceptance of the Big Bang model. The detailed study of the CMB, in particular of the tiny temperature-density fluctuations imprinted in it, their statistical distribution and polarisation spectrum (cf. footnote 25), is the richest source of information and constraints on cosmological models today (see also figure 3.2, sections 3.1 and 3.2, and figure 5.4).

5.3. Primordial nucleosynthesis of light elements

Let us now roll further backward the film of evolution of matter according to the Big Bang model. At about 10^{-6} seconds after the Big Bang the

[10] The volume of space at the time of recombination is about 42 million light years (13 Mpc) in spherical radius and the baryonic matter density about a billion times larger than today.

[11] The energy density of the Universe is essentially all in the form of radiation until then.

temperature of the Universe from its extremely hot beginnings has fallen to around 200 MeV (2.6 trillion kelvin, 170 thousand times hotter than the core of the Sun), which triggers a transition from the state of a quark–gluon plasma to the hadronic phase (see section 14.2.1) where composite particles emerge including protons and neutrons. With the temperature further falling to a few MeV, the typical nuclear binding-energy regime is reached and first nuclear structures appear and survive in the heat bath. This is the period of *primordial nucleosynthesis*, lasting from, approximately, the first tens of seconds to several minutes after the Big Bang. Light nuclei such as helium-4 (^{4}He) a very stable nucleus with a binding energy of 28 MeV were created. The formation process goes through two steps: first the production of deuterium d (that is hydrogen ^{2}H), whose binding energy is only about 2 MeV, through fusion of a proton and a neutron ($p + n \to d$); followed by the fusion process $d + d \to {}^4$He. The deuterium production is in competition with neutron decay, as a free, unbound neutron has a lifetime of only about fifteen minutes (figure 5.2), whereas a neutron bound in a (stable) nucleus is stable. The neutrons thus rapidly disappear from the scene, as the production process in the medium $p + \nu_e \to n + e^+$ is an endothermic reaction requiring input energy of about 1.5 MeV.[12]

By the end of the nucleosynthesis phase, most of the hadronic matter is made of protons (hydrogen ^{1}H), about 75%, whilst helium (^{4}He) makes up about 25%. The other light elements synthesised, such as deuterium (^{2}H), tritium (^{3}H), helium-3 (^{3}He), lithium (^{7}Li) or beryllium (^{7}Be) are present only at trace levels, amounting to fractions between 10^{-5} and 10^{-10} in comparison to hydrogen ^{1}H (figure 5.2). It is possible to determine experimentally the primordial nuclear composition through the study of, for example, very ancient stars or celestial bodies such as quasars,[13] whose formation occurred early after the nucleosynthesis period.[14] Stars, galaxies, and quasars began to form a few hundred million years after the Big Bang.

[12]The mass difference between neutron and proton is 1.3 MeV, which allows the neutron to decay into a proton plus an electron and an electron-antineutrino via nuclear β decay.

[13]The present understanding of a quasar (quasi-stellar source) is that of a compact region surrounding a very massive black hole at the centre of a massive young galaxy. The quasar activity peaked approximately three to four billion years after the Big Bang. Quasars are the most luminous objects in the Universe.

[14]These measurements of primordial nuclear compositions must be corrected for the subsequent modifications induced by stellar nucleosynthesis (see following section).

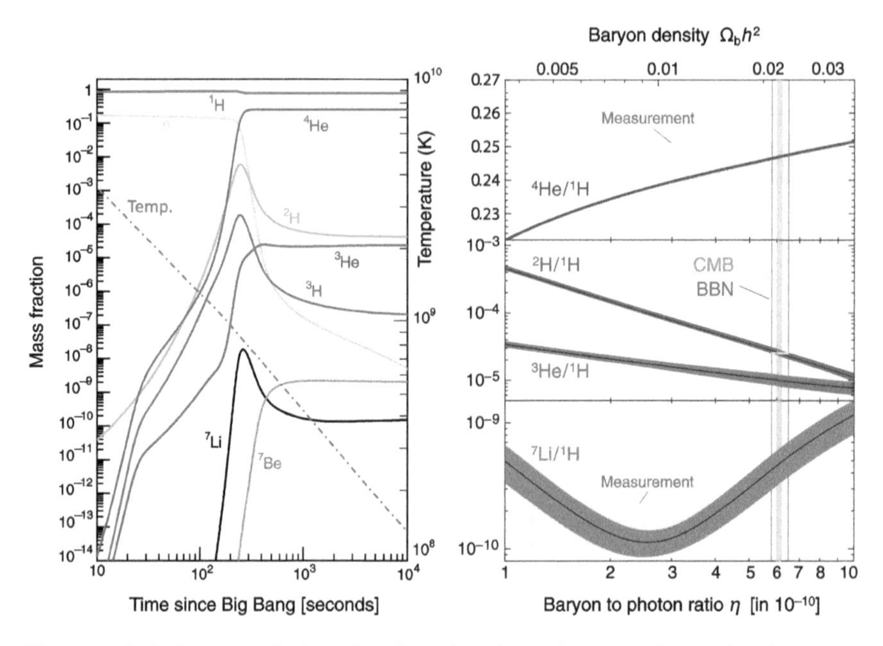

Fig. 5.2. Left figure: evolution of nuclear abundances in terms of mass fraction versus time elapsed since the Big Bang during the primordial nucleosynthesis era. The red-dotted monotonically decreasing curve shows the variation of temperature (ordinate on the right). As soon as the temperature drops below 2×10^8 K, about an hour after the Big Bang, nucleosynthesis freezes out and stops. Hydrogen ^{1}H is the most abundant element (about 75% in mass), followed by ^{4}He (about 25%). The other isotopes or nuclei are present only at trace levels. Figure on the right: predicted abundances of light elements relative to ^{1}H as a function of the ratio of baryon-to-photon densities $\eta = N_b/N_\gamma$ (bottom abscissa) and the baryon density (top abscissa). The yellow rectangles show the astrophysical measurements. Taking into account experimental uncertainties, the various measurements are mutually compatible with a common value of η (pink vertical band). Furthermore, they are in agreement with the value derived from cosmic microwave background measurements (blue vertical band labelled CMB). *Images: A. Coc et al., Phys Rev D. 76, 023511 (2006); Particle Data Group, Prog. Theor. Exp. Phys. 2020, 083C01 (2020); figures modified by authors.*

Theoretical predictions of nucleosynthesis and the primordial nuclear composition depend on the ratio of the number of baryons to photons, N_b/N_γ, during the process of nucleosynthesis, where N_b is the small post-matter–antimatter annihilation residue.[15] This ratio governs the cooling and expansion rate of the medium and thus the probability for the nucleons to encounter and fuse. For $N_b/N_\gamma \sim 6 \cdot 10^{-10}$ the theoretical calculations

[15]As discussed in section 2.4, it depends also on the number of light neutrino species.

match well the astrophysical measurements of the nuclear composition. It is also consistent with the result determined in a totally independent way from studies of the cosmic microwave background (figure 5.2, on the right).

During the primordial nucleosynthesis epoch the temperature of the Universe falls so rapidly that it does not permit the formation of more massive nuclei. The other elements of Mendeleev's table are synthesised later from the nuclear fusion of lighter elements in the cores of stars, a process denoted *stellar nucleosynthesis.* Elements heavier than iron or cobalt up to uranium are produced through rapid neutron capture in the explosions of supernovae and more generally in explosions or stellar mergers (neutron stars) marking the final stages in the life of massive stars. This is the way through which stars in their death-throws feed the interstellar medium with the variety of chemical elements required for the creation of life where conditions are adequate, as was the case with our solar system and the Earth.

5.4. Accelerating expansion of the Universe

Our understanding of the Universe has significantly evolved since 1998 thanks to new astrophysical and cosmological measurements that have changed and much refined the observations discussed in the preceding sections. The concordance among measurements from three independent methods of the Universe's curvature and its matter-energy density has led to what is called the *Standard Model of Cosmology.* This new vision includes the extraordinary finding of an accelerating expansion of the Universe.[16]

First among these results is the measurement of the luminosity of distant type-Ia supernovae as a function of distance. These supernovae are used as

[16]What would be the long-term effects of an accelerating expansion of the Universe? According to a scenario called the *Big Rip*, the density of dark energy would increase without limits (becoming a ghost energy), provoking the break-up of all the objects bound by gravity: planets, stars, solar-type systems, galaxies. But this is not the full story. The expansion would eventually become so rapid that even the electromagnetic and strong forces would not be strong enough to prevent that molecules, atoms and even nuclei are ripped apart. The expansion would end up in a singular state of matter in which all composite matter would be broken up into its elementary constituents. This catastrophic scenario is just one among the possible destinies envisaged for the Universe. Its hypothetical realisation depends on the ratio ω of the dark energy pressure to density, which should be less than -1. The experimental data, including the most recent Planck satellite results, are not yet precise enough to differentiate between $\omega < -1$, $\omega = -1$, and $\omega > -1$. A value of $\omega = -1$ would correspond to a constant dark energy density and an exponentially increasing scale.

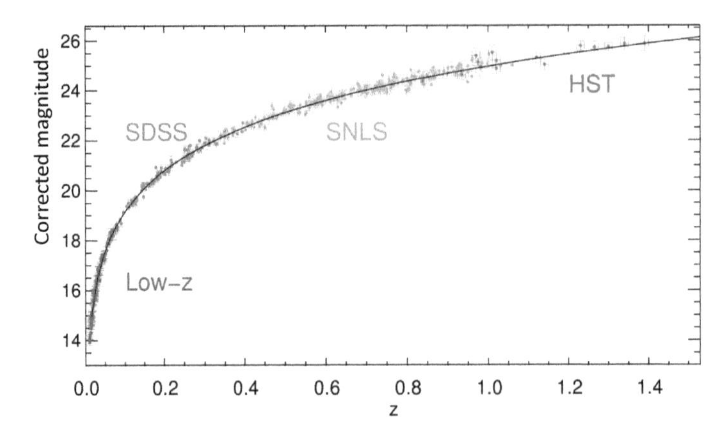

Fig. 5.3. Apparent magnitude of type-Ia supernovae as a function of redshift z obtained from several surveys. The apparent magnitude is a logarithmic function of the apparent luminosity defined such that it is larger for smaller observed luminosity, i.e. more distant objects. *Image: Betoule et al., A&A 568, A22 (2014).*

"standard candles" as it is possible to deduce their distance directly from their luminosity, owing to the good understanding of the thermonuclear mechanism responsible for the stellar explosion. As discussed in section 5.1, the distance can also be determined from the redshift z of a supernova.[17] By noticing that the luminosity of distant supernovae is weaker than expected (figure 5.3), Saul Perlmutter, Brian P. Schmidt, Adam Reiss and collaborators showed in 1998 that our Universe is in a phase of accelerating expansion, due to some sort of dark energy accommodated in the equations of general relativity, thus resurrecting Einstein's cosmological constant Λ. But this is just a phenomenological description of the observation.[18] Some models inspired by particle physics, such as *quintessence*, suggest a scalar particle field, while others propose modifications of general relativity to explain the phenomenon. The exact nature of dark energy is so far unknown, but more and more ambitious experiments and surveys are underway or in preparation to address this question.

[17]The measurement of z can be done from the spectral redshift of the host galaxy containing the supernova, or from the supernova spectrum itself.

[18]Recently, in 2019, some cosmologists put in question the acceleration of the expansion of the Universe and thus the need for dark energy, attributing the observed effect to anisotropies of the "bulk flow" in the local Universe which extends to much larger scales than was expected. This would introduce an observation bias not fully accounted for in the study of distant supernovae.

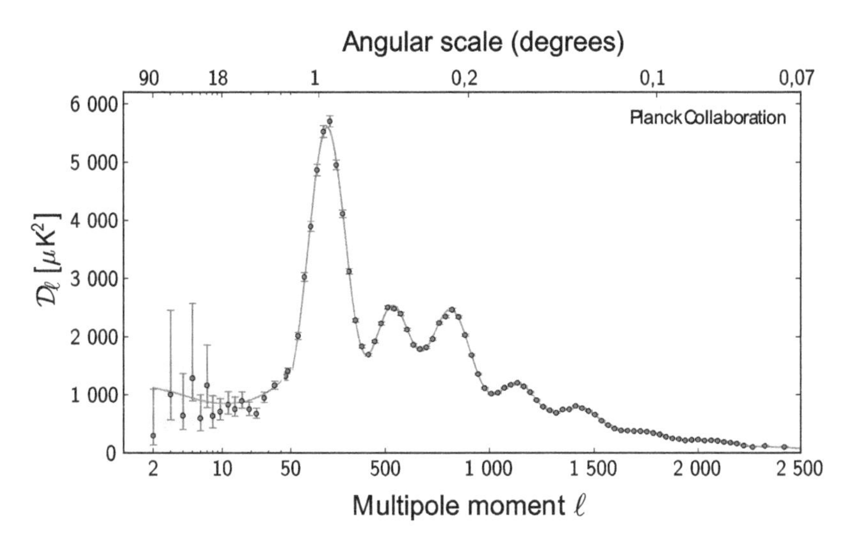

Fig. 5.4. Power spectrum of temperature fluctuations in the cosmic microwave background as a function of angular scale, as measured by the Planck satellite (see figure 3.2). The angular scale on the celestial sphere is given in terms of multipole moments ℓ, large values of ℓ corresponding to small angular sizes. Just for comparison, the angular size of the Sun or the Moon is about half a degree. The position of peaks is related to space curvature, the heights to matter density, the relative heights to baryonic density, etc. *Image: Planck Collaboration, A&A 571, A16 (2014).*

The second observation has to do with the structure of tiny temperature fluctuations observed in the cosmic microwave background (figure 3.2). Figure 5.4 shows the intensity spectrum of these fluctuations as a function of the angular scale at which they are measured. The shape of this spectrum contains a host of information on cosmological parameters characterising the structure of the Universe, such as its spatial curvature, the present expansion velocity H_0, the abundance of baryonic matter, of dark matter and dark energy, and more. For example, the presence of peaks, called *acoustic peaks*, indicates that there are favoured scales whose origin goes back to the separation of matter and radiation 380 000 years after the Big Bang. Before this time the Universe is filled with a plasma made of light nuclei, free electrons, photons and likely WIMPs too (see section 4.1). This plasma is not perfectly homogeneous, but rather has *density inhomogeneities*, presumably originating from primordial quantum fluctuations emerging in the post-inflation phase. The regions of higher density attract through gravity matter from surrounding regions of lower density. This effect is countered by the radiation pressure of photons, which tends instead to push apart electrons and protons. The presence of these two contradictory effects gives rise to instabilities resulting in oscillations between denser and hotter regions and more diluted and colder ones.[19]

These pressure oscillations are similar to acoustic waves in a gas. They froze out at the time of formation of atoms and the separation of light from matter. The distribution between wavefronts corresponding to higher temperatures and the colder ones is visible today through the presence of acoustic peaks in the angular spectrum of temperature fluctuations. The size and position of these peaks are the result of a complex, but well-understood physics depending on a number of cosmological parameters. For example, the spatial curvature of the Universe determines the position of the peaks. The density of matter (dark and baryonic) largely determines their heights. The density of baryons determines the relative heights of these peaks. Up to the first data delivered by the Planck satellite in 2013, the limited precision of previous measurements did not allow to determine simultaneously all the relevant cosmological parameters. In particular, with the data from the WMAP satellite alone it was not possible from the power spectrum of fluctuations to measure separately the abundance of matter and dark energy without making the assumption of a flat Universe (i.e. of zero spatial curvature), as the geometrical degeneracy introduces a strong negative correlation between the measurements of these two quantities (see orange shaded line in figure 5.5).

The third measurement exploits so-called *baryonic acoustic oscillations* (BAO), which offer yet another observationally and methodologically independent way to probe the abundance of (baryonic) matter. Just as the acoustic peaks in the CMB, they are due to fluctuations in the density of the plasma before the recombination of nuclei with electrons into atoms. Here we are, however, interested in the propagation of acoustic, i.e. mechanical waves. These propagate through the plasma with a velocity close to the velocity of light until the freeze-out occurs when matter and radiation decouple about 380 000 years after the Big Bang. At that moment the distance between density crests was of the order of 150 kiloparsecs. This has left an imprint in the cosmic microwave background and in the

[19]The amplitudes of temperature and density fluctuations in the CMB are very small, of the order of 10^{-5} relative to the average values (see figure 3.2). In magnitude this is comparable to waves on an essentially calm sea as seen from a plane flying at an altitude of ten kilometres.

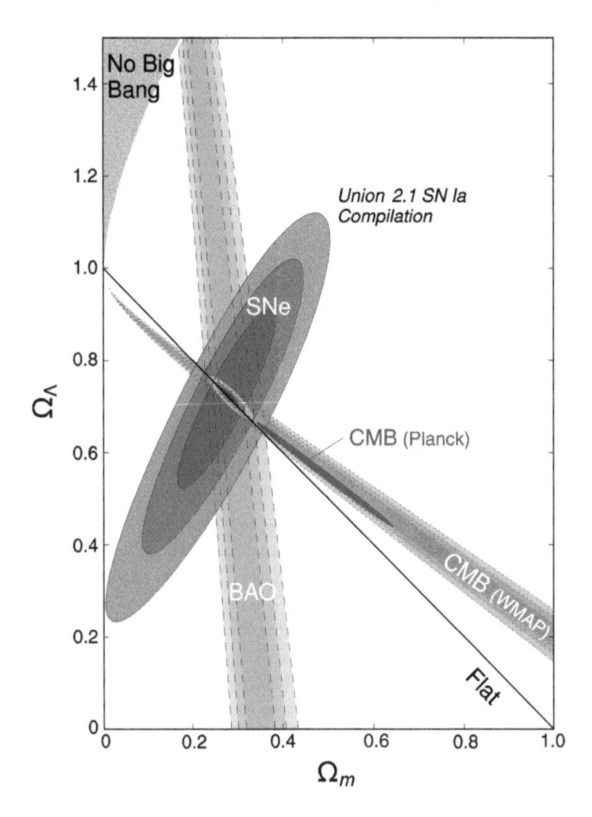

Fig. 5.5. Results of measurements of the correlation between the present density of dark energy Ω_Λ (so labelled to recall the connection with the cosmological constant Λ, which is the simplest form of dark energy) and the overall matter-density (dark matter plus baryonic matter) Ω_m, obtained from three complementary methods: supernovae (SNe), CMB and BAO (see text). The areas labelled CMB correspond to the results obtained from the analysis of WMAP (in orange) and Planck data (95% confidence level, in purple). These densities are given relative to the critical density of the Universe, the one leading to a flat, zero-curvature, Euclidean geometry Universe. The energy density contained in radiation, Ω_{rad}, which was dominant before recombination, is negligible in the present-day Universe ($\Omega_{\text{rad}} < 10^{-5}$). A flat Universe thus corresponds to $\Omega_\Lambda + \Omega_m = 1$. The three measurements are mutually compatible, and show that the Universe is compatible with being flat, with relative densities of matter and dark energy in accordance with those shown in figure 3.3. *Images: Suzuki et al., Ap. J. 2011; Planck collaboration, A&A 641, A6 (2020); figure modified by authors.*

distribution of ordinary baryonic matter (galaxies, hydrogen clouds, etc.). The distance between regions of accumulation of matter is a privileged ruler that is expected to evolve with the distance scale of the Universe. When we observe today the distribution of mutual distances between galaxies, it shows an excess around 150 Mpc (figure 5.6).[20] It is the manifestation of the original privileged distance scale in a Universe, which in the meantime suffered a dilatation by a factor z of about 1100. The evolution of this scale as a function of the spectral redshift z, from the decoupling epoch until today, provides information on dark matter and dark energy. The use of different probes (galaxies or quasars, which are the farthest away objects that can be observed) allows the measurement of the expansion rate of the Universe as a function of the time elapsed from the Big Bang, as shown on figure 5.7.

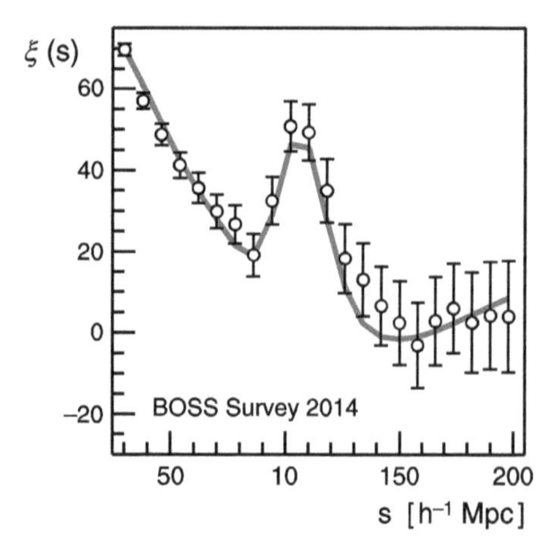

Fig. 5.6. Distribution of mutual distances among remote galaxies for all pairing combinations (two-particle correlation function), as measured by the SDSS-III BOSS experiment. The excess at $s = 100\ h^{-1}\,\mathrm{Mpc} \simeq 150$ Mpc ($h \simeq 0.71$ is the reduced Hubble constant, i.e. H_0 in units of 100 km/sec/Mpc) is attributed to the fossil presence of baryonic acoustic (mechanical) waves. *Image: Anderson et al., MNRAS, 441, 24A, 2014.*

Each one of these methods leads to correlated measurements of the baryonic plus dark matter density Ω_m, contributing to the slowdown of the expansion of the Universe, and the dark energy density, Ω_Λ, responsible for the acceleration of the expansion. The results are shown in figure 5.5.[21] It is remarkable that the three measurements based on very different processes have a region of mutual compatibility, all in agreement with the hypothesis of a flat Universe (of zero-curvature, and thus described by Euclidean geometry) in the framework of the present-day cosmological paradigm called ΛCDM. The acronym stands for the cosmological constant (Λ) cold dark matter version of a hot Big Bang model, where dark matter

[20]In the 2017 Sloan Digital Sky Survey (SDSS-III) survey, more than a million galaxies where included. The most recent results of SDSS-III and SDSS-IV show the BAO clustering effect also for intergalactic hydrogen clouds and for quasars.

[21]Thanks to the measurements using gravitational lensing effects alone, Planck data alone allow to remove the geometrical degeneracy and therefore the simultaneous determination of both the curvature of the Universe and the quantities Ω_m and Ω_Λ, without using data from supernovae surveys and BAO.

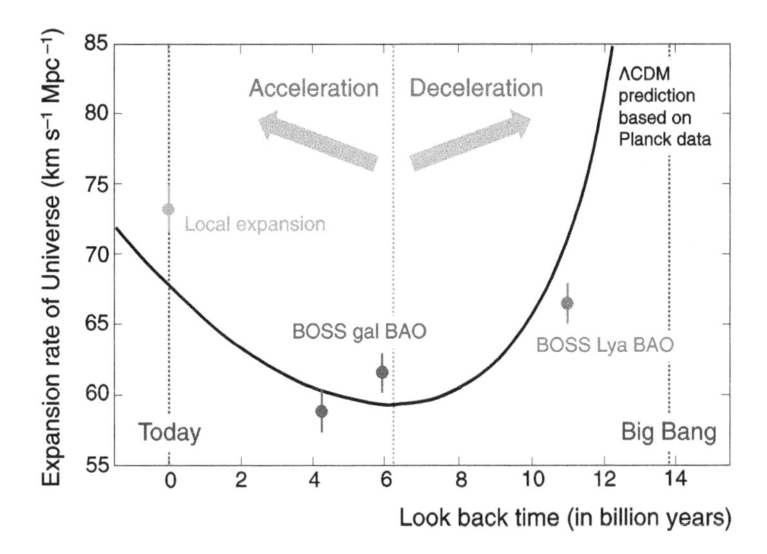

Fig. 5.7. Expansion rate in kilometres per second and per megaparsec (Hubble velocity) as a function of time looking back from today to the Big Bang. The thick line shows the prediction obtained in ΛCDM cosmology after a fit to Planck satellite CMB data. The blue data point corresponds to a result based on quasars from the BOSS experiment, the red points are BAO measurements using galaxies, and the green point on the left corresponds to a local measurement of the Hubble constant. About 6 billion years ago, when dark energy became the dominant energy component of the Universe, the expansion turned from decelerating to accelerating.

is labelled cold as it has become nonrelativistic since the early Big Bang times previously discussed. The new measurements of the time evolution of the cosmological expansion velocity as a function of look-back time and interpreted within this flat universe ΛCDM model are shown in figure 5.7. The Universe is subject to an accelerated expansion over the past six billion years.

Most recent results (2018) from the Planck experiment constraining H_0 and Ω_m (figure 5.8) show the compatibility within ΛCDM of the Planck CMB power spectrum data with the results from BAO, Big Bang nucleosynthesis and distant supernovae surveys. But it also shows the increased tension (almost 4σ, already visible in figure 5.7, most recent values in 2018 give 4.4σ) between the measurement of H_0 coming from the ΛCDM fit based on these sources and the most recent direct measurement based on nearby supernovae — right-hand side of the plot (see also footnote 4 on page 148).

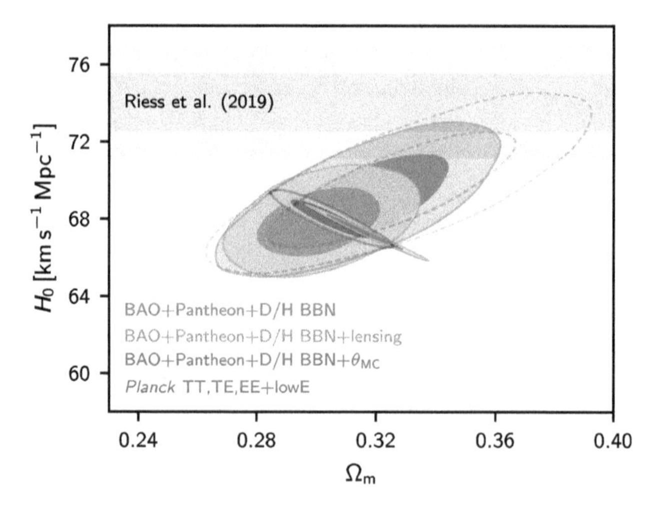

Fig. 5.8. Final constraints from the Planck satellite on the Hubble constant versus Ω_m plane (blue ellipse) within the ΛCDM model. Results from a combination of BAO, Big Bang nucleosynthesis and distant supernovae experiments correspond to the green and grey areas. The figure also shows the grey band coming from the direct measurement of H_0 using nearby supernovae. *Image: Planck Collaboration, A&A 641, A6 (2020).*

5.5. The Big Bang, particle physics and the LHC

The laws of elementary particle physics govern the evolution of the Universe at times earlier than primordial nucleosynthesis, which is the epoch explored with particle accelerators, most particularly by the LHC. Present-day theories do not allow for a quantitative discussion of the period preceding the Planck time, that is the time before 10^{-43} seconds, as this would require a theory of quantum gravity and such a theory does not exist yet (section 3.6). Around 10^{-43} seconds after the Big Bang elementary objects have energies of the order of the Planck mass, 10^{19} GeV, and might interact through a single unified interaction. The temperature falling with the expansion, this interaction soon breaks up in two, the gravitational force on one hand and the unified strong-electroweak force on the other. Quantum fluctuations should then give rise to first inhomogeneities in the distribution of matter-energy, those at the origin of the temperature fluctuations subsequently imprinted in the cosmic microwave background. These inhomogeneities seed what will eventually become the large structures in the Universe, filaments, clusters, galaxies. In a generally assumed scenario, the grand unified strong and electroweak interactions subsequently break apart

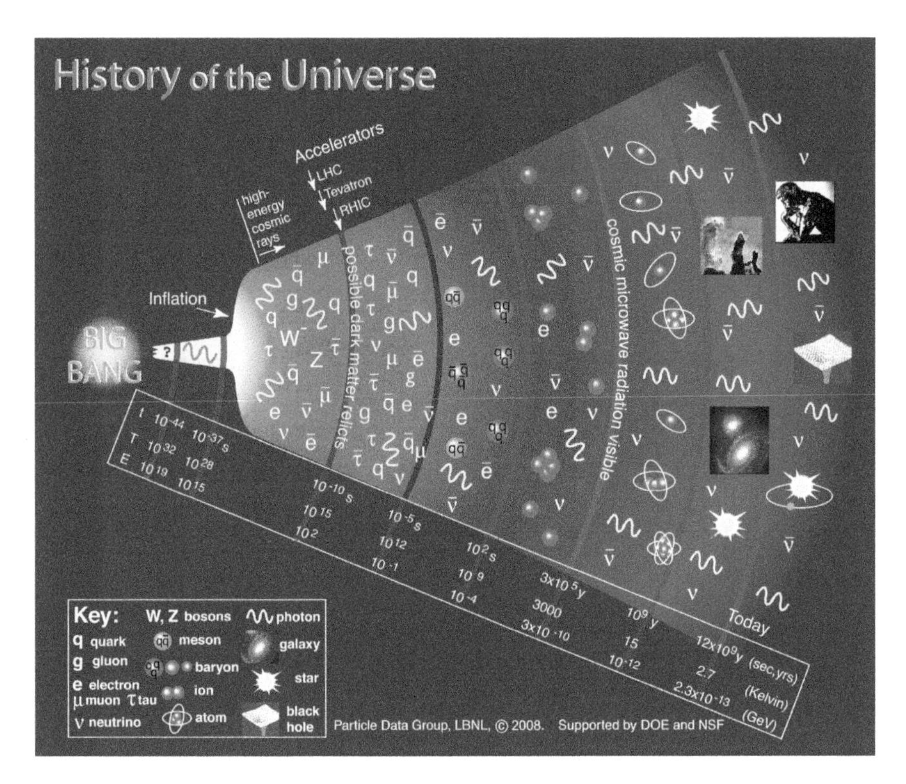

Fig. 5.9. History of the Universe as a function of time, temperature and the average energy of photons. *Image: public domain.*

at around 10^{-35} seconds, after which the Universe enters the inflationary epoch (between about 10^{-33} and 10^{-32} seconds, see figure 5.9).

According to still speculative models, the expansion during this period proceeds at a speed much faster than the speed of light, allowing the Universe to inflate by a very large factor, 10^{26} or likely even more.[22] The idea of cosmological inflation was introduced in the years 1980 by the physicists Alan Guth (USA) and Andreas Linde (USA and Russia)[23] to solve the

---

[22]The dramatic expansion velocity of the Universe is not in contradiction with Einstein's law of relativity. One of the consequences of the exponential expansion during inflation and the continuous and even accelerating expansion since then, is that the present size of the observable Universe, whose geometry seems to be Euclidean for its visible part, could be more than 90 billion light years. In either case, the size of the observable Universe should be much larger than its age (13.8 billion years) multiplied by the speed of light.

[23]Another name that should be associated with the concept of inflation is François Englert, one of the fathers of the Brout–Englert–Higgs mechanism.

horizon and flatness problems of the standard Big Bang cosmology model. The first problem, horizon, reflects the fact that the Universe at very large distance scales looks extremely uniform (as is manifest with the microwave background at same temperature in antipodal directions), despite the fact that these regions could not have been in causal contact in the past. This is the case for distances that are larger than what could be traversed with a speed of light within the age of the Universe. The second problem, flatness, is related to the observation that the matter-energy density observed today agrees so well with a zero-curvature (flat) Euclidean Universe; achieving this would have required an extreme amount of fine-tuning of cosmological parameters at 10^{-35} seconds after the Big Bang, a situation most disliked by physicists. The inflationary Big Bang model provides a natural solution to both problems.[24] Inflation also provides the right magnitude for the emerging fluctuations ($\sim 10^{-5}$) that are found in the CMB and that are believed to be at the origin of all large-scale cosmic structures. It moreover dilutes the magnetic monopoles expected in grand unification schemes consistent with present levels of non-observation. Despite the strong phenomenological arguments in favour of inflation, direct observational evidence is still wanted. Indications were released early 2014 by the BICEP2 telescope near the South Pole from the study of CMB polarisation,[25] but not confirmed in subsequent studies by the Planck Collaboration. An important ground and space born experimental programme is currently underway or planned to tackle this most fundamental among the CMB measurements.

After the period of inflation, at around 10^{-32} seconds (a rather uncertain moment), the Universe enters a better understood epoch in the evolution of matter, to a good extent described by the particle physics we know today, up to theoretical speculations concerning grand unification, neutrino mass generation as well as any form of other new physics. It is likely that

---

[24]The observable Universe is then only a small part of the entire Universe, the part being entirely in causal contact and flattened by the expansion.

[25]The detection of so-called *B-mode* polarisation from primordial gravitational waves (GW) could provide an experimental hint in support of cosmic inflation. B-mode polarisation is weak and very difficult to observe. The primordial B-mode signal must be distinguished from dominant GW and non-GW B-mode polarisation background due to, e.g., CMB gravitational lensing effects (induced by the so-called *E-modes*) and rotating interstellar dust, respectively. The measurement of B-mode polarisation is expressed in form of a power spectrum, where the signal from inflationary gravitational waves is expected to broadly peak at around 120 GHz. This allows the experimenters to separate it from the B-mode background. So far, no experiment had sufficient sensitivity to observe primordial B-modes.

this phase is responsible for the observed matter–antimatter asymmetry of one part in a billion, surviving the big annihilation drama, to which we owe our existence (cf. sections 3.3 and 5.3). In the interval between 10^{-32} and 10^{-13} seconds after the Big Bang the temperature fell by a factor 10^{10} to about 1 TeV. At this time the Universe contains all our present-day Standard Model particles, possibly also additional yet unknown particles like the supersymmetric neutralino $\tilde{\chi}_1^0$ (the lightest stable particle of the MSSM, a possible realisation of a WIMP, see section 4.1). Near the end of this period, between 10^{-13} and 10^{-12} seconds, the WIMP might begin to evolve independently, separately from the rest of matter. Gradually, under the influence of gravity, it would have accumulated in regions of larger density generated previously, and would have given rise to the dark matter accumulations — the halos of today's galaxies and galaxy clusters, within which galaxies evolve with the minority component, the baryonic matter, as foam on an agitated dark matter sea.

When the temperature falls to somewhere between 1000 and 100 GeV the electroweak phase transition takes place, spontaneous symmetry breaking due to a sudden non-zero vacuum expectation value of the Higgs potential, resulting in the separation of electromagnetic from weak interactions, and providing mass to elementary particles. With the temperature falling below 100 GeV, the W and Z bosons, and the Higgs boson become too heavy compared to the available thermal energy in the medium, and are thus not any more regenerated in collisions, they fall out of thermal equilibrium, decay very rapidly in a few 10^{-25} seconds and disappear definitively as on-shell physical particles from the world scene, until again produced[26] at CERN in 1982 and 2012!

When the temperature is around 100 GeV the matter content of the Universe is made of quarks and leptons as we know them today. The top quark is just disappearing, since its mass is too large (173 GeV) to remain in equilibrium and its lifetime too short (about 5×10^{-25} seconds). If the dark matter is made of stable neutralinos or other forms of WIMPs, they are also around as cosmological fossil particles and would even represent the majority form of matter. When the temperature of the Universe falls to few GeV, it is the turn for b and c-quarks to fall out of equilibrium and decay via the weak interactions with lifetimes of about 10^{-12} seconds, as well as the τ-leptons (lifetime of 3×10^{-13} seconds). All that is now gone,

[26]In fact, some are continuously being produced in cosmic ray interactions with the Earth's atmosphere, well before the appearance of humans and their technology.

beside (possibly) the WIMPs, light quarks (u, d, s), electrons, muons (lifetime of 10^{-6} seconds), neutrinos, gluons and photons, where the photons exceed the number of quarks by about nine orders of magnitude! There remains only one major transition in the evolution of matter, occurring at around 10^{-6} seconds, when the temperature of the Universe falls to about 200 MeV. This is the *QCD phase transition* during which all matter subject to strong interactions, which was up to that point in the form of a plasma of quarks and gluons, condenses into hadrons, ultimately into protons (baryons made of uud quarks) and neutrons (ddu), as further described in section 14.2. At a temperature of around 1 MeV it is the turn of stable particles subject to weak interactions only such as the neutrinos to separate from other matter with which there is practically no further interaction (the exothermic reaction $p + e^- \rightarrow n + \nu_e$ requires about 1 MeV activation energy). This background of cosmological neutrinos, frozen-in and evolving independently 1–2 seconds after the Big Bang, is now 13.8 billion years later still around, cooled down to a temperature of about two kelvin (neutrino energies of 10^{-4} eV). Their number density throughout the Universe is presently about 80 neutrinos per cubic centimetre for each neutrino species. The period of primordial nucleosynthesis described earlier in section 5.3 takes place after this, between tens of seconds and minutes after the Big Bang (figure 5.2). Then, about 380 000 years later, follows the formation of first atoms (figure 5.9). The Universe becomes transparent to light and falls into darkness, all electrons being captured in stable atomic orbits. About half a billion to a billion years later, under the influence of gravity acting relentlessly on baryonic and dark matter, first galaxies are formed and first stars begin to shine — *post tenebras lux* — burning with nuclear fusion fires the nuclei inherited from primordial nucleosynthesis and now synthesising in their cores all the heavier nuclei in Mendeleev's table, up to iron (the most tightly bound nucleus). At the end of their lifetime, with all nuclear fuel spent, massive stars return these nuclei through cataclysmic explosions (during which nuclei from iron to uranium are synthesised, mostly through rapid neutron capture) to the intergalactic medium and thus seed galaxies with all the chemical elements that will be needed for the development of life when conditions of appropriate temperature and stability are met.

What can the LHC bring to the understanding of the primordial Universe? Through high-energy collisions of heavy-ions (see chapter 14) the experiments study properties of the quark–gluon plasma and address questions related to whether it is an *ideal fluid* with no viscosity for example, and explore the QCD phase transition occurring about a microsecond after the Big Bang. With proton–proton collisions at 7 up to 14 TeV centre-of-mass energies, the LHC allows to directly explore the *particle content* of the Universe at the epoch when thermal energies were about a TeV per particle, just before the electroweak symmetry-breaking transition occurs. There are many questions one may ask: is there a Higgs boson as predicted by the theory of spontaneous electroweak symmetry breaking, does it have the predicted properties, are there possibly more of them? Has the Brout–Englert–Higgs potential the simplest (*ad hoc*) form of a Mexican hat, or more complicated features? Can dark matter particles be directly produced in proton–proton collisions? Are there additional families of quarks and leptons, are there supersymmetric particles, too, and if so, what supersymmetry scenario was realised by Nature? Are there other gauge bosons such as heavier W', Z' carriers of still unknown interactions? Are there right-handed neutrinos at the TeV mass scale, partners of massive left-handed ones that would shed light on the oscillation phenomena observed in the neutrino sector? Are there new forms of interactions moderating the hierarchy problems of the Standard Model?

It should be emphasised, however, that collisions at the LHC give rise to an extremely localised excitation of the quantum vacuum only, provoking the creation of the Higgs boson for example. Even if these collisions allow to recreate particles existing just before or at the scale of electroweak symmetry breakdown, the *thermodynamical equilibrium* corresponding to a temperature $k_B T$ of about a TeV is not realised. It is thus not (yet) possible to study the phase where the particles have not yet acquired their masses via the Brout–Englert–Higgs mechanism. The particles observed, the W and Z bosons, the top quark, etc., carry their non-zero masses acquired in the broken phase.

Chapter 6

The LHC

January 1983, discovery of the W boson, April 1983, discovery of the Z boson. October 1984, Nobel Prize in physics to Carlo Rubbia and Simon van der Meer. In these two years the preeminence in particle physics passed from the USA to Europe. The US physicists, dominating the particle physics scene since the end of World War II, decided to launch a colossal project, the *Superconducting Super Collider* (SSC). With such a machine they would crush all competition, they would surpass by far amounts the physics discovery potential of the CERN proton–antiproton collider Sp$\bar{p}$S at 0.6 TeV, the physics reach of the 1.8 TeV Tevatron then in a commissioning phase at Fermilab, near Chicago, and of the large electron–positron collider (LEP) in construction at CERN. At the same time in a most manifest and definitive way they would outdistance the USSR and its UNK project[1] in this race for scientific and technological supremacy. The SSC, a proton–proton collider of 40 TeV centre-of-mass collision energy, would be installed in an 87-kilometre circumference tunnel to be located in Waxahachie close to Dallas in Texas. This sets the scene on which the LHC will appear.

6.1. Historical development of acceleration technologies

Before describing the LHC it should be useful to briefly review some of the acceleration technologies developed and employed. Let us begin

[1]A projet for a 21-kilometre circumference Accelerator and Storage Complex (UNK for the Russian acronym), located at Protvino, near Serpukhov, south of Moscow, which aimed at producing 6 TeV centre-of-mass energy proton–proton collisions. Excavation work started in 1983 and the tunnel was completed in 1994. However, with the collapse of the USSR, the subsequent economical crisis, and the ongoing construction of the LHC, it was finally abandoned in 1998.

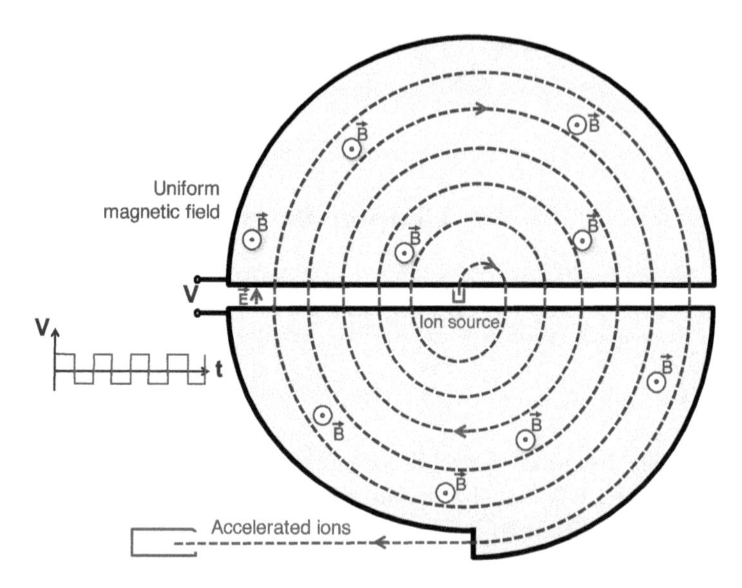

Fig. 6.1. Functioning principle of a cyclotron. The direction of magnetic field is perpendicular to the page. Charged particles are bent by the Lorentz force in perpendicular direction to the magnetic field and their flight direction towards the centre of the cyclotron, counterbalancing the centrifugal force.

with a few words about direct ancestors of the LHC, the circular proton accelerators.[2]

6.1.1. *Cyclotrons and synchrocyclotrons, early ancestors of the LHC*

In a *cyclotron*, electromagnetically charged particles (protons, deuterons, α particles, etc.) are first produced by an ion source at the centre of a gap between magnetic poles of a large static dipole magnet with a constant (time-independent) magnetic field. Particles are then accelerated in a hollow metallic vacuum chamber shaped in form of two D's located between the two polar pieces of an electromagnet (figure 6.1). Inside the hollow electrodes, particles are subject only to the perpendicular magnetic dipole field $\vec{B}$ bending their trajectories onto circular paths.

Let us consider a particle of mass m with charge e and velocity $\vec{v}$ in a plane perpendicular to a uniform magnetic field $\vec{B}$. Under the influence

[2]Historically, in parallel to the development of proton accelerators, there are the developments of electron (positron) accelerators, betatrons, synchrotrons, linear accelerators, e^+e^- colliders, but describing these equally fascinating projects is beyond the scope of this book.

of the Lorentz force $\vec{F}_L = e\vec{v} \times \vec{B}$, the particle follows a circular path of radius R. The force $\vec{F}_L$ counterbalances the centrifugal force $\vec{F}_C = -m\frac{v^2}{R^2}\vec{R}$ ($\vec{R}$ is the radius vector giving the position of the particle relative to the centre of curvature). The modulus of particle momentum p is thus

$$p = mv = e \cdot B \cdot R,$$

where B is the magnitude of the field $\vec{B}$. The particle is accelerated when traversing the gap between the two D's by the electric force $\vec{F}_E = e\vec{E}$, where $\vec{E}$ is the electric field between the D's at the moment when the particle is crossing the gap. The period T of one revolution:

$$T = \frac{2\pi R}{v} = \frac{2\pi m}{eB},$$

is constant, as long as we remain in a non-relativistic kinematic regime, i.e. the mass m is unchanged by relativistic effects. The electric field in the gap changes direction at each passage of the particle so as to be always accelerating. The oscillation frequency f of the electric field is fixed so as to coincide with the revolution frequency of the particle: $\omega = 2\pi/T = 2\pi f$. As the particle is acquiring a higher and higher velocity v, it spirals on a path of increasing radius R:

$$R = \frac{mv}{eB},$$

until it reaches the edge of the pole piece, whose radius can be up to one or two metres.

Ernest Lawrence invented the cyclotron in the years 1930 at Berkeley, USA (figure 6.2). The first machine designed in 1929 and tested in 1931 had a diameter of thirteen centimetres, fitting into the palm of a hand. The protons were accelerated up to 80 keV. Within few years Lawrence constructed a machine with 28-centimetre diameter

Fig. 6.2. Ernest Orlando Lawrence received the Nobel Prize in physics in 1939 for the invention of the cyclotron which he initially dubbed *proton merry-go-round*. In his hand his first device. A rather modest ancestor for the 27-kilometre circumference LHC. *Images: public domain.*

providing an energy exceeding 1 MeV. The first cyclotrons were used to produce radioactive isotopes and in 1937 the first artificial element was produced, *technetium*, thanks to a machine with a diameter of almost one metre.

Cyclotrons made it possible to accelerate protons up to energies of 25–30 MeV. The limitation in energy was due to the fact that, whilst the magnetic field $\vec{B}$ and the accelerating frequency f were fixed, the accelerated particle of *rest mass*[3] m_0 has its *relativistic mass* m gradually increasing with velocity v according to the relation $m = m_0/\sqrt{1-(v/c)^2}$. Taking into account this increment of mass with velocity, the circulation period T is not constant any more, the particle gradually falls out of phase with fixed-frequency oscillations of the accelerating field between the two Ds, thus limiting the maximal energy attainable in a cyclotron.

This limitation was removed by synchronising the accelerating frequency f with the circulation period T; this is the functional principle of *synchrocyclotrons*. In this manner, it was possible to increase the proton energy up to about 600 MeV, as reached by CERN's very first accelerator that came in operation in 1957. The limitation now became the size of the pole pieces of the dipoles that reached several metres and weighed thousands of tons.

6.1.2. *Synchrotrons*

The next chapter in the development of accelerators was the introduction of *synchrotrons*. Instead of the gigantic pole pieces required by cyclotrons and synchrocyclotrons, particles are now accelerated in a circular vacuum chamber of fixed radius located inside a ring of dipole magnets (figure 6.3). The radius of the device can now reach tens if not hundreds of metres without the cost of iron exploding.

In a synchrotron the magnetic field B varies during the acceleration cycle. At injection of beam particles, when their energy is low, the magnetic field B is also low and gradually increases as the energy of circulating particles increases under the influence of accelerating electric fields. The magnetic field must be present over most of the trajectory so that the beam particles follow the path of the circular vacuum chamber. The oscillating

[3]The rest mass m_0 is an intrinsic property of a particle, independent of its state of motion. The particle can be elementary or composite like a proton or a nucleus. The basic components of the Standard Model, quarks, leptons, gauge bosons, acquire their rest mass due to coupling to the Higgs field, while the majority of the mass of composite particles stems from their binding energy. The relativistic mass m, which depends on a particle's velocity, is also the coefficient appearing in the famous equation $E = mc^2$.

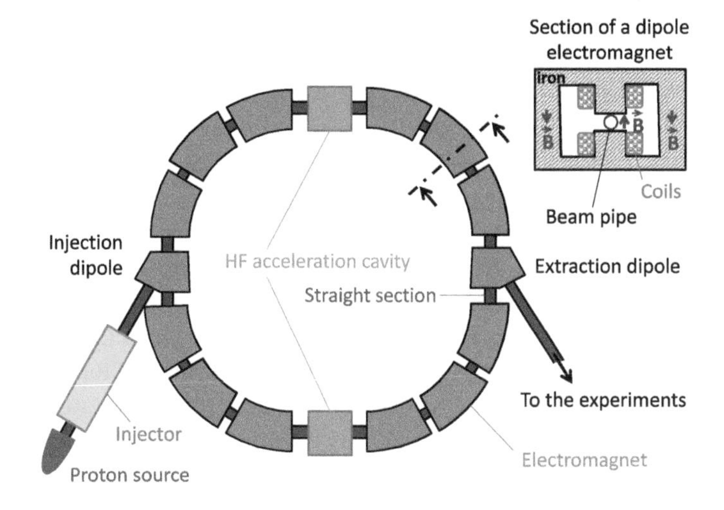

Fig. 6.3. Scheme of a synchrotron.

Fig. 6.4. Left: control room of the Cosmotron, the first large synchrotron whose construction at the Brookhaven National Laboratory (BNL, Long Island, USA) started in 1948; it operated until 1966. Right: Saturne, the synchrotron built at CEA (Saclay, France), started operation in 1958. Several times upgraded and modified, it was in operation up to 1997. *Images: BNL and CEA.*

electric field is localised at one or few places along the circumference of the machine. It is generated by *radio-frequency cavities*, whose frequency also varies during the acceleration cycle to be in phase with the increasing particle velocity. The particles circulate, not as a continuous beam, but grouped into bunches following closely the same trajectory. We nonetheless talk of particle beams when referring to these bunches.

The first large *proton synchrotron*, the Cosmotron at the Brookhaven National Laboratory (BNL, figure 6.4) was commissioned in 1953. It could accelerate protons up to the energy of 3300 MeV, i.e. 3.3 GeV, and was used to bombard proton and nuclear targets allowing the discoveries of the Σ^0 and $\Sigma^\pm$ hyperons (see inset 2.2 on page 75). In 1956 began the operation of another large synchrotron, the Bevatron at Berkeley, designed to produce antiprotons. The antiproton ($\bar{p}$) was not yet observed, but theory was suggesting[4] that it could be produced through the reaction $pp \to pp + p\bar{p}$. The idea was therefore to produce a beam of protons and hit a proton target with it. A simple relativistic calculation (see inset 6.1 on antiproton production) shows that the kinetic energy of incident protons must be at least 6 GeV. This energy was reached at the Bevatron allowing the first observation of antiprotons in 1956 and of antineutrons the following year.

Large synchrotrons were also built in Europe, for example, Saturne (figure 6.4) in Saclay near Paris started operation in 1958 accelerating protons to 3 GeV, Nimrod in the UK reached 7 GeV in 1964 and the 10 GeV Synchrophasotron in Dubna (USSR) commissioned in the 1960s (its operation ceased only in 2003). Despite these successes the development of synchrotrons, was plagued with a technical problem, the accelerated particles had a tendency to oscillate laterally relative to the ideal trajectory. In this first generation of synchrotrons these transverse deviations were sizeable, requiring large vacuum chambers (apertures of up to one-metre diameter for the Synchrophasotron!) and thus huge air-gaps between the polar pieces of the dipole magnets. The cost of magnets, the return yokes, became prohibitive limiting the size of the rings, the value of magnetic fields and the achievable energies.

It was the time when the *strong focusing principle* was invented independently by the Greek physicist Nicholas Christophilos in 1949 and, slightly later in 1953, by Ernest Courant and co-authors at BNL, USA, as well as Vladimir Veksler in the USSR. The key idea was to separate the two functions of the magnetic field: bending the trajectory of accelerated particles so as to keep them on an approximately circular paths, and focusing the particles, i.e. keeping them on trajectories close to the ideal one. In first-generation synchrotrons, this was obtained (but not very efficiently) by shaping the polar pieces so as to produce a restoring force towards the ideal central trajectory. The strong focusing idea consisted in having two types of magnets: to the bending dipoles were added quadrupole-type

[4]The antiproton is a fermion and thus necessarily produced in particle–antiparticle pairs (see inset 2.2 on page 75).

Fig. 6.5. Left: face-on view of a quadrupole magnet. Right: combination of two sextupole magnets with a dipole in between. Multipolar magnets, sextupoles, octopoles, etc., are used in accelerators to correct chromatic aberrations. *Images: public domain.*

magnets[5] playing the role of magnetic lenses. The synchrotrons were constructed as a sequence of *bending dipole* and *focusing quadrupole* magnets (figure 6.5). This configuration reduced significantly the transverse beam oscillations allowing smaller diameter vacuum chambers of the order of ten centimetres and thus a large reduction in the size, weight and cost of the magnets, whilst permitting much larger acceleration energies.

The first two machines of this novel type were the Proton Synchrotron (PS) at CERN commissioned in 1959 with a diameter of two hundred metres and an energy of 25 GeV (figure 6.6), and the *Alternating Gradient Synchrotron* (AGS) at BNL in 1960 of

Fig. 6.6. Aerial view of the 25 GeV Protron Synchrotron at CERN in 1965. *Image: CERN.*

[5]The simplest form of a quadrupole magnet is a set of four magnetic bars configured symmetrically around a centre such that their north and south poles alternate.

about same size and operating at 32 GeV.[6] In 1967 a proton synchrotron of 70 GeV started operation in Protvino in the USSR, and for a few years it was the most powerful machine in the world. In 1972 Fermilab put into operation a still larger proton synchrotron of 6.3 km circumference and 200 GeV energy (subsequently raised up to 500 GeV in 1976). CERN countered with a Super Proton Synchrotron (SPS) of 450 GeV with a circumference of nearly seven kilometres. In the 1980s Fermilab installed superconducting magnets in the synchrotron which allowed to increase the proton energy to almost 1000 GeV, thus justifying its name Tevatron.

Inset 6.1: To produce antiprotons, start with protons

Let us consider the reaction $pp \to pp + p\bar{p}$. Energy–momentum conservation in the formalism of special relativity kinematics requires that the sum of four-vectors of initial state particles is equal to the four-vector sum of the three protons and antiproton in the final state. Expressed in the laboratory frame where the target proton is at rest, the initial four-vector sum is:

$$(E_{\text{inc}}, \mathbf{p}_{\text{inc}}) + (m_p, \mathbf{0}),$$

where E_{inc}, $\mathbf{p}_{\text{inc}}$ are the energy and momentum of the incident proton. At reaction threshold all final state particles are at rest in the centre-of-mass of the reaction and in this reference frame the final-state four-vector sum is

$$4(m_p, \mathbf{0}),$$

[6]Several more specialised synchrotrons were built over the years around the world, some of them already dismounted by now. Among these were a 4 GeV machine in Gatchina, Leningrad, the 10 GeV Zero Gradient Synchrotron (ZGS) at the Argonne National Laboratory (near Chicago, USA), the 4 GeV Princeton-Pennsylvania Accelerator (PPA) rapid-cyclying synchrotron. More recently two very successful machines were built: a 1.6 km circumference 50 GeV proton synchrotron, construction launched in 2001 and operational since 2009, at the Japan Proton Accelerator Research Complex (J-PARC); and a 6.3 km circumference (up to) 920 GeV proton synchrotron operating from 1992 until 2007 at the DESY laboratory in Hamburg, Germany (the HERA project), largely devoted to nucleon structure studies through electron–proton deep inelastic collisions.

where we use that proton and antiproton have the same mass. The equality of the moduli of these two four-vectors, expressing energy–momentum conservation, which can be applied in any reference frame, gives

$$2\,m_p^2 + 2\,m_p E_{\rm inc} = 16\,m_p^2,$$

and that is

$$E_{\rm inc} = 7\,m_p.$$

The kinetic energy of the incident proton at threshold for antiproton production must thus be $E_{\rm inc} - m_p = 6\,m_p$, which is about 6 GeV.

6.2. Hadron colliders

6.2.1. *The first proton–proton collider: the ISR*

In the years 1960 CERN undertook the construction of the first proton–proton collider called the *Intersecting Storage Rings* (ISR). It started operation in 1971 and consisted of two distinct accelerator-storage rings, with two vacuum beam pipes, two systems of bending dipoles and focusing quadrupole magnets. Two beams of protons circulated in opposite directions and crossed at specific intersection points (figure 6.7). The ISR were the most direct precursor of the LHC.

Protons were accelerated in the PS, injected in the ISR where they were circulating for hours colliding at well-defined intersection points where detectors were located. The centre-of-mass energy of the colliding protons could vary from about 20 to 62 GeV. The electric current of protons amounted to several amperes (A), with a maximum of 20 A, providing high luminosity (see inset 6.2). This machine, unique of its type, was a real *tour de force*, both technically and technologically, demonstrating the mastery achieved by CERN accelerator designers and engineers. It also served as a test bench of the following machine, the proton–antiproton collider at CERN. The ISR operated as a research device until 1984. It held the luminosity record for hadron colliders until 2004. Among the primary physics results, experiments at the ISR confirmed the quark structure of protons uncovered a few years earlier at SLAC (section 2.1). The lack of appropriate transverse coverage of the first detectors in operation at the ISR intersection points prevented CERN from discovering the charm and bot-

Fig. 6.7. View on one of intersection points in the ISR at CERN. The picture shows the two distinct sets of magnets corresponding to two counter-rotating proton beam lines. *Image: CERN.*

tom quarks via the observation of the heavy J/ψ and Υ resonances, or hadronic jets, although these were abundantly produced in the ISR proton collisions.[7]

[7] In the early 1970s, the ISR committee and CERN management were still guided by the dominant paradigm of soft hadron physics with collision products emitted preferentially at small angle with respect to the colliding beam directions. Not enough attention was paid to the new hard scattering physics pioneered by the SLAC experiment with the study of high-energy electron–proton scattering (see section 2.1), which predicted quark or quark–antiquark pair production at large angle in hadron collisions. The few visionary proposals of detectors with a 2π azimuthal coverage at large angle were rejected or postponed in favour of experiments with a small solid angle coverage, either around the beams or at larger angle, which could not measure leptons pairs that would have been the signature of heavy charmonium and bottomium resonance decays. The lack of experience with collider physics and focus on fixed target experiments at CERN have played a role in these unfortunate decisions.

6.2.2. CERN's proton–antiproton collider $Sp\bar{p}S$

As discussed in section 2.3, a central challenge in experimental particle physics in the 1970s was to observe the W and Z bosons whose masses were expected to be in the 80–100 GeV range. The usual experimental procedure in those years consisted in bombarding with an accelerator beam an external fixed target. Despite beam energies that were reaching about 450 GeV, this collision mode did not allow to exceed proton–proton centre-of-mass energies of 30 GeV.[8] Even with the innovation introduced by the ISR, where the centre-of-mass frame was equal to the laboratory frame of the colliding protons, and where proton–proton collision energies up to 60 GeV were attained, W and Z bosons remained beyond reach.

In 1976, David Cline, Peter McIntyre and Carlo Rubbia, at the time all three working in the US, proposed to convert one of the existing large proton synchrotrons, the Tevatron at Fermilab or the SPS at CERN, into a proton–antiproton collider. The idea was to have proton and antiproton beams circulating in the same ring and same beam pipe, but in opposite directions. As these two particles are antiparticles of each other, the electromagnetic forces acting on them via the magnets and accelerating structures will be of exactly the same magnitude but opposite direction in the accelerator-collider. The two beams brought to collide head-on would provide energies in the proton–antiproton centre-of-mass of several hundreds of GeV. The principle difficulty of this set-up was, of course, the production of an adequate number of antiprotons.

Only a fraction of the proton collision energy, about 1/6 on average, is really available in collisions at the level of fundamental constituents of protons and antiprotons, i.e. at the quark–antiquark level. It was already known in the 1970s that about half of the momentum of a proton is carried by gluons (and these do not participate directly in the creation of W and Z which do not carry colour charge), the remaining half being mostly shared by the three valence quarks in the proton or valence antiquarks in the antiproton. With two beams of about 300 GeV circulating in opposite direction this would give in the quark–antiquark centre-of-mass collision energies of 50–200 GeV, sufficient to produce the longed-for weak bosons.

[8] For an incident energy $E_{\rm inc}$ much larger than the proton's rest mass, the energy in the proton–proton centre-of-mass, usually denoted $\sqrt{s}$, in a collision with a static proton target, is equal to the modulus of the sum of energy–momentum four-vectors: $\sqrt{s} = |(E, \mathbf{p}_{\rm inc}) + (m_p, \mathbf{0})| = \sqrt{2m_pE}$. For $E = 450$ GeV one finds for the centre-of-mass energy $\sqrt{s} = 30$ GeV.

The central difficulty in the proton–antiproton collider project was how to obtain a sufficiently intense antiproton beam to yield the luminosity needed for the production and observation of W and Z bosons. The expected production cross-section, multiplied by the decay probability of these bosons into lepton pairs — the decay mode required to observe them in the midst of expected hadron backgrounds is rather small, 0.5 nb for the W and ten times smaller still for the Z. The antiprotons could be obtained by bombarding a fixed target, for example with the 26 GeV beam from the PS at CERN, at sufficient production rate of about one antiproton per million proton–target collisions. However these antiprotons were produced at the target with a broad range of energies and production angles and thus could not be directly captured, collected and accelerated to obtain a beam of required size and intensity.[9] These antiprotons are said to be *hot.* The problem was solved thanks to the method of *stochastic cooling* proposed originally for the purpose of damping beam oscillations in the ISR by Simon van der Meer at CERN in 1972 (see inset 6.3).

Confident in the quality of its accelerators, CERN was first to engage into the endeavour of converting its recently commissioned SPS into a proton–antiproton collider. The two beams were accelerated in the SPS, which became known as the Sp$\bar{\text{p}}$S, up to about 300 GeV and then brought into collision at the centre of two large detectors, UA1 and UA2. In the years from 1981 until 1983 the collisions were at centre-of-mass energies of 540 GeV, reaching, after addition of further magnet cooling, 630 GeV until 1990. The discovery of the W and Z bosons described in section 2.3 was a triumph for the scientists involved in the project and for CERN.

Inset 6.2: Luminosity of a collider

Luminosity is a measure of the collision rate and is, together with the collision energy, the single most important quantity to be optimised in a particle collider. It is essentially determined by two factors: the number of protons circulating in the two beams (the beam currents) and the focusing at the collision point, which means the

[9]The phase space volume occupied by the produced antiprotons was about a hundred million times above that required to fit into the SPS.

concentration of particles in the plane perpendicular to their motion (transverse plane) at the point where they are colliding head-on. The instantaneous luminosity of a collider can be expressed as follows

$$L = \frac{f n_b N_{p,1} N_{p,2}}{4\pi \sigma_x \sigma_y},$$

where $N_{p,1/2}$ are the numbers of protons per bunch in beams 1 and 2, n_b is the number of colliding bunches at each revolution, f the revolution frequency of bunches in the ring (the ratio of light speed to ring circumference), and $\sigma_{x/y}$ are the dimensions of the beams at the crossing point in the two directions x, y in the plane transverse to the beam direction. These quantities are determined by the beam focusing strength at the interaction point, characterised by a focusing length denoted β^*.

After beam injection the luminosity decreases exponentially with time through gradual loss of beam particles participating in the collisions ("burn off") and particles that leave stable beam orbits. A characteristic quantity is the *luminosity lifetime*, defined as the time after which the instantaneous luminosity has decreased by $1/e = 0.37$, which can be as short as an hour or as long as a day. Another important quantity is the *integrated luminosity*, L_{int}, which is the integral of the instantaneous luminosity over the duration of data taking, $L_{\text{int}} = \int L\,dt$. Knowing the production cross-section σ_i of a specific process i, the number of events of that process is given by

$$N_i = \sigma_i \cdot L_{\text{int}}.$$

As the dimension of a cross-section is a surface, the instantaneous luminosity is expressed in units of $\text{cm}^{-2}\,\text{s}^{-1}$. The unit used for the integrated luminosity is barn^{-1}, where $1\ \text{barn} = 10^{-24}\ \text{cm}^2$. The following sub-multiples are particularly useful for the LHC: $1\ \text{pb}^{-1} = (10^{-12}\,\text{barn})^{-1} = 10^{36}\,\text{cm}^{-2}$ and $1\,\text{fb}^{-1} = (10^{-15}\,\text{barn})^{-1} = 10^{39}\,\text{cm}^{-2}$. A collider operating at an average instantaneous luminosity of $10^{34}\ \text{cm}^{-2}\,\text{s}^{-1} = 0,01\ \text{pb}^{-1}\,\text{s}^{-1}$ during 24 hours yields therefore an integrated luminosity of $860\ \text{pb}^{-1}$ or $0.86\ \text{fb}^{-1}$.

Inset 6.3: Stochastic cooling

Stochastic cooling at the SPS consisted in capturing the maximum number of antiprotons produced at around 3 GeV average momentum from the 26 GeV PS beam hitting a stationary target and circulating them in a special ring called *antiproton accumulator*, where corrective electrical impulses were applied to reduce their initial dispersion in momentum and direction.

The trick was to measure the average deviation of the barycentre of the antiproton bunch at one place in the ring and, while this bunch was travelling a half circle through the ring, send on the opposite side of the ring an electrical signal to correct for the measured deviation, just in time for the passage of this same bunch in front of the electrodes. Repeating this procedure for hours it was possible to cool the antiproton beam, that is, to gradually reduce the motion and momentum of the beam particles in transverse directions. This way one obtained a beam of *cold* antiprotons of sufficient density and adequate dispersion, both longitudinally and transversely, to be re-injected from the accumulator ring into the PS.

The time required to accumulate a sufficient number of antiprotons was about half a day. During this time about a billion antiprotons were collected at 3 GeV, cooled and stacked, ready for injection onto the PS where they were accelerated up to 26 GeV, and then finally injected into the Sp$\bar{\text{p}}$S and accelerated up to about 300 GeV, with protons circulating in the opposite direction. The luminosity lifetime was of several hours, up to half a day, with the beams crossing and colliding at the UA1 and UA2 detector positions. The SPS accelerator has become the proton–antiproton collider Sp$\bar{\text{p}}$S.

6.2.3. *The Tevatron collider at Fermilab*

At about the same time when van der Meer proposed his method of stochastic cooling, Gersh Itskovich Budker, eminent physicist at Novosibirsk (USSR) at the origin of a series of electron–positron colliders (VEPP) still continuing today, proposed another method of antiproton cooling. This method, called *electron cooling*, consisted in producing an intense beam of

Fig. 6.8. Aerial view of Fermilab's proton–antiproton accelerator and collider complex towards the closure of its operation in 2011. The Main Injector and Recycler ring is seen in front and the 6.3-kilometre circumference Tevatron collider ring in the background. On the left the iconic Wilson Hall, Fermilab's central laboratory building. Close-by are the proton source and the Booster. Tevatron's main detectors CDF and DØ are placed to the left and right along the ring, respectively. *Image: Fermilab.*

electrons surrounding, and co-moving with the hot antiprotons with same velocity as the antiproton beam. The component of random motion of the antiprotons was thus gradually transferred to electrons thereby cooling the antiprotons. It took a long time for this method to be mastered and it is only at the end of the 1990s that it could be implemented for the second phase of operation of the proton–antiproton collider at Fermilab.[10]

The Tevatron collider (figure 6.8) started physics operation with experiments in 1987. Thanks to the use of superconducting magnets providing 4.4 tesla bending dipole fields, the centre-of-mass collision energy rose to 1.8 TeV, outperforming the CERN collider whose operation was therefore stopped in 1990. The Tevatron collider energy was subsequently increased up to 1.96 TeV for the years from 2001 till 2011. It has been in operation for

[10]The first level of cooling at the Tevatron was still obtained through stochastic cooling, while electron cooling was applied at a second level at higher acceleration energy.

almost 20 years with constantly improving luminosity. Two large detectors, CDF and DØ operated at the Tevatron (CDF since 1987 and DØ joining in 1992). Among their many physics achievements, two results stand out until today: the discovery of the top quark in 1995, and later in 2006, during the second phase of the Tevatron operation after massive detector and accelerator upgrades, the first observation and measurement of the rapid oscillation between the neutral B_s^0 and $\overline{B}_s^0$ mesons containing each a b quark and an s quark (see section 14.1). The final years of operation were characterised by intense searches for the Higgs boson, whose discovery was, however, only accomplished at the LHC. The Tevatron collider stopped operating in September 2011, in part due to budgetary difficulties but also because of the excellent performance of the LHC in 2010 with which the Tevatron collider could not compete.

6.3. A new hadron collider

The idea of installing a hadron accelerator and collider in the LEP tunnel was put forward since the beginning of the LEP project in the late 1970s. Visionary physicists in a working group initiated by the European Committee for Future Accelerators (ECFA) in 1978–1979, chaired by the Italian physicist Antonino Zichichi,[11] and Herwig Schopper, CERN Director General between 1981 and 1988, had pushed for the largest achievable tunnel of sufficient diameter when LEP was approved in 1981. In April 1983, two accelerator experts, Stephen Myers and Wolfgang Schnell, published a note (LEP Note 440) describing the concept. The following year in March 1984 Giorgio Brianti and Maurice Jacob organised a workshop with the title "Large Hadron Collider in the LEP Tunnel" in Lausanne, where physicists and machine experts discussed the principles of a proton and heavy-ion collider (see figure 6.9). Many consider this event the unofficial launching of the LHC. A Long Range Planning Group, chaired by Carlo Rubbia, was established in 1985 by CERN Council. The group's report recommended a 13–15 TeV proton–proton collider with 8–10 tesla bending magnets and a luminosity of 10^{33} cm^{-2} s^{-1} as next option for CERN, with first collisions possible in 1995, if the decision to proceed would be taken in 1989! A 2 TeV electron–positron collider was considered to have comparable reach, but not

[11]The ECFA-LEP working group report mentioned a "tunnel with 27-kilometre circumference and a diameter of five metres, with a view to the replacement of LEP at the end of its activities by a proton–proton collider using cryogenic magnets".

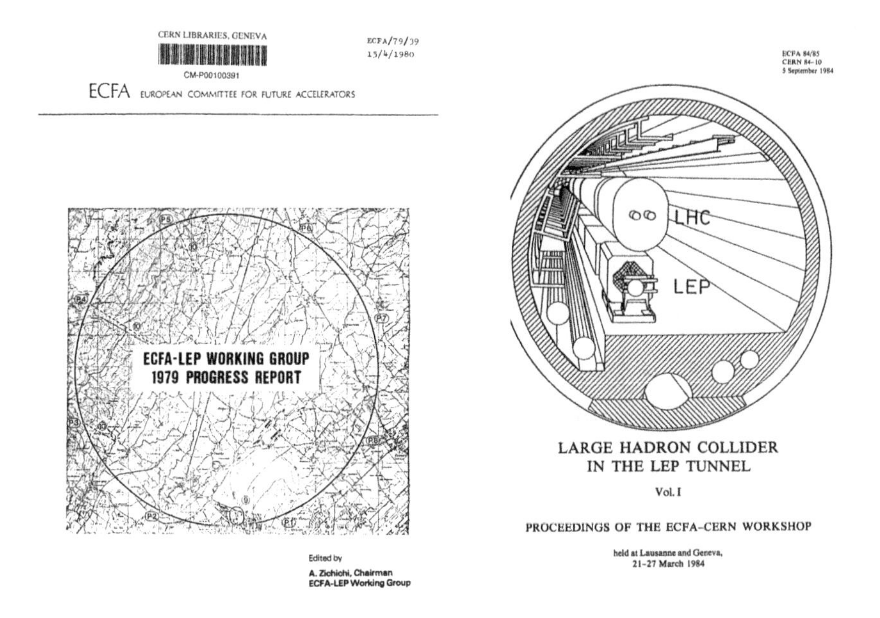

Fig. 6.9. Front pages of the 1979 ECFA-LEP working group report (left), where the option of a large tunnel to host a cryogenic hadron collider was considered, and of the 1984 Lausanne workshop proceedings (right). The image shows one proposal from the workshop (that was later dismissed), where the LHC was added on top of the existing LEP machine.

technically feasible. The report further explained that lesser energy of LHC relative to the SSC could be partially recovered with the help of a larger luminosity.

6.3.1. *LHC versus SSC*

In the meantime the USA had proposed a competing project, the SSC. Thanks to its gigantic tunnel of 87 kilometres in circumference this machine could fairly straightforwardly reach a centre-of-mass collision energy of 40 TeV, dwarfing what could be envisioned for the LHC with its 27-kilometre tunnel.[12] When the SSC was officially approved and launched in 1987, the idea of an LHC was seriously questioned. With lower energy and much riskier technological choices, its final approval and construction was decided through a long and difficult procedure.

[12]The exact circumference of the LEP/LHC tunnel amounts to 26.659 kilometres.

With its much longer tunnel, the SSC could be built with more conventional accelerator techniques. The bending magnets had to be superconducting to provide a field of six to seven tesla (6.7 tesla in its final configuration), substantially lower than the 8.3 tesla dipole field required to achieve 14 TeV collision energy at the LHC. Likewise, the SSC was conceived to operate with two beam tubes (one on top of the other) in a tunnel of similar inner diameter (3.7 metres) as the LEP tunnel, two independent magnetic rings (as in the ISR), whilst due to the limited cross-section of the LEP/LHC tunnel, only a single set of magnets could be accommodated there.

The SSC was supposed to begin operation at the beginning of the years 2000. The cost of the project, as estimated at its launch in 1987, was about five billion US dollars. By 1991 it had already escalated to about eight billion and in 1993, when the project was cancelled by the Clinton administration complying with a decision by the US Congress, it had reached an estimated cost of at least eleven billion US dollars. By this time about 24-kilometre tunnel had been excavated including 17 access shafts and several large buildings erected on site. A total of two billion dollars had been spent on civil engineering activities, development of superconducting magnets for the machine, and SSC detector developments.

The apparent lack of control over the costs has surely played a major role in the cancellation of the SSC project, but this was not the only reason. The international context had changed much within few years: with the end of the cold war, the desire of the USA for asserting its technological superiority over the USSR, which was very beneficial to such *big science* projects in the past, was not so prevalent anymore. Another factor was the strong opposition of eminent US physicists from domains other than particle physics to such gigantic projects depriving them from funds, in particular as very similar physics could be done in Europe at a much lower cost to the US taxpayer. Finally, there was competition for funding with the development of the International Space Station, which included the Johnson Space Center and other NASA undertakings in Texas. The big advantage of the SSC over its European competitor, the fact that it would be much easier to increase its energy and luminosity in a later upgrade, was not enough to help the US project prevail.

6.3.2. *1989: launch of the LHC project*

If the LHC project had survived during the SSC years, according to CERN accelerator physicist Lyn Evans, who served as LHC project manager during eighteen years, it is thanks to the tenacity and resolve of Carlo Rubbia. In September 1989, recently nominated Director General of CERN, Carlo Rubbia initiated a detailed study of the physics potential of a proton–proton accelerator-collider to be located in the LEP tunnel operating at a 17 TeV centre-of-mass energy with an instantaneous luminosity of 10^{34} cm^{-2} s^{-1}. The W and Z bosons had been discovered a few years earlier, the next goal of high-energy particle physics was clearly the search for the Higgs boson or whatever mechanism might be responsible for the W, Z and fermion mass generation. On the program was, however, much more than the Higgs boson search, namely the investigation of physics beyond the Standard Model. Prominent examples among these were supersymmetry or new heavy gauge W', Z' bosons indicating a new force at the TeV scale. The key idea for LHC was to install bending magnets with a *two-in-one* magnetic scheme in the LEP tunnel as suggested by Robert Palmer from BNL already at the 1984 Lausanne meeting, with maximum achievable field at the time. It was hoped to reach fields close to ten tesla, which would allow accelerating protons up to 8.5 TeV per beam. Most importantly, however, the LHC needed to compensate its two-to-three times inferior energy compared to SSC by a ten times larger luminosity.

The LHC was also required to be capable of accelerating heavy ions, in particular lead nuclei. The goal was to have lead–lead collisions at a nucleon–nucleon centre-of-mass energy of about 5 TeV, which is about a factor of 25 beyond that provided by the 3.8-kilometre circumference Relativistic Heavy-Ion Collider (RHIC) then being planned in the USA at BNL.[13] The initial LHC project also considered the possibility of having collisions between the LHC proton beams at about 8 TeV and an electron or positron beam with 50–100 GeV from LEP, supposed to be kept in the same tunnel. The only electron–proton collider then in construction at the DESY laboratory in Hamburg (Germany), called HERA, was planned to operate with 800 GeV protons colliding with 30 GeV electrons or positrons. That machine has operated successfully from 1990 until 2007 providing detailed studies of the proton structure, QCD and also performing electroweak physics measurements.

[13]RHIC was conceived in 1983, R&D began in 1986, its construction in 1991, and physics operation started in 2000. A number of different nuclei, such as gold, copper, as well as polarised protons were brought into collision. Among the main physics achievements of the four experiments using RHIC were detailed measurements of quark–gluon plasma properties.

6.3.3. *Study of the LHC's physics potential*

The evaluation of the discovery and measurement potential of the LHC in all modes of operation originally considered, proton–proton, lead–lead, and electron–proton, occurred in several workshops following the 1984 Lausanne meeting, in particular at La Thuile, France in 1987 and Aachen, Germany in 1990. The preparations of the Aachen workshop, which was organised by ECFA, took about one year and involved around 250 physicists from all over the world. The main challenges in proton–proton physics were to identify the most promising production and decay modes for a Higgs boson discovery, and the most appropriate channels and methods in the search for supersymmetric particles or new heavy gauge bosons. To quantitatively estimate the LHC discovery potential it was also necessary to evaluate the level of backgrounds, which are processes either leading to the same final state particles as for the desired search channels, or of sufficiently similar final states that they could confound and mislead the interpretation of the selected events. The Aachen workshop has been considered by many as a founding moment of the LHC, as all the aspects of the project were presented and discussed: feasibility of the machine, conceivable searches and expected discoveries, and ideas concerning possible detectors. From this moment, the work on all these issues proceeded at full strength and uninterrupted.

One of the main results of these studies was the detailed evaluation of the search and discovery potential for the Higgs boson, whatever its mass could be, and in those years the possible mass range extended from about 100 to about 1000 GeV. It became rapidly evident that several decay modes were particularly suitable: decays into two photons ($H \to \gamma\gamma$), into two Z bosons ($H \to ZZ$) and into two W bosons ($H \to WW$). The first channel just required the detection and precise measurement of two energetic photons. When the Higgs decays to two Z bosons, each of these can decay into two leptons, e^+e^-, $\mu^+\mu^-$ or $\tau^+\tau^-$ (each with a probability of 3.4% among all possible Z decays), or into a pair of quarks. While it was not easy to detect and precisely measure a pair of τ-leptons which further decay (not to even talk about quarks, which form jets of hadrons in the detector), muons and electrons represented a very promising signature. One of the Z bosons can also decay into a pair of neutrinos, $\nu\bar{\nu}$ (20% decay probability), which escape the detector unseen. With a hermetic detector it is nonetheless possible to detect the missing energy in the final state thus signalling the presence of neutrinos. This is also the way to look for $H \to WW$ decays when

W bosons decay each into a lepton plus a neutrino. Similar to the Z boson, the W boson also decays in about 2/3 of all cases to a quark–antiquark pair, whose hadronic jets lead to much larger confusion with background processes than the leptonic W decays.

These most promising decay modes of the (still hypothetical) Higgs boson have to a very large extent influenced the design of the two large experiments ATLAS and CMS (see chapter 8). The electromagnetic calorimeters had to satisfy challenging constraints on precision, linearity in response, and granularity required in the face of large expected backgrounds. The internal spectrometers for the detection of charged particles, and external spectrometers for muon detection, as well as magnetic systems needed to measure charged particle momenta through curvature, were largely determined by requirements for the channels $H \to ZZ \to 4\mu$ or $H \to ZZ \to 2e2\mu$ (as well as the capability of reconstructing very high-momentum (TeV-scale) electrons and muons from decays of putative heavy Z' and W' bosons, see discussion further below). As it turned out to be beneficial a few years earlier to the UA1 experiment, great care was also taken to design the detectors as hermetic as possible to detect missing energy (see inset 7.5) carried away by otherwise undetectable particles such as neutrinos, possible neutralinos or other elusive candidates for dark matter.

Beyond these aspects related to searches for the Higgs boson, the full physics potential of LHC was reviewed in Aachen. Production and decay modes of the top quark,[14] detected only later in 1995 at the Tevatron, were studied in detail and the expected precision on its mass measurement

[14]Several years before the top quark was discovered, its mass was only indirectly constrained to be above approximately 90 GeV from early LEP measurements. Its width, $\Gamma_t \propto m_t^3$ for $m_t \gg m_W$, dominated by the decay to Wb, rises fast with the top mass. It ranges from $\Gamma_t \simeq 0.09$ GeV at $m_t = 100$ GeV to $\Gamma_t \simeq 2.5$ GeV at a mass of 200 GeV, with corresponding top-quark lifetimes of 7×10^{-24} and 3×10^{-25}, respectively. At large mass, the lifetime is shorter than typical strong-interaction timescales. A $t\bar{t}$ *toponium* bound state would therefore decay through individual top-quark decays rather than the hadronisation of the bound state itself. The toponium width is approximately given by $\Gamma_{t\bar{t}} \sim 2\Gamma_t$, so that toponium should not exist beyond about 120 GeV. At a hadron collider, toponium would be dominantly produced in a 0^{-+} state, which would allow it to decay with significant branching fraction into two photons, if m_t lies between about 90 and 110 GeV. Discovery of toponium through the di-photon decay was discussed in the 1992 Letter of Intents of ATLAS and CMS. More discussion of these considerations is provided in section 12.3.1.

evaluated.[15] It was also estimated that strongly interacting supersymmetric particles could be found provided their mass would not exceed about 2–2.5 TeV, and Z' and W' bosons could be discovered up to masses of around 4 TeV. Searches for additional spatial dimensions or quantum micro-black holes were not considered as these phenomena appeared only towards the end of 1990s in the LHC physics program, as a consequence of new theoretical ideas to solve the hierarchy problem (see section 3.7).

The key message from the 1990 Aachen workshop was that the LHC, in spite of its smaller collision energy, would nonetheless be competitive with the SSC for a large part of the physics programme, provided a ten times larger luminosity (in the range of $10^{34}\ \mathrm{cm}^{-2}\,\mathrm{s}^{-1}$) could be achieved. Furthermore, it was thought that it could be operational earlier, possibly by the year 2000, and — most importantly — it should be much cheaper.

6.3.4. *LHC approval*

CERN Council[16] was the decision taking body that had to approve the LHC project. After hearing in 1993 the plan to build the LHC within about ten years, Council, albeit being positive but concerned about the implications on CERN's budget, approved the LHC project in 1994 in a staged "missing-magnet" version with initially only two-third of the required number of dipole magnets, allowing the LHC to reach only 10 TeV centre-of-mass energy. As in the following CERN's Director General Christopher Llewellyn Smith and LHC project responsible Lyn Evans succeeded in securing additional contributions from CERN non-member states, Council approved in 1996 the complete 14 TeV LHC with an envisaged completion date of 2005.

---

[15] Prospective studies estimating the potential for measuring CP symmetry violation in $B^0 \to J/\psi K^0_S$ or in $B^0 \to \pi^+\pi^-$ were also performed, but for these studies the LHC was overtaken by the e^+e^- B factories which began operating at SLAC, Stanford, USA and KEK, Japan towards the end of the 1990s.

[16] The *CERN Council* is the highest authority of CERN, which controls all important scientific, technical and administrative activities. It also adopts the budgets and reviews expenditure. Each of the (in 2020) 23 member states of CERN sends two official delegates to the CERN Council, one representing the government and the other representing national scientific interests. Each member state has a single vote. Although most decisions require a simple majority, Council traditionally aims for a consensus as close as possible to unanimity.

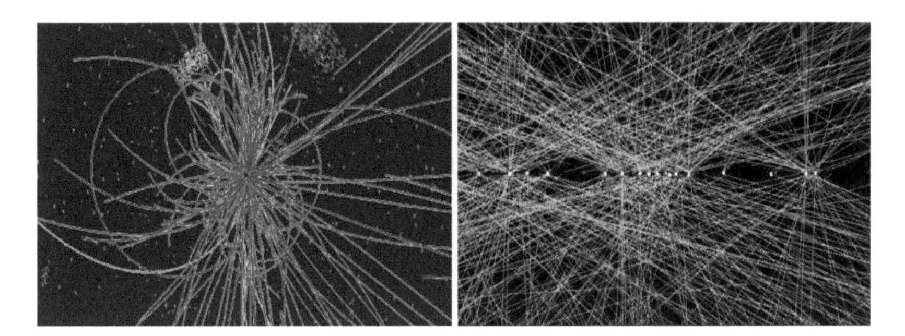

Fig. 6.10. Left: simulation in the CMS tracker of a high-luminosity event with about twenty simultaneous proton–proton collisions resulting from the crossing of two bunches. The figure shows a transverse view with beam proton trajectories perpendicular to the page. Right: longitudinal view of about ten-centimetre length along the beam line around the crossing point, for a real collision event recorded during the early 2011 data-taking period. Individually reconstructed proton–proton collision vertices are clearly visible. *Images: CMS Collaboration.*

6.3.5. *Challenges for the detectors*

To reach the very high luminosity required, the LHC was designed such that the 2800 bunches of protons making up the beams, spaced by 7.5 metres, cross and collide every 25 nanoseconds at the centre of detectors. With each bunch containing more than a hundred billion protons, on average about twenty simultaneous proton–proton collisions were expected per bunch crossing. Such simultaneous collisions are called *pile-up* . The rate of pile-up collisions is a crucial parameter in the operation of the LHC: a large pile-up rate allows for high luminosity but increases the difficulty of distinguishing between tracks or energy deposits in the detectors and assigning them to the correct proton–proton collision. Indeed, only a small fraction of these collisions, less than one in ten million, produce an interesting event with, for example, a top quark, W, Z, or just a Standard Model process that one wishes to study in greater detail. Collisions containing possibly a Higgs boson, W', Z' or supersymmetric particles are much more rare still (see section 10.1).

The mode of operation of the LHC has imposed very hard constraints on the design of experiments and detector concepts. On one hand, detectors were required to have a very rapid response; on the other hand, they had also to be very granular, i.e. made of a very large number of elementary cells (channels), so as to be able to distinguish and resolve the multitude of particles produced at the same instant due to the pile-up collisions coming with each bunch crossing (figure 6.10). The particle tracker of CMS is thus made of 66 million silicon pixels and ten million silicon microstrips, while the ATLAS tracker contains 80 million pixels and six million microstrips. Furthermore, the detectors themselves and the directly mounted electronic readout systems had to be radiation hard. The data selection, trigger and online readout systems needed to be extremely powerful, fast and selective, with data processing requirements on a scale never encountered before.

6.4. The long way towards the LHC start-up

6.4.1. *The LHC magnets*

The research and development studies on LHC magnets started immediately in 1990. The requirements were to design dipole bending magnets to maintain proton beams of highest achievable energy, hopefully 10 tesla so as to have 8.5 TeV beams, in the LEP tunnel, and to conceive quadrupole magnets to keep the beams focused. Two types of low temperature superconductors were initially considered, niobium-titanium (NbTi) and niobium-three-tin (Nb_3Sn) alloys. As the technology of NbTi magnets was much more advanced and better mastered, rapidly all efforts were concentrated on this option for the superconductor. Niobium-tin could in principle allow higher magnetic fields, but the material is very brittle and difficult to handle. The development of this superconductor into a workable solution for industrial production would have required much more time, which the LHC team could not afford in the race with the SSC.[17] Led by Giorgio Brianti since 1986, then after 1993 by Lyn Evans, CERN took the main responsibility of developing the dipole magnets, the key elements of the LHC. Under the presidency of the renowned physicist Nicola Cabibbo, the Italian nuclear physics organisation INFN made an exceptional financial contribution to the LHC magnet research. INFN helped develop the superconducting cables, which were manufactured in industry, and prototype magnets. The French institutions, CEA/IRFU and CNRS/IN2P3, designed in collaboration with and support from CERN the quadrupole magnets. Special interaction-region dipoles and focusing magnets were developed by the USA and Japan, and beam optics corrector magnets by Great Britain and India. Russia contributed various magnets (about 400 in total) to transfer protons from the SPS to the LHC.

[17]The development of Nb_3Sn magnets continued in the meantime in the USA and Europe, and it is only in 2016 that a first workable prototype two-metre long, 14 tesla dipole has been manufactured, see discussion in chapter 15.

The R&D effort on the NbTi dipoles took ten years and three generations of prototypes before finally producing satisfactory magnets. First generation prototypes were of reduced size, about a metre long. Second generation ones were ten-metre long, while the final ones are a little above fourteen metres. These magnets use superconducting technology operating at 1.9 kelvin (−271°C), cooled with superfluid helium. During the development phase it became clear that only 8.3 tesla fields (corresponding to an electrical current flow of 11 000 amperes) could be achieved for stable operation and not 10 tesla as initially hoped. For comparison, 8.3 tesla is 130 000 times the flux density of the Earth's magnetic field at its strongest intensity of 65 micro-tesla, south of Australia and in central Siberia). The beams could thus be accelerated up to 7 TeV, the proton velocity reaching 0.999 999 991% of the velocity of light, just short by 10 km/h in comparison.[18] The centre-of-mass proton–proton collision energy would thus be 14 TeV. In April 1994, a ten-metre long bending magnet prototype with a 8.3 tesla field was operated for the first time. It was a decisive step in the long and difficult development phase of these magnets which continued up to 2002 when the very last design modifications were implemented.

The main innovation in accelerator design and technology that made the LHC construction possible was the two-in-one magnet concept[19] illustrated in figures 6.11 and 6.12. What is this about? In preceding proton–proton colliders, for example the ISR, the two proton beams circulated in opposite directions in the machine and were accelerated and kept in orbit in two distinct vacuum tubes, each beam pipe at the centre of a complete and independent set of bending and focusing magnets (figure 6.7). The SSC was designed according to these same principles.

The two-in-one scheme required twice less magnets as the two vacuum beam pipes are incorporated in the same magnetic structure and a single cryogenic enclosure. The price to pay for this is that these magnets are much more complicated to build and the operation of the accelerator is more delicate, less flexible in the manipulation and tuning as the two beams are fully coupled. Nonetheless, these challenges did not impede the performance and robustness of the LHC, which featured amazing flexibility in its various collider modes, proton–proton, lead–lead, and even proton–lead and xenon–xenon over the years 2010 to 2018 as discussed in chapter 9.

---

[18]This number can be compared to the velocity of protons and antiprotons in the Tevatron, which is about 480 km/h smaller than the velocity of light.

[19]This novel concept was first proposed by Brookhaven, USA, but dropped for use in the SSC.

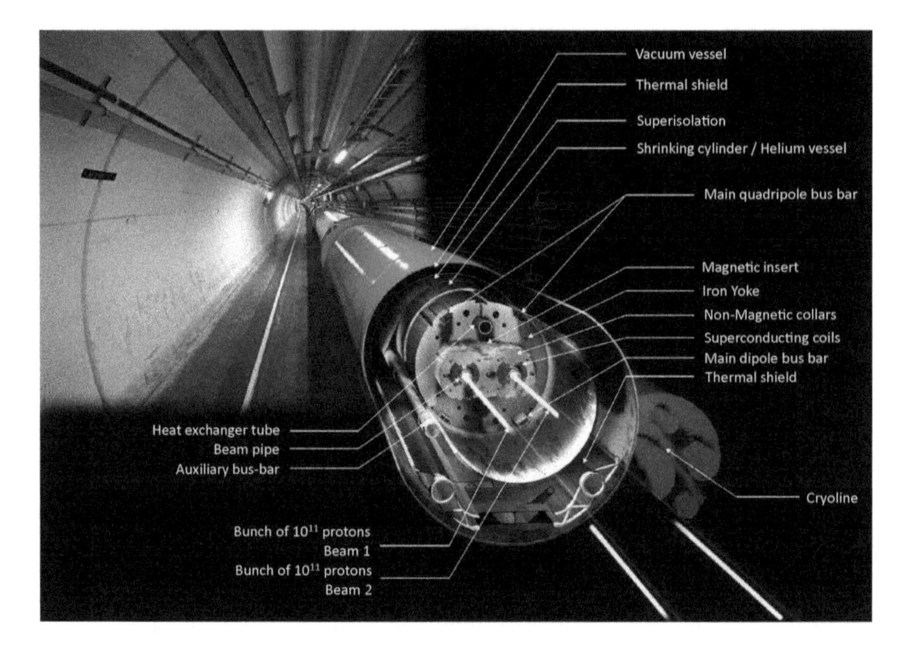

Fig. 6.11. Schematic drawing of a bending (dipole) magnet of the LHC. At the centre the two beam pipes in which the proton beams circulate in opposite direction. They are surrounded by the superconducting coils and separated by 18 centimetres within rigid steel collars providing good thermal contraction and magnetic permeability. The direction of the dipole magnetic field points bottom-up in the left-hand tube, and top-down in the right-hand one (see figure 6.12). *Image: CERN.*

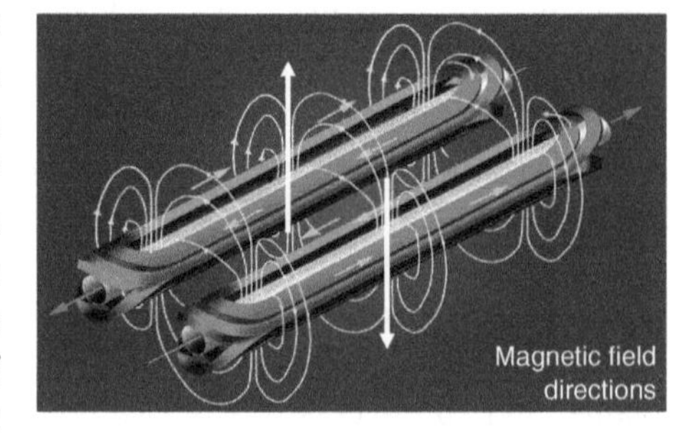

Fig. 6.12. Scheme illustrating the two-in-one concept of the LHC dipoles. *Image: CERN.*

The two-in-one scheme was in fact the more realistic option for the LHC, as with the 3.8 m diameter cross-section of the LEP tunnel (in the arcs, the straight sections of the tunnel have 4.4 m or 5.5 m diameter), it would have been difficult to install two sets of magnets side-by-side. It should not be forgotten that the tunnel must also accommodate over its entire length a complex and voluminous cryogenic system to feed the magnets with superfluid helium, and possibly also an electron ring, as initially planned. This whole enterprise was risky, requiring audacity in view of the understanding of magnet technologies and accelerator dynamics at the time, and its undertaking was owing to a large extent to the daring of Carlo Rubbia as well as Giorgio Brianti.

In 2002, with the required performance finally achieved with third-generation prototypes, the production of the 1232 dipole magnets and about forty spares was launched. Each dipole magnet is 14.3-metre long, weighs 35 tons, and costs about one million Swiss francs. The manufacturing orders were distributed over three large European electro-industry companies specialised in magnet production: in France the consortium Alstom MSA–Jeumont Industries, Ansaldo Superconduttori in Italy, and Babcok Noell Nuclear in Germany. Engineers from these companies spent a year at CERN working with local researchers to master the fabrication procedures[20] before starting mass production at their home plants, a very good example of technology transfer from CERN towards key European industries.

Quadrupole magnets with a field gradient of 223 tesla per metre were a bit less difficult to develop. They are again superconducting magnets operating at 1.9 kelvin with coils made of NbTi in a two-in-one scheme. About 400 main quadrupole magnets of various types were needed. Their lengths vary from three to twelve metres with a weight of up to 18 tons. Two prototypes were built in Saclay and tested before launching final production by ACCEL Instruments in Germany. The LHC construction also required the production of about five thousand smaller cryogenic multipole corrector magnets, which were built in the UK and India. In the years between 1994 and 1998 a first accelerator cell made of prototype magnets was put in place and operated at CERN to test and validate electrical, mechanical and cryogenic aspects of the various components.

During the years 2003 through 2007, while industrial magnet production was ongoing, assembly and quality assurance tests were performed on various sites in the Gex county around CERN, mysterious long blue cylinders could be seen aligned all around (figure 6.13). These were the machine dipoles, tested and waiting to be mounted in the tunnel as soon as they could be accommodated. The physicists from the LHC experiments were following with greatest interest the delivery and performance of these magnets as everything depended on their infallible quality.

[20]The production of the NbTi conductor required the highest standards involving hundreds of individual superconducting strands assembled into a cable that had to be shaped to fit the geometry of the magnet coil.

Fig. 6.13. Left: aligned LHC dipoles in the Gex county region (France) waiting for installation in the LHC tunnel. Right: lowering of the first magnet dipole into the tunnel on March 7th, 2005. *Images: CERN.*

6.4.2. *The eight octants of the LHC and the experimental areas*

The LHC tunnel is located about 100 metres underground (figures 6.14 and 6.15). It is subdivided into eight octants each about 3.5-kilometre long and containing about 350 large magnets (figure 6.16). Access to the tunnel and experimental areas is reached through eight shafts, one per octant. The cryogenic and electrical powering systems are also subdivided into eight mutually independent sectors going from a middle of one octant to the middle of the next one. Thus we talk about sector 7–8 when referring to the one that goes from the middle of the 7th to middle of the 8th octant.

The two large omni-purpose experiments ATLAS and CMS are installed in pits 1 and 5 at opposite locations around the ring. The ALICE detector, designed specifically for the study of heavy-ion collisions, is installed in pit 2. The LHCb detector in pit 8 is specialised for the study of relatively low-mass forward going particles such as B, D and other hadrons. Octants 3 and 7 are dedicated to the cleaning and collimation of the beams.[21] Octant 6 is equipped with the system required to discharge the beams when needed (beam dump), while the radio-frequency acceleration system is located in octant 4.

[21] *Collimators* are jaw-like movable blocks of robust material positioned close to the beam to absorb stray particles, which could leave the beam orbit and generate backgrounds in the LHC and its experiments. More than a hundred such devices are placed around the LHC. They are arranged in chains of primary, secondary and tertiary collimators of different material and distance to the beam. Collimators are among the most complex elements of the LHC.

Fig. 6.14. View of the ATLAS cavern in July 2002 during the phase of floor consolidation. The cavern is 53-metre long, 30-metre wide and 35-metre high. The LHC beam line is at centre of the opposite-side cavern wall. *Image: ATLAS Collaboration and CERN.*

Fig. 6.15. Superconducting magnets aligned in the LHC tunnel. The blue cylinders are external envelopes of the vacuum enclosure of the bending dipoles. *Image: CERN.*

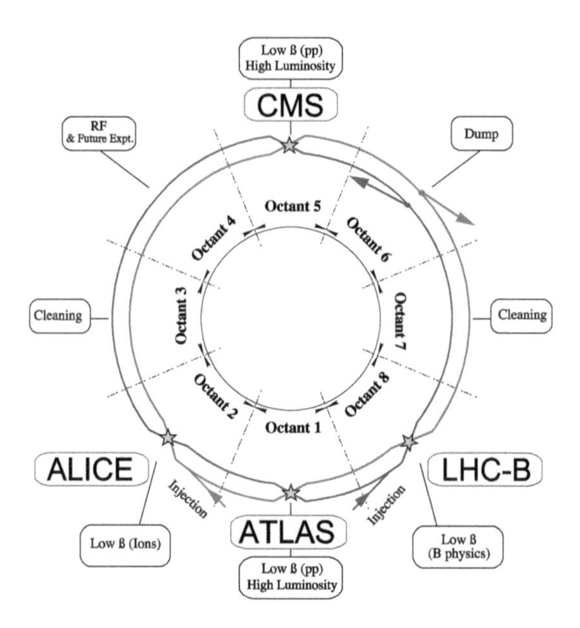

Fig. 6.16. Segmentation of the LHC in eight octants, with locations of experiments and the main accelerator facilities. *Image: CERN.*

Large underground cavities had to be excavated to house the ATLAS and CMS experiments. The ATLAS cavern, with its 53-metre length, 30-metre width, and 35-metre height (equivalent to a 10-storey building), is one of the largest underground excavations ever built (figure 6.14). During the excavation, the ceiling, with its 1380 square metre surface, was suspended through a system of cables anchored in the rock from galleries excavated on the sides. While the cables still exist, the ceiling now rests on the concrete cavern walls. The ATLAS cavern was inaugurated in June 2003. During the excavation of the CMS cavern it was necessary to freeze at liquid nitrogen temperature (−196°C) the soil over a depth of 40 metres, as an underground river was going through the area. These very major civil engineering activities were completed in February 2005 and the cavern fully equipped by October 2006. The excavation cost for each of these two caverns amounted to about 80 million Swiss francs. The ALICE and LHCb experiments reused two of the already existing LEP experimental caverns. Once the civil engineering work on the access tunnels and the experimental areas completed, the cryogenic system was put in place in the 27-kilometre long tunnel. In March 2005, the first dipole magnet was lowered (figure 6.13), and by April 2007 the installation of the 1232 dipole magnets was accomplished. Each one of these dipoles as well as the quadrupole and additional corrector magnets had to be positioned and aligned with a precision of about 100 micrometres (about the thickness of human hair). Such a precision was required along the entire LHC circumference for a perfect mastery of the magnetic field, allowing to rapidly set-up proton beams of the required lifetime in order to reach the instantaneous luminosity goals. With 65% of the 27-kilometre LHC machine length occupied by superconducting magnets operating at superfluid helium temperature of 1.9 kelvin,[22] the LHC is the largest refrigerator in the world and probably the coldest place in the Universe — unless there exists a technologically more advanced civilisation than ours. Furthermore, in the vacuum tubes where the beams are circulating, the vacuum is at a quasi-interstellar level.

6.4.3. *Cost of LHC*

The use of the two-in-one scheme for the LHC magnets has allowed substantial cost savings. Even if the magnets and the system as a whole are more complex, the reduction by a factor of two in the number of required magnets is estimated to have reduced the cost by about 600 million Swiss francs. Furthermore, the reuse of the already existing LEP tunnel has saved about one billion Swiss francs in civil engineering cost. Another saving of about one billion Swiss francs is owed to the use of the already existing CERN infrastructure, the entire injection and acceleration system, linear accelerator (Linac2), booster, PS, SPS (figure 6.17).[23] These cost-saving arguments were successfully used in the years 1990/91/92 by CERN's Director General to attract participation of large CERN non-member states such as Japan, India, Russia, Canada and others to the then nascent LHC project.

[22]The full cooling procedure of the LHC magnets, consisting of three different stages, takes weeks to complete. During the first stage, helium is cooled to 80 kelvin using liquid nitrogen heat exchangers, and then to 4.5 kelvin with the help of turbines. In the second stage it is injected into the cold masses of the magnets, before being cooled to a temperature of 1.9 kelvin in the third stage. Below 2.17 kelvin, fluid helium becomes superfluid with very high-thermal conductivity making it a perfect refrigerant.

[23]No accelerator technology had a dynamical range that would allow the direct acceleration from zero to the maximum energy required here, it had to be done in steps. Before injection into the LHC, protons extracted from a hydrogen ion source are injected and accelerated through a sequence of accelerators: first in the Linac2 where they are accelerated to 50 MeV, then into the venerable — but regularly renovated — PS booster (acceleration to 1.4 GeV and packaging of protons into bunches) and PS (further acceleration to 26 GeV, splitting into 72 bunches), followed by the SPS (acceleration to 450 GeV, accumulation of 4 times 72 bunches) to be finally injected in counter directions into the LHC ring where they are accelerated up to 7 TeV (see figure 6.17).

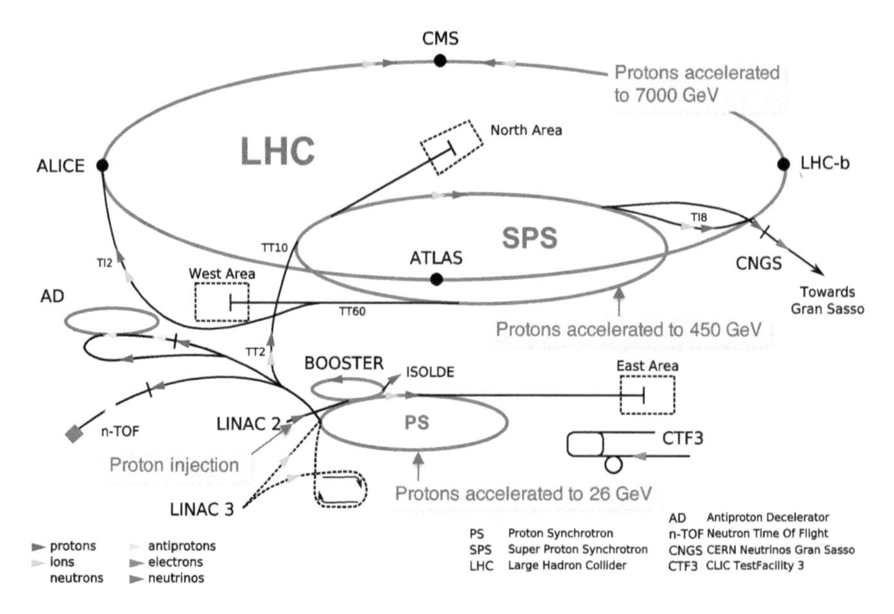

Fig. 6.17. Accelerator-injector complex at CERN (see text and footnote 23 for details). *Image: CERN.*

The cost of the machine itself, not including the experiments, was initially estimated at 3.5 billion Swiss francs, about one third of the cost of the SSC in 1992. The start-up was (somewhat optimistically) thought to be possible by the year 2000, i.e. before the SSC. Ultimately the cost rose to 4.8 billion Swiss francs, including the upgrade of the entire injection-acceleration system upstream the LHC (figure 6.17). Out of these, about 3.8 billion were spent on material costs and the remaining billion on human resources (researchers, engineers, technicians, administration). CERN member states carried 85% of the total expenditure and non-member states contributed about 500 million Swiss francs (USA: 180 million, Japan: 120 million, Russia: 90 million, Canada: 20 million, India 20 million), while the host countries France and Switzerland have added 200 million on top of their regular contribution as member states of CERN. The subdivision of cost according to the various technical items is given in figure 6.18 showing that the cost of magnets alone is just about half the total material cost of the machine.

After the cancellation of the SSC in October 1993, universities and laboratories from the USA joined the LHC experiments *en masse*. In the ATLAS and CMS collaborations, for example, US physicists today

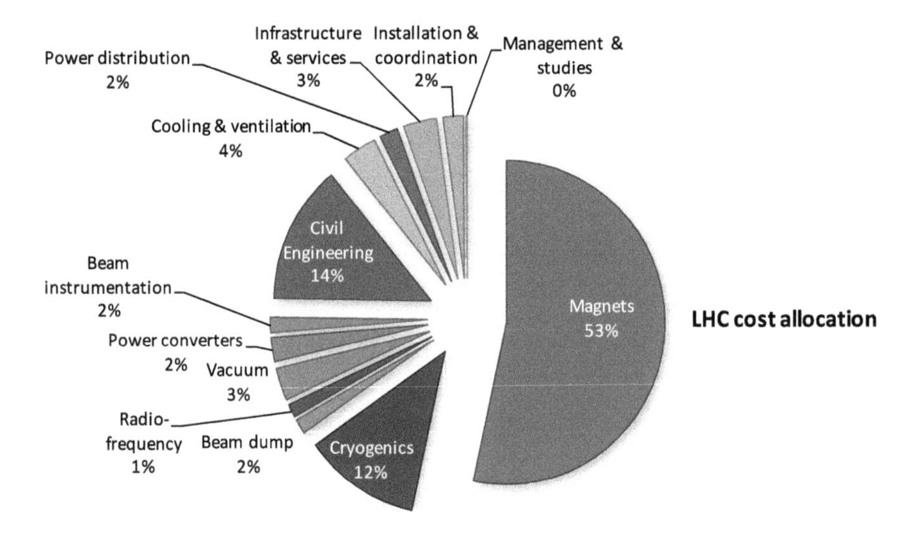

Fig. 6.18. Distribution of the total LHC material cost. The largest fraction is for magnets, followed by civil engineering (tunnels, caverns, pits) and cryogenics (cooling system for superconducting magnets). The cost of studies and management amounts to only about 0.1%.

represent the largest contingent per nationality. The USA contributed to 20% of the overall construction cost of the CMS detector, similarly to ATLAS it is the largest financial contribution by country. In view of the total number of US physicists taking part in the various experiments at the LHC, CERN is today the largest US particle physics research laboratory. Thanks to CERN, Europe succeeded in reversing the brain drain, at least in the field of fundamental physics, that worried so much European politicians during the 1960s and 1970s.

The construction and commissioning of the LHC took almost ten years longer than initially hoped for. This is largely because the development and production of the dipole magnets was more difficult and has taken more time than initially expected. Budget pressure on CERN, which had to realise this colossal project with a fixed budget, also slowed down its completion.

In March 2010, twenty years after the Aachen Workshop, first proton–proton collisions at a new world-record centre-of-mass energy of 7 TeV were produced. To fully accomplish this incredible bet and endeavour it was still necessary to reach design centre-of-mass energy of 14 TeV and $10^{34}\ \mathrm{cm}^{-2}\,\mathrm{s}^{-1}$ instantaneous luminosity, as we will discuss in the following.

Inset 6.4: The LHC in numbers

The LHC is made of 9300 magnets, including

- 1232 superconducting bending dipoles, each 14.3 metres long and weighing 35 tons; an electric current of 11 600 amperes generates a magnetic field of 8.33 tesla; the dipoles operate at superfluid helium temperature of 1.9 kelvin and altogether fill about 18 kilometres out of the 26.6 kilometres total LHC circumference;
- 392 superconducting quadrupoles for beam focusing with a field gradient of 223 tesla per metre;
- 2464 sextupoles, 1232 octupoles and about 4000 other smaller magnets for beam orbit corrections, most of them cryogenics;
- Winding of the superconducting cables: the cable (1.5-centimetre diameter) is made of 36 braids, each made of 6400 niobium-titanium (NbTi) filaments of seven-micrometre diameter; the total amount of superconducting material at 1.9 kelvin amounts to 1200 tons; the overall length of filaments in the 7600 kilometres of cable equals ten times the distance Earth to Sun; the niobium-titanium used in the LHC represents about 28% of the world production of superconducting raw material over a five-year period.

Energy stored in the beams: two times 2800 bunches, each bunch with more than 10^{11} protons at 7 TeV correspond to a stored energy of almost a gigajoule. This is equivalent to the energy of a 400-ton high-speed train at a velocity of 250 km/h, or of 200 kilograms of TNT explosive, or the energy required to melt 1.5 tons of copper. The energy stored in the magnets is about thirty times larger still! Despite all these large numbers, only about two nanograms (10^{-9} grammes) of protons originating from a hydrogen bottle are accelerated each day. The pressure in the vacuum tube where protons circulate is 10^{-13} atmospheres, ten times less than the pressure on the surface of the Moon.

Cryogenics: the LHC is the largest refrigerator in the world, 40 000 tons of material are kept at 1.9 kelvin, cooled down by 120 tons of superfluid helium. This is also probably the coldest place in the Universe (the cosmic microwave background heats homogeneously the Universe to 2.7 kelvin).

Chapter 7

What is a Particle Detector?

7.1. Particle detection techniques

We have seen in the preceding chapter how the development over the years of more and more powerful accelerators allowed to measure particle properties and uncover new phenomena thus advancing the understanding of Nature. To be able to *see* and study results of particle collisions occurring at ever increasing energy and rate, it was necessary to develop in parallel with accelerators ever more powerful particle detection techniques.

Some major advances marked the progress over the years, as the discovery of the *bubble chamber* in 1952 by the American physicist and (later) neurobiologist Donald A. Glaser (Nobel Prize in Physics, 1960). Reduced to essentials, a container is filled with liquid hydrogen or a noble gas in liquid state, the liquid being at same time the target material and serving as a detector. Just before the passage of a particle beam (typically few tens of particles) through the detector, the liquid is decompressed for a brief moment not allowing transition into a gaseous state and bringing it thus into a metastable state of an *overheated liquid.* Particles produced in the collision of a beam particle with a target nucleus in the liquid seed the growth of tiny bubbles, up to about a hundred micrometres in diameter along particle trails, thus materialising particle trajectories, a little as for the condensation trails left behind high-flying airplanes (see figure 1.12, or figure 2.1, for example). A photograph is systematically taken at every traversal of the beam through the chamber, and the chamber is immediately recompressed to prevent boiling of the liquid throughout the volume. Photographs at a rate of one every few seconds could be taken depending on the operating rate of the bubble chamber and accelerator beam cycle. A famous and

Fig. 7.1. Left: the heavy liquid bubble chamber Gargamelle during installation on the PS beam at CERN in 1970. Right: "scanners" measuring particle trajectories on bubble chamber photos projected onto dedicated scanning tables and selecting potentially interesting collisions on basis of their topologies.

pioneering hydrogen bubble chamber was the Alvarez[1] 1.6-metre chamber at the Bevatron, where most of the initial hadronic resonances were discovered. Another famous bubble chamber was Gargamelle (figure 7.1), an almost 5-metre long and 2-metre diameter heavy-liquid (freon) chamber. It allowed the discovery of weak-interaction neutral currents in a neutrino beam at CERN in 1973 (see section 2.2).

Even if bubble chambers remain up to this day unsurpassed in terms of precision in the reconstruction of particle tracks[2] (nuclear emulsion

[1]Luis Walter Alvarez was a particularly creative American experimental physicist who received the Nobel Prize in Physics in 1968 for the discovery of a large number of resonance states using his hydrogen bubble chamber at the Bevatron. He published in 1951 the idea that led to the development of the Tandem Van de Graaff accelerator, an electrostatic accelerator capable of delivering continuous or high-intensity pulsed ion beams in a wide range of ion species at various energies. In the late 1960s he carried out pioneering cosmic ray muon imaging in the Chephren pyramid using spark chambers as detectors (see inset 7.6). This *muon-tomography* method is nowadays applied at many sites worldwide. In 1980, together with his son, a geologist, and along with the nuclear chemists Frank Asaro and Helen Michel, Alvarez proposed the asteroid-impact theory to explain the large over-abundance of iridium in a clay stratum of the Cretaceous–Tertiary (dinosaurs and much else) extinction boundary, a thin rock crust layer marking the end of the Cretaceous Period 66 million years ago.

[2]Every bubble can be measured with a precision of about a hundred micrometres, but the ensemble of all bubbles along a trajectory results in a track measurement precision of about a micrometre.

detectors achieve similar precision), their slow operation is a serious shortcoming: few seconds are required after every expansion cycle until the chamber is operational again. Furthermore, data acquisition is not selective but systematic as a picture has to be taken at every beam passage, irrespective of whether or not an interesting interaction occurred in the chamber. In the golden days of bubble chamber physics, the years from around 1960 until around 1975, in several tens of laboratories and universities in the USA, Europe, USSR and Japan, tens of thousands of pictures were visually inspected day and night by specialised crews (named "scanners" or also "scanning girls" in the male dominated particle physics science of that time, see figure 7.1). However, with the gradual improvement in the understanding of the internal structure of hadrons, the emergence of quarks as actual elementary particles and not anymore as purely mathematical concepts, progress required investigating rarer and rarer processes. It thus became necessary to select pictures at the very moment when an interesting interaction took place, and the bubble chamber technique did not allow *online triggering and selection*. Another limitation was that the chamber liquid must also act as the target material. The development of colliders, hadronic or electron–positron ones, required totally different detector configurations and the development of other detection techniques.

Due to these drawbacks of bubble chambers, purely electronic detectors were developed since the 1960s, whose extremely rapid evolution was largely owed to the evolution of ever faster electronic circuits, becoming also less and less expensive. Such systems allowed the immediate selection of data online (i.e. before storing the recorded interaction) by selecting on the desired event topologies or shapes. This was a very major advance in particle physics. The *spark chamber* was first among this type of detectors. Just as in a bubble chamber, a particle trajectory is materialised through sparks occurring along its track and a picture is taken. However, thanks to the fast response of two scintillators[3] placed in front and behind the chamber, the decision whether or not to take a picture can be taken electronically almost instantaneously (figure 7.2). Spark chambers had their moment of glory with the discovery of the muon-neutrino (and proof of two distinct neutrino types, ν_e and ν_μ) at Brookhaven in 1962. The spatial

[3]A scintillator is a detector made of specific transparent material that emits a light flash (luminescence) when hit by radiation from an ionising particle. The light is measured by a detector such as a photomultiplier tube which converts and amplifies incident photons into an electrical signal.

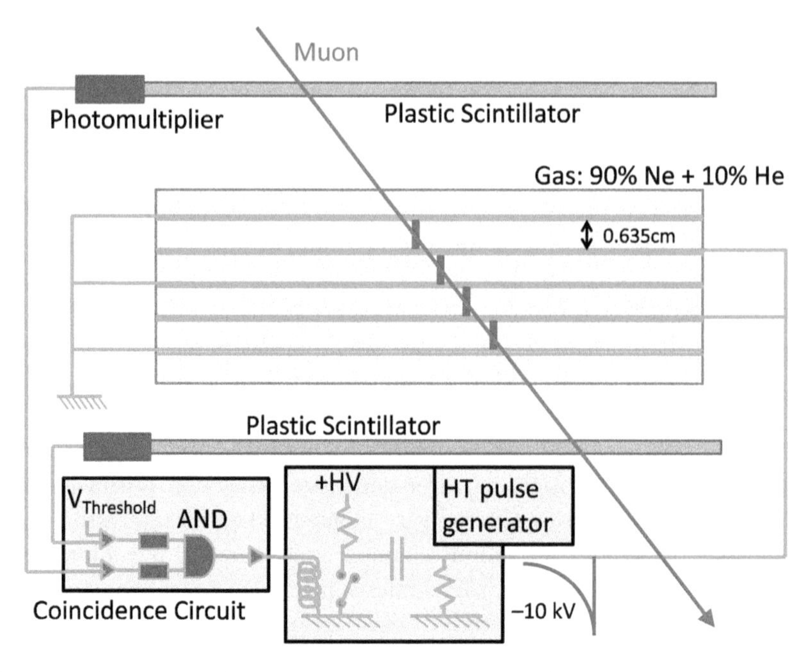

Fig. 7.2. Basic principle of a spark chamber triggered by two scintillator counters. A number of parallel metallic plates are inserted in a gas enclosure. Every second plate is kept at a high potential while the other ones are connected to ground. If a charged particle, say a muon or a pion, crosses the chamber, it ionises gas along its trajectory by extracting electrons from atoms and leaves a trace of electrons and positively charged ions. The acceleration and multiplication of charges between the plates gives rise to sparks in the gaps. The two scintillator counters on top and bottom of the chamber give a very rapid signal indicating the passage of a particle through the chamber resulting in the application of a high voltage pulse on the plates. These counters also trigger data acquisition, a photo or of a purely electronic signal. *Image: CERN.*

resolution was significantly less precise than in bubble chambers, but spark chambers were much faster and the dead time between two triggers was much reduced to an order of a millisecond. However, even this turned out too slow in face of the ever mounting demands of experimentalists for the study of ever rarer phenomena.

All these limitations were overcome with the invention of *multi-wire proportional chambers* (figure 7.3) of which the Polish-born French physicist Georges Charpak (figure 7.4, Nobel Prize in Physics 1992) was one of the main inventors and promoters. Particle detection in this case, too, depended on ionisation of a gas by a charged particle passing through. It induced a small electrical signal recorded on

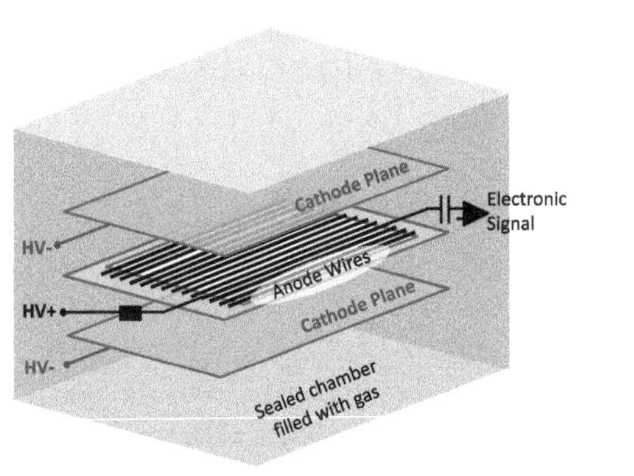

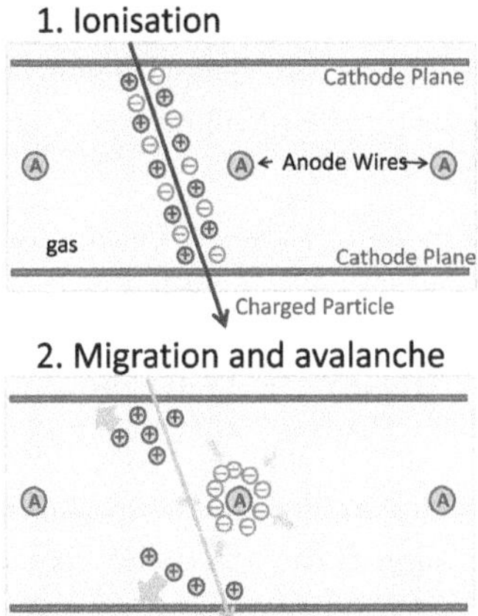

Fig. 7.3. Functioning principle of a wire chamber. A gas enclosure is filled with a noble gas, argon for example. Wires, strung in a plane and kept at a high positive voltage, make up the anode planes. In between these planes are inserted conductive cathode planes at ground potential. A charged particle traversing this arrangement ionises the gas, electrons are attracted towards anode wires and positive ions towards cathode planes. The very high-electric field present in the immediate vicinity of wires gives rise to an *avalanche ionisation* phenomenon generating thousands of ionised atoms. An amplifier circuit is connected to the end of wires allowing to measure the electrical signal generated by this avalanche. The electrical signal recorded on each wire is then analysed to reconstruct the particle trajectory as it would be done in a bubble chamber, but it is now purely automatic, handling only fast electronic signals.

a nearby wire immersed in the gas. With a large number of wires strung in the chamber it is possible to reconstruct the particle's trajectory through the active detector volume. This proved to be a very fast detector with a response time on a scale of a microsecond and with sufficient spatial precision of the order of 100–300 micrometres. The detector can take various shapes for example that of a hollow cylinder surrounding a beam pipe, which makes this type of detector very suitable for use at particle colliders. By the end of the 1960s all major electronic detector experiments were equipped with such multi-wire proportional chambers or of their subsequent evolution, the *drift chambers* (see inset 7.1).

Fig. 7.4. Georges Charpak.

During the years 2000, multiwire or TPC type gaseous detectors at colliders were gradually replaced by *silicon microstrip* detectors (see inset 7.2), even faster, with dead times of order hundred nanoseconds, and more precise, with spatial resolutions of a few tens of micrometres. This is the type of inner tracking detector mostly used today in experiments at the LHC.

Inset 7.1: Drift chambers

In a typical multi-wire proportional chamber, a large number of wires are strung in a gas volume crossed by ionising particles, and the position of the wires that give a signal are used to reconstruct the trajectory of those particles. In a *drift chamber*, the time it takes for the ions to drift through the gas of the detector is also taken into account, allowing greater spatial precision and a larger separation between wires, thus reducing the number of electronics channels and cost.

The electrons produced by primary ionisation drift towards the nearest wire. The signal induced on the wires results essentially not from the primary ionisation electrons, but rather from the drift of ions coming from the secondary ionisation of gas atoms or molecules due to the strong electric field in the immediate vicinity of the wire. The signal is recorded as in a classical wire chamber.

This concept allowed experimentalists to build larger chambers and to measure longer trajectories, yielding more precise momentum measurements (the resolution of the momentum reconstruction of a track in a magnetic field goes with the inverse of the square of the measured trajectory length). A good example of a large drift chamber, with a quasi bubble-chamber-like pattern recognition and measurement precision capability, is that of the UA1 experiment (figure 2.7).

The most recent developments of drift chambers are the so-called *time projection chambers* (TPC), where ionisation electrons drift over metres in large volumes of gas or ultra-pure liquid noble gas like argon or xenon. Examples of gaseous TPCs are the inner trackers of the ALEPH experiment at LEP (see section 2.4) or the ALICE experiment at the LHC (chapter 14), whereas liquid TPCs are now used in large experiments looking for neutrino or dark matter interactions, or proton decay.

Inset 7.2: Semiconductor detectors

With the advent of the LEP experiments in the late 1980s, detectors based on *semiconducting crystals* came into widespread use for the precise measurement of charged-particle trajectories. These counters are based on a p–n junction between two semiconducting regions doped p^+ and n^+, with opposite electric polarity, so as to increase the zone depleted of free charge carriers, neither electrons nor holes.

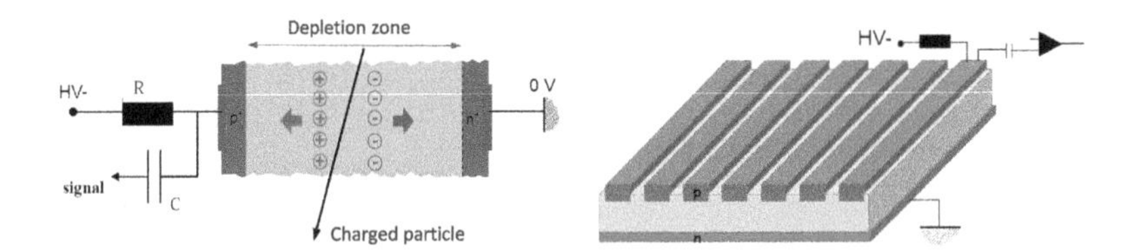

When a charged particle crosses the depletion region, it produces through ionisation electron–hole pairs in the crystal. Under the influence of a large (external) electric field present in this region, the free charges move towards the corresponding electrodes depending on their charge. During this drift, they induce a voltage pulse on the resistance R coupled to the semiconductor, which is proportional to the energy loss of the particle in the medium. As the drift velocity of electrons or holes is high, the generated electrical pulse is very fast, with a duration of the order of tens of nanoseconds.

In practice, the thickness of the semiconductor between the electrodes is few hundreds of micrometres and the voltage applied hundred to two hundred volts. The readout electrode is often shaped as microstrips few tens of micrometres wide (above picture on the right) or in the form of rectangular pixels. The precision in the measurement of a particle position is of the order of tens of micrometres in the direction orthogonal to the microstrips.

7.2. Design principles of a collider detector

The particle physics detectors located at the beam collision points (called *interaction points*) of a collider (e^+e^- or hadronic, pp or $p\bar{p}$, or even $e^{\pm}p$) are made of a set of subdetectors embedded into one another — a little bit in the manner of Russian dolls or onions — each subdetector fulfilling a well-defined and complementary role. A typical scheme for a collider

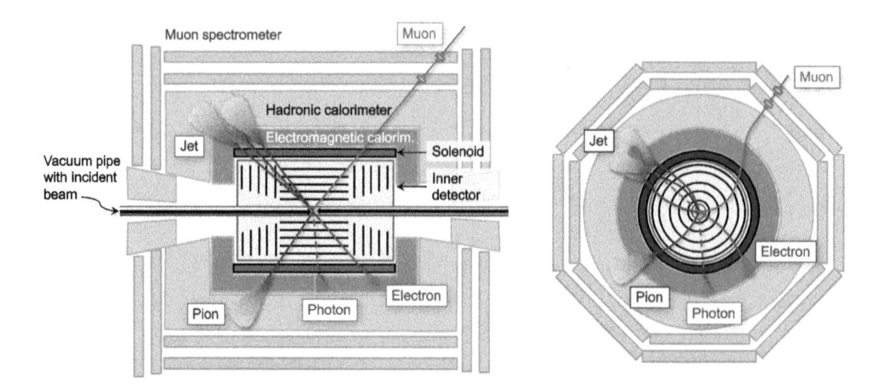

Fig. 7.5. Schematic design of a typical particle detector at a collider, with the longitudinal cut along the beam line on the left and the transverse cut to the beam line on the right (the beam points into the sheet). The physics objects that have to be detected, that are identified and measured, are: electron — produces a track in the inner tracking detector and a dense shower contained within the electromagnetic calorimeter (dark orange); muon — penetrating particle that leaves a track in the inner detector, little energy in the calorimeters and reaches the outer parts of the detector, the muon spectrometer (blue); photon — similar to electrons, produces a dense shower in the electromagnetic calorimeter, but leaves no trace in the inner tracking detector; charged hadron such as pion, kaon, proton, etc. — produces track and showers in both the electromagnetic and hadronic calorimeters; neutral hadrons have no tracks but similar showers as charged hadrons; jet of particles resulting from the fragmentation and hadronisation of quarks and gluons (the jets are therefore macroscopic manifestations of quarks and gluons) — produces a broad shower captured and recorded in the electromagnetic and hadronic calorimeters. In proton–proton collisions occurring at the LHC, on average about 700 charged particles cross the inner tracking system at each bunch crossing, which nominally occurs every 25 nanoseconds (except for some empty bunches).

detector is given in figure 7.5, showing the longitudinal and transverse views relative to the vacuum pipe through which the two beams are circulating, with the collision point located at the centre of the detector. The central part, called the *barrel*, is approximately of cylindrical shape. It is closed at both ends with perpendicular detector planes, called *end-caps*. A particle produced in the collision encounters successively the inner tracking detector, then the electromagnetic calorimeter, followed by the hadronic calorimeter, and — if not yet absorbed by the detector material — the outer muon spectrometer.

7.2.1. *Inner tracking detectors*

The inner tracking system serves the purpose of reconstructing all charged particle trajectories (pions, kaons, muons, electrons, etc.) emerging from the interaction point and usually called just *tracks* (see, for example, figure 2.7). This tracking device is most often located inside a solenoidal magnet whose axis coincides with beam direction. The particle trajectory is thus bent in a plane perpendicular to the magnetic field B and the particle's direction. By measuring the curvature R of the trajectory (or rather its sagitta, see inset 7.3), one obtains the momentum p for a particle with electric charge e: $p = e \cdot B \cdot R$ (see section 6.1). To achieve this, the particle trajectory is sampled with fast electronic detectors functioning either on basis of gas ionisation or as semiconductor devices. As a typical value, for a metre long track measured in a uniform magnetic field of two tesla, the sagitta for a 100 GeV momentum track will be about 0.75 millimetres. With three equidistant measurements along the trajectory with point-measurement precision of 50 micrometres (0.05 millimetres), the particle's momentum is determined with 5% accuracy. With increasing momentum, the curvature and thus the sagitta diminishes and the relative momentum resolution degrades.

Inset 7.3: Momentum measurement in a tracking detector

A particle of momentum p and electric charge e traverses over a length L a region of constant magnetic field B perpendicular to the trajectory. The radius of curvature R of the trajectory is related to the particle momentum by the relation $p = e \cdot B \cdot R$. The goal is to measure R as precisely as possible through the measurement of the sagitta f of the trajectory as illustrated in the following sketch.

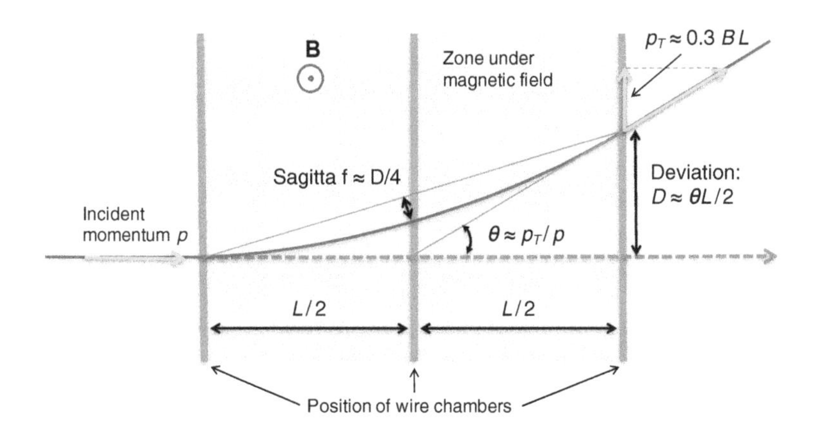

When the deviation D at the exit of the magnetic field zone is small, it is possible to approximate the circular trajectory by a parabola. The particle, whose velocity is close to the velocity of light, acquires a momentum component p_T perpendicular to its direction of incidence equal to:

$$p_T \simeq 0.3 \cdot B \cdot L$$

with p_T given in units of GeV, B in tesla and L in metres. The sagitta f of the trajectory is then given by:

$$f = \frac{D}{4} = \theta\frac{L}{8} \simeq \frac{0.3 \cdot B \cdot L^2}{8p},$$

where θ is the angle between the incident and exit directions from the magnetic field region. Let us assume that the trajectory is sampled at three equidistant points. The relative precision on the momentum measurement is then:

$$\frac{\delta p}{p} = \frac{\delta f}{f} \simeq \sqrt{\frac{3}{2}} \cdot \frac{8 \cdot \varepsilon \cdot p}{0.3 \cdot B \cdot L^2}.$$

Here ε is the point measurement precision assumed to be the same at all three points. The relative momentum error ($\delta p/p$) thus increases linearly with momentum.

As an example, let us consider the characteristics of the muon measurement spectrometer of ATLAS, a track length of five metres, a magnetic field of 0.5 tesla. For a particle of one TeV momentum, the sagitta is 470 micrometres. With $\varepsilon = 40$ micrometres, the momentum resolution then amounts to about 10%.

7.2.2. *Electromagnetic calorimeters*

The electromagnetic calorimeter is primarily used to measure the energy of light particles undergoing electromagnetic interactions. These can be charged ones such as electrons or positrons, or neutral ones such as photons. These particles, impinging onto the calorimeter, interact with the absorber material of the calorimeter and produce a large electromagnetic shower made of electrons, positrons and photons. As these particles interact preferentially with elements of high atomic number (Z), absorbers are usually made of heavy elements such as lead, tungsten or heavy crystals. It is these shower particles — that can number in the millions at shower maximum — that are detected and allow to estimate the energy of the incident parent-particle.

There are two main categories of electromagnetic calorimeters: sampling calorimeters with passive absorber planes interleaved with active detector planes where the deposited shower energy is measured, and homogeneous calorimeters in which the absorber material is such that it also allows the direct measurement of the shower energy. As we are going to see in what follows, the ATLAS electromagnetic calorimeter is of sampling type, whilst the CMS one is of the homogeneous type (see inset 7.4).

Inset 7.4: Energy measurement in a calorimeter

To measure the energy of a particle or of a jet of particles, it is necessary that the entire shower be contained within the calorimeter, be it electromagnetic only or the sum of electromagnetic and hadronic ones.

The main quantity that characterises an electromagnetic calorimeter is the *radiation length* X_0. It is material dependent and corresponds to the average distance traversed by a high-energy electron after which it has lost all but the fraction $1/e\,(\ln e = 1)$ of its initial energy. The energy is lost through bremsstrahlung, a German term designating an electromagnetic radiation produced by an electrical charge (electrons or positrons in the present case) moving through a medium with external electrical fields (the atomic nuclei of the medium) and thus slowing down. The spectrum of bremsstrahlung photons is continuous. Photons of sufficiently high energy can also convert in the medium into electron–positron pairs in the field of nuclei. Bremsstrahlung and photon conversion are the mechanisms through which an electromagnetic shower develops. As an example, the radiation length of lead is about six millimetres. For an electromagnetic shower to be fully contained in a calorimeter, it must have a depth of about 20 X_0.

For a hadronic calorimeter, the characteristic quantity is the *nuclear interaction length* λ_{int}. It is the average distance a high-energy hadron travels through a medium before having a strong-interaction-mediated collision with a nucleus of the medium. Iron, for example, has a nuclear interaction length of about 17 centimetres. The required depth for a hadronic calorimeter to fully contain a hadronic shower is typically 10 λ_{int}.

The precision $\delta E/E$ in the calorimetric measurement of energy E can be expressed as the quadratic sum of three terms

$$\frac{\delta E}{E} = \sqrt{\left(\frac{A}{\sqrt{E}}\right)^2 + \left(\frac{B}{E}\right)^2 + C^2},$$

where $\frac{A}{\sqrt{E}}$ is the so-called *stochastic* or *sampling term* and is related to fluctuations in energy depositions in the calorimeter owing to the statistical nature of the shower development, $\frac{B}{E}$ represents the effect of electronic noise, and C, the so-called *constant term*, reflects the effect of mechanical imperfections and other non-uniformities in the calorimeter construction, granularity and response. For electromagnetic calorimeters in use by the LHC experiments, with energy given in GeV units, A is typically between 5 and 15%, B is smaller than 20% and C is between 0.5 and 2%; at high energy it is this last term that dominates the resolution. For hadronic calorimeters, the stochastic term dominates with values for A ranging from 50 to 120% because hadronic showers are much less dense and exhibit larger fluctuations in their development in depth than electromagnetic showers. With increasing energy the resolution improves and is ultimately also dominated by the constant term C.

7.2.3. *Hadronic calorimeters*

The purpose of hadronic calorimeters is to measure (with additional help from the electromagnetic calorimeter) the energy of hadrons, i.e. particles subject to strong interactions such as charged or neutral pions and kaons, protons and neutrons. Hadronic calorimeters are particularly suitable for the measurement of jets, clusters of particles resulting from the materialisation of quarks and gluons. Most hadronic calorimeters are built as sampling calorimeters. Because the hadronic showers are much more irregular and less dense than electromagnetic ones, hadronic calorimeters are very deep and have a smaller segmentation in depth (fraction of active relative to passive layers, typically few percent) than electromagnetic calorimeters (see inset 7.4). Hadronic calorimeters are also much bulkier and voluminous than electromagnetic ones. In detectors at colliders, care is taken to design electromagnetic and, in particular, hadronic calorimeters with large acceptance, i.e. having large coverage of space angle around the collision point. The detector is then said to be *hermetic*, allowing to measure *missing transverse energy* as described in inset 7.5.

Inset 7.5: Missing transverse energy

In a symmetric collision as seen in the overall centre-of-mass, the vector sum of momenta of all particles produced must be equal to zero owing to momentum conservation between initial and final state. However, if particles that do not interact with the active detector material are produced, such as neutral particles that are not subject to strong interaction, the vector sum computed from the visible particles alone will not be zero as the invisible particle has taken away momentum. Examples of such particles are neutrinos, which are produced copiously in LHC collisions, or the putative dark matter particle or particles. In the picture below on left, we show a view in the plane transverse to the beam of an event with W^+ production, the W^+ decaying into a positron (shown here in yellow), identified by its energy deposition in the electromagnetic calorimeter (here from the ATLAS detector), and a neutrino which escapes the detector without interacting. On the right-hand side, we have a transverse view of an apparently badly unbalanced event measured in the CMS detector, with three jets all in the same hemisphere!

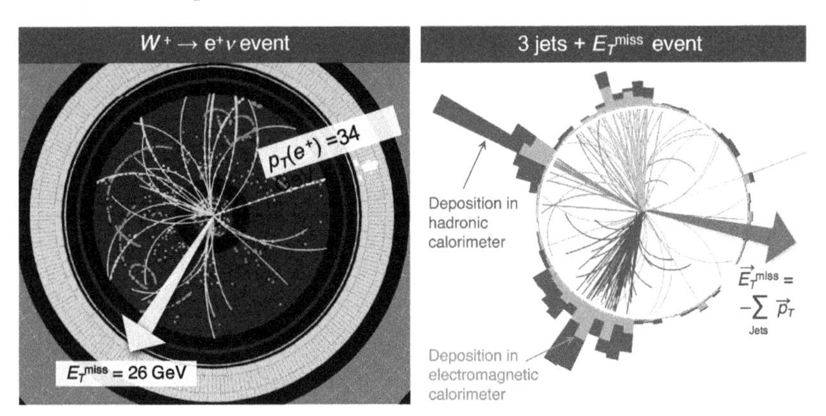

The signature of these non-interacting particles thus appears in the form of unbalanced momentum of the collision products. The neutrino momentum can be deduced as being opposite to the vector sum of all detected particles in the collision. However, in case of proton–proton colliders (or proton–antiproton colliders), a large fraction of the energy of the collision is in fact carried away by debris of the two incident protons flying close to the initial beam lines and are therefore not traversing the detector. Thus longitudinal momentum balance cannot be verified and moreover, because the hard collision occurs not directly amongst the protons but its constituents which carry only a fraction $x < 1$ of the proton momentum, the longitudinal centre-of-mass is usually different from the laboratory frame, i.e. there is no longitudinal balance of final state momenta even if they could be measured.

Fortunately, balance in the plane perpendicular to the beams can be accurately verified and therefore reveal possibly escaping particles and allowing to reconstruct their transverse momentum vector. To achieve this one of the main requirements in the design and construction of modern detectors is to make them as hermetic as possible, in order to measure with minimal loss in the transverse plane all electromagnetically and strongly interacting particles.

The *missing transverse momentum* is defined as the two-component vector built as the vector opposite to all measured momenta of an event in the transverse plane of the collision. At LHC collision energies, where the difference between momenta and energies is negligible (rest masses are small in comparison) and much smaller than the measurement precision, the magnitude of the missing transverse momentum vector is sometimes called missing transverse energy, E_T^{miss}. The measurement of this quantity is an essential analysis ingredient to reconstruct events with neutrino production and to search for invisible particles.

7.2.4. *Muon spectrometers*

Surrounding the hadron calorimeters, muon spectrometers are usually the outermost elements of collider detectors. Their purpose is to identify muons and to contribute to the measurement of their momenta. Muons, thanks to their relatively high mass (compared to electrons), do not suffer much from bremsstrahlung and, not being subject to strong interactions, they can penetrate calorimeters without being stopped, loosing only a small fraction of their energy through ionisation.

Muon spectrometers at the LHC are based on wire chambers sampling the muon trajectories. They provide a measurement of the curvature if embedded in a magnetic field, or of the flight direction otherwise (see for example figure 7.5). Both ATLAS and CMS experiments have chosen to have a muon spectrometer system embedded in a magnetic field, allowing a relatively precise measurement of muon momenta with a 1–10% momentum resolution ($\delta p/p$) depending on momentum and production angle. The muon spectrometers also play a key role in the triggering system of LHC experiments, as pattern recognition for muons is simple and specific elements of muon detectors can be sufficiently fast.

Inset 7.6: Pyramid imaging with cosmic ray muons using particle physics tracking detectors

The use of cosmic ray muons[a] for the imaging of buildings or structures on the Earth's surface is nowadays applied at many sites worldwide. The idea to use muon tomography to search for hidden chambers in the ancient Egyptian pyramids was pioneered in the 1960s by the American particle physicist Luis Alvarez (without success at the time, cf. footnote 1 on page 202). Muon tomography exploits the differential absorption of cosmic ray muons as a measure of the material thickness crossed. A hidden void would appear as a localised excess of recorded charged tracks coming from a well-defined direction. Just like computer tomography composes a three-dimensional view of a person's body from X-ray measurements taken at different angles, a structural image can be reconstructed by measuring muons that crossed an object in different ways.

In 2017, the *ScanPyramid collaboration*, consisting of three teams from Nagoya University (Japan), KEK (Japan), and CEA/IRFU (France) who developed and exploited particle detectors, announced the discovery of a new void in the heart of the Khufu (also known as Cheops) pyramid. This new cavity, which has a length of at least 30 metres, was first observed with nuclear emulsion films installed in the Queen's Chamber (by Nagoya University) and with a scintillator telescope installed in the same chamber by the KEK team. The discovery has been confirmed with high-resolution gaseous detectors, called MicroMegas (which stands for Micro-MEsh Gaseous Structure), installed outside the pyramid by the CEA team, allowing the physicists to refine the location of the void.

The picture below shows the layout of the telescope formed by four 60×60 cm^2 planar chambers. The top left insert illustrates the working principle of a MicroMegas chamber: ionisation electrons produced by the muon in the drift space traverse a metallic mesh and generate avalanches in the high-electric field region between the mesh and an anode, which is segmented in strips. The signals are collected on the strips along two perpendicular directions, allowing a precise two-dimensional measurement (within about 400 μm resolution) of the track position. The telescope then provides a measurement of the cosmic ray direction. As only muons with energy above a few tens of GeV can traverse the pyramid, the exposure time required to accumulate sufficient tracks to be able to identify a void of that size amounts to several weeks or even months.

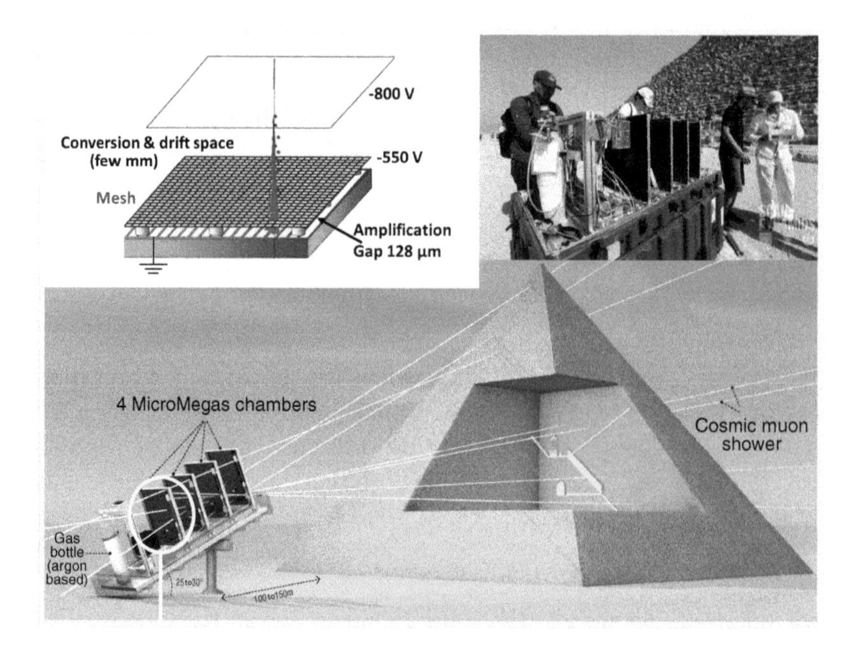

The figure below shows the results in terms of the measured muon flux as a function of the tangent of the two angles θ and ϕ defining the track direction. The big red area are muons that did not penetrate the pyramid. The histograms show the flux along the tangent of the θ direction corresponding to the two rectangles sketched with dashed lines in the two-dimensional flux distribution. The excesses (peaks above the continuum) correspond to the new void (orange cone) and a known chamber (red cone). The results are given for the telescope position labelled G2. The orange cone confirms the triangulation positions already given by the team from Nagoya University.

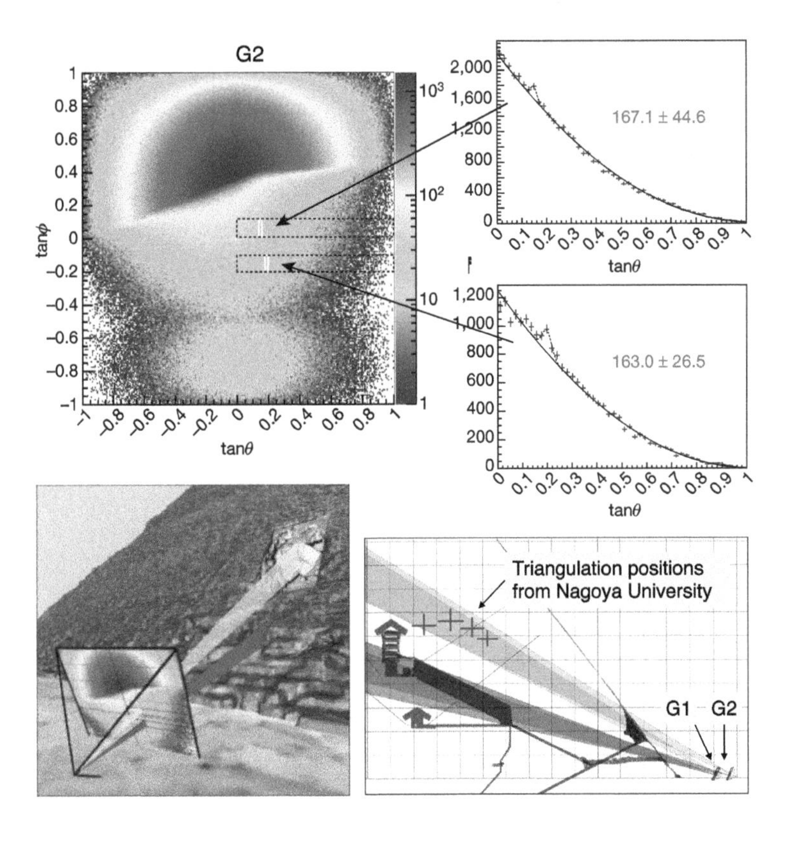

This is the first time that an instrument has detected a cavity located inside a pyramid from the outside. It is also the first discovery of a new and major internal structure of one of the great pyramids since the Middle Ages. *Source: K. Morishima et al., Nature 552, 386 (2017).*

---

[a]Cosmic ray muons are the decay products (together with neutrinos) of pions and kaons produced by the interaction of energetic cosmic particles (mostly protons) with the nitrogen and oxygen nuclei of the Earth's upper atmosphere. The total muon rate at sea level is about 10 000 per square metre and per minute. Their energy spectrum peaks at a few GeV and is falling with energy roughly with a power law $E^{-\alpha}$, with $\alpha \sim 3$ for muons coming from a zero zenith angle. As muons in this energy range loose about 1 GeV per metre of rock, only the high-energy muons reach the detectors after crossing the pyramid.

Chapter 8

The ATLAS and CMS Experiments

We are now going to discuss in some detail the two large general-purpose LHC experiments ATLAS and CMS. They are called *general purpose* as they were designed so as to be able to study essentially all the physics questions that can be addressed with the LHC. Two other large detectors, LHCb and ALICE, are installed on the LHC ring (see figure 8.1). They are optimised for a more detailed study of two specific domains of investigation: CP violation, and beauty and charm-quark physics for LHCb; heavy-ion collision physics for ALICE. Smaller-size detectors installed very close to the beam, TOTEM and LHCf, were designed to investigate "soft" proton–proton collisions with debris in forward direction. These special purpose experiments will be discussed in chapters 14 and 12, respectively. A seventh experiment, MoEDAL (which stands for Monopole and Exotics Detector at the LHC), is installed next to the LHCb detector. It consists of an array of plastic nuclear-track detectors with the aim to take a "photograph" of possible magnetic monopoles, which are hypothetical particles carrying magnetic charge.

8.1. Proto-collaborations EAGLE, ASCOT, L3P, CMS

As a result of the Lausanne, Switzerland workshop in 1984, where the excitement about the LHC grew significantly to mark it as the real starting point of activities, and the series of follow-up meetings, including the 1987 workshop in La Thuile, Italy and the Aachen, Germany conference of

Fig. 8.1. The four large experiments at the LHC and their location around the ring. *Image: CERN.*

October 1990 (see section 6.3.3), collaborations[1] began to form: a number of physicists knowing well each other through previous collaboration, sometimes over many years, as for example in experiments like UA1 or UA2, decided to pool their knowledge and experience in the design of experiments to be set up at the LHC. Thus the origin of the proto-collaborations EAGLE, ASCOT, L3P and CMS is first and foremost human. The LHC experiments, their concept and designs have very much profited from the experience gained in preceding experiments. For example, initiators of CMS (*Compact Muon Spectrometer*) were physicists from the UA1 experiment,

[1]Throughout this book we use the, by now customary, expression *collaboration* to indicate a group of people working together on the design, construction, operation and exploitation of a detector, trying to get the best possible physics results out of it, and signing altogether the scientific publications based on results obtained. There is some overlap and interchangeability in the usage of the words collaboration, experiment and detector, although collaboration refers more to the organisational aspects, experiment to the activities, instrumental and computational ones essentially, and detector to the hardware, the instrument itself.

Michel Della Negra from CERN, Jim Virdee from Imperial College London, and Daniel Denegri from CEA/IRFU (France), plus few others, all from UA1. On the other hand, EAGLE (*Experiment for Accurate Gamma, Lepton and Energy measurements*) was initiated by a number of physicists coming mainly from UA2 (several others coming from UA1, too) and led by Peter Jenni from CERN. The ASCOT experiment (*Apparatus with a Super-COnducting Toroid*) led by Friedrich Dydak (CERN), was gathering a number of physicists from fixed-target experiments at CERN with either neutrino or muon beams, or from experiments at LEP, in particular from the ALEPH experiment. A number of physicists working on the L3 experiment at LEP proposed the L3P project under the leadership of the Nobel Prize winner Samuel Ting. The four detector concepts were presented in Expressions of Interest at a meeting entitled "Towards the LHC Experimental Programme", organised by ECFA and CERN in March 1992 in Evian, France.

The physics of proton–proton collisions does not differ much from that in proton–antiproton collisions with which the physicists now engaging into CMS and EAGLE designs were familiar with. It was an opportunity to apply lessons learned from the ten-year long exploitation of the UA1 and UA2 detectors, correct the most obvious weaknesses of these two experiments, and adapt them to the much more challenging operational requirements of the LHC. These were indeed tremendous: a very fast detector response was required to face a proton bunch collision rate of 40 MHz, detectors and front-end electronics needed to withstand a harsh level of radiation, and extreme requirements on energy and momentum resolution and on detector granularity were set to be able to distinguish and select among the hundreds of particles produced in every collision of two proton bunches.

The starting point in the detector design was to draw lessons from the recently discovered W and Z bosons and apply those to the search for the Higgs boson, which was the obvious next task facing particle physicists and, besides searches for the unknown, the main reason for building the LHC. At the time of the design of the LHC detectors, the upper limit on the Higgs boson mass was about one TeV (derived from theoretical arguments), thus a very broad range of possible decay modes had to be addressed. In UA1 it was the detector hermeticity owing to its almost complete solid angle coverage as well as the detection of electrons at both calorimetric and particle-track reconstruction level that brought ultimate success. However, in 1990 it was not clear whether electron detection and identification was possible at

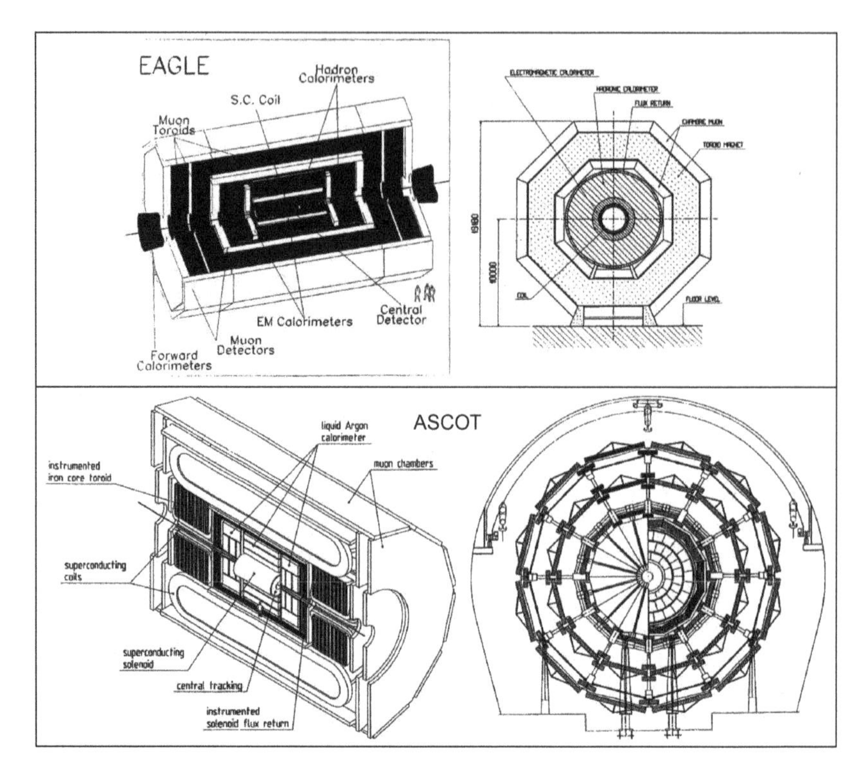

Fig. 8.2. Schemes of the EAGLE and ASCOT detectors, as presented at the Evian, France Workshop in 1992.

all in LHC running conditions. With a nominal luminosity of 10^{34} cm^{-2}s^{-1}, ten thousand times larger than at the CERN proton–antiproton collider, or the Tevatron at the time, it was not clear whether any tracking was feasible at all! It seemed that the best, if not the only way, was to concentrate on muon detection. This is still visible today in a "fossil manner" in the name CMS (*Compact Muon Spectrometer*), even if the detector design over the following two years gradually evolved to incorporate calorimeters and ultimately a very sophisticated high-performance tracking detector, together with a high-field solenoidal magnet (see figure 8.3). On the other hand, the UA2 detector was designed almost exclusively for high-precision calorimetric detection of high transverse energy electrons, photons and jets, disregarding particle momentum measurements with a central spectrometer, as well as the detection of muons. The EAGLE design (figure 8.2) corrected these weaknesses, whilst keeping the emphasis on excellent calorimetric detection of electromagnetic and hadronic showering particles.

Thanks to detailed physics, detector and digital detector response simulation, using also freshly acquired experience with detector prototypes exposed to test beams, physicists rapidly understood that constraints on detectors imposed by the LHC operation requirements could be met, provided that the detectors are of very high granularity, fast and radiation hard. It was the first time in particle physics that detector optimisation relied so deeply on detailed simulation studies benefitting from modern simulation programs (named GEANT and FLUKA) that were co-developed at CERN.

Launching the LHC experiments required the design of high performance detectors capable of addressing the physics questions and coping with the experimental challenges at the LHC. Both, the potential and the challenges were on a scale without precedent and unheard of in the history of particle physics. It was also necessary to form the collaborations, which meant being able to attract with an appealing detector concept a sufficient number of experienced as well as early career physicists and engineers with the competences and enthusiasm required to build the detectors. Moreover, of course, the collaborations needed the financial means to cover construction costs of the detectors and everything that goes with it. The CERN directorate imposed a ceiling of 475 million Swiss francs on the cost of each of the general purpose detectors. Whilst the construction cost of the LHC itself was essentially covered by the CERN budget, the funding of the detectors had to be covered by collaborating laboratories and countries taking part in the experiment and these were not limited to CERN member states. Thus a very important activity during these initial years consisted in looking for collaborators all over the world, find appropriate and satisfying tasks and responsibilities for each group in the common endeavours, ensure the support of the corresponding funding agencies,[2] and thus build a credible collaboration both in terms of competences and from the financial point of view.

After the Evian workshop in 1992 it became quickly clear that the EAGLE and ASCOT projects should fuse as they shared a number of common features: liquid argon for electromagnetic calorimetry and an external muon spectrometer based on a toroidal magnet system which could measure muons independently of the inner tracking detector. This is how the ATLAS (*A Toroidal LHC ApparatuS*) project was born, with Peter Jenni

[2]Funding agencies of the LHC experiments are usually public research organisations or competent ministries. There is at least one funding agency per collaborating country, in larger countries there may be several.

and Friedrich Dydak as co-spokespersons for about two years, then continued by Peter Jenni as spokesperson over fifteen years. The ATLAS Letter of Intent was submitted, together with those of CMS and L3P, on October 1st, 1992.

In 1993 CERN management wished to reduce the number of potential LHC experiments. As both CMS and L3P concepts were based on large solenoidal magnets, it was natural to look for an agreement leading to a merger of the two projects. The discussions among the leaders of the two groups lasted for several months, but no agreement could be found, even if the designs were globally similar. CERN management finally opted for the CMS project as it was respecting the given cost limit, which was not the case for L3P whose design was more ambitious. Subsequently, a large influx of physicists, engineers, and of financial means from L3P reinforced significantly the CMS collaboration.

8.2. The CMS experiment

8.2.1. *A detector built around its magnet*

The key idea in the CMS detector concept was to have as strong a magnetic field as feasible to measure precisely muons in a fully external spectrometer — as initially no inner tracking was thought possible at the expected LHC luminosity. Particles other than muons were to be filtered out in two ways: on one hand by the strong magnetic field bending softer particles, thus not allowing them to reach the spectrometer; on the other hand, by the very structure of the spectrometer, with muon chambers inserted at various depths within the iron return-yoke absorbing all particles except for muons. Thus, whatever could be the complexity and multiplicity of particles produced in collisions at the centre of the detector, this arrangement would allow detection, identification and measurement of muons, thus guaranteeing discovery of the Higgs boson essentially through $H \to ZZ^{(\star)} \to \mu^+\mu^-\mu^+\mu^-$.[3]

The key element of the CMS detector design is therefore a large superconducting solenoid, thirteen-metre long and six metres in diameter

[3] Rapidly it was understood that central tracking would nonetheless be feasible and that at least track segments in the outer parts of an inner tracker could be connected with muon track segments in the spectrometer, thus much improving the momentum resolution. It also gradually became clear that not only hadronic but also electromagnetic calorimetry was feasible and necessary.

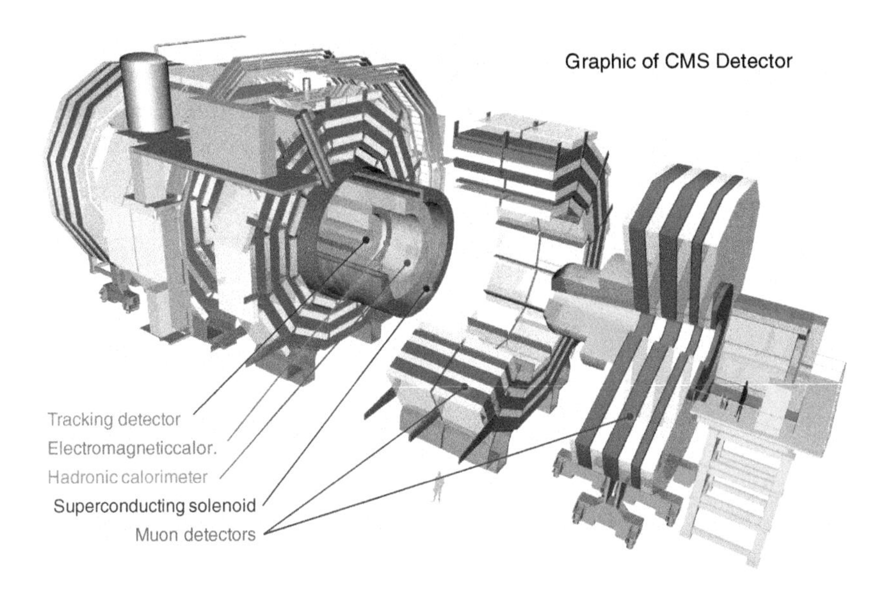

Fig. 8.3. Three-dimensional schematic view of the CMS detector in its open configuration to show the different constituent parts or sub-detectors. In operation condition the detector is closed and these parts are regrouped ensuring the hermeticity of the detector. The three-metre radius of the inner wall of the cryostat of the superconducting solenoidal magnet, itself resting on the central wheel-shaped module, is sufficient to house the central tracking detector, surrounded by full barrel electromagnetic and hadronic calorimeters. There are five wheels, each weighing about 2000 tons, equipped with muon chambers, organised in four layers to ensure redundancy. The end-cap calorimeters are supported by a forward-protruding structure called the *nose* and located on a front-end of the three-layered muon end-cap detector wall. All layers are equipped with muon chambers over their entire surface. Beyond the end-caps are the very forward calorimeters designed to withstand very high radiation doses. *Image: CMS Collaboration.*

(inner bore), a weight of 12 500 tons (by comparison the Eiffel tower is about 8000 tons), producing a uniform 4-tesla field, with axial field lines parallel to the beam axis (figures 8.3 and 8.6). The operating temperature of the superconducting magnet is four kelvin and the current in the magnet amounts to 19 000 amperes. At the beginning of the 1990s, such a magnet represented a major technological challenge, as no other large particle-physics magnet had ever been built with a field exceeding 2 tesla. It became possible thanks to several technical innovations, both at the level of the superconducting cable (figure 8.4) and its winding (in four layers, whereas previous magnets had two layers, with a total cable length of 53 kilometres), and at the level of the cryogenic system using thermosiphon cooling to safely maintain the low temperature.

Figure 8.5 shows a charged-particle trajectory in CMS, first bent by the 4-tesla field,[4] while inside the solenoid, and then reverse-bent when going through the outer spectrometer embedded in the iron return-yoke of the solenoid where field lines are in the opposite direction to close the magnetic loops. The field strength in the return-yoke is about two tesla.

The CMS solenoid is the largest in the world. The energy stored in its magnetic field is about three billion joules. Its construction is the very example of the international nature of particle physics undertakings: the magnet was designed at the CEA/IRFU (Saclay),[5] the coil was wound in Italy at Ansaldo (Genova), the superconducting cables were a co-production of Switzerland, Finland, Japan and France. The huge room-temperature bus-bars (the current leads) were produced in Croatia. The mechanical structure, including the iron return-yoke and its feet, was designed and built in a collaborative effort managed by CERN with laboratories from Germany, Italy, Russia, the Czech Republic, Switzerland and Pakistan for the barrel part, and the USA, Japan and China for the end-caps. The final magnet assembly and tests were performed at CERN. The magnet project is also a good example of control in costs: in 1992 the cost was evaluated at 120 million Swiss francs and when closing the financial construction books in 2008, the final accounting was 125 million Swiss francs.

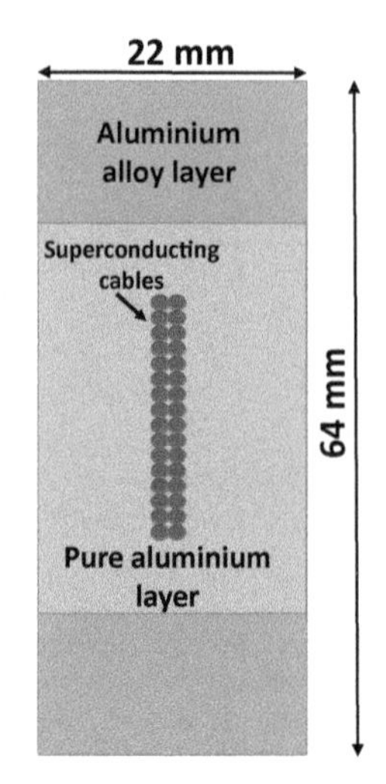

Fig. 8.4. Cross-section of the three-component reinforced superconducting cables of the CMS solenoid. The so-called Rutherford type superconducting cables are surrounded by a pure aluminium layer for thermal and electric stabilisation, ensuring conduction in case of a quench. The outermost aluminium alloy layers are reinforcements to withstand the magnetic forces.

8.2.2. *Inner tracker*

The most complex detector within CMS is the inner tracker (figures 8.5 and 8.7). To be able to reconstruct individual tracks among the many hundreds

[4]To allow a safety margin, the operational field strength eventually chosen during physics runs is 3.8 tesla (current of 18 000 amperes).

[5]Chief designers were P. Desportes, A. Hervé and J.C. Lottin.

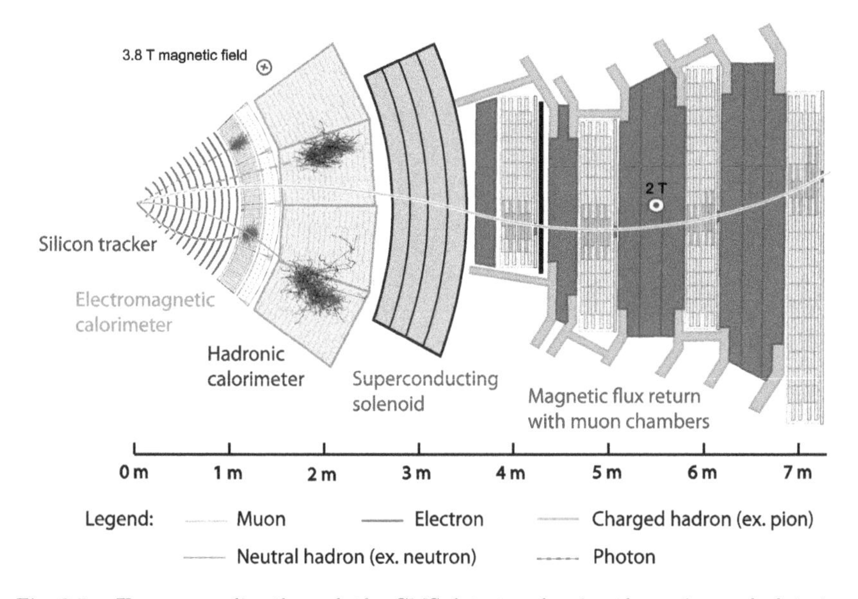

Fig. 8.5. Transverse slice through the CMS detector showing the various sub-detectors and their functions. The signals from different types of particles recorded in the sub-detectors are also shown (tracks in inner tracking detector, showers in electromagnetic and hadronic calorimeters). The muons, having little interaction with matter, are the only ones (except for neutrinos and other non-interacting particles) to reach the outer spectrometer. Their trajectory is bent one way inside the solenoid and the opposite way when traversing the iron return-yoke where the magnetic field lines are reversed. The four-layered structure of the magnet winding is also indicated. *Image: CMS Collaboration.*

of tracks produced in every collision, it is necessary to have very small elementary detectors and in very large numbers. This is a bit equivalent to the finesse of the grains on a Roman mosaic: the finer the grains the finer the details. Granularity, be it for trackers or calorimeters, is at the heart of the pattern recognition capability. The instrumentation option initially chosen by CMS was *microstrip gaseous detectors* (MSGC), the latest evolution of Charpak's multiwire proportional chambers. MSGC were a brand new detector type at the time. The ionisation produced in a gas layer is detected by microstrip electrodes 300-micrometre wide and six-centimetre long. The cost of such a detector was then about ten Swiss francs per square centimetre, and about 200 square metres were needed for the whole detector. This meant an overall price for the tracker of several tens of millions of Swiss Francs, not yet including the cost of mechanics and electronics. Another option considered at the time were silicon microstrip detectors

Fig. 8.6. Lowering in the underground experimental cavern of the central module of the mechanical structure of CMS, fully equipped with all its muon chambers and carrying the solenoid in its cryostat. This central module weighs 2500 tons, and about twelve hours were needed for this element to be carefully lowered from the surface to a depth of hundred metres, as the access shaft of 24-metre diameter was only about 20-centimetre wider than the module. *Image: CMS Collaboration.*

(see inset 7.2). In this type of device, the ionisation takes place in the 300-micrometre thickness of silicon and the overall number of ion pairs is about twenty times larger than for the same thickness of gas. This produces larger and more easily detectable electrical signals. In 1992, the price of such a silicon detector was about 50 Swiss francs per square centimetre, which was prohibitive. However, with the exponential growth of integrated circuit techniques in the 1990s and the exploding demand for mono-crystalline silicon on world markets, the growth of production resulted in a collapse of prices down to the level of nine Swiss francs per square centimetre in 1995.

Despite all the efforts, studies on prototypes and expenses in various laboratories of the collaboration to develop the MSGC, the CMS management took the decision to drop the MSGC option and switch to constructing

Fig. 8.7. Picture of the inner tracking system of CMS. The outer envelope is a cylinder six-metre long and 1.3-metre in radius, surrounding the interaction region. It contains about ten million silicon microstrip channels, with strip widths varying from 80 to 180 micrometres and lengths from 6 to 20 centimetres. The microstrip modules are organised into eight cylindrical shells spanning radii from 20 to 110 centimetres and eleven disc-like structures at each end of the cylinders (only the cylindrical central part of the tracker is visible on the photo). The measurement precision of a point on a trajectory of a particle going through this detector is about 30 micrometres, which is essential for a precise momentum measurement. For example, for transverse momenta not exceeding 100 GeV, the momentum resolution of the overall tracking system (microstrip plus pixel detector) is better than 1% at normal incidence and 7% for tracks almost parallel to the beams. The uniformity of the magnetic field throughout the volume of the tracker is better than one permil. *Image: CMS Collaboration.*

a silicon-based tracker. Such a decision, justified a posteriori on scientific and technical grounds, generated a deep crisis in the CMS collaboration among the laboratories that had spent so much time and resources on MSGC developments. With some diplomacy and distributing the construction of the silicon tracker over many countries, including Belgium, Finland, France, Germany, Italy, Switzerland, USA, UK, India, Taiwan, etc., the problem was overcome, and in the years between 1998 and 2005 CMS succeeded in building the most ambitious tracking detector among the LHC experiments (figure 8.7).

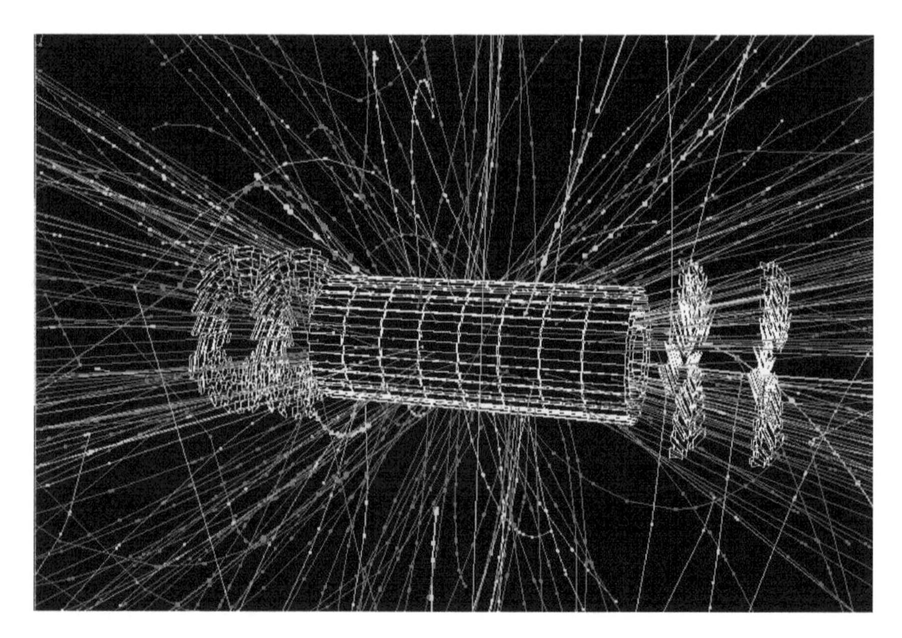

Fig. 8.8. Early simulation of a $H \to ZZ^{(*)} \to e^+e^-\mu^+\mu^-$ event in the CMS pixel detector. *Image: CMS Collaboration.*

Around 1995, in view of the importance that the identification of b quarks (*b-tagging*) took for the detection of top quarks at the Tevatron and was expected for the search for supersymmetry,[6] CMS decided to add at the centre of its inner tracker, as close to the interaction region as possible, a high granularity silicon-pixel detector. The 66 million pixels, $100\,\mu\text{m} \times 150\,\mu\text{m}$ in size, are organised into three cylindrical barrel layers at radial distances from the beam lines of four, seven and eleven centimetres, respectively, with two forward discs[7] closing the cylinders in the end-caps (figure 8.8). With this arrangement, the precision in the measurement of a particle trajectory at the level of the interaction point, the so-called *interaction vertex*, is about ten micrometres. The pixel detector is also essential in seeding tracks in the pattern recognition process. The key role in the design and construction of the CMS pixel detector was played by Swiss and US laboratories for, respectively, its barrel and end-cap parts.

[6]Today we know that b-tagging is also a crucial tool for Higgs boson physics.
[7]For the runs of 2018, the CMS pixels system was reinforced with a third forward disc in the end-caps.

8.2.3. *Electromagnetic calorimeter*

The design and construction of the electromagnetic calorimeter was the most delicate and expensive part of the CMS detector. This sub-detector plays a key role in the Higgs boson search, as it gives access to two essential Higgs boson decay channels: two photons, and two Z bosons with one or both Z decaying to an electron–positron pair.

In its initial design, the electromagnetic calorimeter was to be made of elements in the shape of oblong rectangular towers themselves made of alternating millimetre-thick slabs of lead and scintillator over a depth of about 40 centimetres and readout through optical fibres. The aim was to achieve a 1.5 to 2% precision on the energy measurement of electrons and photons in the 30 to 2000 GeV range. This so-called *shashlik* configuration was, however, not totally satisfactory: it was not providing sufficiently high granularity (the towers had $4 \times 4\,\mathrm{cm}^2$ front-face cross-section), the energy resolution was marginal, and, most importantly, there were serious worries about the long-term radiation hardness of scintillators.

In 1994, CMS physicists learned about a scintillating crystal, lead-tungstate (PbWO4), a rapid scintillator of ten nanoseconds scintillation lifetime, very high density of 8.3 g/cm^3 and therefore short radiation length of nine millimetres and a Molière radius (quantifying the lateral spread of a shower) of two centimetres. This would allow to build a higher granularity and more compact calorimeter with a crystal depth of about 20 centimetres, thus giving more radial space for the tracker and thus even better momentum resolution.

However, at the time, there existed just about one cubic centimetre of this crystal on the entire planet and it did not have the required transparency, radiation hardness nor mechanical properties. The full electromagnetic calorimeter needed about eleven cubic metres of crystals and one had to be ready for data taking in 2003 or 2004. This conundrum was at the origin of an incredible scientific, political and industrial endeavour. From 1995 on, prototype crystals of constantly increasing performance were produced at Bogorodijsk in Russia and the Shanghai Institute of crystals in China and shipped to CERN for testing. The choice of Russia or China rather than US or European suppliers was due to the very competitive prices initially proposed, and the available production capacities in Russia. By the year 2000, after five years of R&D, prototype crystals meeting all requirements were finally obtained. Their shapes were $2.2 \times 2.2 \times 23$ cm^3 in the barrel part of the detector, and slightly larger for the end-caps (figures 8.9 and 8.10). Mass production of 76 000 crystals was to be launched.

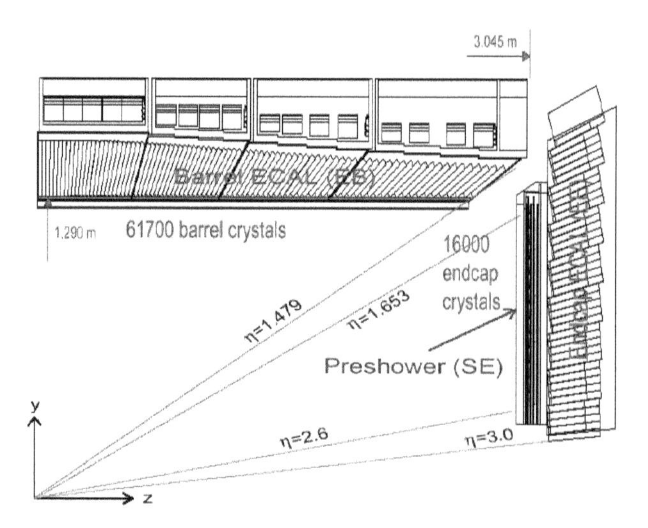

Fig. 8.9. Transverse cut through a quadrant of the CMS electromagnetic calorimeter. The mechanical mounting is such that all crystals point towards the centre of the apparatus (collision point). In front of the end-cap section is a pre-shower detector to improve the separation of a single isolated photon from two nearby photons originating from a $\pi^0 \to \gamma\gamma$ decay at high energies. *Image: CMS Collaboration.*

Fig. 8.10. Optical control of PbWO4 crystals before insertion in the electromagnetic calorimeter of CMS. *Image: CMS Collaboration.*

At that time first beams in the LHC were expected in 2005. A race with time thus started, with all the crystal-growing capabilities available at Bogorodijsk put into use. This production order was a benediction for the Bogorodijsk factory, as it formerly provided crystals for the needs of the Red Army, but orders became scarce after the collapse of the Soviet Union. If initially the crystals costed about two US dollars per cubic centimetre, a rapid adaptation to capitalism let the price rise at every new delivery to finally reach more than seven US dollars per cubic centimetre! By 2005 about 80% of the crystals were produced. CMS, fearing that it would be late despite a postponement of the LHC start-up by two more years to 2007, launched a second production line for crystals in China, where ultimately about 15% of the crystals were produced.

By summer 2008 all the crystals were produced, equipped with their photodiodes[8] to measure the scintillation light, and mounted on their mechanical structures (figure 8.9). Modules were tested and pre-calibrated either in a test beam or with cosmic rays after installation. The calorimeter of CMS was ready just in time for the expected data taking in September. What a race!

Thanks to a laser calibration system that allows to follow and monitor the effects of irradiation on the crystals (loss of transparency during irradiation and its recovery during periods without beam),[9] the calorimeter provides today high-precision measurements of photons and electrons in the 1–2.5% range.

The main institutes responsible for the construction of the electromagnetic calorimeter, including the mechanical structure and the electronics, came from Switzerland, France, USA, UK and Italy. The pre-shower detector installed in front of the end-cap, equipped with silicon-readout channels (figure 8.9), and helping with the identification of photons, was a collaborative effort of several institutes mainly from Taiwan, Greece, the UK and Armenia.

[8]Specific avalanche photodiodes (APD) were developed for CMS needs by the Japanese company Hamamatsu Photonics, which has a long and successful record of cooperation with experimental particle physics. A total of 120 000 APDs were produced for CMS. The same APDs and their electronics readout system are now used in the industrial production of TEP (electron–positron tomography) cameras for medical imaging. The first TEP devices were using 12 000 APDs, and the most recent ones employ more than 100 000 APDs per readout system.

[9]This calibration system was developed by physicists from CEA/IRFU in France and CalTech in the USA.

8.2.4. *Hadronic calorimeter*

The downstream component of the CMS calorimeters, the hadronic calorimeter, follows a more traditional conception and design. It is of a sampling type with alternating 4–5-centimetre thick brass plates and 4-millimetre thick scintillator plates. It is subdivided into 5500 towers of 10×10 cm^2 cross-section pointing towards the centre of the detector. The calorimeter is located inside the solenoid magnet to avoid too much non-instrumented material — the thickness of the magnet plus its cryostat — between electromagnetic and hadronic calorimeters, which would degrade the energy and missing transverse energy resolution. The hadronic calorimeter is a co-production mainly of the USA, Russia, Ukraine, Bulgaria, and India.

An anecdotal aspect in the construction of this calorimeter was that Russian members of CMS managed to recover at low cost 800 tons of brass from discarded naval artillery shell casings (figures 8.11) from the Soviet Black Sea fleet. The brass shells were melted in Bulgaria, cut and shaped in a shipyard in Spain according to engineering drawings from Fermilab in the USA. The scintillator equipping the calorimeter was produced in Ukraine and the readout and calibration systems in the USA. Another nice example of particle physics trans-national cooperation.

The calorimeter precision in the energy measurement of a 100 GeV hadron is of the order of 10%, dominated by a stochastic uncertainty component of 100% (see inset 7.4). The central hadronic calorimeter covers the barrel and part of the end-cap regions. It is complemented in the two forward regions (figure 8.3) by special quartz-fibre calorimeters, built and equipped cooperatively by the USA, Turkey, Hungary, Russia and Iran. Quartz-fibre technology was chosen for its exceptional radiation hardness needed in this most exposed area near the beam. The collected signal in such a fibre calorimeter is produced by Cherenkov photons (see inset 14.1 on page 475) emitted by hadronic shower particles propagating in the fibres and not by scintillation light.

A special effort was made at the design stage of CMS to avoid a vertex-pointing junction or "crack" (11-centimetre wide and 2-centimetre clearance) in the transition between the barrel and end-cap regions (figure 8.12). Through this gap all the cables and pipes of the inner detectors services had to exit, and great care had to be taken to ensure maximal hermeticity and minimal dead zones in the detector. The CMS calorimetric system has an almost complete coverage in terms of space-angle of the interaction region,

Fig. 8.11. The hadronic calorimeter of CMS (photo on the right) is made of brass plates interspersed with scintillator plates. The brass comes from recycling 800 tons of naval artillery shell casings discarded by the Russian Black Sea Fleet (left). *Image: CMS Collaboration.*

down to a polar angle of 0.8 degrees with respect to the beam direction, thereby ensuring a good measurement of missing transverse energy.

8.2.5. *Muon spectrometer*

The outermost part of the CMS detector is instrumented by a muon spectrometer. It has to fulfil several tasks and for that purpose several detection techniques have been implemented. Firstly, it has to measure the muon trajectory as precisely as possible to provide an accurate measurement of muon momentum, but it also serves to very rapidly identify the passage of a muon through the detector that should trigger the data acquisition of signals throughout the CMS detector.

Two different techniques are used for the precision measurement of muon trajectories. In the central, barrel detector the iron return-yoke wheels are equipped with chambers containing dense layers of *drift tubes* (DT),[10] which provide a position measurement precision of about 250 micrometres per tube (and about 100 micrometres per muon chamber of twelve layers) and an angular precision (transverse to the beam) of about one milliradian per muon-chamber track segment. In the two end-cap regions (figure 8.12)

[10]Drift tubes are simplified wire chambers. They are made of individual tubes about one-metre long and a few square centimetres cross-section. The tubes are filled with gas and have a single stretched anode wire in the centre and a cathode on the opposing sides of the tube. From the drift time measurement of the ionisation signal and known drift velocity one can obtain a relatively precise position measurement. However if several particles cross the same tube along its length, there are ambiguities in the signal determination. A total of 172 000 DTs are used in CMS.

proportional *cathode strip chambers* (CSC) are employed.[11] This type of detector has initially been developed for the needs of the SSC experiments (see section 6.3.1) and is more suitable for the larger muon rates and rapidly varying magnetic field encountered in the CMS end-cap region. The precision of the CSC measurements is comparable to that of the drift tubes.

Drift tubes and CSCs are also used to trigger the data acquisition system in case an interesting collision event occurred as their time resolution is in the five- to ten-nanosecond range. This system is complemented by a third one based on *resistive plate chambers* (RPC). These devices provide a worse position measurement, of order a few millimetres to a centimetre per point, but they are extremely fast, with response time[12] of two nanoseconds. This is very important to be able to distinguish the individual proton bunch collisions following each other at a rate of one every 25 nanoseconds.

Thanks to the double triggering scheme, heritage of its initial concept as a Compact Muon Spectrometer, the CMS muon measurement system is very robust. Altogether, combining the track measurement in the inner tracker with the one in the outer muon spectrometer, a precision ($\delta p/p$) on the muon momentum measurement of about 1% is obtained for central muons with momenta up to 100 GeV, degrading to 7% at 1 TeV. It is worth mentioning that, owing to the large silicon tracker and the 4-tesla field of the CMS magnet, with a proportionally large track bending and sagitta, it was sufficient to require a 70–100 micrometres position measurement precision on the muon tracking devices, and about the same on the alignment system, which are less severe requirements than those of the ATLAS muon spectrometer.

The construction of the huge CMS muon system was a large cooperative effort among many institutes. The barrel DT sections, including mechanics and electronics, were mainly provided by institutes from Italy, Germany, Spain, Russia, China; for the end-cap CSCs, mainly institutes

---

[11] Cathode strip chambers are also detectors based on ionisation with multiplication in gases, but the cathodes are now segmented into strips orthogonal to planes of anode wires. This allows distinguishing without ambiguities multiple-particle crossings of the same chamber layer. A total of 500 000 CSC channels are used in CMS.

[12] In resistive plate chambers two planar electrodes placed in a gas enclosure and spaced by few millimetres are put at a high voltage difference. The passage of a charged particle gives rise to a spark and the induced signal is recorded on back-plate strip electrodes. In contrast to drift tubes and CSCs, where ion drifts can take hundreds of nanoseconds, the readout of RPCs is almost immediate. The RPC detector covers a very large surface of about 5000 square metres.

Fig. 8.12. CMS detector during the final assembly stages, after installation of the beam pipe. On the right-hand side is protruding the so-called *nose* carrying end-cap calorimeters and a layer of CSCs; in the final detector closure configuration, this nose fits inside the barrel part of the detector visible on the left, providing a hermetic detector arrangement. *Image: CMS Collaboration.*

from the USA, Russia, China, and JINR-Dubna,[13] whilst for the RPCs the effort was made by institutes from Italy, Korea, Pakistan, India, Bulgaria, Poland, Belgium, Brazil and Egypt. The laser-based alignment system of the entire muon system, ensuring a better than 100-micrometre precision, was mainly under the responsibility of Spain. For cost reasons the full CMS system was not implemented during the initial running period of the LHC at lower energy (see section 9.4). It was completed during the 2013–2014 LHC technical stop.

[13]The Joint Institute of Nuclear Research (JINR) is an international research organisation with eighteen member states (as of 2020) from part of the former USSR plus Bulgaria, The Czech Republic, Cuba, Mongolia, Poland, Romania, Slovakia and Vietnam, some countries being also member of CERN. Associated member states are Egypt, Germany, Hungary, Italy, Serbia, and South Africa.

8.2.6. *Financial matters*

The overall cost of CMS at closure of the accounting books during the LHC start-up in 2008, including also infrastructure expenditures, installation and commissioning, amounted to 540 million Swiss francs. For a project whose ceiling cost in 1994 was set at 450 million Swiss francs by CERN management, this is rather remarkable. The most expensive subsystems were, as expected, the solenoid (125 million Swiss francs including the return-yoke) and the electromagnetic calorimeter (117 million Swiss francs). In chapter 15, we discuss the required upgrade and related costs for CMS running up to about 2035 during the high-luminosity phase of the LHC.

8.3. The ATLAS experiment

With its 46-metre length and 25-metre height, ATLAS is the largest particle physics detector ever built (figures 8.13 and 8.14). The sub-detectors making up ATLAS have same functions as we saw in CMS, but the specific technical choices for many of them are quite different.

8.3.1. *Inner tracking detectors*

The inner tracking detector adopted by the ATLAS collaboration follows the design proposed for the EAGLE project. It is located inside a thin-walled and transparent (4.5 centimetres, 0.66 X_0, see inset 7.4) superconducting solenoidal magnet[14] of 2.4-metre bore and 5.3-metre length, designed at KEK in Japan. This solenoid provides an axial magnetic field of 2 tesla through a 7600-ampere current with a single coil winding. This choice was dictated by the desire to minimise the amount of material in front of the electromagnetic calorimeter and thus minimise fluctuations in initial shower developments. The price to pay is, however, a reduced momentum measurement precision for charged particles by a factor of almost two compared to CMS, for particles emitted in the central region and with high momentum for which degradation from multiple scattering is small. For particles produced at smaller angles relative to beam lines, the difference between ATLAS and CMS diminishes notably. The impact of this design choice on the majority of the physics is limited as the energies of electrons and particle jets, mostly composed of pions, kaons and photons, are more precisely measured in the calorimeters.

[14]The magnet is housed in the same cryostat as the electromagnetic barrel calorimeter.

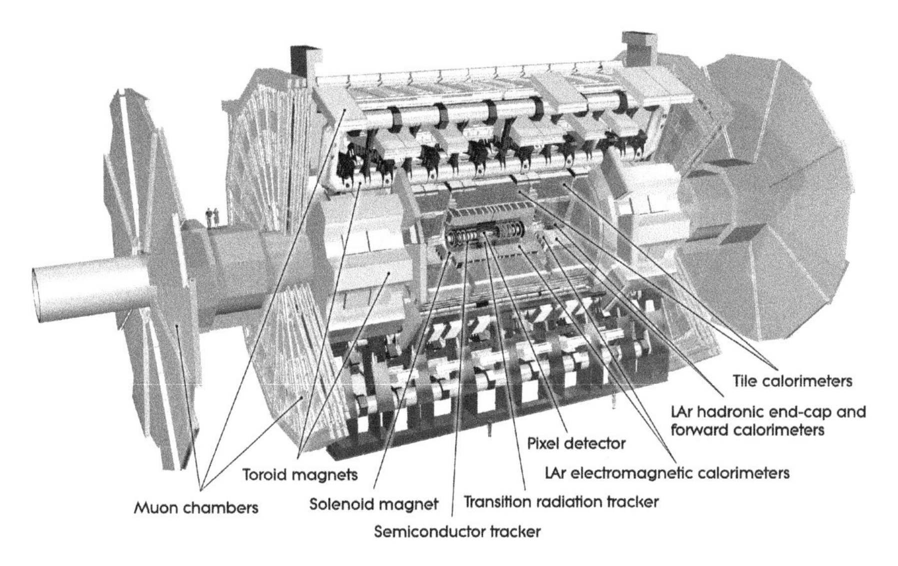

Fig. 8.13. The ATLAS detector in the configuration at the LHC start-up in 2008. *Image: ATLAS Collaboration.*

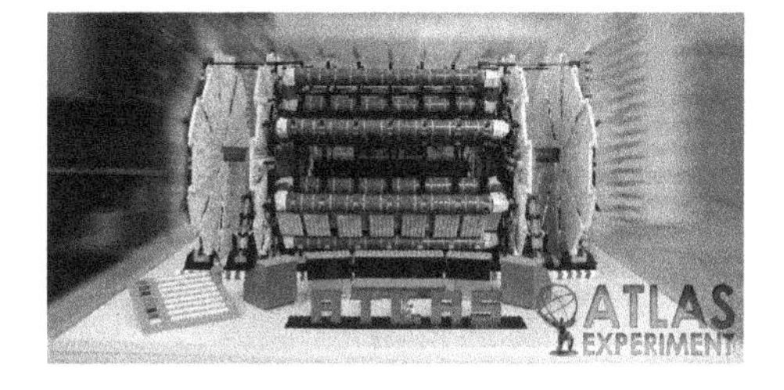

Fig. 8.14. A Lego model of ATLAS.

The tracker (figures 8.15 and 8.16) in its inner part, for radii less than 50 centimetres, is made of concentric cylindrical layers of silicon detectors: first, three layers of pixels, surrounded by eight layers of silicon microstrips. This was clearly not the obvious choice at the beginning of the 1990s. At the time it was thought that the high-radiation levels to which the innermost detector layers would be exposed at full LHC luminosity would not allow the use of silicon pixels. However, subsequent studies and advances made in the design and construction of pixel detectors greatly improved their radiation hardness, leaving no doubt on the applicability of this detection technology.

During the technical stop of the accelerator in 2013–2014, a new cylindrical layer of pixel detectors with a very small radius, $R = 3.3$ centimetres to the beam, was inserted inside the existing set of cylindrical pixel layers. The purpose of this new detector, called the *Insertable B-Layer* (IBL), was to

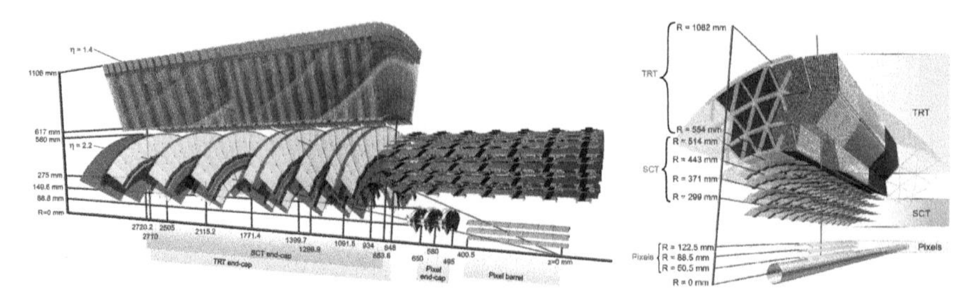

Fig. 8.15. Scheme of the ATLAS inner tracking detector. Left: cut through central and forward parts showing the three types of detectors used, silicon pixels at smallest radii, silicon microstrips at intermediate radii, and a transition radiation tracker (TRT) at largest radii. Right: a cut through the central part of the tracker. *Image: ATLAS Collaboration.*

improve the reconstruction of secondary vertices, mostly due to b hadrons, and thus improve b-tagging. With a lifetime of about 1.6 picoseconds (in their rest frame), b hadrons[15] decay within a few millimetres from the collision point where they are produced (the primary vertex). Such decays are difficult to detect especially when there are many proton–proton collisions occurring in a single bunch crossing (see, for example, figure 10.3), a situation experienced during most of the LHC run periods. The identification of b-hadron decays is of primordial importance for the study of Higgs bosons, whose dominant decay mode is into a pair of b quarks, but also for all processes involving top quarks, which decay almost exclusively into a b quark and a W boson. The silicon part of the ATLAS inner detector is the fruit of a collaborative effort among many institutes[16] from several countries with (in alphabetical order) Australia, Czech Republic, France, Germany, Italy, Japan, The Netherlands, Norway, Poland, Russia and JINR,[17] Slovenia, Spain, Sweden, Switzerland, UK, and USA as the main contributors.

Surrounding the silicon-detector layers, ATLAS has an original tracking detector made of thin (4-millimetre diameter) reinforced Kapton *straw tubes* and equipped with a single 30-micrometre thick anode wire in the centre.

[15]Hadrons carrying a b quark are called b hadrons. These may, for example, be a B^- meson, made of a b quark and a $\bar{u}$ antiquark, or a Λ_b^0 baryon, made of udb quarks.
[16]For most sub-detectors, physicists, engineers and technicians from CERN were involved at one or several stages of the development, from the design to construction and commissioning. CERN contributions are thus implied in the lists of countries provided in this chapter.
[17]The specific feature of JINR as an international inter-governmental organisation motivated a number of institutions from countries of the post-Soviet area to join ATLAS: Armenia, Azerbaijan, Belarus and Georgia.

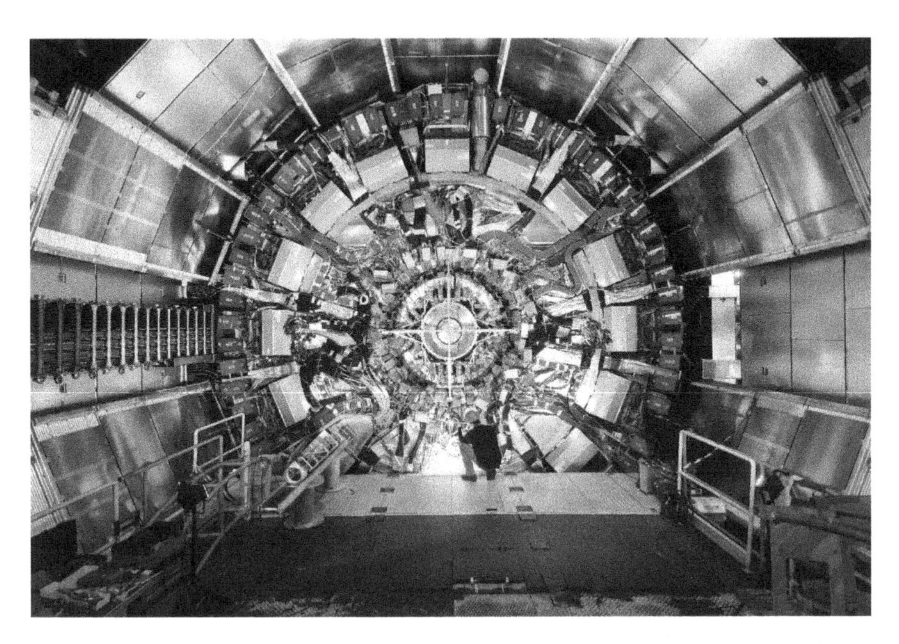

Fig. 8.16. View of the ATLAS inner detectors. At the centre, the inner tracking detector with a radius of about one metre. It is surrounded by the inner solenoid producing a uniform 2-tesla field throughout the volume of the tracker. The magnet itself is located inside the cryostat of the liquid argon electromagnetic calorimeter, whose electronic readout boxes are visible, covering in part the hadronic calorimeter located at larger radius. On the outer part, the first layers of muon chambers, part of the muon spectrometer, subdivided into sixteen azimuthal sectors, are seen. *Image: ATLAS Collaboration.*

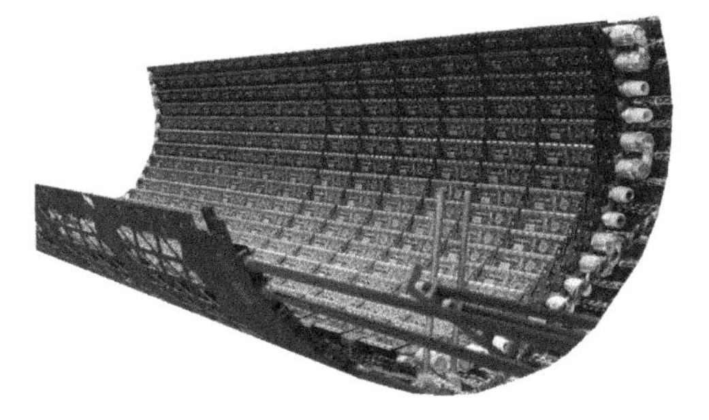

Fig. 8.17. A half-cylindrical silicon pixel module of ATLAS, with its readout electronics. *Image: ATLAS Collaboration.*

The straws are filled with a special gas mixture containing 70% of the noble gas xenon for good X-ray absorption. Each straw is in fact a small drift chamber with spatial resolution of about 170 micrometres per measurement point (figure 8.17). Thanks to the 50 000 axial straws in the barrel part and 320 000 radial ones in the forward parts, the trajectory of each charged particle is sampled about 30 times for a precise measurement of the transverse track coordinates.

The remarkable granularity of the ATLAS tracking detector makes of it a sort of bubble chamber, allowing to distinguish the many particles produced at each collision in the LHC. Furthermore, 20-micrometre thin polypropylene fibres are placed between straw tube layers. As polypropylene has a high-refraction index, a charged particle traversing it emits so-called *transition radiation.* The lighter the particle (an electron for example), the more intense is the radiation. The photons from the transition radiation can, in turn, ionise the xenon gas in the straws and thus be detected. This type of detector, called a *Transition Radiation Tracker* (TRT), allows by the amount of detected transition radiation to distinguish electrons from all the other, much heavier charged particles in a broad energy range.

At nominal LHC luminosity, when proton bunch collisions follow each other at a very high rate of 40 MHz, about 20% of the straw tubes nearest to the beam are hit and thus occupied. Pattern recognition among the many charged particles is then rendered more difficult. Nonetheless, compared to a tracker fully equipped with silicon detectors, the TRT tracker is significantly cheaper. It also provides many more measurement points per track, simplifying pattern recognition (the correct reconstruction of tracks from measured points). The choice of ATLAS has thus been to combine this TRT tracker in the outer regions of the tracker with a silicon detector technique for the regions closer to the beams (figure 8.15). In addition to CERN, which carried a leading role in the development of the TRT detector, the main contributions came from Denmark, Poland, Russia and JINR, Sweden and USA.

8.3.2. *Calorimeter systems*

The powerful calorimetric system of ATLAS, both its electromagnetic and hadronic components (figure 8.18), follows also the original EAGLE proposal, inheriting, among others, from the experience gained with the earlier experiment UA2. A strong point of UA2 compared to the UA1 detector was the geometric arrangement of the calorimeter cells in pointing direction towards the collision point. For a sampling calorimeter this is not so simple to realise as the most straightforward way is to form absorber plates along cylindrical surfaces. In such an arrangement, however, the angle of incidence of a particle on the calorimeter entrance surface changes with the production angle of the particle at the collision point relative to the beam line. The thickness of absorber plates traversed thus also varies with

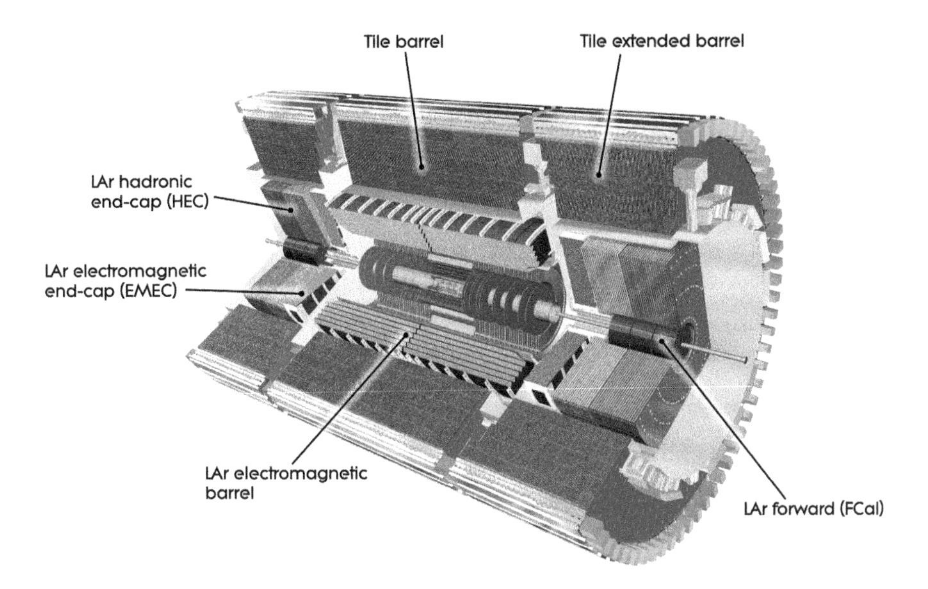

Fig. 8.18. Calorimeters of the ATLAS experiment. *Image: ATLAS Collaboration.*

the particle incidence angle, rendering the calibration of the device more difficult. With calorimetric cells of projective geometry, one does not have such problems as all particles approximately impinge at normal incidence.

For the ATLAS electromagnetic calorimeter this is solved in a rather astute way, thanks to the so-called *accordion geometry* of both lead absorber plates and copper-Kapton electrodes (figure 8.19), a design initially proposed by Daniel Fournier from LAL (*Laboratoire de l'accélérateur linéaire*), a CNRS/IN2P3 laboratory in Orsay near Paris.[18] The calorimeter is based on the liquid argon technology, pioneered by William Willis and Veljko Radeka from BNL. The absorber lead plates are immersed in liquid argon which provides the ionising medium. Such a calorimeter requires therefore a cryogenic system cooling the argon below the boiling point of 88 kelvin. The technology has the advantage to be immune against ageing due to the large radiation exposure at the LHC, encountered in more traditional approaches with plastic scintillators (or crystals for homogeneous calorimeters). Furthermore, the accordion geometry allows a rapid extraction of the electrode signal (within 450 nanoseconds), which is directly readout

[18]In 2020, LAL and four other laboratories in Orsay were joined into the new *Laboratoire de physique des deux infinis Irène Joliot-Curie (IJCLab).*

outside the detector. In practice, the readout is done by sampling the measured electronic pulse shape in multiples of successive bunch crossings (25 nanoseconds) and integrating over it. The height of the pulse is proportional to the amount of ionisation of the liquid argon, which itself is proportional to the energy deposited by the traversing particle.

The choice of a sampling calorimeter has also allowed ATLAS to introduce, by means of etching the readout electrodes with readout pads, a three-layer segmentation in depth with differing granularity. It is a satisfactory compromise between construction cost, increasing with granularity, and the achievable precision on the measured longitudinal shower shape, which helps identify the particle traversing the calorimeter (electron versus photon, muon or hadron).

This marvel of innovation and engineering, assembled with a mechanical precision of absorber plates and printed circuit boards to better than 200 micrometres, has been designed, constructed and commissioned by institutes from Canada, France, Germany, Italy, Morocco, Spain, Sweden, Russia and JINR, and the USA to cite just the main contributors.

A hadronic calorimeter surrounds the electromagnetic one (figure 8.19). During the design phase it was considered to use again liquid argon and an accordion geometry with lower granularity than for the electromagnetic part. However, in view of the cost of such a detector, an alternative solution, which offered a better performance to cost ratio, was chosen for the extended central part of the detector, while the end-caps and very forward parts of ATLAS remained with liquid argon technology to ensure radiation hardness. The central hadronic calorimeter maintains the projective tower geometry, but employs plastic scintillator plates as active elements and interspersed iron plates as absorber material. The 500 000 scintillating tiles (therefore the name *tile calorimeter*, see figure 8.20) are readout using fibres collecting and guiding the light towards photomultiplier tubes, where it is transformed into an electrical signal. Thanks to its location outside the inner solenoid, the radial depth of the hadronic calorimeter could be chosen large enough to fully contain hadronic jets up to highest expected energies at the LHC, thus providing the best hadronic energy resolution. The tile calorimeter extends radially between 2.3 and 4.3 metres, providing together with the electromagnetic calorimeter an overall depth between ten and fourteen interaction lengths (see inset 7.4). On the contrary, the hadronic calorimeter of CMS is located inside the solenoid coil, and with an outer radius of three metres its depth varies from seven to eleven interaction lengths, requiring the presence of an outer hadronic calorimeter beyond

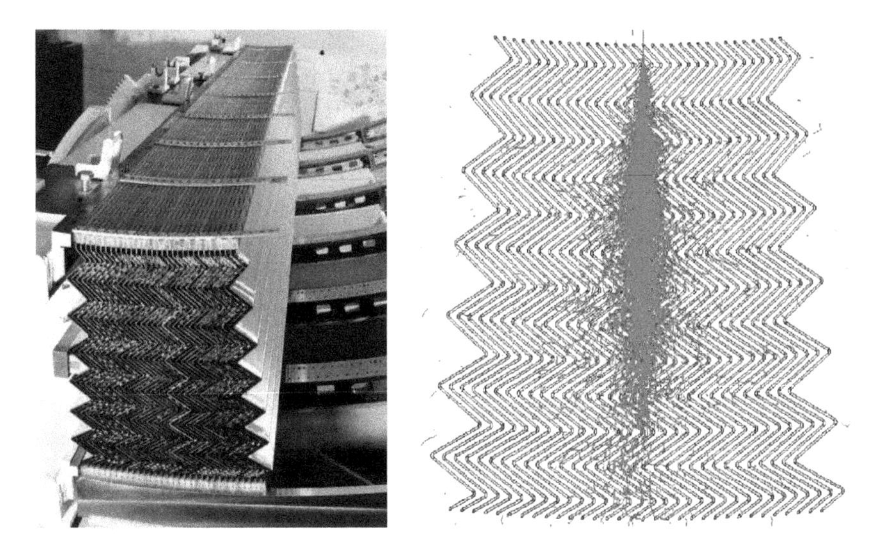

Fig. 8.19. Left: photo of the first module of the assembled ATLAS electromagnetic calorimeter. The accordion structure is clearly visible, with lead plates sandwiched between two steel foils providing mechanical rigidity. Between two absorber layers are located printed circuits to collect the ionisation signals from the liquid argon, as well as insulating spacers ensuring constant thickness of the gap structure. Right: simulation of a 50 GeV electromagnetic shower development in the accordion structure of the calorimeter. *Image: ATLAS Collaboration.*

Fig. 8.20. A complete ATLAS tile calorimeter module covering 1/64 of the full azimuth. *Image: ATLAS Collaboration.*

the solenoid as *tail catcher* for full shower containment. The ATLAS tile calorimeter was mainly constructed by institutes from the Czech Republic, France, Italy, Portugal, Romania, Russia and JINR, Spain, Sweden, and USA. The liquid argon hadronic part (end-caps and very forward calorimeters) were built by institutes from Canada, Germany, Russia and JINR, Slovak Republic, and USA.

8.3.3. *ATLAS muon spectrometer*

The muon spectrometer of ATLAS is inherited from the original ASCOT detector concept. Its design, in particular of the magnet system (figures 8.21 and 8.23) owes a lot to Marc Virchaux, a physicist from the CEA/IRFU in Saclay (France) who passed away prematurely in 2004. It is based on a magnet made of eight superconducting coils producing a toroidal field in air of up to 4 tesla close to the winding, 0.5 tesla in average in the barrel part, and 1.5 tesla in the end-caps. When compared to the iron toroid initially proposed by EAGLE, the advantage of the air-core toroid in terms of momentum measurement precision is considerable. It is based on either the measurement of the track curvature, thanks to three planes of precision drift tube chambers located inside the regions of magnetic field, or on the measurement of the angle of deviation of the trajectory exiting from the field region measured thanks to two additional planes of muon chambers. With a dense core material such as iron, the quality of the position measurement would be strongly degraded due to multiple scattering of muons on the nuclei of the traversed medium. With an air-core toroid, only the material in the mechanical structures, the cryostat of the magnetic windings and of the wire chambers contributes to multiple scattering. The resulting standalone momentum measurement precision is 4% for central 20 GeV muons, degrading gradually to about 10% at 1 TeV.

Achieving a mean field of 0.5 tesla in such a large air-core toroid requires an electric current of 20 400 amperes in the 120 superconducting cable turns that are wound inside each coil. With an iron toroid, on the other hand, thanks to the ferromagnetic effect amplifying by a factor of one thousand the magnetic field, a current of just about 1000 amperes would have been sufficient for a field of 1.8 tesla. Despite the much larger cost of the superconducting coils when compared to the ones for an iron toroid (about 150 million Swiss francs compared to 100 million), the large advantage of the air-core toroid in terms of momentum resolution was considered well worth the investment.

Fig. 8.21. Iconic view of the central part of the ATLAS toroid with its eight cylindrical cryostats housing the superconducting coils, just after the assembly. Each coil is 5-metre wide and 25-metre long. The cryostats are interconnected by aluminium structures to resist the magnetic compression forces reaching up to 200 tons per metre on the inner parts of each coil. These forces acting on the cold mass at 4.2 kelvin are transferred to the aluminium structures at room temperature through eight titanium mechanical connections, so-called struts. *Image: ATLAS Collaboration.*

In the initial design of ATLAS, the air-toroid magnet was made of twelve coils providing an almost twice larger average magnetic field, but this was termed too expensive. The total number of coils was thus reduced to eight, and the current lowered too. This choice has further increased the inhomogeneity of the field, which complicates the track reconstruction requiring more computation.[19] In view of the expected exponential increase in computing power with time, this problem was not considered insurmountable, and this option was finally retained. The design and drawings of the magnet were finalised in 1997, largely at CEA/IRFU (France) for the central barrel part, and at the Rutherford Laboratory in England

[19]With an infinite number of coils (ideal toroid) the field lines would be circular and the field strength would vary inversely proportional with the distance from the beam axis. The inhomogeneity of the field is defined as the difference between the real value of the field at a given space point and the value it would have for a perfect toroidal field.

Fig. 8.22. View of the central part of ATLAS during final assembly stages. The wire chambers mounted inside and around the toroid magnet cryostats are well visible, as is part of the end-cap hadronic calorimeter before its insertion inside the magnet. *Image: ATLAS Collaboration.*

for the two end-cap toroids. The construction and final assembly of the barrel toroid in the underground experimental cavern took place in 2005.

As for the CMS muon spectrometer, several detection techniques are employed by ATLAS. The precision measurements are obtained with three planes of wire chambers dubbed *monitored drift tubes* (MDT), made of six to eight layers of drift tubes.[20] The wires at the centre of each tube have been positioned for each of the 1200 chambers with a precision of 20 micrometres. An alignment system consisting of more than 10 000 CCD

[20]The tubes, three centimetres in diameter, are made of aluminium. They are filled with a mixture of argon and CO_2 at a pressure of three bar, with a central wire kept at a high voltage of 3000 volts. As in traditional drift chambers (see inset 7.1), the ionisation signal is amplified through an avalanche mechanism in the vicinity of the wire and the measurement of the drift time provides information of the distance between the particle trajectory and the wire. The high gas pressure increases the primary ionisation yield and thus allows reaching a position measurement precision of about hundred micrometres.

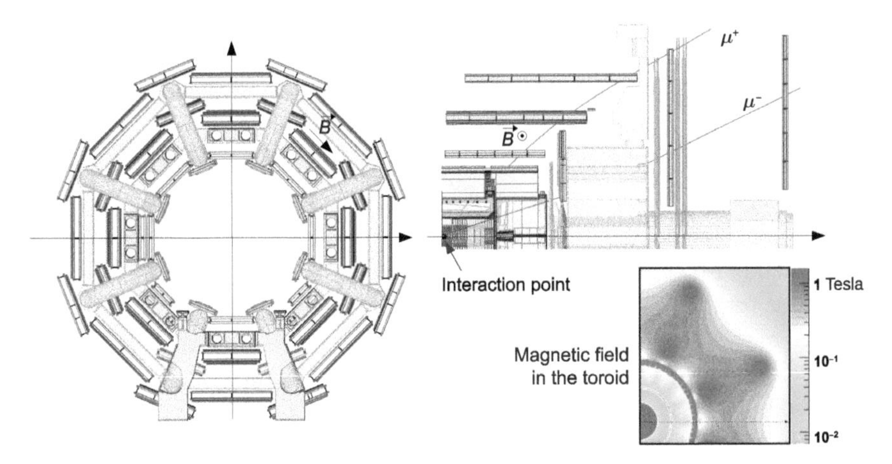

Fig. 8.23. Scheme of the ATLAS muon spectrometer. On the left is a transverse view and on the right a longitudinal cut through the upper quadrant. Trajectories of two muons with 5 GeV transverse momenta are simulated and shown in red. The muon momentum is obtained from curvature measurements with the trajectory sampled in the three drift tube chambers. The figure on the bottom right shows the magnitude of the magnetic field in the barrel part of the muon spectrometer. This field is strongly inhomogeneous. It varies roughly in inverse proportion of the distance to the beam axis with an average value of about 0.5 tesla. The field must be measured with magnetic probes to a relative precision of a few parts per thousand. *Image: ATLAS Collaboration.*

sensors has been developed to monitor the relative positions of the chambers with 30-micrometre precision.[21]

The MDT system is complemented by two other sets of chambers characterised by a fast response: in the barrel part RPCs, as in CMS, and in the end-cap regions conventional wire chambers (i.e. without drift-time measurement), called *thin gap chambers* (TGC), capable of sustaining a very high particle rate. These two types of chambers, of rather mediocre spatial resolution (few millimetres for TGCs and one centimetre for RPCs) have a very fast response and are used essentially by the trigger system to signal the presence of one or more muons in a collision. Finally, as in CMS, CSCs are used for precision coordinate measurements in the forward region where the background conditions are harsher.

[21]The space coordinate of a muon traversing six tubes of a chamber is determined with a precision of $100/\sqrt{6} = 40$ micrometres, and the sagitta of the trajectory from the measurements in three chambers with $40\sqrt{3/2} = 50$ micrometres precision (see inset 7.3).

A big challenge of the ATLAS muon spectrometer was to ensure the quality of construction spread among ten institutes from Germany, Greece, Italy, Netherlands, Russia and JINR, and USA. The 1200 MDT chambers cover a total area of 5500 square metres (almost a soccer field) and are made of 370 000 tubes assembled with a mechanical precision better than 30 micrometres each. The CSC chambers were developed and built by institutes from the USA. The RPC and TGC trigger chambers extend over an area of 6650 square metres, but were less demanding in terms of mechanical precision. Their construction was initially shared among four institutes from three countries: Italy for the RPCs, Israel and Japan for the TGCs. Around the year 2000, several institutes from China joined the muon collaboration and contributed to both MDT and TGC chambers.

Fig. 8.24. External wall of the surface building at CERN containing the access shafts to ATLAS. The building on the right next to the stairs hosts the ATLAS control room.

8.4. Trigger and data acquisition systems of ATLAS & CMS

When at the beginning of the 1990s physicists and engineers started to work on the design of detectors for the LHC, their experience was the one acquired at the proton–antiproton colliders Sp$\bar{\text{p}}$S at CERN and Tevatron at Fermilab. However, at the Sp$\bar{\text{p}}$S the collisions occurred at most every four microseconds, at the Tevatron already every 396 nanoseconds, whilst collisions at the LHC were planned to occur every 25 nanoseconds. On top of that, in order to reach the nominal luminosity of 10^{34} cm^{-2}s^{-1}, it was expected to have about 25 proton–proton collisions occurring simultaneously at each bunch crossing, resulting in the production of about a billion inelastic proton–proton collisions per second. The rate of really interesting events[22] is, of course, much smaller. Let us consider for example the weak bosons W and Z decaying into leptons, which can be identified by the trigger system thanks to the presence of a muon in the muon spectrometer or an electron in the calorimeter. At full LHC luminosity, their production rate is about 400 $W \to \ell\nu$ events per second and ten times less for $Z \to \ell\ell$ ($\ell = e$ or μ). Ten top–antitop quark pairs are expected every second. For the Higgs boson, with a mass of 125 GeV, the expected production is a bit more than one every two seconds, and among these only one out of 8300 is expected to decay into the easily recognisable four-lepton (electrons or muons) mode.

At every recorded bunch crossing, the total number of electronic signals produced and recorded by the detectors amounts to about one megabyte.[23] Recording 40 million bunch crossings per second would lead to a data rate of 40 terabyte per second, which is prohibitive and also not needed given the expected rates of interesting events. It is therefore mandatory to be able to make an extremely rapid and efficient selection to reduce by a factor of about 40 000 the number of proton-bunch collisions to be recorded, keeping about 1000 of the potentially most interesting events every second for permanent storage and a subsequent detailed investigation.

The information technology of the early 1990s, the period when ATLAS and CMS were designed, was totally unable to face this situation. In the design of these detectors it was thus necessary to project into the future and conceive triggering, data acquisition and recording systems based on

---

[22]As already mentioned in chapter 2 (see footnote 8 on page 79), the term "event" is, like so many other terms, unfortunately, experimental particle physics jargon. An event contains the full detector recording during one collision among two proton bunches. Interesting events usually contain one *hard* proton–proton collision, giving rise to the production of high-momentum particles measured in the detector and selected by the trigger system, plus several (typically around 30, but reaching up to 60) additional parasitic *soft* proton–proton collisions (pile-up) producing low-momentum particles in the detector.

[23]This is the price to pay for the granularity provided by the ATLAS and CMS detectors, exceeding by orders of magnitude that realised in their predecessor experiments. For example, the tracking detector of UA1, a projective drift chamber, a technological masterpiece in its days, contained about 6000 active wires. In comparison, the CMS tracker has 10 million silicon microstrips and 70 million silicon pixels, each readout individually, an increase in granularity by 10^4. The electromagnetic calorimeter of UA1 was readout by altogether 756 photomultipliers, whilst CMS has 76 000 individually read crystals and ATLAS has 170 000 independent read-out channels for its liquid argon calorimeter.

electronics and computing capabilities that would be available more than ten years later. This was possible trusting Moore's law according to which the computer chip performance roughly doubles every year and a half.[24] Altogether, the data acquisition and processing capabilities of the LHC experiments are among the great successes of this colossal project.

In both ATLAS and CMS the online event selection proceeds through two levels.[25] The first-level selection is obtained electronically, with fast custom-designed circuits that have to take a decision in few microseconds on whether to accept an event or reject it. This is done on the basis of simple topological or energy-deposition criteria applied on the information given by calorimeters and the muon spectrometer. The first selection accepts for further study about one event in 400, reducing thus the data rate from 40 MHz to a maximum of 100 kHz, usually a bit less. All information on rejected events is erased from the memories of the electronic boards of the detector systems, where they were stored in so-called *pipeline memories*, several-microsecond deep, while waiting for the trigger to decide on their lot.

An event that has passed the first selection level is then readout from the on-detector pipeline memories, and sent to a large computer farm counting several tens of thousands of processors for a more fine-grained analysis. The available latency at this second selection level is about half a second per event. Events are fully reconstructed using the information provided by all detector systems, but not yet with the full precision; this will only be possible during offline analysis when more computing time per event can be accommodated (see section 10.1.1). Ultimately, the second-level online trigger and data acquisition system accepts about one thousand events per

[24]Note that since the end of the years 2010, this law is not valid anymore as the chip performance shows signs of saturation. This is counterbalanced by increased parallelisation and the use of heterogeneous computing resources.

[25]The chief architect of the CMS trigger and data acquisition systems was CERN physicist Sergio Cittolin, while Nick Ellis and Livio Mapelli (both also CERN) are among the principal designers of the corresponding ATLAS system. On the ATLAS side, the main contributors are institutes from Austria, Denmark, Germany, Israel, Italy, Japan, Netherlands, Portugal, Switzerland, Turkey, UK, and USA. For CMS, the contributing institutions are from Austria, Finland, France, Greece, Korea, Lithuania, Poland, Portugal, Switzerland, UK and USA. In principle, one should here also mention the contributions from all the institutes which developed the online and offline reconstruction software that is running in the high-level trigger processors and the prompt offline processing of the triggered events.

second in average, which are transferred to permanent storage and offline reconstruction at the CERN computing centre.[26]

The flux of data produced corresponds to about 10 000 terabytes, that is 10 petabytes or 10^{16} bytes per year for each ATLAS and CMS. The amount of data is more than doubled once all corrections and calibrations are applied to the data, they are fully reconstructed and necessary replications of events throughout the worldwide computing grid have been done, so that the data can be analysed for physics by the collaborating institutes. To this must be added the even larger amount of data resulting from event simulations required for data analysis. Altogether this leads to an overall amount of recorded data on hard drives exceeding 200 petabytes (200×10^{15} bytes) per experiment in 2020.

8.5. Organisation of the large international collaborations in particle physics

Both ATLAS and CMS collaborations have more than 3000 physicists and engineers, affiliated to 180 to 200 member institutions in 40 to 50 countries.[27] The ALICE experiment is also a large international collaboration, with about 1500 international collaborators. LHCb is the smallest among these, though still sizeable with about 900 international collaborators.

How to organise the work of so many people, employed by a variety of research institutions and universities, themselves depending on one or more funding agencies even in a single country?[28] While CERN is the host laboratory and in charge of the LHC, it is neither the prime contractor nor main responsible for the experiments, and the universities and national

---

[26]Before offline reconstruction, which is a very computing intensive operation, the data are processed to take into account calibration corrections, alignment corrections, specific aspects at data taking (for example, if there were some problems with high voltage in some detector channels meaning that these specific data are less reliable), etc. We shall discuss these aspects in more detail in chapter 10.

[27]The number of nationalities present in the collaborations largely exceeds that of the member countries. ATLAS, for example, has members from 103 different nationalities.

[28]For example the Department of Energy (DOE) and National Science Foundation (NSF) in the USA, the *Centre national de la recherche scientifique* (CNRS) and *Commissariat à l'énergie atomique et aux énergies alternatives* (CEA) in France, the *Bundesministerium für Bildung und Forschung* (BMBF), *Max-Planck-Institut* (MPI) and the *Helmholtz-Gemeinschaft* in Germany, the *Istituto Nazionale di Fisica Nucleare* (INFN) and Ministry of Science and Education for Italy.

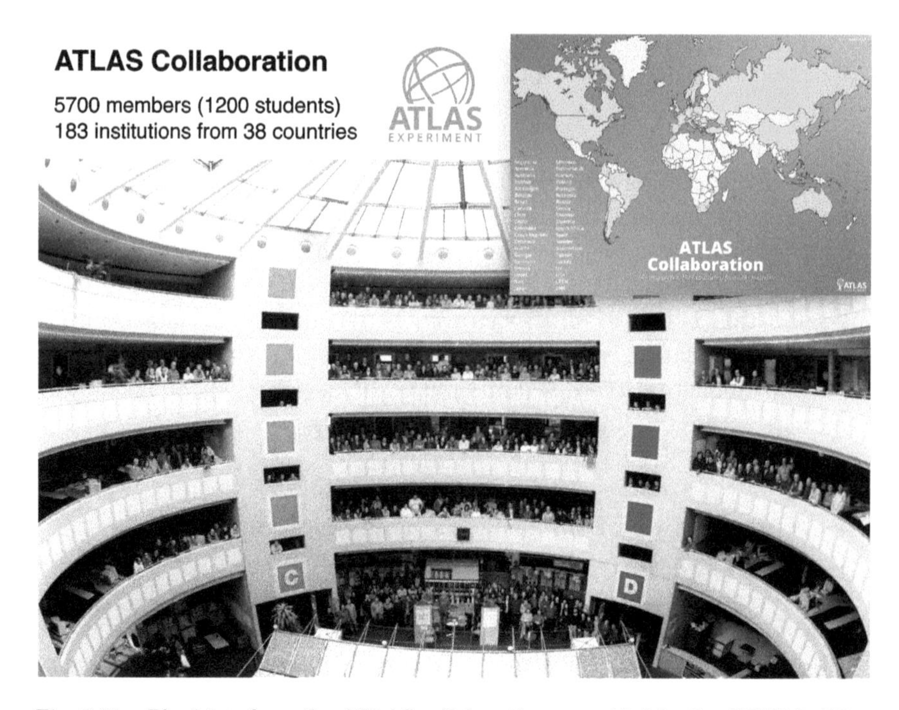

Fig. 8.25. Physicists from the ATLAS collaboration assembled in the CERN building where their offices are in part located. The numbers of collaborators and collaborating institutes are in steady flux, but overall constant since data taking started (before which they strongly increased). *Image: ATLAS Collaboration.*

research organisations are not sub-contractors.[29] The term *collaboration* reflects rather well the democratic and consensual ways these large enterprises are run (see figures 8.24–8.27).

Both ATLAS and CMS have written *constitutions*, based on similar general principles, to define the various aspects of operating the detectors and managing the collaborations. They build upon the foundations and experience of previous particle physics experiments. At the centre of the organisation is the *spokesperson* of the experiment, whose task and responsibility is to make the synthesis of the various opinions expressed by members of the collaboration and to take appropriate decisions. The spokesperson

[29]The local CERN groups in the four LHC experiments are nevertheless the largest in number and, owing to their size, experience and local presence, they carry more responsibility for the experiments than most other groups.

is assisted by a *technical coordinator* overseeing the operation and maintenance of the detector systems, the infrastructure and experimental area, as well as providing a link to the CERN host laboratory, and a *resources coordinator* in charge of contacts with funding agencies of all countries participating in the experiment, whether or not member-state of CERN.

The spokesperson is working and interacting with three structures. The *Collaboration Board*, which is a sort of collaboration parliament, each full member institution being represented by one member whether it has fifty or just two researchers taking part in the experiment. It is a similar principle as for example in the European Union, where large and small countries have, in principle, equal weight. The second structure is somewhat equivalent to a government of the experiment, either called *Executive Board* or *Management Board*. Its members are the leaders of the various detector systems and of other activities such as operations, trigger, software and computing, physics, as well as the Collaboration Board chair. While the policy making and voting Collaboration Board meets typically three or four times a year, the Executive or Management Board meets much more often, typically once a month or more when required to take executive decisions. The third structure is the most direct manifestation of democracy: the plenary meetings of the collaboration where everybody can freely express themselves. At least three times a year there are *collaboration weeks* attended by a large fraction of collaboration members, and every year one of these general meetings takes place outside CERN. In addition, there are several other large meetings, such as physics weeks, upgrade weeks, trigger weeks, software and computing weeks, etc. The large number of resulting meetings sometimes exasperates collaboration members, but it is the price to pay for such a large and democratic endeavour, where transparent communication and decision taking is key for a trustful and collaborative spirit.

This is the organisational structure that allows to take all major decisions, as far as possible based on consensus. However, not all decisions concerning the experiments are treated at this level. The organisations are further subdivided in sectors according to the sub-detectors or other activities, each organised as a smaller collaboration with its own leadership levels. The spokesperson and most of the top-level coordinators are elected by the Collaboration Board. Other leaders are appointed by search committees after open calls for nomination.

To organise the work on a scale of the entire planet, particle physicists have developed tools allowing to work from remote places. It is worth

Fig. 8.26. A number of CMS physicists assembled in the surface building at the site of the CMS experiment. On the back is a picture of CMS at scale one-to-one, the real detector is located hundred metres underground. *Image: CMS Collaboration.*

remembering that the *World Wide Web* (WWW) was born at CERN[30] precisely out of this need to share among many collaborators automatic monitoring results, documents, plans, physics results with colleagues all over the world. Likewise, all meetings, even small ones with few participants only, can be followed at distance by audio and video communication means. Only the time differences limit the physicists and engineers in their desire to reduce distance, as it is not easy to find the hour of the day for a meeting that would simultaneously suit well collaborators in Asia, Europe and the Americas.

[30]The web has been invented in 1989 at CERN by the British scientist Tim Berners-Lee, and was refined by him in collaboration with the Belgian system engineer Robert Cailliau the following year. They outlined the principal concepts and terms behind the WWW, including "hypertext documents" that could be viewed by "browsers". The world's first web server running on a NeXT computer at CERN had the address info.cern.ch, and the first website was http://info.cern.ch/hypertext/WWW/TheProject.html.

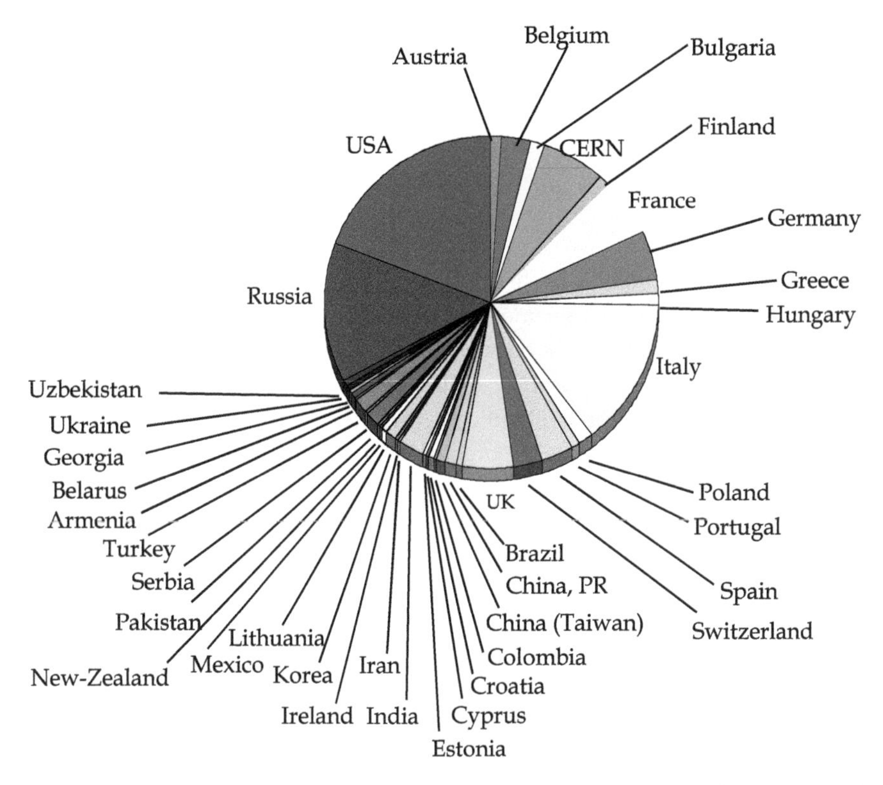

Fig. 8.27. Composition of the CMS collaboration, with the numbers of physicists and engineers involved from each participating country shown in proportion. This pie chart shows the situation in 2012, the year of the Higgs discovery; in the meantime many new countries have joined CMS: Egypt, Thailand, Malaysia, Indonesia, Equator, Oman, Saudi Arabia, Lebanon, Montenegro, etc. *Image: CMS Collaboration.*

A key role in the collaborations is the *physics coordinator* who defines research priorities and organises the data analysis work from which physics measurements are obtained, ultimately leading to scientific publications. The physics coordinator works in conjunction with group leaders organised according to the major research activities: search for new phenomena (such as supersymmetry), Higgs-boson physics, top-quark studies, electroweak and strong interaction studies, B-physics studies, heavy-ion studies, etc. Within each of these groups physicists confront their ideas, studies and results. There is no *a priori* distribution of the research topics among the collaboration. Individuals and groups are free to contribute to whatever study they are interested in. This leads to competition within a collaboration on some high-profile analyses, which allows to compare and to mutually confirm results obtained. Physicists are imaginative, they like to use novel techniques and perform creative analyses. Therefore, instead of simply reproducing somebody else's work, people working on a similar topic approach it in different ways. The research thus progresses through cooperation among physicists, seasoned with an appreciable dose of competition inside a collaboration and among different experiments.

A healthy tradition in particle physics, valid for both experiment and theory, is that publications list the name of co-authors in alphabetical order.[31] Furthermore, experimental collaborations include all their members as authors on their physics papers, which means thousands of signatories on each publication of ATLAS and CMS![32] Indeed, everyone's contribution is considered significant, be it ensuring that the detector operates and takes good data, preparing the software and computing to reconstruct the data, up to analysing them and writing the paper. This partnership is an important reason for the successful realisation of the large particle physics experiments. It is, however, sometimes challenging to explain this culture to other science fields and funding agencies, as the value of each individual contribution to the experiment and to a given publication is not visible from the author list, but needs to be conveyed through other means.

Beyond their internal organisation, the collaborations interact with the management of CERN and with scientific organisations of countries taking part in the experiments. Indeed, they must be informed and follow all major technical decisions concerning the construction, running and eventually upgrading the detectors, as these political and financial bodies *in ultima linea* fund the experiments. This is achieved through the *LHC Resource Review Board* (RRB) where, twice a year over the past more than two decades, the management of each experiment, for half a day in turn, presents to the representatives of funding bodies and ministries of all countries involved the status of the experiment, the major physics results obtained, the yearly expenditures of construction, operation, maintenance, cost of detector upgrades needed, and upcoming expected expenses. This is the place where, under the supervision of CERN, material, personnel and

[31]In other research fields, such as biology, author lists are composed according to the importance of the individuals' contributions. While this may in some cases be more equitable as not everyone contributes at the same level and intensity, it obstructs the formation of larger collaboration where it becomes difficult to form such a hierarchy.
[32]In early 2020, ATLAS, CMS, ALICE and LHCb papers are signed by 2953, 2318, 1008, and 907 co-authors, respectively.

computing expenses are carefully scrutinised, but also that each country gets a fair share of orders or investment in its home industries and institutions, as a return for its involvement with CERN. This mode of operation proved to be very effective for all stake holders. Another important body is the *LHC Committee* (LHCC), an overseeing committee appointed by CERN. It is composed of experienced physicists, not part of the experiments, with membership renewed every few years. The committee meets four times a year with the management of the LHC experiments and, for a full day, review the status of operation, physics and detector upgrades. Additional sub-bodies of the RRB and LHCC review in detail computing usage and costs as well as the plans and execution of detector upgrades.

These innumerable meetings of evaluation and cost-scrutiny committees, whose aim is to guarantee the optimal technical choices for detectors and their most efficient and productive exploitation for physics, while keeping the needed level of cooperation and minimising hierarchical aspects, appears rather substantial if not ponderous. It is a criticism often put forward by detractors of this field of *Big Science*, but also by some experimentalists who have experienced different times. While finding the right level of scrutiny is itself a question requiring deep study, one should recognise that, also thanks to these procedures, detectors with technological complexity, performance and operational efficiency without historical comparison have been successfully realised at the LHC.

Chapter 9

LHC Start-Up and Data Taking

9.1. A promising start

In January 2007 all magnets in sector 7–8 were connected, both electrically and the cryogenics. The vacuum was established and it was possible to begin cooling down to 1.9 kelvin this first sector. This took about three months. One after another, all eight sectors had been cooled down by May 2008. The cryogenic system of LHC is, and by far, the largest in the world. It requires more than 50% of the world's supply of helium. The magnets represent 1200 tons of superconductor and the overall cold mass to be maintained at 1.9 kelvin is 40 000 tons, over five times the mass of the Eiffel Tower! With such a huge thermal inertia, several months are required to cool the whole machine starting from room temperature. Thus, while previous non-cryogenic machines were operated with cycles of three to six months of data taking, followed by few months for machine maintenance and upgrading, the operation mode of the LHC is totally different: working cycles of three to four years with short technical interruptions in the winter, followed by longer machine stops of at least one year.

To accelerate the beams from 450 GeV, the energy at injection from the SPS, to the nominal energy of 7 TeV per beam, the current in the magnets must be gradually raised from 700 to almost 12 000 amperes. By September 2008 seven out of eight sectors were fully tested. Due to lack of time, magnets of the eighth sector had their magnets tested for currents up to about 5000 amperes, as the date of September 10th had been fixed a long time ago to inaugurate the LHC, with the world's media and officials invited to attend (figure 9.1). The machine and experiments were ready for

Fig. 9.1. On September 10th 2008 LHC inauguration in the control room of the LHC, the CERN Control Centre (CCC): the five CERN Directors General during the LHC construction phase. From left to right: Robert Aymar, Luciano Maiani, Christopher Llewellyn-Smith, Carlo Rubbia and Herwig Schopper. *Image: CERN.*

first beams and everyone expected to see first collisions about three weeks to a month later.

On September 10th at around 9:30 in the morning a bunch of 3×10^9 protons of 450 GeV was transferred from the SPS to the LHC (figures 9.2–9.4). After about two hours of efforts by the machine operators, the beam[1] made a first full circle around the ring[2] and could be maintained circulating for several minutes. This was the first huge success of the day and was

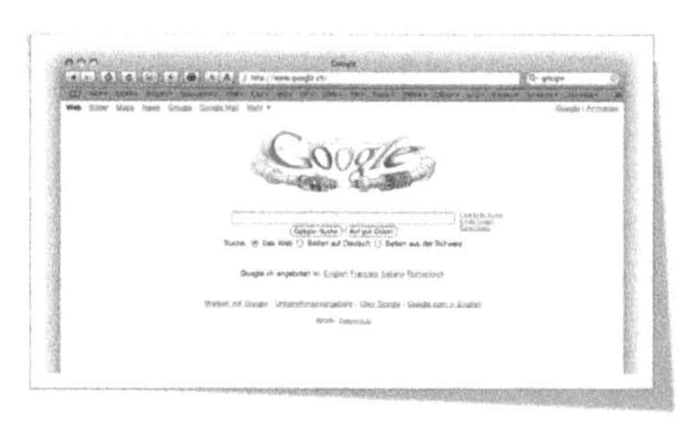

Fig. 9.2. Google page on September 10th, 2008.

[1]At the beginning of the injection procedure, the beam, made up of just a single bunch of protons, would be stopped at the collimators (see footnote 20 on page 194) sector by sector, until all eight sectors of the machine, corresponding to almost 27 kilometres, have been crossed.

[2]It takes 89 microseconds for a proton to make a full turn over the 27-kilometre circumference of the LHC. Within a second a proton can accomplish 11 200 turns.

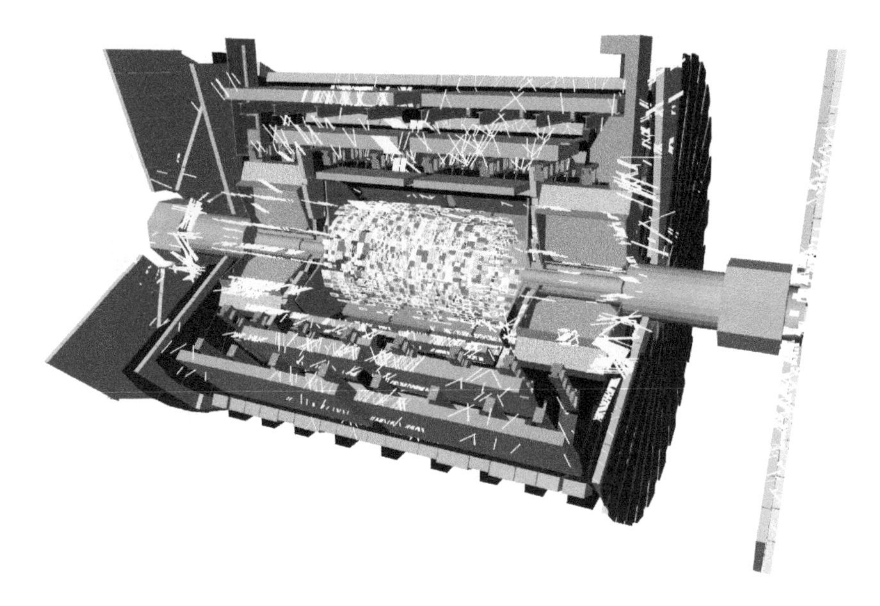

Fig. 9.3. ATLAS on September 10th. While the LHC was being tuned, the beam was directed onto the collimators just preceding the detector. This produced a huge shower of particles, called "beam splash" by the physicists, lightening up almost the entire detector. This type of event already brought much information to physicists on the functioning of the various detector elements and their timing with respect to the LHC beams. *Image: ATLAS Collaboration.*

accompanied by a tremendous applause in the LHC control room. That same day, at around 15:00 in the afternoon, an attempt was made to inject the second counter-rotating beam. After two hours of tuning the magnet lattice, this beam too performed its first full LHC turn.

This was a real tour de force and a case without precedent in the history of accelerators. With hadron machines of previous generations this operation would take days, weeks, months and sometimes could even never be brought to a successful completion. Compliments arrived from all over the world, including all particle physics research centres, where the events were followed in real time. The following day the two beams were already simultaneously circulating in the machine.

Fig. 9.4. A cake sent by physicists from BNL in the USA to CERN on September 10th, 2008.

Fig. 9.5. A junction between two magnets contains about ten interconnections between superconducting cables operating at 1.9 kelvin, all immersed in a super-fluid helium bath. Altogether there are about 10 500 such connections along the LHC ring. *Image: CERN.*

9.2. The incident of September 19th, 2008

The LHC operators worked first on achieving stable beams at injection energy of 450 GeV, that is, being capable of circulating the beams on stable, low-background orbits for long periods. The next task was to gradually increase the beam energies to the nominal 7 TeV per beam.

In the morning of September 19th, however, while testing the magnet powering up to 9300 amperes, an incident occurred in sector 3–4 when the current in the dipoles reached 8700 amperes. A defective interconnection between superconducting cables (figure 9.5) of one dipole and one quadrupole was at the origin of what would turn out to become a major accident. Its resistance, in principle of order of 10^{-9} Ω (ohm), was in reality two hundred times larger. With increasing current and the local heating induced by this faulty connection, the superconducting cable made a sudden transition from superconductivity with zero electrical resistance to

Fig. 9.6. A junction between two magnets after the incident of September 19th, 2008. On the right, the same junction after removing the bellows. *Image: CERN.*

normal conductivity with finite resistance. This type of transition is called a *quench*. As the connection among the massive copper conductors surrounding the superconducting cable, supposed to take over the current in case of a quench as part of the LHC's quench protection system, was also deficient, the current could not discharge through them as it should have done in a normal quench. This gave rise to a local electric arc (discharge) at the level of the superconductor, heating up the super-fluid helium and provoking a transition to the gaseous state, which resulted in a large local overpressure followed by an explosive release of around two tons of liquid helium, before detectors triggered an emergency halt. A further four tons of liquid helium leaked at lower pressure. Security valves for this type of catastrophic mishap had not been not foreseen. The shock wave generated mechanical deformations and displacements in a whole set of 10-ton magnets nearby, dipoles and quadrupoles (figure 9.6). A total of 53 magnets were damaged. Furthermore, the formation of the initial electric arc provoked the melting of cables, insulators and metallic tubes nearby, causing the deposition of soot in the beam pipes over a length of about three kilometres.

To repair the damage caused by the accident 40 magnets had to be removed from the tunnel and replaced with spare ones. To avoid a similar blow to happen again, CERN teams undertook radiography and electric resistance measurements (at the nano-ohm level) of about 10 500 connections in the machine looking for other weak spots. Indeed several were found. It was thus necessary to remove the defective junctions and devise a faster and more effective alarm system in case other incidents of similar nature occurred. Security valves of larger diameter, able to release the helium pressure before damage could occur, were installed. A new quench detection and protection system was put in place. The full analysis of the incident and following corrective actions took over a year to complete.[3] LHC physicists, some of whom were looking forward to the collider start-up for about twenty years, had to be patient once more. Nevertheless, had this incident occurred later on, after substantial running and irradiation of the machine components, the analysis and repair would have taken more time still.

9.3. Successful collisions

In November 2009 the machine was ready again for a new start. However, it was too perilous to right away raise the beam energy (and thus magnet currents) to full 7 TeV. It was therefore decided to limit the energy initially to 3.5 TeV per beam (7 TeV proton–proton collision energy). First proton bunches were injected into the LHC on November 20th. Within few hours both beams were circulating stably at injection energy of 450 GeV. The next step was to bring the two beams into collision with 900 GeV in the proton–proton centre-of-mass. First collisions were observed on November 23rd at 14:22 local time in the ATLAS detector and, after some adjustments, at around 19:00 in CMS. When the first events appeared on monitoring screens in the control rooms of the experiments the excitement of the physicists was at its maximum (figure 9.7).

During the following days the main task of the LHC team was to raise the luminosity by increasing the number of protons per bunch and the number of circulating bunches. Both rapidly rose to 10^{10} protons per bunch and four circulating bunches, respectively. These were still very modest numbers compared to what was needed to achieve the nominal instantaneous luminosity of $10^{34}\,\text{cm}^{-2}\,\text{s}^{-1}$, namely more than 10^{11} protons per bunch and 2800 bunches, but they were important milestones before the LHC

[3]This system was further reinforced during the scheduled LHC shutdowns in 2013 and 2014 as well as in 2019 and 2020.

Fig. 9.7. ATLAS, November 20th, 2009. Top: the first bunches passing through the interaction point in ATLAS. Middle: a spectacular beam-splash event on the left, and one of the first recorded proton–proton collision events on the right (taken on November 23rd). Between this first day of collisions in 2009 and the end of the first LHC run in 2013, the intensity of collisions had increased by a factor of about fifteen billion. Bottom: physicists rejoicing in the control room at the sight of a first beam-splash event. *Image: ATLAS Collaboration.*

took off towards records. In December 2009 the protons were accelerated up to 1180 GeV per beam in the LHC and were brought into collision. With 2360 GeV centre-of-mass collision energy the LHC had overtook the Tevatron and become the most energetic collider of all time.

9.4. The LHC Run 1 (2010–2013)

First collisions of a single 3.5 TeV proton bunch against another 3.5 TeV bunch, thus giving 7 TeV in the proton–proton centre-of-mass, took place on March 30th, 2010. The peak luminosity then was just about 10^{27} cm^{-2} s^{-1}, ten million times smaller than the expected nominal one. During this first day of data taking, the ATLAS and CMS experiments already collected several hundreds of thousands of events that were immediately reconstructed and analysed to commission the proper operation of detectors.

During the following weeks the progress on beam lifetime and machine luminosity was spectacular. By end of June the peak instantaneous luminosity already exceeded 8×10^{29} cm^{-2} s^{-1} with three colliding proton bunches in the ATLAS and CMS interaction points. In October, just before the end of LHC running in the proton–proton collision mode, the luminosity reached 2×10^{32} cm^{-2} s^{-1} with 348 colliding bunches in ATLAS and CMS, a five orders-of-magnitude increase in six months. The integrated luminosity delivered to ATLAS and CMS during this first period of data taking in 2010 was 48 pb^{-1}. It allowed the LHC experiments to measure a number of already known particles and physics processes (figure 9.8), and use these to evaluate the detector and data reconstruction performance.[4] For example, the production cross-section for a Z boson in proton–proton collisions at 7 TeV is about 30 nanobarns (nb), which means that the overall number of Z bosons produced was $30\ \text{nb} \cdot 48\ \text{pb}^{-1} = 1.4$ million.[5]

[4]A careful reader has probably noticed that figure 9.8 is labelled "preliminary". It indicates that the result was not taken from a published article in a peer reviewed scientific journal, but rather from a scientific note released by an experiment (here CMS) for the purpose of presentation at a conference or workshop. Many particle physics results are first presented in preliminary form, and, after all checks and corrections have been carried out (which usually results in minor modifications only), they are published.

[5]Only Z bosons decaying to a pair of muons or a pair of electrons stand out clearly above the background — as seen in figure 9.8 — whilst having good detection efficiency, these leptonic decays represent barely 7% of all Z decays.

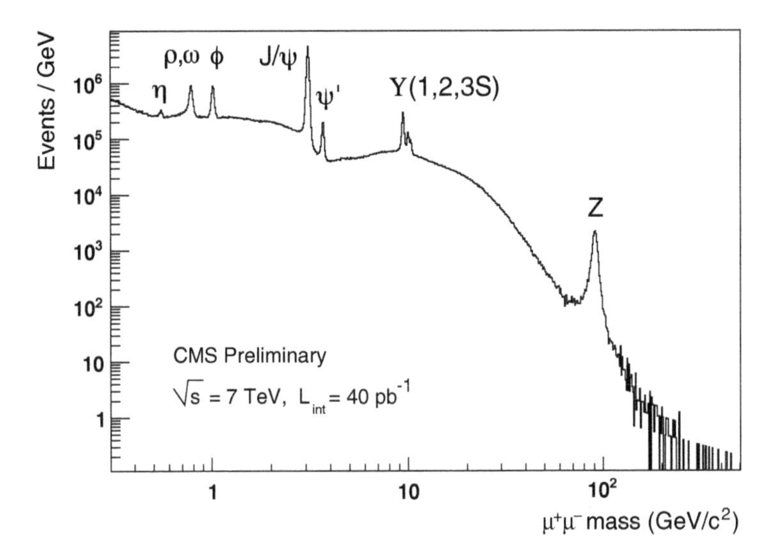

Fig. 9.8. Mass spectrum of muon pairs measured with the CMS detector in 2010. A number of important discoveries in the history of particle physics were made through the search for particles decaying to a pair of charged muons or electrons (leptons). New particles appear as spectral lines on a continuum of lepton-pair masses. The particle zoo thus gradually acquired the ϕ (bound state made of $s\bar{s}$ quarks) in the early 1960s, the J/ψ (bound state of $c\bar{c}$) in 1974, the Υ (bound state of $b\bar{b}$) in 1977. The decay to a pair of electrons allowed the UA1 and UA2 experiments to discover the Z boson in 1983 (see section 2.3). Just a few months after the beginning of data taking, the LHC experiments were able to re-observe all these particles which wrote science history, demonstrating the good quality of the detector, data as well as reconstruction and analysis software already achieved. (See footnote 4 for an explanation of the "preliminary" label.) *Image: CMS Collaboration.*

Despite the modest amount of data collected during this first full year of LHC operation, the 3.5 times larger collision energy than that reached at the Tevatron proton–antiproton collider opened the possibility to observe new heavy particles if their production cross-section were large enough. Indeed, even a very small integrated luminosity of 315 nb^{-1} was sufficient for ATLAS to present in July 2010 results of a first search for new physics through the study of final states with two particle jets (figure 9.9). Already this very early study allowed ATLAS to establish a lower limit of 1.3 TeV on the mass of hypothetical excited quarks, which could exist for example if quarks were composite objects. This limit exceeded the one previously established by the Tevatron experiments by 400 GeV in mass.

With the successful proton–proton run in store, the LHC switched in November 2010 to heavy-ion collisions using lead nuclei (Pb–Pb) with a

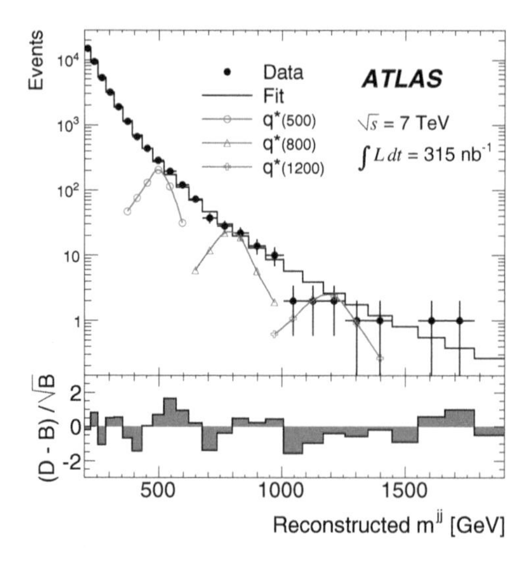

Fig. 9.9. Two-jets invariant mass distribution measured on a very first set of data recorded in 2010 by the ATLAS detector (points with statistical uncertainty bars). The solid line histogram represents the expectation in the absence of any resonance. The coloured curves show putative signals of excited quarks with varying assumptions on their mass. This study excludes their presence. *Image: ATLAS Collaboration, Phys. Rev. Lett. 105, 161801 (2010).*

nucleon–nucleon centre-of-mass collision energy of 2.76 TeV. The run lasted for about a month before closing the CERN accelerator complex for the traditional end-of-year technical stop. A total integrated luminosity of about 10 inverse microbarn (μb^{-1}) was delivered by the LHC. Sufficient data to allow exciting new insight into the physics of the hot dense strongly-interacting matter created in these collisions, where quarks and gluons are no longer confined in hadrons. The highlight of this first run was the discovery of *jet quenching* in heavy-ion collisions by observing events with a large (and unexpected) imbalance in the transverse momentum of two jets. The hypothesis is that, in an originally balanced production of a quark–antiquark pair, each fragmenting and hadronising into a jet, one jet has lost significant amounts of energy (quenching) when traversing the medium (or quark–gluon plasma) created in the lead–lead collision. An event display of such a striking event recorded by the ATLAS detector is shown in figure 9.10.[6] We will provide a more detailed discussion of results obtained in heavy-ion collisions in section 14.2.

[6]The operators in the ATLAS control room noticed the occurrence of such events already on the spot while taking lead–lead data in November 2010.

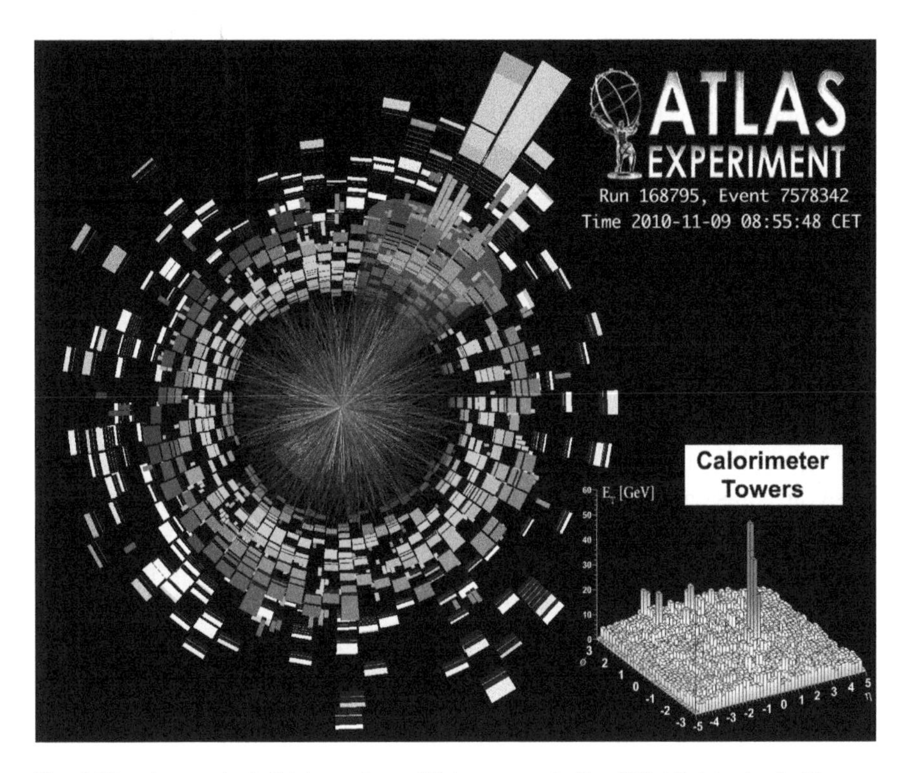

Fig. 9.10. A quenched di-jet event candidate as seen in the ATLAS detector in November 2010. The asymmetry in the jet energies is clearly visible on the bottom right plot, which shows the energy deposits in the calorimeter cells displayed in the (η, ϕ) plane. The quenched jet is hardly visible. *Image: ATLAS Collaboration, Phys. Rev. Lett. 105, 252303 (2010).*

For the data-taking period of 2011, which started in March, engineers and physicists hoped to reach an instantaneous luminosity of $2 \times 10^{33}\,\text{cm}^{-2}\,\text{s}^{-1}$ and an overall integrated luminosity of $1\,\text{fb}^{-1}$. The number of colliding proton bunches for ATLAS and CMS in the machine gradually rose to 1024 end of May, reaching 1331 by the end of the proton run on October 30th, 2011, with bunch crossings every 50 nanoseconds. The average number of protons per bunch reached 1.4×10^{11} at the beginning of an LHC fill and could even be forced up as high as 1.6×10^{11} in special trial runs. The peak luminosity reached $3.7 \times 10^{33}\,\text{cm}^{-2}\,\text{s}^{-1}$. The net outcome of this performance was that by the end of 2011 ATLAS and CMS had each collected not $1\,\text{fb}^{-1}$ as initially targeted but $5\,\text{fb}^{-1}$. It exceeded by a factor of 120 the data sample recorded in 2010. With that amount of data it became possible to look for the Higgs boson!

For about one month at the end of 2011, the LHC operated again as a lead–lead collider. The gain in luminosity was more than an order of magnitude compared to the 2010 run. The new data allowed a much more detailed study of the quark–gluon plasma properties by the dedicated heavy-ion experiment ALICE. The ATLAS and CMS experiments took also advantage of the lead–lead data taking. Thanks to the large space-angle coverage of their calorimeters and muon spectrometers they explored for the first time the so-called *hard probes* of the quark–gluon plasma: jets, hard photons, W and Z bosons. These studies were complementary to those done by ALICE.

As a consequence of a careful and systematic study of the quality of the electrical connections between the magnets, the component that has caused the September 2008 incident, the machine experts were confident enough to raise the collision energy from 7 to 8 TeV for the proton–proton data taking in 2012. It may seem a modest improvement, but it resulted in (depending on its mass) at least a 25% increase in the predicted Higgs-boson production cross-section. For more massive particles such as hypothetical supersymmetric particles or possible additional heavy gauge bosons Z', W', the increase in the cross-section can be even larger. Raising the beam energy is also helpful to reach higher instantaneous luminosity, which, during 2012, gradually increased up to $7.7 \times 10^{33}\ \mathrm{cm}^{-2}\,\mathrm{s}^{-1}$, thus close to the nominal value of $10^{34}\ \mathrm{cm}^{-2}\,\mathrm{s}^{-1}$. The number of colliding bunches in 2012 did not exceed 1368, which is less than half the design value because the distance between filled bunches remained 50 nanoseconds, twice as large as designed. A significant improvement in 2012 represented the stability and regularity of the LHC operation. By the end of June, after barely three months of operation, the integrated luminosity reached already $5\ \mathrm{fb}^{-1}$, equalling what was accumulated during the entire 2011 data taking (figure 9.11). The data available in June 2012 were sufficient to make another attempt, which proved to be decisive, to uncover the elusive Higgs boson (see chapter 11).

Nominal LHC operation would require 2808 proton bunches in each beam with bunches crossing every 25 nanoseconds.[7] For the 2012 running the decision was taken (as in 2011) to limit the number of colliding bunches in ATLAS and CMS to just below 1400 with 50-nanosecond crossings. Switching to the nominal operation regime may seem a simple way to double

[7]The LHC circumference of 26.659 kilometres allows in principle for 3564 25-nanosecond bunches, but not all bunches can be filled with protons. Gaps without protons are needed to allow so-called "kicker" magnets to dump the beam, and for other technical reasons.

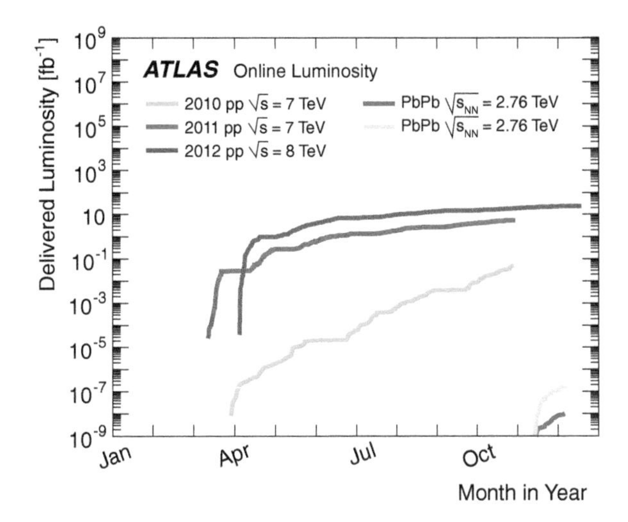

Fig. 9.11. Integrated luminosity delivered to the ATLAS experiment as a function of months in the years 2010 to 2012. The corresponding curves for CMS are very much the same. The ordinate (vertical) axis is in a logarithmic scale, so as to allow showing also the proton–proton data-taking period of 2010 as well as the lead–lead periods. Only the short proton–lead data-taking period beginning of 2013 is not shown in this figure. *Image: ATLAS Collaboration.*

the luminosity, but in reality it is a rather complicated undertaking and was postponed for the running in 2015, after the first long technical stop of the accelerator complex. The main goal for 2012 was to accumulate as much luminosity as possible, so as to find the Higgs boson (or whatever new physics there might be), as the Tevatron was still in the game for the Higgs hunt. Working with 1400 bunches was minimising the problems of beam instability due to electromagnetic beam–beam interactions (that is, the perturbation of the beams as they cross the opposing beam), and allowed to cram more protons per bunch, yielding ultimately a higher luminosity. The price to pay, however, was a larger number of simultaneous proton–proton collisions per bunch crossing, the so-called pile-up (section 6.3.5): there were on average 21 simultaneous collisions, reaching sometimes up to 40, in 2012 compared to an average of nine pile-up interactions during the 2011 running. It obviously complicated the data analysis (see figure 6.10), but this was the only way to record a sufficient number of events giving hope to still observe the Higgs boson in 2012. All in all, a record integrated luminosity of 22 fb^{-1} was collected by ATLAS and CMS in 2012, and the Higgs boson was found!

The first run of LHC data taking ended beginning of 2013 with one last month, not of lead–lead collisions as in previous years, but with proton–lead collisions. This last phase once again showed the great competence of the CERN engineers and machine experts, not only in terms of spectacular performances, but also of the flexibility of the LHC, capable with little preparation to collide protons with protons, ions with ions, and finally also protons with ions.

To get a feeling for the step forward in technology that the LHC represents, the above-mentioned values should be compared with those for the Tevatron: 10 fb^{-1} of integrated luminosity at 1.96 TeV recorded in over ten years of data taking. In terms of numbers of collisions the LHC numbers are astronomical: three million billion (3×10^{15}) inelastic proton–proton interactions have been looked at by ATLAS and CMS during the year 2012 alone![8]

9.5. Long LHC shutdown 1 and high-energy restart

The first long LHC shutdown (LS1), from mid February 2013 to March 2015, was devoted to reinforcing the machine in view of a rise of the proton–proton centre-of-mass collision energy to 13 TeV. This meant: strengthening all the 10 500 magnet interconnections — where the 2008 incident occurred — through clamping connections operating at liquid helium temperature; installing the second half of valves of larger diameter on the LHC magnets, not yet mounted during the 2008–2009 repair phase; preparing and strengthening the machine protection system with additional collimators and absorbers.

It was also necessary to adapt the radio frequency system to run with a 25-nanosecond bunch separation instead of 50 nanoseconds as in previous years. The 25-nanosecond running mode was a persistent demand by the ATLAS and CMS experiments as it allows halving the rate of pile-up interactions (the same integrated luminosity is produced by twice as many proton bunches), thus simplifying the reconstruction of the interesting collision debris and improving physics analysis. However, going from a 50-nanosecond bunch crossing rate to a 25-nanosecond one comes with difficulties in the machine operation, in particular as it increases the so-called *electron-cloud* effect: electrons stripped from residual atoms in the beam pipe by the circulating beams. These electrons are then accelerated towards

[8]Of course, only a small fraction of these were recorded, as we discussed in section 8.4 in relation to the experiments' triggers.

the beam-pipe walls where they produce *secondary-emission electrons*, attracted in turn by the circulating proton beam with the effect gradually amplifying. The effect is much more pernicious at 25-nanosecond than at 50-nanosecond crossing rates, introducing instabilities and possibly too large energy depositions in the cold sectors of the machine provoking an increased heat-load on the cryogenic system of the LHC. The effect is mitigated by using reduced secondary-emission materials in the beam pipe, and by the *beam-scrubbing* procedure, which is a sort of *baking* of the beam pipe with a controlled electron bombardment technique of its surface, during few days at the beginning of major running periods.

During this first LHC long shutdown, not only was the collider prepared for operation at 13 TeV, but also the experiments went through a completion and upgrade stage. For example, CMS was upgraded with 72 additional CSC and 144 additional RPC-type chambers in its end-cap parts (see section 8.2.5). These chambers were planned from the beginning in the design of CMS, but their construction and mounting was postponed to save money during the initial lower luminosity running periods where they were not yet absolutely necessary. The tracker of ATLAS was augmented by a fourth silicon-pixel layer as close as 3.3-centimetre radius from the beam line (see section 8.3.1) — much improving its capability of identifying jets initiated by b quarks. The trigger systems of ATLAS and CMS were also adapted and upgraded in expectations for 13 TeV collision energies, 25-nanosecond crossings and higher luminosities. The LHCb detector, discussed in chapter 14, received a new pixel VELO detector and new tracking stations in front of and behind the magnet.

The main aim of the 2015 LHC running was to prepare gradually the machine for a stable long-term operation at 13 TeV collision energy and to approach the nominal $10^{34}\ \mathrm{cm}^{-2}\,\mathrm{s}^{-1}$ luminosity with 25-nanosecond bunch crossings. The several month-long *training* of LHC magnets was completed by March 2015, first beam were injected on April 5th, and on April 10th a 6.5 TeV circulating beam was obtained for the first time. First collisions at 13 TeV were observed by the detectors on May 21st, 2015. These were one-bunch operations with a bunch of 1.4×10^{11} protons. First injection in the machine of 40 bunches of nominal intensity was achieved few days later, on May 25th, with beams declared *stable* on June 3rd.

It is worth re-emphasising that machine protection is the first and absolute priority, as the energy content of one LHC beam at the nominal 7 TeV energy and at full intensity is 360 megajoules (MJ), two orders of magnitude above that of any previous accelerator. In the 2012 run at 4 TeV the peak energy stored in each beam was 140 MJ. The most delicate machine

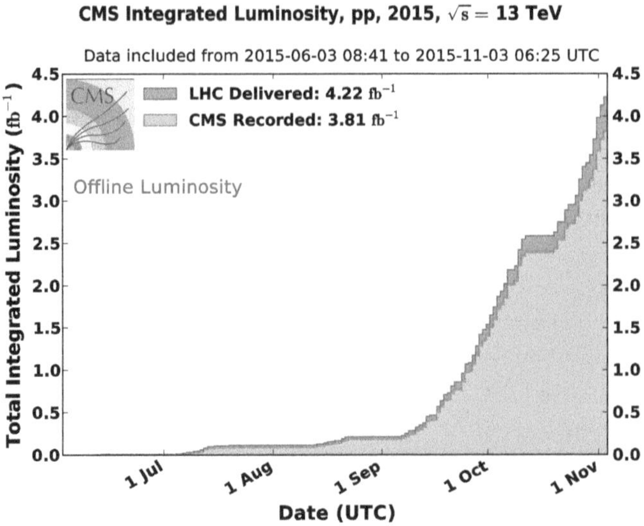

Fig. 9.12. Cumulative growth of the LHC proton–proton luminosity during 2015. The picture shows both the delivered luminosity by the collider and that recorded by CMS. The greater than 90% efficiency achieved was tremendous for such complex detector systems. It was further improved throughout the years to reaching 96% operational efficiency (ATLAS in 2018). *Image: CMS Collaboration.*

operation is during beam injection, which is therefore always performed with low intensity beams. Every year of LHC running from 2010 to 2012, on dozens of time, the beams were accidentally dumped during injection phase, most likely because of interaction with dust particles left in the beam pipe. When the origin of such dumps was uncertain, it was called UFO, standing for *Unidentified Falling Objects*! Because UFO-initiated dumps occurred especially around the kicker magnets used during injection, new ceramic linings were put in place for the 2015 start-up to minimise such beam losses.

In July 2015 the operation towards high luminosity at 25-nanosecond bunch crossing rate started, first with 3 on 3 circulating bunches, then 48 on 48, then 150 on 150, followed by 219 on 219, etc. In the early days of November 2015, the LHC was operating with 2244 bunches in each beam and a peak luminosity of $5 \times 10^{33}\ \mathrm{cm^{-2}\,s^{-1}}$. It delivered up to 250 $\mathrm{pb^{-1}}$ of integrated luminosity per day (figure 9.12), allowing ATLAS and CMS to collect 4 $\mathrm{fb^{-1}}$ integrated luminosity for the year. The goal for the rest of the year was then to have a month-long period of lead–lead collisions at an almost doubled equivalent nucleon–nucleon collision energy of 5 TeV, compared to ion run in 2011, and to accumulate 0.5 $\mathrm{nb^{-1}}$ of integrated luminosity.

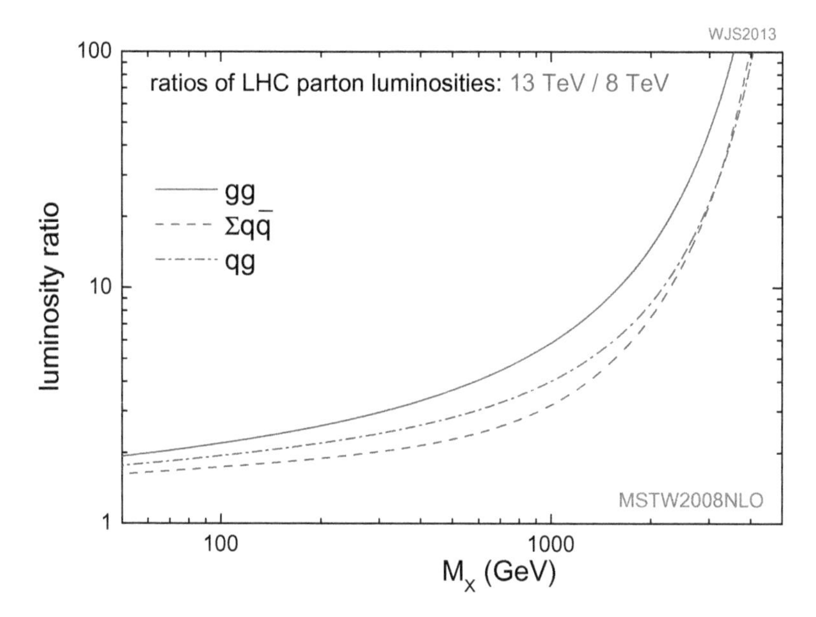

Fig. 9.13. Ratio of the various parton (gluon–gluon, quark–antiquark, gluon–quark) luminosities versus parton–parton centre-of-mass energy M_X, comparing proton–proton collisions at 13 and 8 TeV centre-of-mass collision energy. The parton luminosity combines the parton distribution functions of the two partons involved in the process, and therefore reflects the probability of having a parton collision at centre-of-mass energy M_X, for any fractions of proton energy x_1 and x_2 carried by each parton. It depends on the total proton–proton centre-of-mass collision energy (here 8 or 13 TeV). The advantage of higher energy for production of heavy — or a pair — of heavy objects is evident. *Image: W.J. Stirling, private communication.*

The year 2015, essentially devoted to the running-in of the LHC at higher energy, could be termed as very successful, both operationally for the LHC and experiments as well as from the point of view of the physics perspectives it offered. In the search for possible new heavy particles, despite a much smaller accumulated luminosity than in 2012 (23 fb^{-1}), the 60% increase in collision energy provided a large benefit as shown in figure 9.13. For example, the production cross-section of a pair of supersymmetric particles with mass of 1.5 TeV each would be increased by a factor 35 (see chapter 13, for more details)! Therefore, there was tremendous excitement and hope to discover new physics. However, surprise came not from SUSY searches but from the analysis of the photon–photon mass spectrum where both ATLAS and CMS observed a three standard deviations excess at 750 GeV. Nonetheless, the excess was not confirmed by the 2016 data.

9.6. The LHC Run 2 (2015–2018)

If 2015 was mainly seen as a recommissioning year after the long shutdown to probe the 13 TeV energy regime, 2016 was very successful in terms of accumulating data at an unprecedented rate. First beam injection took place end of March and first stable beams, at low intensity, were obtained in May. The aim was first to operate 6.5 TeV proton beams at 25-nanosecond bunch crossing rate with 1.2×10^{11} protons per bunch, then squeeze the beams harder in the ATLAS and CMS interaction regions to reach a peak luminosity of $10^{34}\,\mathrm{cm}^{-2}\,\mathrm{s}^{-1}$, and accumulate 25 fb^{-1} per experiment integrated over the year. A month-long period of proton–lead collisions at 5 TeV was scheduled towards the end of the year.[9] The year 2016 was thus planned as a *production year* for physics.

During the start-up of the 2016 run there was an anecdotal incident that is worth mentioning here. On April 29th at 5.30 pm all LHC operations were interrupted by a severe electrical perturbation, which affected all LHC accelerators and during which most LHC magnets performed fast aborts. Apparently a weasel sneaked through the fence into the protected area on the surface next to LHCb's experimental area and produced a short circuit on a 66 kV transformer (short to ground fault on the 18 kV cable terminations on top of the transformer, a 1991 model made back in the USSR). Figure 9.14 shows a relative of the culprit and damaged cable terminations. Better protection was subsequently introduced to prevent future intruders.

More serious was a problem with insufficient cooling of the SPS beam-dump system which limited the number of bunches per injection into the LHC to 96. Thus the entire 2016 running was performed with maximally 2208 colliding bunches instead of the nominal 2740. This cooling problem was remedied during the few-month shutdown scheduled at the end of the year.

Despite these incidents, there was a reasonably quick ramp-up in the number of bunches, the peak luminosity reaching $5 \times 10^{33}\,\mathrm{cm}^{-2}\,\mathrm{s}^{-1}$ around May 29th and design $10^{34}\,\mathrm{cm}^{-2}\,\mathrm{s}^{-1}$ beginning of July. It gradually increased to reach $1.4 \times 10^{34}\,\mathrm{cm}^{-2}\,\mathrm{s}^{-1}$ in October, i.e. 50% above the design value.

At such a peak luminosity the pile-up rate seen in the detectors amounted to about 50 simultaneous proton–proton interactions per bunch crossing. The maximum beam intensity reached 1.3×10^{11} protons per

[9] In addition there was a short proton–proton run at 5 TeV centre-of-mass energy, to be used for comparison with proton–lead and lead–lead collisions at the same energy.

Fig. 9.14. A relative of the weasel that sneaked through the fence and provoked a short circuit on 18 kV cable terminations, with the damage induced. *Images: CERN.*

bunch. The total integrated luminosity delivered to ATLAS and CMS was 39 fb^{-1} per experiment, well in excess of the planned 25 fb^{-1} for 2016, and the two detectors took data with efficiencies exceeding 92%.

Figure 9.15 compares the (exceptional) LHC performance over the years in terms of peak (left) and integrated luminosity (right) in proton–proton runs from 2011 to 2018. As the dipole magnets behaved well at 6.5 TeV, at the end of 2016 during two weeks, dipoles of two sectors of the machine were trained to reach 7 TeV beams to eventually reach 14 TeV collision energy when all sectors will be trained accordingly. Prior to this test, the last month of LHC physics operation in 2016 was devoted to proton–lead running at nucleon–nucleon collision energies of 5 and 8 TeV, the LHC delivering 0.5 nb^{-1} at 5 TeV (used for calibration purposes) and 190 nb^{-1} at 8 TeV, respectively.

In 2017, the LHC has been mainly running in proton–proton collision mode at 13 TeV at an average peak luminosity up to 1.5×10^{34} $cm^{-2}\,s^{-1}$. At the beginning of August, due to an accidental air leak during cool down in one of the LHC sections, leading to a deposit of frozen air in an interconnection between magnets in the arc connecting the interactions points of ATLAS with ALICE,[10] the number of colliding bunches had to be reduced

[10]The problem was addressed during the 2017–2018 technical stop by warming up the arc to 80 kelvin, which helped eliminate most, although not all of the ice. A few beam dumps due to remaining ice still occurred during 2018.

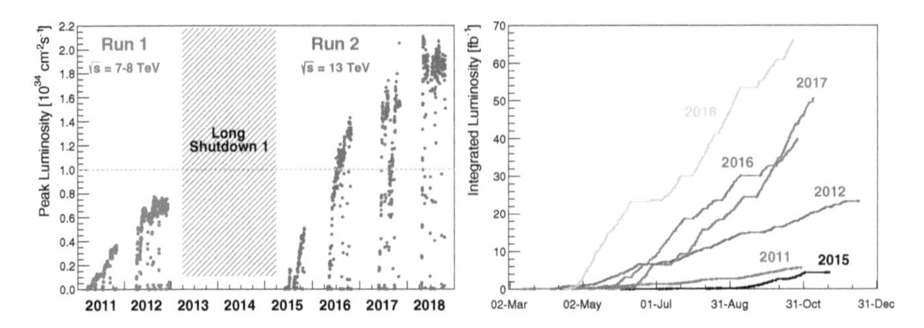

Fig. 9.15. Left: peak instantaneous luminosity delivered to (each) ATLAS and CMS for the proton–proton runs from 2011 till 2018. In 2018 the peak luminosity is regularly around 2×10^{34} cm^{-2} s^{-1}. Right: comparison of the rise of the integrated luminosity in proton–proton runs from 2011 to 2018 as a function of time during the year. *Images: J. Wenninger, CERN-ACC-NOTE-2019-0007 (2019).*

from 2544 to about 1900 for the rest of the year. Despite this setback, by increasing the number of protons per bunch up to 1.35×10^{11} and stronger focusing the beam through decreasing the β^* down to 30 centimetres (for a design value of 80 centimetres), the LHC operators managed to recover the luminosity and even set a new peak luminosity record of 2.1×10^{34} cm^{-2} s^{-1}, twice the design luminosity. However, the price to pay for the experiments was a huge amount of up to 80 pile-up interactions. The instantaneous luminosity was then intentionally levelled at 1.5×10^{34} cm^{-2} s^{-1} to keep the pile-up (60) and trigger rates under control. The delivered integrated proton–proton luminosity for 2017 was 50 fb^{-1} per experiment (figure 9.15).

In 2018, the peak instantaneous luminosity reached 2×10^{34} cm^{-2} s^{-1} with 2544 colliding bunches throughout the year (figure 9.15), limited by the cryogenic cooling capacity in the ring, with a new record integrated luminosity of 63 fb^{-1} for each ATLAS and CMS. The total delivered luminosity for the 13 TeV runs (2015–2018, the so-called Run 2) amounts to 157 fb^{-1} per experiment. The corresponding recorded integrated luminosity for ATLAS is 147 fb^{-1} (94% data taking efficiency) among which 139 fb^{-1} (95%) are good for physics analysis with all detector systems working nominally. The numbers for CMS are similar.

As has been the case since the inception of the LHC physics program, almost every year the last month of running is devoted to heavy-ion physics. Figure 9.16 shows the rise of the integrated luminosity for lead–lead collisions delivered in 2018 for all four LHC experiments. This corresponds to about twice the integrated luminosity delivered in the previous lead–lead runs. The record peak luminosity was 6.2×10^{27} cm^{-2} s^{-1}. Due to its

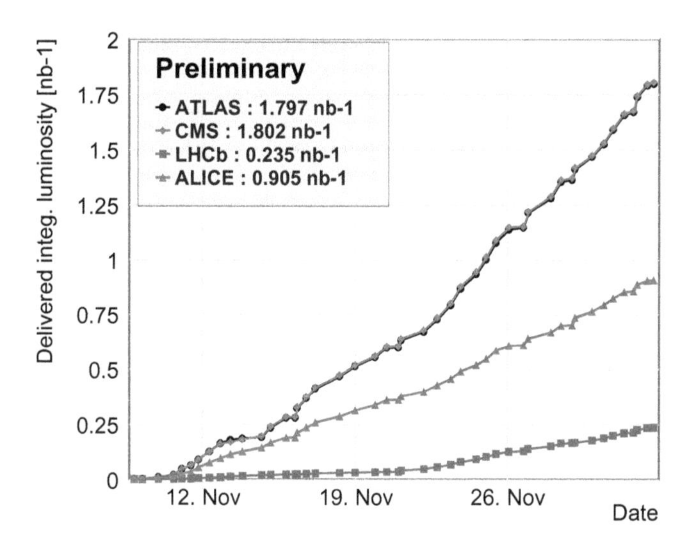

Fig. 9.16. Integrated luminosity versus time for lead–lead collisions delivered in November 2018 to the four LHC experiments. *Image: CERN.*

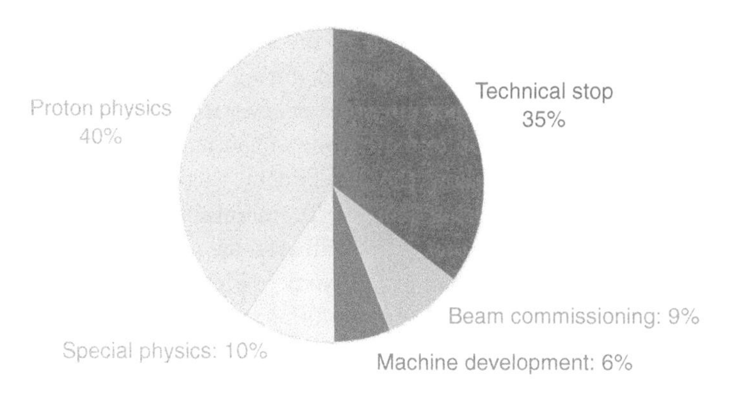

Fig. 9.17. Pie chart showing the time spent for the various LHC machine conditions for the Run 2 production years (end of the year shut-down not included). Special physics includes the heavy-ion runs.

large gaseous tracking system, beams are transversely separated in ALICE to level the instantaneous luminosity at constant $1.0 \times 10^{27}\ \mathrm{cm^{-2}\,s^{-1}}$.

Figure 9.17 shows the remarkable availability of the CERN accelerator complex during the Run-2 production years 2016–2018, which has been able to deliver stable beam conditions for physics half of the time.

9.7. Long LHC shutdown 2 (2019 to end of 2021)

Since 2019, during the three-year long shutdown 2 (LS2), the main LHC activities consisted of extensive consolidation work on the superconducting magnets, checking about 1400 interconnections, 8000 electrical quality assurance tests, 2500 leak tightness tests, and the replacement of 22 cryomagnets, more specifically 19 dipoles and 3 quadrupoles (figure 9.18 shows the lowering through the access shaft and positioning in the LHC tunnel of a replacement dipole) and the installation of up to four new-technology (Nb_3Sn) 11-tesla magnetic cryo-assemblies (figure 9.18, lower part), allowing to reduce the 15-metre length occupied by present 8.3-tesla magnets so as to liberate space to insert additional collimators for Run-3 heavy-ion runs and in view of future High-Luminosity LHC (HL-LHC) running.[11] The LHC injector complex also underwent significant upgrade for improved beam brightness and operational reliability. In addition, civil engineering work was performed in preparation of the HL-LHC.

The four large LHC experiments also underwent important upgrades. ATLAS refined the trigger selection in view of Run 3 and the HL-LHC with better first-level trigger granularity in the calorimeter and a new muon detector replacing the current small muon wheel (with nevertheless ten-metre diameter!). CMS already renewed during Run 2 the inner silicon-pixel tracking detector. Further upgrades of several systems and consolidation work were made. LHCb executed one of the most ambitious upgrade programmes that should allow the experiment to take five times higher luminosity and pile-up during Run 3. To maintain the performance of detector under these conditions, almost all systems needed to be upgraded. New tracking detectors and upgraded particle identification systems were installed. A remarkable feature is a new 40-megahertz rate all-software trigger (the Run-2 hardware trigger allowed only 1.1-megahertz rate), exploiting new computing technologies such as graphical processing units. Finally, the ALICE experiment also performed a very ambitious upgrade with as main theme a trigger-less readout allowing it to record 50 to 100 times more soft collisions with a readout frequency of 50 kilohertz (improved from 1 kilohertz during Run 2). Among the many system upgrades, a new inner tracker using silicon pixel technology and forward muon tracker was installed, with altogether 13 billion pixels and using pioneering monolithic (CMOS) technology.

[11]To reduce technical risk it was decided not to install 11-tesla dipole magnets during LS2.

Fig. 9.18. The upper images show the replacement of an old-type cryomagnet: positioning in the tunnel and lowering through the access shaft. The centre images show a cryomagnet built with Nb_3Sn superconductors, which will operate at 11 tesla and prefigures what could be the magnet technology used in future hadron colliders (see section 15.2.2). The left-hand image shows a coil after impregnation. The lower photo shows HL-LHC civil engineering work next to the CMS experimental cavern. *Images: CERN.*

After LS2, physics production resumes in 2022 with an instantaneous luminosity remaining at a level of $2 \times 10^{34}\ \mathrm{cm}^{-2}\,\mathrm{s}^{-1}$, but for a longer time, possibly an entire 10 hours run owing to the improved beam brightness. The LHC operators will also attempt to achieve design energy of 7 TeV per beam, that is, 14 TeV proton–proton collision energy. This period, termed Run 3, will last until 2025 when LS3 starts and machine and detectors are upgraded to the HL-LHC phase starting in 2027. The HL-LHC is designed to deliver an instantaneous luminosity of up to $7 \times 10^{34}\ \mathrm{cm}^{-2}\,\mathrm{s}^{-1}$ (up to 200 pile-up events) and 3000 fb^{-1} integrated luminosity by 2037 (see chapter 15).

Chapter 10

Data Analysis

With 5 fb^{-1} of 7 TeV proton–proton collision data recorded in 2011, 21 fb^{-1} at 8 TeV in 2012, and 150 fb^{-1} at 13 TeV until end of 2018, the ATLAS and CMS physicists had plenty of data to analyse. During all these years, the work schedules in both collaborations were often determined by the occurrence of the large international conferences on particle physics, each collaboration striving to present in the shortest possible time the most important and original physics analyses, based on all the available data.

Before considering specific physics analyses, for example the search for the Higgs boson, a graviton, or a new heavy W' or Z' gauge boson, or a more precise measurement of the W boson or top-quark mass, a long and careful preparation of the data is required. Indeed the quality of this preparatory work determines the quality, precision and reliability of the final physics result. The data being barely recorded have to be processed taking into account latest detector calibrations and alignment measurements, and verifying the quality of the data. The *event reconstruction* allows to transform the recorded detector signals into signatures of particles produced in a collision. These can be used to perform physics measurements or undertake a search for yet unknown particles. Depending on the complexity and priority of the physics study, the analysis can involve just a few physicists up to several tens. Once the work is performed, a scientific note summarising the analysis procedure and the results is written. It is submitted to the scrutiny and criticism of the many experts within the collaboration before it is "approved": the result is then ready to be presented at a conference or published in a physics journal. How many nights and weekends have been spent with great passion[1] to get results ready and released!

10.1. What does data analysis involve?

We described in the preceding chapters what spectacular instruments the LHC collider and its particle detectors are. Less visible, but not less spectacular, are the chain actions required to analyse the wealth of collision data recorded by the detectors, as we will show in the following.

From the electronic signals recorded by detector elements, one has to first extract the physical quantities such as the nature, the energy and/or momentum, and the direction of the particles produced in the collision. On the basis of these quantities, the experimenters select among the multitude of collisions those that correspond to the intended analysis. Most of the time this looks like the proverbial search for a needle in a hay stack.

Let us first give a look at figure 10.1. When the LHC is at its nominal working point, with centre-of-mass collision energy of 14 TeV (dotted line) and a luminosity of 10^{34} cm^{-2} s^{-1}, about a billion proton–proton collisions per second take place in ATLAS and CMS. Among these, about 300 events contain a Z boson, which is almost easy to detect through its instantaneous decays $Z \to e^+e^-$ and $Z \to \mu^+\mu^-$. However, the decay probability for each of these modes is only 3%, meaning that one has to find about twenty events out of a billion produced each second. The task is much harder in case of a Higgs boson search: according to Standard Model expectations and for a mass around 125 GeV, it should be produced at a rate of about one event every two seconds (figure 10.1). However, the cleanest decay modes, $H \to ZZ$ and $H \to \gamma\gamma$, are only at the 0.1 and 2 permil level, respectively, taking into account that the two Z bosons have again to be identified through the e^+e^- or $\mu^+\mu^-$ decay modes. Altogether there is on average one Higgs boson event followed by a four-lepton decay[2] only once in 15 000 billion collisions! The $H \to \gamma\gamma$ channel may then look easier and more favourable for a discovery, as its ratio is just one in 1000 billion. But the difficulties turn out to be comparable as we will see later.

---

[1] To be fair: sometimes also great frustration in view of technical obstacles or never ending reviews.

[2] It may be worth recalling that only muons and electrons and their antiparticles are considered here. The τ lepton is not included, as it decays promptly and its experimental signature is complicated by many possible decay modes. Talking of the four-lepton decay mode of the Higgs boson is thus an abuse of language, often used in the following.

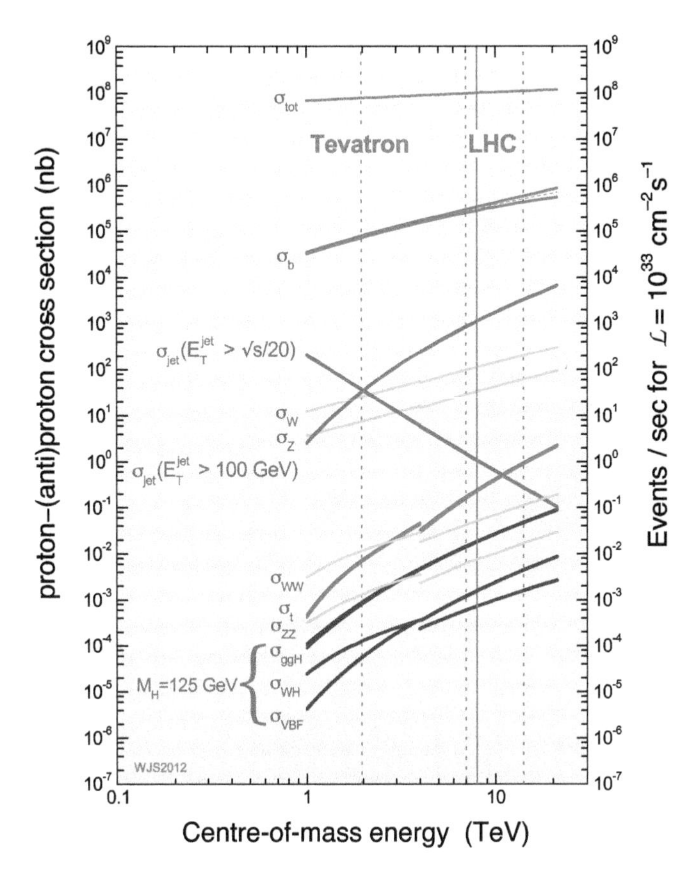

Fig. 10.1. Production cross-sections (left ordinate) and event rates (right ordinate) for an instantaneous luminosity of 10^{33} cm^{-2} s^{-1} versus the proton–proton centre-of-mass energy $\sqrt{s}$ (the notation $E_{cm} = \sqrt{s}$ is for historical-theoretical reasons). The rightmost vertical dotted line corresponds to the nominal LHC energy (14 TeV). There is a strong dependence of the Higgs boson production rate on the collision energy. Running at half the nominal energy reduces the Higgs production cross-section by a factor slightly larger than three for a Higgs mass of 125 GeV. One notices also the huge difference, by a factor of 10^9, between the overall proton–proton interaction rate given by σ_{tot} and that of Higgs boson production, through the various production processes gluon–gluon fusion (ggH), associated production with a weak boson (WH, ZH not shown), and vector boson fusion (VBF), all discussed in chapter 11. *Image: W.J. Stirling, private communication.*

Finding a specific type of event among thousands of billion collisions is in fact possible. This is exactly the scope and purpose of the online triggering system already discussed in section 8.4, and of the offline data analysis, which is described in this section. Let us look step-by-step how the data analysis proceeds.

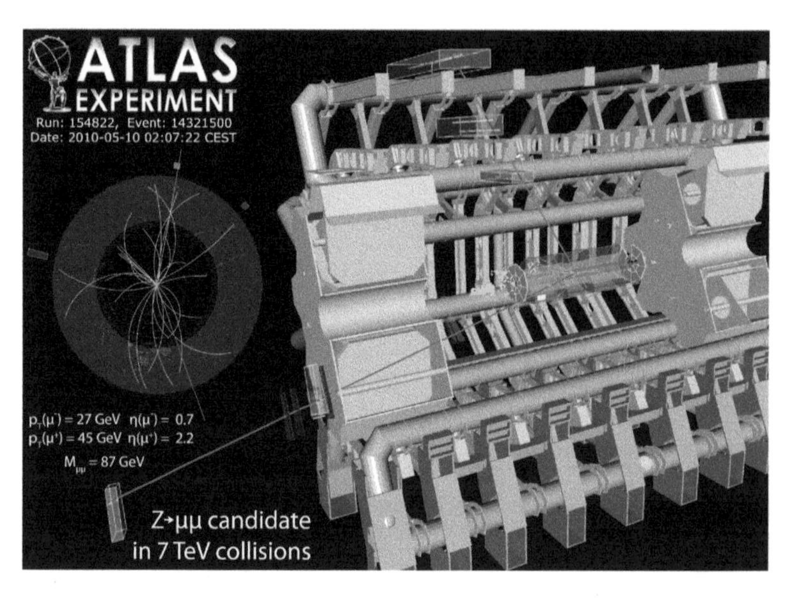

Fig. 10.2. Proton–proton collision event recorded with the ATLAS detector showing a Z-boson candidate, which subsequently decayed to two muons. *Image: ATLAS Collaboration.*

10.1.1. *Event reconstruction*

Figure 10.2 shows a computer view of the ATLAS detector with a 7 TeV collision event recorded during the first data-taking period of 2010. In the small transverse view on the left-hand side of the picture, as if taken from the centre of the detector looking down the beam directions, one sees a multitude of yellow points representing position signals ("hits") recorded by the central tracker. Some of the points can be connected into charged particle trajectories emerging from the beam interaction point at the centre. Knowing the value of the ambient magnetic field, the reconstructed curvature of trajectories (barely visible for the red lines here) determines the particle momenta (the straighter the line, the higher the momentum). Some of the hits could not be associated to trajectories, for example because they are fake signals due to electronic noise in the detector, or because they are badly measured. On the three-dimensional view one notices that some of the tracks reconstructed in the inner tracker can be related to track segments reconstructed in the outer part of the ATLAS detector, more specifically in the muon spectrometer. It is thus very likely that in this specific collision two muons were produced.

Fig. 10.3. Event recorded in the ATLAS detector featuring a Z-boson candidate decaying into two muons together with 25 simultaneous proton–proton collisions. Thanks to the presence of these two high-energy muons, the trigger system could identify the event as potentially interesting and took the decision to record it. During the subsequent full event reconstruction it was possible to distinguish the many collision vertices, all produced during the same bunch crossing, as visible on the lower part of the figure showing the individual vertices and associated tracks. *Image: ATLAS Collaboration.*

Every event recorded by the LHC experiments is processed by a reconstruction software program, whose task is exactly what we have just described for charged particles: from the detected electronic signals reconstruct hits and trajectories, and thus obtain the charge, direction and momentum of particles produced in the collision (figure 10.3). In case of a neutral particle, the inputs to the reconstruction software are instead the signals registered in the calorimeters to measure their energy. Particle reconstruction is a sophisticated task that is essential for all subsequent analysis work. In ATLAS for example, the reconstruction of a single collision event requires on average about ten seconds of computing time on standard processors. After the online event selection by the trigger system, the rate of events retained for final recording on disc (and tape) amounts to several hundreds per second at the beginning of data taking to more than one thousand per second towards the end of Run 2. All those recorded events are first reconstructed at CERN, usually within two days. This task requires the usage of several ten thousand processors to keep-up with the rate of data taking. A second round of event reconstruction of the same data may be performed towards the end of a data taking year, when the best calibration and alignment information is available. Reprocessing the entire stock of data recorded during a year usually takes several weeks. It requires information technology means, processors and disc space, well beyond what is available locally at CERN, as discussed in section 10.2 about the worldwide LHC computing grid.

10.1.2. *Calibration and alignment of detector systems*

Before proceeding with the event reconstruction it is necessary to estimate the conversion factors allowing to transform digitised signals provided by various detector systems into quantities to be used by reconstruction programs, such as the energy deposition in calorimeters or the precise geometrical location of a hit in the tracking system, to name a few. In a liquid argon calorimeter, as in the ATLAS experiment, the signal corresponds to the electric charge generated by the charged particles from the electromagnetic shower ionising the liquid argon placed in the gaps between electrodes and passive absorber material. For calorimeters based on scintillation, as for the CMS calorimeters or the ATLAS hadron calorimeter, the signal results from the transformation of the intensity of the scintillation light pulse into an electric pulse, as performed by photomultipliers or semiconductor diodes. This is also the case in calorimeters based on the Cherenkov effect (see inset 14.1 on page 475) in quartz fibers, as for the CMS forward hadron calorimeter. For all those calorimeters, the conversion factors are referred to as *calibration coefficients*, while for tracking detectors the usual term is *alignment coefficients*. In general, the term alignment describes the entire procedure allowing to determine the exact position in space of a silicon microstrip or pixel detector element, or of a measuring wire in a drift chamber.

Calorimeter calibration usually proceeds in several steps depending on the nature of the calorimeter. A first calibration is performed in the years or months preceding the installation of the detector in the experimental cavern. It consists of exposing modules to a beam of particles (denoted "test beam") of known energy and composition, and measuring the calorimeter response. Later, when the detector is operating and taking data, calorimeter calibration is done *in-situ* at regular intervals to correct for time drifts. In case of liquid argon, a charge pulse of shape and duration similar to the one generated by a particle passing through the calorimeter is injected upstream of the readout electronics. In calorimeters based on scintillating crystals (or on Cherenkov light) instead, a well-controlled light pulse of known amplitude, for example from a laser or a radioactive source, is injected directly in front of the crystals, in order to calibrate the full readout chain. In CMS, for example, this allows to continuously estimate and correct for the variation with time of the light captured by the readout elements, avalanche photodiodes, due to a gradual loss of transparency of the crystals[3] from irradiation occurring during data taking. The final stage in calibration relies on proton–proton collision data themselves. For example, the very abundant $\pi^0 \to \gamma\gamma$ decays can be used to inter-calibrate the calorimeter response over all its elements, and thanks to the very precisely known Z-boson mass (measured at LEP), it is possible to use $Z \to e^+e^-$ data to calibrate the calorimeter's absolute energy scale. Altogether, it is possible to achieve a very uniform and precise response of the electromagnetic calorimeter over its entire acceptance with a precision better than 1%.

The alignment of wire chambers used in the muon spectrometers is done through an ingenious system of optical probes exploiting the straightline propagation of light (figure 10.4). In the case of ATLAS, ten thousand triplets composed of a diode laser, a lens and a CCD readout element are used. These triplets are positioned with a precision better than 30 micrometres relative to the detection wires, allowing one to align the 1200 drift chambers with a precision better than 50 micrometres. As a

[3]The transparency of the lead tungstate crystals forming the CMS electromagnetic calorimeter is affected by irradiation during data taking. The larger the instantaneous luminosity, the larger the transparency loss. During intervals with no collisions a spontaneous but incomplete recovery occurs. Over one year of data taking the permanent transparency loss is only a few percent in the less-exposed barrel part, but can reach up to 40% for the end-cap crystals closest to the beam line. The crystals must therefore be constantly monitored.

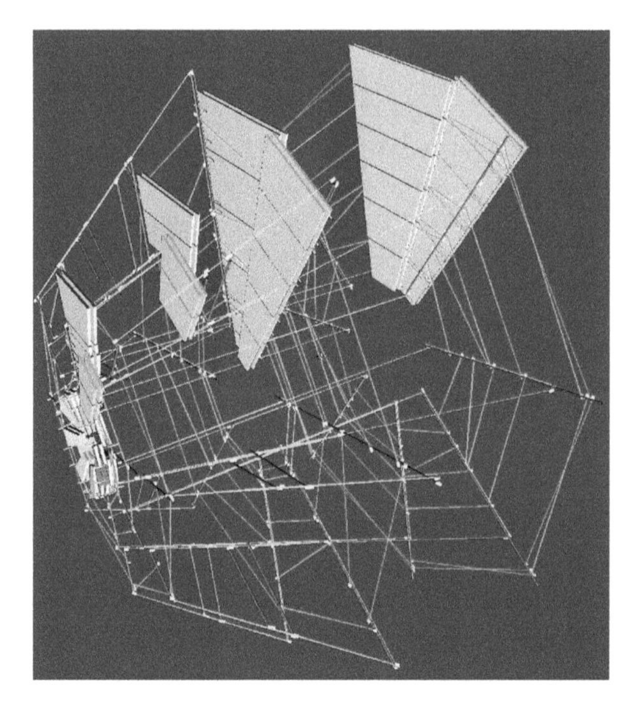

Fig. 10.4. Scheme of the optical alignment system in one of the two end-cap muon spectrometers of ATLAS. A system of light rays (in green and blue on the picture) allows reconstructing muon chamber positions with accuracy better than 50 micrometres. For clarity only two sectors of muon chambers are shown in the picture. *Image: ATLAS Collaboration.*

result, the momentum of a 1 TeV muon can be measured with a 10% precision. The data provided by the CCD elements are analysed several times a day to follow potential displacements of chambers due to temperature changes.

The alignment of the inner tracking devices uses tracks of high-energy particles produced in the proton–proton collisions. This is possible as many hits per track are sampled and recorded. Applying a fitting procedure with more than 30 000 free parameters corresponding to the various geometrical degrees of freedom of the tracking detector elements (see figure 8.7), an overall optimisation to match the measured hit positions to reconstructed tracks is performed using millions of tracks. Ultimately, an alignment precision of 10 micrometres or better is obtained. The momentum resolution for high-energy particles is then essentially limited by the space-point measurement accuracy of the detectors.

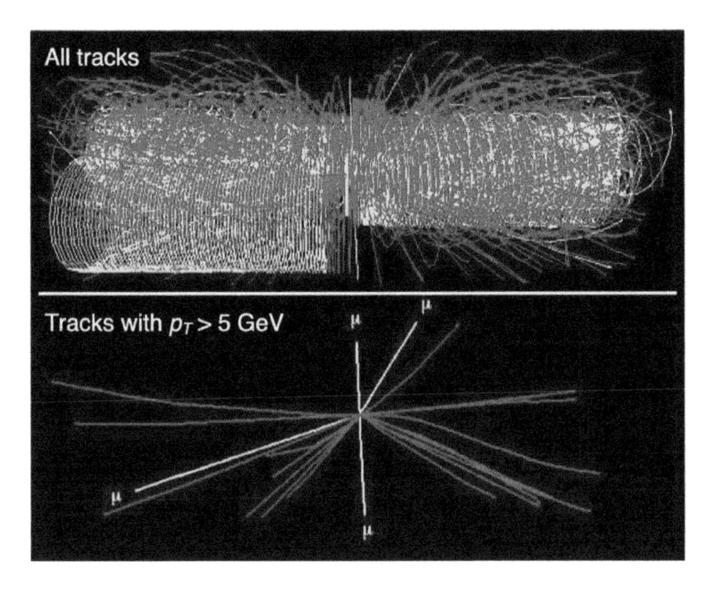

Fig. 10.5. Simulation of Higgs boson production followed by the decay into two Z bosons, each subsequently decaying into two muons. The upper figure shows all tracks reconstructed in the CMS tracking detector, most of which are soft and spiralling in the strong magnetic field. By selecting only tracks with transverse momentum larger than a specified threshold (here 5 GeV) the muons resulting from the Higgs boson decay are well exposed (lower figure). *Image: CMS Collaboration.*

In addition to alignment with proton–proton collisions, a procedure using cosmic-ray muons is performed at the beginning of major running periods, prior to injection of beams into the machine: external muon chambers are aligned with the inner tracker, first with magnetic field turned off, therefore with straight tracks, and then with field on.

10.1.3. *Signal and backgrounds*

Among the physics analysis goals at the LHC is the search for evidence of a signal corresponding to a physics phenomenon of interest, as for example the production of a new particle with given decay modes. For this purpose, events are selected if they exhibit the features of the signal that is sought for. The selection is based on a set of topological and kinematic variables as measured by the detector: momenta measured from the tracks reconstructed in the inner tracking and muon detectors (figure 10.5), energies deposited in the calorimeters, number of leptons and jets in the event, missing transverse energy, invariant mass of a given combination of particles, etc. The most straightforward way is to select events for which the values of these variables fall in specific ranges as expected for the phenomenon of interest. More sophisticated methods, such as multivariate analyses (see inset 10.1) or machine learning techniques combining the available information, are applied when the expected signal is small compared to the background or signal and background events have similar properties making them hard to distinguish.

The sample of selected events is never pure: it always includes events from other phenomena that look alike the one being targeted. This is what is called background. At first sight background looks like the signal, but it is not the signal. Sometimes, or may we say in most of the new physics searches, the whole selected data turn out to be only from background processes. One then concludes that the signal expected on basis of some novel theory or phenomenology was not produced in an amount that is statistically significant. Its production above a certain rate (or cross-section) can therefore be excluded.

After the signal selection procedure, the evidence for or exclusion of signal is derived from the statistical distribution of one or more variables characteristic of the signal looked for: the reconstructed mass, missing transverse energy, or some combination of kinematic quantities attached to final state particles (figure 10.6). The production of a new particle could be signalled by a peak in the mass distribution of a combination of final state particles. This is sometimes called a resonant signal; the Higgs boson, a fourth generation top quark, or a Z' is expected to appear this way. The background mixing into the peak can be reliably estimated from the amount of events seen left and right to the peak (denoted *sidebands*). A signal can also be searched for and established through an excess of events relative to the background expectation without very distinctive kinematic features. This is called a non-resonant signal, and is quite often what is sought for in searches for supersymmetry for example. A non-resonant signal is

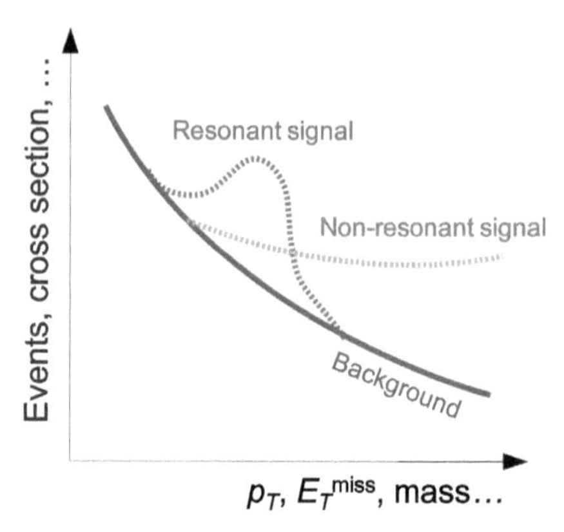

Fig. 10.6. Schematic representation of distributions expected at the outcome of an analysis looking for a new physics signal.

more difficult to extract than a resonant one as it requires precise and reliable knowledge of the properties and abundance of all relevant background processes.

If evidence for a signal has been seen, one can measure its production cross-section (see inset 10.2). If the outcome of the search is negative, that is no signal was found, the result of the analysis consists in deriving an *upper limit* on its production cross-section. Had the signal been larger than this limit, it would have been detected by the analysis. This is also a valuable physics result, as it provides constraints on the underlying theoretical model (section 13.2).[4]

Inset 10.1: Multivariate statistical event classification

A crucial task in every physics analysis consists in selecting a sample of events corresponding to the signal that is sought for. This sample should be as pure as possible, which means it should be as little as possible contaminated by background processes while retaining as many signal events as possible. The selection is performed using a set of N variables x_i (such as transverse momentum, production angle, etc.) on which certain requirements are applied. The variables must be chosen according to their discriminating power to distinguish signal from background events, that is, their shapes should look different.

The figure below illustrates a simple case with $N = 2$ variables. The distribution of variable x_i as a function of a variable x_j is shown for a sample of signal events (in blue) and background events (in red). The requirements applied can be simple rectangular or oblique threshold values ("cuts"), as on the left portion of the figure, but this rarely corresponds to the most effective selection. Often an optimised selection requires nonlinear and complex cuts, as sketched on the right part of the figure. Such a *multivariate analysis* can become rather complex especially when dealing with a large number of variables.

[4] A beautiful example how the non-observation of a process sparked progress is the Glashow–Iliopoulos–Maiani (GIM) mechanism conceived in 1969: to explain the measured suppression of the decay $K_L^0 \rightarrow \mu^+\mu^-$, a new particle, the charm quark, was postulated to complement the strange quark in the doublet forming the second quark generation. The charm quark was experimentally confirmed in 1974.

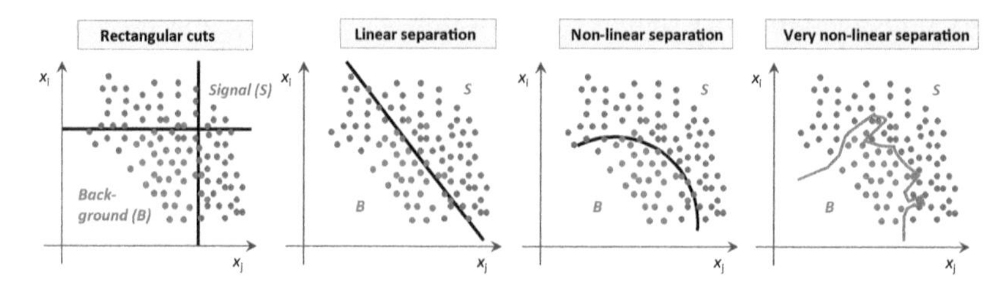

There is a whole arsenal of statistical classification methods to deal with such problems that are also encountered in other domains than particle physics, for example in finance, automation, medicine, etc. All methods have in common that they learn features from the data in an automatic way ("machine learning"). In supervised machine learning, the training exploits the knowledge of what is a signal event and what is background. In particle physics the methods most often used are *maximum likelihood* calculations and fits, *neural networks* inspired by decision-taking processes in biological neural systems, and *boosted decision trees* (BDT). The latter are smart combinations of a set of decision trees, each implementing a sequence of cuts on the x_i variables. Machine learning allows computer systems to perform specific tasks without using explicit instructions. As such, it is a subset of artificial intelligence technology. New industry-driven developments of ever more sophisticated machine learning tools are now being applied in experimental particle physics to more complex problems than signal and background separation: regression algorithms are used to calibrate objects reconstructed in calorimeters, fast neural networks improve online event selections, etc.

The output of the simplest classification algorithms is a real number, often defined to lie between 0 and 1, indicating how much each event is signal or background-like. The closer the output is to 1 (0), the more the event features resemble that of signal (background). The algorithms rely on a set of internal parameters that are optimised during the learning (or training) procedure exercised on samples of signal events (from simulation) and background events (from simulation or from recorded data when possible), to obtain maximum discriminating power for the algorithm output.

The figure on the right shows example distributions of a BDT output for signal (blue) and background events (red) after the training procedure. The signal selection is performed by rejecting events for which the BDT response is below a certain value, chosen according to how pure the final sample must be.

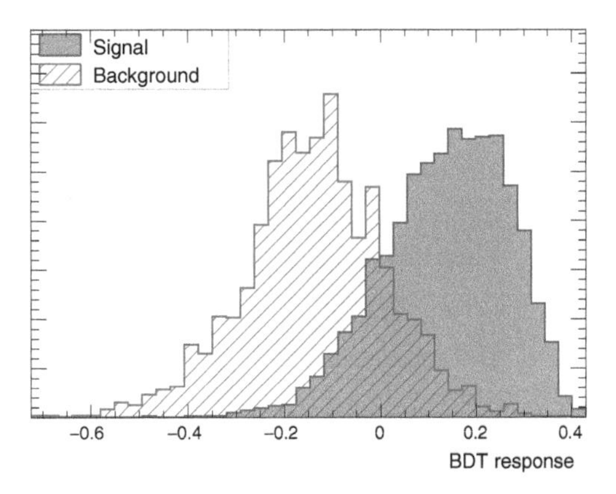

This algorithmic procedure thus provides a quasi-optimal way to reduce an N-dimensional kinematic variable space to a single dimension, with one variable given by the output of the BDT. It goes without saying that the reliability of these selection methods depends crucially on the quality of the simulation used.

Inset 10.2: Cross-section measurement

The study of a particular physics phenomenon with final state f, for example the production of a top-quark pair or a Z boson, most often implies the measurement of its production cross-section σ (see also inset 2.5 on page 93). Experimentally, σ is measured from the number of observed events as follows

$$\sigma(pp \to f) = \frac{N^{\text{sel}} - N^{\text{bkg}}}{L^{\text{int}} \cdot \mathcal{A} \cdot \varepsilon^{\text{trig}} \cdot \varepsilon^{\text{sel}}},$$

where

- $\boldsymbol{N^{\text{sel}}}$ is the number of selected events;
- $\boldsymbol{N^{\text{bkg}}}$ is the number of background events in the sample, estimated either on basis of simulations or from data themselves, for example through extrapolation of an observation in a nearby control region where no signal is expected;

- L^{int} is the integrated luminosity corresponding to the event sample studied;
- $\mathcal{A}$ is the detector acceptance, that means the probability for the particles making up final state f to be within the geometrical coverage of the various detector systems where they can be identified and measured;
- $\varepsilon^{\text{trig}}$ is the triggering efficiency of the detector for events within the geometrical acceptance of the detector, that means the probability that a signal event be recorded;
- ε^{sel} is the selection efficiency defined as the probability for a recorded event of final state f to be actually selected; this also includes the event reconstruction efficiency, as for every particle in the event, electron, muon, jet, etc., there is a certain reconstruction inefficiency (of hardware or software origin), even within detector acceptance.

The integrated luminosity is calculated from the instantaneous luminosity that is measured all along the data-taking period, thanks to dedicated particle counters located close to the beam line in the detector. These counters must be appropriately calibrated and several techniques exist for doing so. The most precise is the so-called *van der Meer beam separation scan*, which consists in displacing the beams during a special low intensity fill relative to each other in the transverse plane to measure the overlapping beam profile (luminous region) from the recorded collision rates. Combining this profile with the precise knowledge of the electric current in each beam allows to estimate the luminosity during the scan and to thus calibrate the response of the luminosity detectors. This tremendously accurate method achieves percent-level precision exceeding the most optimistic expectations of the machine experts!

Monte Carlo simulations (see section 10.1.5) are an essential tool to estimate the signal acceptance. The trigger, selection and reconstruction efficiencies are also determined using simulation, but often corrected with the use of real data by exploiting well understood and controlled final states, such as Z boson production and decay (section 10.1.6).

Inset 10.3: Probability and the particle world

The initiators of the mathematical theory of probabilities, driven largely by the need to develop a theory of games, were in the mid-17th century the mathematician Pierre de Fermat (1607–1665) and the philosopher, physicist and mathematician Blaise Pascal (1623–1662). In 1687 Newton established the laws of motion and the law of universal gravitation in his *Principia Mathematica Philosophiae Naturalis.* These achievements were followed by the developments of mathematics and physics through the 18th century and the more appropriate formulations of Newton's law of motion (Lagrange, Hamilton, Poisson, etc.). In the early 19th century, the most renowned classical physicist of that time, Pierre-Simon de Laplace (1749–1827, figure 10.7), posited that a totally deterministic evolution of matter and thus the Universe is possible, both towards the future and in the past, provided all the laws of forces and initial conditions are known. Probability theory would still be needed, but only to complement technical difficulties and the practical impossibility of knowing the initial conditions of all objects, in particular in large statistical ensembles.

During the second half of 19th century, James Clerk Maxwell (1831–1879) and Ludwig Boltzmann (1844–1906), applying classical Newtonian mechanics and statistical methods on large-population systems as gases or bulk matter to obtain the average physical properties (pressure, temperature, etc.), reduce and incorporate thermodynamics into classical physics. But with the advent of quantum mechanics in the 20th century, with Heisenberg's uncertainty relations, an irreducible uncertainty enters in the description of the microscopic world, and the strict determinism of Laplace becomes a statistical determinism: we cannot determine the outcome of an individual particle collision, but only a distribution of probabilities. Albert Einstein, although being one of the fathers of quantum mechanics, was not accepting Max Born's statistical interpretation of quantum mechanics. As well known he exclaimed "*Gott würfelt nicht!*" ("*God does not play dice!*"), hoping in the restoration of Laplacian determinism, but hidden variables that would bring back determinism through the back door into the quantum world were never found. In the following we address some aspects of the use of statistics in particle physics.

Probability and quantum mechanics. In quantum mechanics, particles are represented by wave functions whose size gives the probability that the particle is found in a given position. The rate at which the wave function varies from point to point gives the speed of the particle. Quantum phenomena like particle reactions occur according to certain probabilities. Quantum field theory allows us to compute cross-sections of particle production in scattering processes, and decays of particles. It cannot, however, predict how a single event will come out. We therefore use probabilistic techniques to simulate event-by-event realisations of quantum probabilities.

Fig. 10.7. Pierre-Simon de Laplace (1749–1827). Fate was considered deterministic and, in principle, predictable if the complete equation of state were known.

Statistics of large systems. Statistical physics uses probability theory and statistics to make statements about the approximate physics of large populations of stochastic nature, neglecting individuals. Heavy-ion collisions at the LHC are modelled using notions of hydrodynamics (the strongly interacting medium behaves like a perfect fluid — with the smallest sheer viscosity ever observed in fluids). Statistical mechanics provides a framework for relating the microscopic properties of individual atoms and molecules to the macroscopic properties of materials that can be observed in everyday life. It thus explains thermodynamics as a natural result of statistics, classical mechanics, and quantum mechanics at the microscopic level. Probability and statistics are fundamental ingredients and tools in all modern sciences. *"Statistics is the grammar of science"*, as the English mathematician and statistician Karl Pearson (1857–1936) is quoted.

Statistics in measurement processes. In addition to the intrinsic probabilistic character of particle reactions, the measurement process through the interaction of particles with active and passive detector materials contributes statistical degrees of freedom leading to measurement errors and to genuine systematic effects (e.g. detector misalignment), that need to be considered in the statistical analysis. For example, track fitting in the LHC environment is challenging: it must deal with ambiguities, hit overlaps, multiple scattering, bremsstrahlung, multiple vertices, etc. Track fitters take Gaussian noise (via the so-called Kalman filter) and non-Gaussian noise (Gaussian sum filter) into account. Fitting is a statistical procedure that interprets and exploits *a priori* known uncertainties.

Measurements and hypothesis testing. From a measured data sample, we want to determine parameters of a known model (e.g. the top-quark mass in the Standard Model), we want to discover and measure missing pieces of the model (e.g. the Higgs boson, neutrino masses), and we want to watch out for the unknown (test the data versus the predictions of a given model), or exclude parameters of suggested new physics models. Due to the intrinsic randomness of the data, probability theory is required to extract the information that addresses our question. Statistics is used for the actual data analysis. The interpretation of probability depends on the statement we want to make:

- for repeatable experiments, probability can be a measure of how frequently a statement is true;
- in a more subjective approach, one could express a degree of belief of a statement.

Repeatable experiments are for example: playing a dice and finding 6; finding a peak of some size or more in fluctuations of a background distribution. *Non-repeatable statements* are for example: the probability that dark matter is made of axions; the probability that the new 125 GeV boson is the Higgs boson.

Random variables. In a statistical context, instead of "data" that follow a distribution, one typically speaks of a "random variable". The data follow an (unknown) distribution of a random variable. Although we plot error bars around observed data points, a given observation does not have an uncertainty. It is fixed. It would be more accurate to plot the error bar around the prediction. By plotting the error bar around the observation, we assume by convenience the observed value to be the truth. This is a delicate issue that sometimes leads to confusion and misinterpretation among physicists.

Statistical distributions. Measurement results typically follow some *distribution*, that is, the data do not appear at fixed values, but are spread out in a characteristic way. Which type of distribution it follows depends on the particular case. It is important to know the occurring distributions to be able to pick the correct one when interpreting the data, and it is important to know their characteristics to extract the correct information.

10.1.4. *Systematic uncertainties*

Systematic uncertainties are the evil (see figure on the right) in every measurement. Well designed experiments minimise systematic uncertainties by achieving maximum phase space coverage, high-measurement precision, response homogeneity and linearity, large calorimeter depth, sufficient longevity of the detector components including resistance against irradiation, etc. The understanding, evaluation and reduction of systematic uncertainties is often the main analysis challenge. A high-quality analysis stands out by its thoroughness on all relevant sources of systematic uncertainty. It is thereby important to distinguish relevant from irrelevant sources, where in doubt a source should be considered relevant. For many uncertainty sources, in particular theoretical ones, estimating a "one-sigma" error is very difficult or simply impossible. In such cases conservative uncertainties should be chosen where possible.

(Reasonably) conservative uncertainty estimates are a must! It is of no use to the scientific endeavour to make over-aggressive statements that one cannot fully trust.

10.1.5. *Event and detector simulation*

The simulation of proton–proton collisions and the interaction of the generated final state particles with the detector material are essential ingredients of physics analysis. For example, the measurement of a production cross-section often relies on simulation for estimating the detector acceptance (see inset 10.2) or the level of residual background in the selected sample. The simulation tools used by the LHC collaborations have reached an extraordinary degree of sophistication.

The simulation technique mostly employed is called *Monte Carlo* as a reminder of the games based on random choices and chance occurrences practised in casinos in the eponymous quarter of the Principality of Monaco. Every physics process, be it production of a given final state, decay of an unstable particle, or the interaction of particles with matter in the detectors, can be reproduced through the selection of random numbers following a particular probability distribution underlying the specific mechanism or feature being simulated.

The simulation of a single collision event most often requires generating thousands of random numbers. For an unbiased physics analysis one needs to generate very large numbers of events, often exceeding the sample size of real experimental data, so as to have a fully representative and detailed comparison sample. However in practice, as the time to generate a single LHC collision event including the detector simulation amounts to several minutes if not tens of minutes, the number of simulated events is always ultimately limited by the available computing resources.[5] Nonetheless, the number of events generated every year by each LHC experiment reaches several tens of billion. Three stages can be distinguished in the event simulation.

(1) **Generation of particles** present in a specific final state of a proton–proton collision, for example: production of a Higgs boson decaying into two Z bosons, or the production of a top–antitop pair, followed by all their decays. Particles produced as left-over debris of the two interacting initial-state protons must also be generated (figure 10.8). The key point here is the simulation of the quantum mechanical probability distribution (the matrix element) for the process under study. Specialised Monte Carlo programs used to generate the complex processes occurring at the LHC, including higher-order loop processes as well as strong-interaction and electromagnetic radiation, are written by

[5]Sometimes a less detailed and thus faster simulation can be used if the best level of accuracy is not needed. Such a fast simulation is obtained by parameterising with simple functions the complex response of the full simulation.

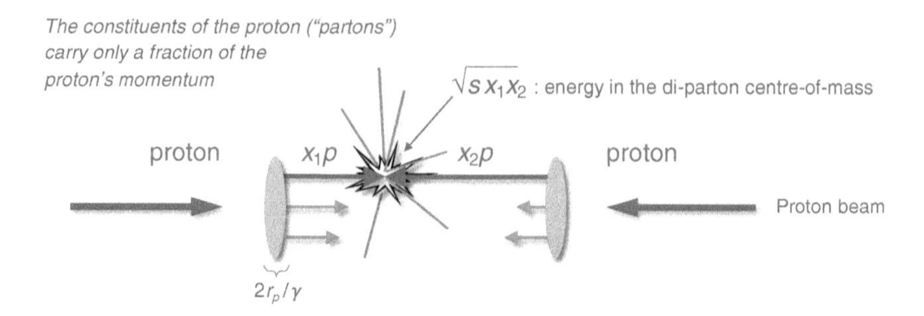

Fig. 10.8. Because protons are composite particles, their high-energy collision actually corresponds to hard interactions among their constituent partons, the quarks and gluons. Each parton takes a fraction x of the incident proton's momentum p. Denoting, as customary, the proton–proton centre-of-mass energy $\sqrt{s}$, the effective centre-of-mass energy in the parton–parton collision is equal to $\sqrt{x_1 x_2 s}$, where the indices correspond to the two colliding protons. The momentum fraction x for each parton in the proton follows a well-defined probability distribution (see inset 10.4). At LHC energies only a very small fraction of collisions has a parton–parton collision with the product $x_1 x_2$ exceeding 0.01. Higgs boson production at $\sqrt{s} = 13$ TeV requires only $x_1 x_2 \approx 0.0001$. The crushed shape of the protons is due to the Lorentz boost. The femtometre radius of the proton is contracted along the beam direction by the Lorentz factor $\gamma = E_{\text{beam}}/m_p \simeq 7500$ at $E_{\text{beam}} = 7$ TeV. The proton's shape resembles a two-dimensional disc.

theorists or phenomenologists and are in widespread use among experimentalists.

(2) **Propagation of particles through the detector**. This stage of the simulation procedure describes the interaction of particles produced in the final state of the collision (electrons, photons, muons, pions, etc.) with the material of the various detector systems. For example, random interactions of strongly interacting particles with nuclei in calorimeter materials giving rise to hadron showers are generated at this stage. Another phenomenon that must be simulated is multiple scattering, i.e. random deviations suffered by charged particles through electromagnetic interactions with atoms and nuclei of the active and passive materials traversed (see figure 10.9). If metastable particles produced in the collision have sufficiently short lifetimes to decay while traversing the detector, these decays must be simulated too. Every time a particle crosses a sensitive part of the detector, be it a silicon microstrip, the gas gap in a wire chamber, or the scintillation layer of a calorimeter, the space coordinates of the traversing particle and the information of the subdetector hit are stored, to be used in the next simulation step: digitisation of the response. The detector simulation is the slowest part

Fig. 10.9. Illustration of the effect of multiple scattering on the measurement accuracy of a detector: just as rain of the windshield of a car blurs the light from incoming cars, the presence of matter on the particle trajectory degrades the momentum measurement.

of the full simulation chain, and among it the simulation of showers propagating through the calorimeter layers dominates by far.

(3) **Digitisation.** This stage describes the interface between the detector response and the data analysis system. The response of every subdetector element the particle hits along its path is simulated. For example, when a charged particle crosses and ionises the sensitive medium of a chamber, the drift time is converted by the readout chain into a number encoded in a limited chain of bits. In the digitisation phase, this is simulated yielding numbers exactly in the same format as for real data. Some detector imperfections, such as dead readout channels, can be inserted in the simulation at this stage, so as to make it as realistic as possible.

The output of the simulation chain after the digitisation step resembles one-to-one that of real detector data, except for one critical difference: the Monte Carlo "truth" information. It contains all the identifiers and properties of the simulated particles. This information allows the analysts to test their reconstruction algorithms and signal selection procedures, and to understand when particles and processes were properly identified and when not. The simulated data are then reconstructed through the exact same chain of programs as the real data collected in the detector. One has thus access to particle momenta (for example) as if they were really measured, with all detector inefficiencies and defects, multiple scattering effects, etc., as for real data.

Inset 10.4: Monte Carlo generators for the simulation of proton–proton collisions at the LHC

How to predict kinematic distributions of particles produced in a proton–proton collision? Quantum field theory describes the interaction between partons, i.e. the quarks and gluons making up the protons (figure 10.8). The quantum probability of the interaction of two partons from each proton is computed first and for each type of relevant parton pair. The result is then convolved with *parton momentum distribution functions* (PDF) inside the proton. These functions describe in a statistical way the incident momentum fraction x of the total momentum of the proton, taken by each type of parton inside the incoming protons. The PDFs cannot currently be predicted from quantum chromodynamics, but they have been precisely measured at deep inelastic scattering experiments where high-energy electron or muon beams scatter on protons at facilities like the HERA electron–proton collider at DESY in Hamburg, Germany. In the figure below, the left-hand picture shows the distribution functions, for example in blue, green and cyan for u, d, and s quarks, respectively, and in red for gluons (divided by a factor of 10 for better visibility). The bands indicate the size of the uncertainties.

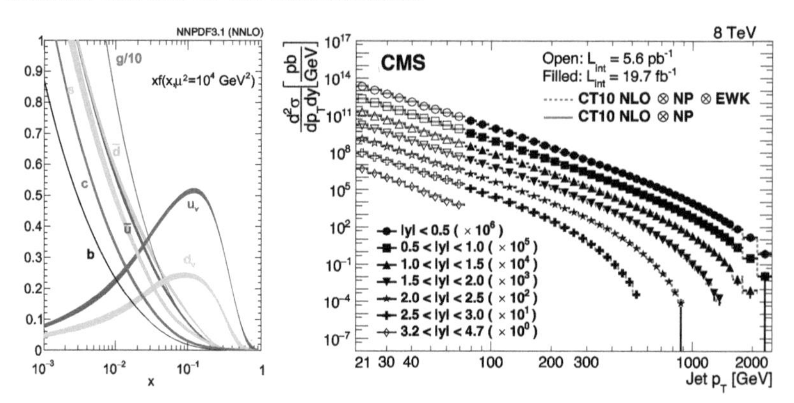

The presence of gluons and strange (s) quarks, and even of antiquarks, inside the proton may surprise. It is due to the fact that, besides the uud valence quarks that define the proton's quantum numbers, the proton also contains gluons and transient quark–antiquark pairs. These are the so-called *sea components* whose existence is allowed for very short time intervals (see section 1.9). The u and d quarks in a proton can be either *valence* or *sea* ones, but s quarks as well as $\bar{u}$ and $\bar{d}$ antiquarks are necessarily from the *sea*, as are the still heavier b quarks. This is the reason why u and d quarks carry a larger momentum fraction than the s quarks. The gluons in a proton carry collectively almost 50% of its momentum, their frequency of occurrence being concentrated at small values of fractional momentum x. The gluons being so numerous at low x, the prevalent parton–parton collision is gluon–gluon. This is the reason why the LHC is sometimes called a *gluon collider*.

The figure on the right shows the production cross-section for jets, originating from either gluons or quarks, at 8 TeV proton–proton collision energy, as a function of transverse momentum p_T of jets. The black points correspond to the experimental measurements in various intervals of jet rapidity y (see inset 10.5). For better visibility, the values of the data points in the different y ranges have been re-scaled. The red coloured histograms correspond to the state-of-the-art theoretical predictions, which depend on the PDF set used. One notices a spectacular agreement between data and theory over more than thirteen orders of magnitude of the measured cross-section! This demonstrates impressively the high level of reliability attained by QCD theory.

The quarks and gluons produced in the collisions must ultimately recombine with parton–antiparton pairs from the QCD vacuum to produce the visible particles, the hadrons, i.e. those that are actually detected (see figure 2.4). This recombination cannot be computed from first principles as it occurs in the non-perturbative QCD regime. It is thus modelled phenomenologically with the help of parameters that are determined from experimental data.

Inset 10.5: Rapidity and pseudorapidy in hadron collider physics

Rapidity is a measure of relativistic velocity. The rapidity y of a particle or a jet is related to the scattering polar angle θ relative to the beam axis. Zero rapidity ($y = 0$) corresponds to scattering perpendicular to the beam axis and large $|y|$ corresponds to forward scattering.

The rapidity expression is derived from the constraint that the difference Δy between the rapidity of two particles is invariant under a Lorentz boost (see inset 1.5 on page 24) along the beam axis

$$y = \ln\left(\frac{E + p_L}{E - p_L}\right),$$

where E and p_L are the energy and longitudinal momentum (projection upon the beam axis) of the particle (or jet), respectively. For a particle with a negligible mass compared to its momentum, y simplifies to the *pseudorapidity* η, which can be directly expressed in terms of the polar angle θ

$$\eta = -\ln\left(\tan\frac{\theta}{2}\right).$$

The figure below on the right illustrates the above relation.

The Lorentz invariance of Δy (or $\Delta\eta$) is an important feature in hadron collider physics, where the colliding partons carry different longitudinal momentum fractions $x_1 \neq x_2$, which means that the rest frame of a parton–parton collision has an (unknown) longitudinal boost.

Similarly, for the azimuth angle ϕ defined in the transverse x–y plane, $\Delta\phi$ is also invariant along a longitudinal boost. In soft, so-called *minimum bias* proton–proton collisions, which makes the bulk of inelastic collisions, the rapidity distribution of the produced hadrons is approximately flat in the interval Δy (or $\Delta\eta$) $= [-4, +4]$ (the so-called *rapidity plateau*), as is the azimuth distribution in the interval $\Delta\phi = [-\pi, +\pi]$.

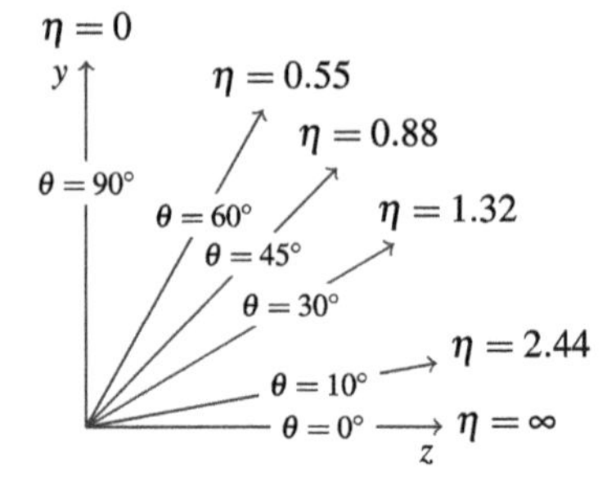

The rapidity is so useful in particle or jet kinematics that often the geometrical coverage of tracking detectors or calorimeters is not expressed in polar angles but rather in rapidity coverage. Pseudorapidity is also often used to define a measure of the angular separation between two particles $\Delta R = \sqrt{(\Delta\eta)^2 + (\Delta\phi)^2}$, which is Lorentz invariant if the involved particles have negligible mass.

10.1.6. *The Z boson: A standard candle to study the detector performance*

A good control of energy measurements with the calorimeters and momentum measurements from trackers and spectrometers are of central importance for most physics studies. This is especially true in case of precise property measurements of particles. Vice versa, an already well-known particle can be used to check and even correct the response of a detector. The best such example is the Z boson whose mass has been precisely measured at LEP to be $m_Z = 91.1876 \pm 0.0021$ GeV. This is a remarkably precise result,[6] better than what can ever be achieved at the LHC.

Figure 10.10 shows as an example the distribution of the muon-pair invariant mass as recorded during the first data-taking period in 2010 by CMS (black points). When the muons result from the decay of a Z boson, the mass of the pair corresponds to the mass of the parent Z particle. The accumulation of data around 91 GeV therefore witnesses the Z bosons in the data sample. The yellow histogram is the Monte Carlo expectation. The agreement between the two distributions proves that the detector response as simulated in the Monte Carlo program is realistic and well understood, and that even the inevitable small defects and inefficiencies were properly introduced in the simulation.

But why don't we measure the exact mass value of 91.1876 GeV for every muon pair? There are two reasons. Firstly, the Z boson decays very fast (within 3×10^{-25} seconds) so that due to Heisenberg's uncertainty relation its mass is blurred leading to the *natural width* of $\Gamma_Z = 2.5$ GeV (see inset 2.1 on page 73). The second reason is the detector resolution: muon momenta are not measured with infinite precision so that the measured invariant mass differs slightly from the real one produced in a given event.

The mass of the Z boson is given to first approximation by the position of the peak of the mass distribution. More precisely, the shape of the distribution is prescribed by relativistic quantum mechanics and therefore theoretically well-known: it is a Breit–Wigner function. This functional shape, after convolution with detector effects such as resolution, acceptance and efficiency, can be fitted to the data leaving free some parameters, among which the peak position. In this way CMS measured on the early 2010 data a mass $m_Z = 91.17 \pm 0.56$ GeV. This result is far less precise than that of

[6]It relied on the extremely precise knowledge of the LEP beam energy that was achieved by exploiting the energy-dependent precession frequency of the electron spin.

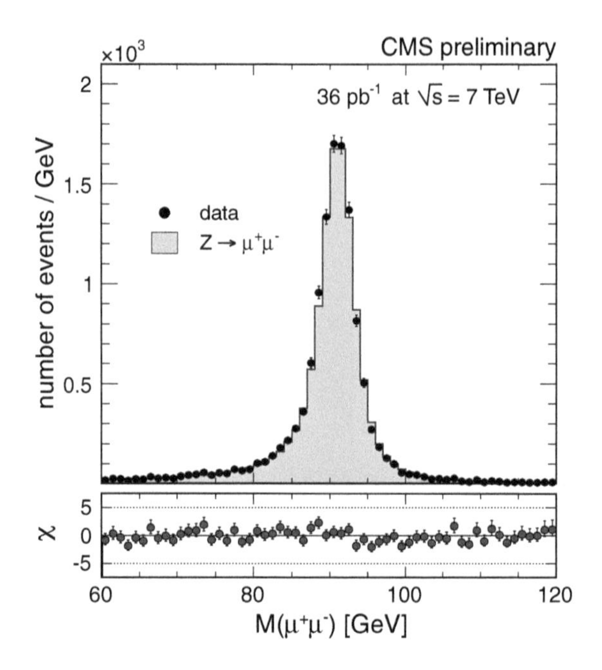

Fig. 10.10. Di-muon invariant mass distribution of events containing two muons registered in the CMS detector in 2010 (black points). The peak at 91 GeV shows a predominant presence of Z bosons in this data sample. The yellow histogram is the Monte Carlo simulation, combining the theoretically expected Breit–Wigner shape with effects due to the measurement procedure. The figure at bottom is the ratio of real data to this expectation obtained by simulation. The quantity χ on the ordinate is a measure of the agreement. *Image: CMS Collaboration.*

the LEP experiments as one should expect after just a few months of data taking. It is however in agreement with LEP, showing that the detector performance was already at this stage understood and reliable. This check was an important step towards digging deeper and beginning searches for new phenomena.

10.2. Information technology challenges and the worldwide LHC computing grid

The amount of data produced by the LHC is colossal. In 2012, for example, the large experiments ATLAS and CMS have recorded each about two billion events of real collision data, and an even larger number of simulated Monte Carlo events were generated. As the size of an event is about one megabyte, the raw data alone (before going through the reconstruction procedure) already required about two petabytes[7] of storage space. This number is at least doubled after the data have gone through event reconstruction. Over the year, adding up the real data produced by the four large experiments and the needed Monte Carlo simulations, the total required storage space corresponded to roughly 40 petabytes. The total amount of disc and tape storage used by the LHC experiments at the end of the Run-2 data taking in 2018 reached about 500 petabytes and 800 petabytes, respectively. Taking into account that these numbers will increase with pile-up and the LHC will be operating for about seventeen more years (although with several scheduled interruptions), these are indeed huge amounts of data requiring large-scale and reliable storage facilities, but also smart ideas to save resources.

The data and Monte Carlo processing is also very demanding in terms of processor power: both ATLAS and CMS need several hundred thousand processing units. The most demanding task is the detector simulation for the Monte Carlo simulation production, which alone consumes about one-third of the required processing power. While the time required to reconstruct one event is between 10 and 20 seconds, the full simulation of a single event may take up to few tens of minutes.

Another constraint (and strength) is the distribution of computing resources. While the experiments are located at CERN, the physicists are working at universities or laboratories throughout the world. They all need access to the data and to sufficient computing power to perform physics analysis. This means that high-speed internet of at least one gigabit per second, better ten, is needed. Most often, research-dedicated networks are used as commercial networks would not fulfil these requirements.

It could have been conceivable to concentrate all the information technological resources at CERN, but the need for fast and simultaneous access to data worldwide would rapidly have generated saturation issues. Furthermore, taking into account the cost of establishing and running such a huge computing centre, it would have been difficult, if not impossible, to convince the various participating countries to invest large amounts of money into buying hardware not located on their own soil, and that could not be used in home countries by scientific communities other than particle physicists.

[7] A petabyte corresponds to one million gigabytes, i.e. 10^{15} bytes.

To circumvent these problems and ensuing constraints, the LHC physicists opted for a *computing grid.* The word grid in the present context was introduced in an analogy with the electrical power grid. Just as when one connects to the electrical grid the consumer does not need to care about where the electrical power is coming from, the idea behind the computing grid is to have access to computing power and storage capacity without knowing and caring about the exact nature and geographic location of the facilities. In practice, the grid is made of hundreds of thousands of processors and storage means distributed around the globe, but it appears to the user as a single computing source.

Fig. 10.11. The CERN data centre which hosts the Tier-0 node of the world wide LHC computing grid. *Image: S. Bennett, CERN.*

This is the working principle underlying the *worldwide LHC computing grid* project (WLCG). It connects about 170 computing sites in 42 countries throughout the world providing round-the-clock support. It is structured into four levels of *Tiers.*

- The computing centre at CERN is the level 0 (called *Tier 0*) node. It is the central hub of this structure, although it provides less than 20% of the grid's total computing capacity. The totality of data recorded by the four large experiments is safely stored here. The Tier 0 is also in charge of performing the first-pass reconstruction of the data, and to distribute the reconstruction outputs to the Tier-1 centres through optical-fibre links working at 10 gigabits per second.
- There are thirteen large Tier-1 centres serving the ATLAS and CMS experiments (most centres serve both experiments — see figure 10.12 for ATLAS). Their responsibility is to store a proportional share of detector and reconstructed data, as well as to perform large-scale data processing, to store the corresponding output and distribute them to the Tier-2 sites. Part of the simulated event production and reconstruction is also done in these centres.

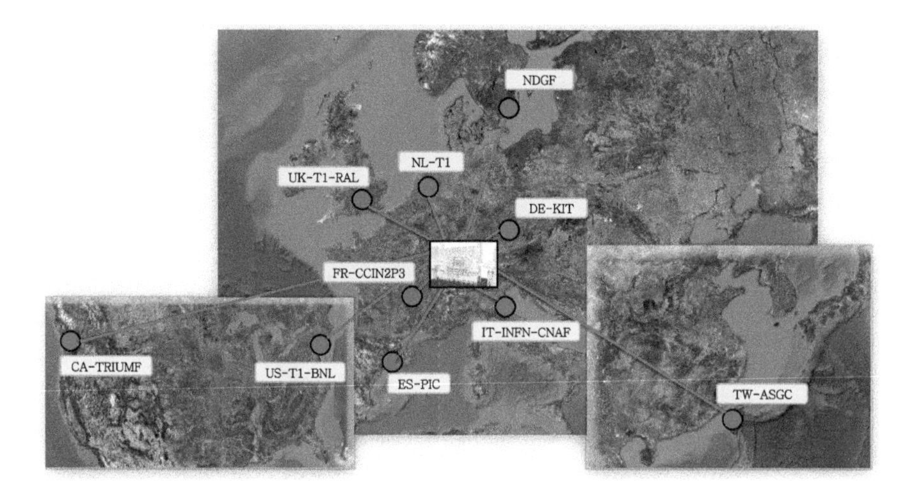

Fig. 10.12. World map of the ten Tier-1 nodes used for data processing by the ATLAS collaboration. CERN is the central Tier-0 node. The CMS computing grid is comparable to the ATLAS one. In total, thirteen Tier-1 computing centres contribute to the WLCG. *Image: WLCG, CERN.*

- The about 155 Tier-2 sites, usually universities and laboratories, provide an aggregate computing power of almost half the full grid,[8] and more than that of the Tier-1 centres. They are mostly used for the production of Monte Carlo simulations as well as for physics analysis.
- The smaller Tier-3 resources in the local research institutions are typically the entry points for the submission of processing tasks to the grid. They are also used to complement physics analyses made locally by physicists at their terminals. The Tier-3 sites do no have a formal engagement with the WLCG.

In practice the WLCG functions as follows: reconstructed data are stored in one, or at most two, Tier-1 centres. The same applies for simulated data in Tier-2 centres. So-called derived data, obtained by filtering and reducing the original reconstructed datasets, are directly accessed by the

[8]It should be said that in this particular field, calculations are relatively simple but repetitive, for example calculating tracks, trajectories from particle hits (signal in detectors traversed), so commercially available processors can do the job. In others domains such as lattice QCD, meteorology, simulation of nuclear explosions, totally different computations are performed requiring gigantic dedicated mainframes.

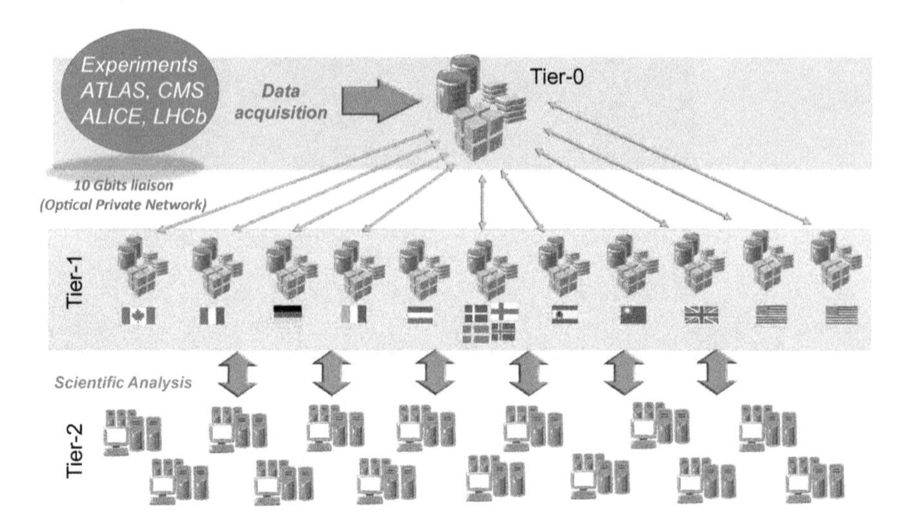

Fig. 10.13. Computing architecture of the WLCG with its Tier-0, 1, and 2 centres (the Tier-3 nodes are not shown). *Image: WLCG, CERN.*

users and must therefore be available at all times and on several sites. To ensure this, copies based on user access patterns are managed automatically by a smart data management software developed by the LHC experiments and the WLCG.

Since data storage is the biggest limitation of LHC computing, all experiments have made efforts to reduce their event sizes and output formats. CMS has pioneered during Run 2 very small derived data formats, reducing the original size of reconstructed data by a factor of almost 400, down to one kilobyte per event for some datasets used for physics analysis. To save processing resources, LHCb uses the online processing of their events directly for some of their offline physics analyses, after a prompt calibration of the data within the online system. ATLAS follows a stable-software-release strategy with long-term planning of updates to reduce the need for data (and consequently Monte Carlo simulation) reprocessing. The collaboration also employs efficient use of slower but cheaper tape resources using an intelligent system of prompt copies from tape to disc buffer for processing.

The computing grids, in particular the networks, have recently made important forward strides and still continue progressing, but the computing speed per processor core is almost saturating (Moore's law is failing). The available computing time is thus becoming a limiting factor and the operating principle of the grid has evolved accordingly. Processing tasks are sent to nodes that have computing time available, and the required data are transferred there through fast network.

The application scheme allows data processing of the type required in particle physics to be easily done with a very large number of cheap, commercially available processors. Each core processes a collision event (real data or simulated) independently of the others as the events are statistically independent. This is one of the reasons why a computing grid is very well adapted to the computing requirements of the field. Thus, expensive supercomputers with massive parallel processing architecture, as needed in meteorological applications and modelling, are not needed. On the other hand, the industry trend to high-performance computing and dedicated processing units with prodigious compute capacity and relatively little memory per unit requires the LHC experiments to adapt. The simulation and reconstruction software has been transformed to run also in environments where even the processing of a single event is shared among several units (threads). It allows the experiments to take benefit from super computers and to process data on graphic processors.

Particle physicists are not the only ones interested in grid computing technologies. Other scientific domains have also large appetite for computing, for example bioinformatics or Earth sciences. As for particle physics, these are sciences with large demands either for simulation or the processing of large quantities of data. The grid can also be of interest to scientists wishing to share data, for example for epidemiological studies. Furthermore, states or international organisations may benefit from this technology when there is need to share data and react quickly and effectively, as in case of large catastrophes. The data management software developed for the WLCG is, like all software written for particle physics experiments, open source and already successfully employed by other parties.

Chapter 11

The Higgs Boson: Search and Discovery

The search for the Higgs boson started well before the LHC. It took about a decade from the formulation of the BEH mechanism in 1964 until the first comprehensive phenomenological analysis of Higgs boson production and decay was published in 1975 by the CERN physicists John Ellis, Mary K. Gaillard, and Dimitri V. Nanopoulos, at the time still considering the full mass range down to MeV scale. In the 1980s a number of experiments nearby Zürich (Switzerland), at CERN, and at Cornell (USA) looked for the then already famous Higgs boson in decays of mesons containing strange, charm and bottom quarks. As a result, at the start-up of the electron–positron collider LEP at CERN, a Higgs boson with a mass smaller than about 4 GeV was already practically excluded, although theoretical predictions on production rates and decay modes in this mass region are rather uncertain. Thus thirty years later, when the searches at LHC were already in full swing and the allowed mass range quickly shrunk, some physicists suggested to explore anew this low mass region, in case it is not found at the LHC. As by now well known, the outcome of the story is different, but before telling it all, we want to recall the many years of Higgs boson searches at LEP and the Tevatron, as the analysis techniques used by ATLAS and CMS in the finally successful search owe a lot to these earlier efforts.

11.1. The LEP era

The LEP era marks the beginning of the systematic and intensive search for the Higgs boson. According to the Standard Model, the main production

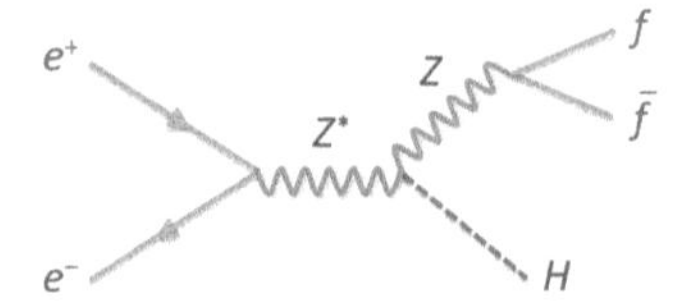

Fig. 11.1. Feynman diagram for the production of a Higgs boson through the reaction $e^+e^- \rightarrow ZH$, followed by the Z boson decay into a fermion–antifermion pair. The intermediate state is denoted as Z^* to indicate that the Z boson is off-mass-shell. This was the case during the LEP-2 phase, when the centre-of-mass energy was practically twice the Z mass. During the LEP-1 phase, however, it is the final-state Z that was off-mass-shell.

mode in an electron–positron collision is through the so-called *Higgsstrahlung* process with a Higgs boson radiated by a Z (figure 11.1); the term Higgsstrahlung was chosen in analogy with the QED process of bremsstrahlung, which is the radiation of a photon by an electron in presence of an external field. During the first data-taking period between 1989 and 1995 (LEP-1), the centre-of-mass collision energy was chosen to be equal to the Z mass at 91 GeV, the Z boson was thus produced at rest. In these conditions, and if the Higgs boson mass is smaller than the Z mass, it is more appropriate to talk about a Z decaying to $H + Z^*$, where Z^* stands for an off-mass-shell Z boson. Such a process heavily penalises the decay rate and hence the Higgs production cross-section. During this LEP-1 period it was nonetheless possible to thoroughly — but unsuccessfully — explore the Higgs mass range from a few GeV to about 60 GeV.

We saw in section 2.4 that it is possible to deduce from a precise measurement of the W mass a constraint on the mass of the Higgs boson through its contribution to radiative corrections calculable within the Standard Model. Radiative corrections sensitive to the Higgs boson mass affect also a number of other measurable quantities related to Z production or decay. Such measurements have been performed with a remarkable level of accuracy at LEP-1. Confronting the Standard Model calculations with the precise LEP data[1] made it possible to indirectly constrain the Higgs boson mass. By the end of the 1990s the result was that it should be lower than about 160 GeV and larger than about 50 GeV. The preferred value,

[1]The SLD experiment at the linear e^+e^- collider at SLAC (Stanford, USA) also contributed to establishing these constraints during the 1990s. Thanks to the ability to polarise electron beams it was possible to measure the Z production asymmetry as a function of left versus right polarisation of the incident electrons. This asymmetry is calculable in the Standard Model and is also sensitive to the Higgs boson mass.

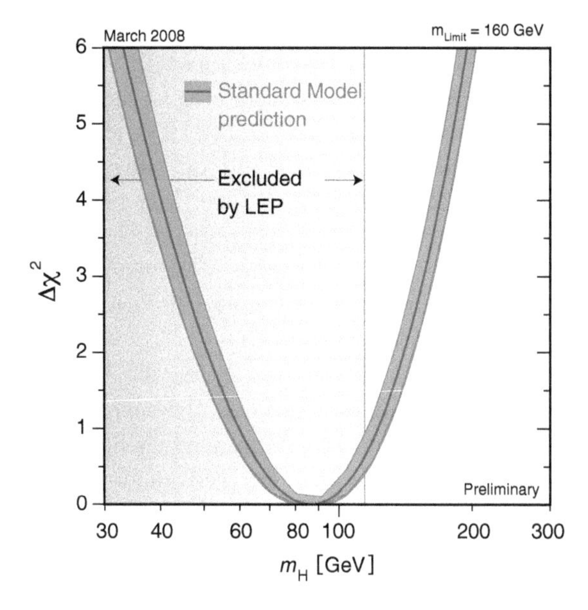

Fig. 11.2. Constraint on the Higgs boson mass m_H in the year 2008 resulting from precision measurements performed at LEP (essentially the W mass, and a number of quantities related to Z production and decays), SLC (Z production with polarised electron beam), and the Tevatron (top-quark mass and W mass). The quantity $\Delta\chi^2$ indicates the level of agreement of the Standard Model predictions with the experimental measurements when varying the Higgs boson mass. The best agreement is obtained for $\Delta\chi^2 = 0$ corresponding to $m_H \sim 90$ GeV, a value excluded by the direct Higgs boson searches at LEP-2, represented by the grey area. The value $\Delta\chi^2 = 1$ gives the limits around the preferred value within which m_H should be expected with 68% probability. Similarly $\Delta\chi^2 = 4$ gives the limits within which m_H should be expected at a 95% confidence level. It is thus possible to say that, within the Standard Model, m_H should be lower than 160 GeV at a 95% confidence level. The abscissa on the plot is given in a logarithmic scale, as the radiative corrections themselves depend logarithmically on m_H. *Image: LEP Electroweak Working Group, 2008.*

i.e. the one giving best agreement, was 90 GeV (figure 11.2), suggesting that the Higgs boson should be within the reach[2] of an electron–positron collider with twice the energy of LEP-1.

[2]It should be clear that, as always in particle physics, this is a statistical result: it means that in view of the data and given the Standard Model, the true Higgs boson mass should lie within these limits, although masses slightly above 160 GeV are not totally excluded. These LEP limits nonetheless represented huge progress remembering that at the time of launching the LEP and LHC projects, at the beginning of the 1990s, the only upper limits were theoretical constraints based on unitarity and these were of the order of $m_H < 1000$ GeV.

Achieving this was one of the goals of the second, high-energy LEP phase (LEP-2) launched in 1996. The collision energy was greatly increased thanks to the installation of additional accelerating radiofrequency cavities. The energy was first raised to 140 GeV end of 1995, then reached 161 GeV mid-1996 (the threshold to produce W boson pairs), 172 GeV in October 1996, 183 GeV in 1997, 189 GeV in 1998, and 192 GeV in 1999. The LEP engineers and machine experts further pushed the machine to its limit to reach 209 GeV in May 2000. Beginning of November 2000 LEP finally bowed out, as it was necessary to dismantle the accelerator and free the LEP tunnel for the installation of the LHC.

LEP-2 data allowed an effective search for the Higgs boson directly produced by Higgsstrahlung (figure 11.1), as the energy was now sufficient to produce both the Higgs and the Z on-mass-shell, provided the Higgs mass does not exceed 115 GeV. Searches were made for the Higgs decay $H \to b\bar{b}$ expected to be dominant in this mass range. Indeed, if the Higgs mass is smaller than 115 GeV, the decays into W or Z pairs, or into a top–antitop pair are heavily suppressed or excluded. About 60% of the decays occur into a $b\bar{b}$ pair (figure 11.5).

The thorough searches performed at LEP-2 did not yield any positive results. In the last months of data taking, however, mainly one of the four LEP experiments (ALEPH) found several events that could be interpreted as candidates for a Higgs boson. As it is often the case in an analysis, selecting events corresponding to a particular final state is not sufficient to eliminate all sources of background. In this specific case, the pair production of Z bosons with one Z decaying to $b\bar{b}$ represents a source of background to mimic a Z accompanied by a Higgs boson. Since Z pair production is well known, it is possible to simulate it precisely and thus estimate the number of false candidates faking $Z + H$ events. In the year 2000 the ALEPH experiment observed a bit more of these $Z + b\bar{b}$ events than expected purely from $ZZ^{(*)}$ background. Figure 11.3 shows one of these events. The problem was that the other three LEP experiments did not confirm the excess (the DELPHI experiment even observed a deficit of such events).

Not being able to confirm the observation of the Higgs boson, LEP physicists had to content themselves with a (strong) lower limit of 114.4 GeV for the Higgs boson mass, a result that made a significant contribution to future searches. Nevertheless, LEP went into retirement with a deep sentiment of frustration among some of the LEP Higgs hunters, who felt that they had spotted the beast at 115 GeV, at the very limit of LEP's energy reach.

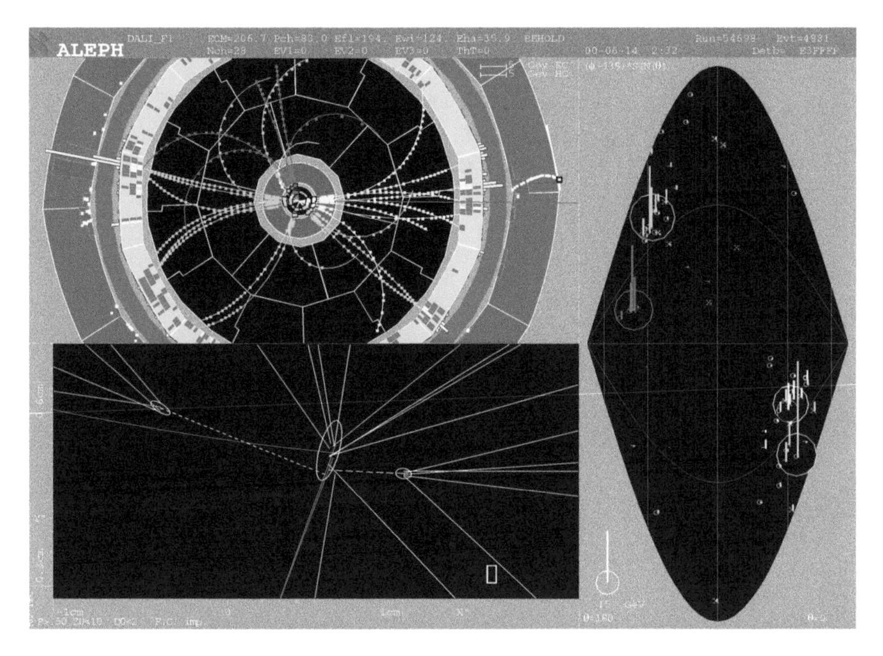

Fig. 11.3. Event seen in the ALEPH detector in an e^+e^- collision at centre-of-mass energy of 206.7 GeV that could be interpreted as the production of a Higgs boson according to the process $e^+e^- \to ZH \to 2\text{jets}+b\bar{b}$. The final state thus contains four jets. Two are coming from b quarks (yellow and green), which can be identified by the displaced decay vertices of the weakly decaying b hadrons, as indicated in the bottom-left inset. Their invariant mass is 114 GeV. The other two jets, in red and blue, have an invariant mass close to 90 GeV and are thus compatible with originating from a Z decay. In view of the subsequent discovery at LHC of the Higgs boson at a mass of 125 GeV, we can now say with certainty that these candidates were indeed background events. *Image: ALEPH Collaboration.*

11.2. Tevatron times

With the shutdown of LEP in 2000, the hope to find the Higgs boson crossed the Atlantic towards the Tevatron, although it was clear for everyone that this would be a challenging task. Just as at LEP, the Higgs could be produced by Higgsstrahlung. In this case, however, it would not anymore be electron and positron that fuse to produce a (off-mass-shell) Z^* boson radiating the Higgs boson, but rather a quark from the proton and an antiquark from the antiproton producing the Z^* (figures 11.1 and 11.4(d)), which would then emit the H. An analogous process, not allowed at LEP, also came into play: an up-type (down-type) quark and a down-type (up-type)

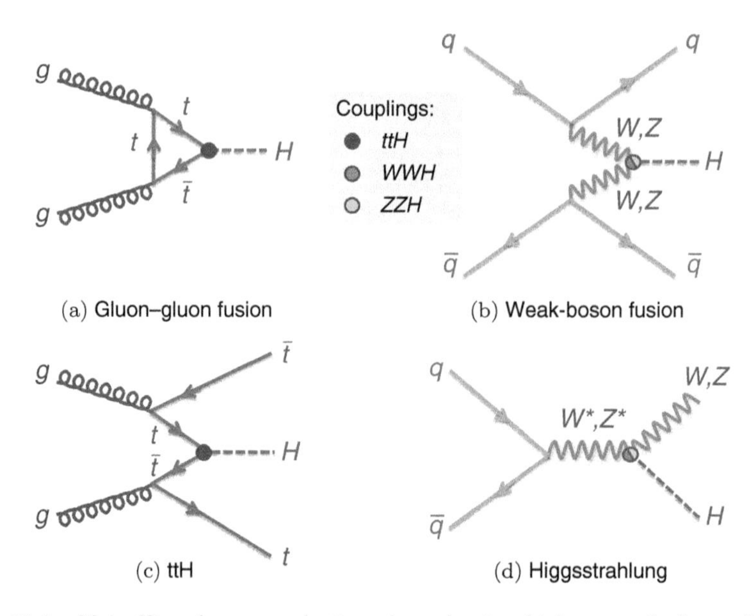

Fig. 11.4. Main Higgs-boson production channels at a high-energy hadron collider: (a) gluon–gluon fusion through a top-quark loop is the dominant production mode at both Tevatron and LHC; (b) weak or vector-boson fusion (VBF); (c) associated production with a top–antitop quark pair; and (d) Higgsstrahlung from a W or Z boson. The latter three production channels, despite having much smaller cross-sections, are nonetheless very useful in searches for Higgs boson decay modes plagued by large backgrounds (as is the case for $H \rightarrow b\bar{b}$, which is the dominant decay mode for $m_H < 130$ GeV) thanks to the associated production of either two forward jets (VBF), two top quarks (ttH), or a W or Z boson, allowing to select events online in the trigger and to suppress backgrounds. At Tevatron energies, the gluon-fusion and Higgsstrahlung channels were the most useful modes.

antiquark with differing electrical charges can produce a $W^{+(-)}$ boson, which subsequently emits a Higgs boson just as a Z does. One should thus look for final states containing either a Z or a W boson. The most abundant Higgs production mode at the Tevatron was however gluon–gluon fusion (figure 11.4(a)). Owing to the high-gluon density at low momentum fraction contained in the proton and antiproton (see inset 10.4), the gluons can mutually interact through a loop of quarks, which fuse into a Higgs boson. Because the Higgs boson couples to mass, this mechanism is more likely for heavier virtual quarks circulating in the loop and is thus dominated by top quarks. From five to ten times as many Higgs bosons were expected to be produced by gluon fusion than by Higgsstrahlung at the Tevatron.

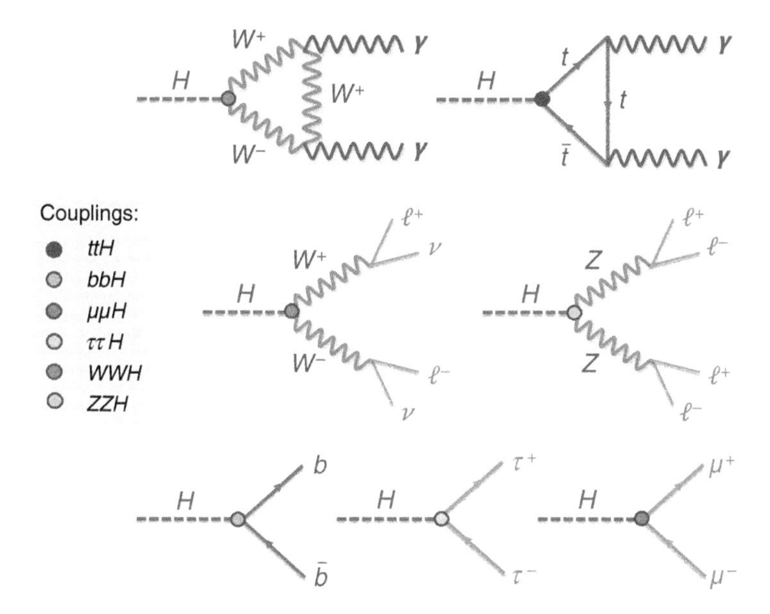

Fig. 11.5. Main Higgs-boson decay modes. On top, the two diagrams contributing to decay into two photons, which mainly proceeds through W or top loops. One of the final state photons may also be replaced by a Z boson. Middle row, decays into a pair of W or Z bosons (one of them possibly off-mass-shell, depending on the Higgs mass), with subsequent decays of these into a lepton–antilepton pair. Bottom row, decays into fermion–antifermion pairs.

Now that we know how the Higgs boson could be produced in proton–antiproton collisions, let us see how it would decay. Figure 11.5 shows a number of possible decay modes into bosons (upper two rows) and fermions (bottom row). The decay rates into each of these modes depend on the Higgs mass. We have already seen that the Higgs boson preferentially decays into the heaviest particles allowed by kinematics. If the Higgs is lighter than twice the mass of a particle, it cannot decay into two real, on-mass-shell ones, but it may still decay into off-mass-shell particles with, however, suppressed rate. The decay rates to lighter, kinematically allowed decays are then augmented, as visible through the interplay of the various decay modes versus the Higgs boson mass in figure 11.6.

When searching for the Higgs boson in a mass range just above the LEP limit, say $m_H < 120$–$130\,\mathrm{GeV}$, there is no major change in strategy, the most favourable decay mode remains the one into a pair of bottom quarks. However, in a search at a hadron collider, be it pp or $p\bar{p}$, one must

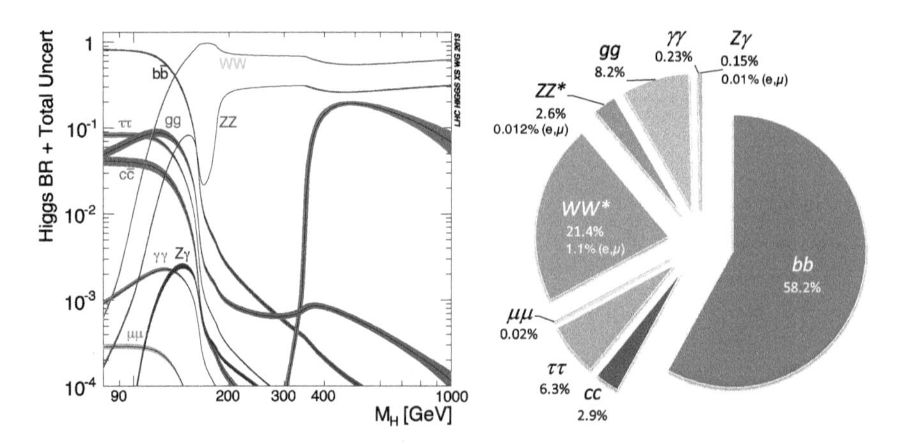

Fig. 11.6. Left: predicted Higgs boson decay branching fractions as a function of the Higgs boson mass on a logarithmic ordinate scale. When approaching 160 GeV the decay rate into two (then on-mass-shell) W bosons increases rapidly, while the decays to lighter particles, bottom and charmed quarks, and *a fortiori* gluons and photons, decrease in proportion. Similar features occur when approaching the di-Z-boson and the di-top-quark masses of about 180 GeV and 350 GeV, respectively. Today, we now know that the Higgs boson mass is 125 GeV, but at the time of the search at the Tevatron and in particular during the design stages of the LHC experiments in the 1990s, all the possible masses, decay modes and corresponding search strategies had to be considered. Right: Higgs boson branching fractions for a mass of 125 GeV. *Left figure: LHC Higgs Cross Section Working Group.*

now take into account the direct production of these pairs through strong-interaction $q\bar{q} \to b\bar{b}$ and $gg \to b\bar{b}$ processes, a background almost a billion times larger than the targeted Higgs signal (cf. figure 10.1)! There is no way to distinguish a $b\bar{b}$ pair resulting from a Higgs boson decay from one produced in $q\bar{q}$ or gg interactions. However, in the first case one would expect the $b\bar{b}$ mass to accumulate around the Higgs mass, but it is difficult to reconstruct precisely from jets and a resolution of only about 15 GeV is expected. Few events in a peak distributed with such a broad mass resolution on top of the huge number of $b\bar{b}$ pairs produced by gluons and quarks are totally undetectable. To reduce this background, the Tevatron experiments have restricted their search to events produced through Higgsstrahlung by requiring a Higgs boson produced in association with a leptonically decaying W or Z boson. This reduces by a factor of about 25 the signal rate, but eliminates most of the directly produced $b\bar{b}$ background events. Other backgrounds still remain, but these are only few tens of times larger than the signal!

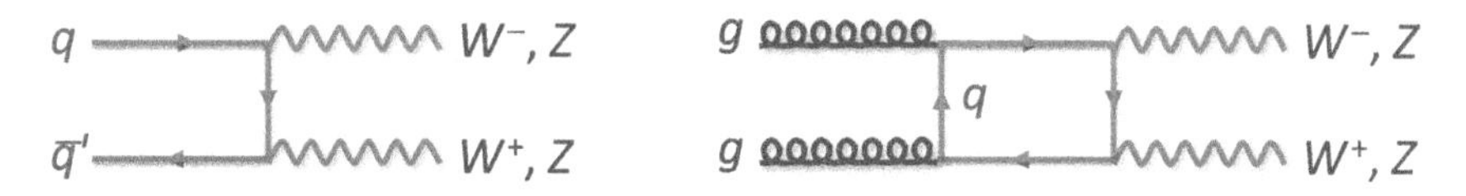

Fig. 11.7. Feynman diagrams describing the direct production of a W or Z boson pair from quark and gluon interactions. These are main background processes in the search for Higgs boson decays to $ZZ^{(*)}$ or $WW^{(*)}$.

If one looks, however, for a slightly more massive Higgs boson, other channels open up. Even if LEP results favoured a light Higgs boson, there was no certainty and the whole accessible mass range had to be experimentally explored. If the Higgs boson were in the range 160–180 GeV, it could decay into a pair of W or Z bosons, which greatly simplifies the search. If the mass is below threshold, one of the bosons in the pair can be off-mass-shell as seen in figure 11.6. The decay rate is then suppressed but not forbidden, and looking for a final state with a leptonically decaying $WW^{(*)}$ or $ZZ^{(*)}$ pair (even with one member off its mass shell) is much more favourable from the point of view of background rejection. A pair of W or Z bosons can also be produced directly in the proton–antiproton collisions, without creation of a Higgs boson, as can be seen in figure 11.7. However, other properties of the events make them distinguishable from Higgs boson production. For example, because the Higgs boson is a spin-zero (scalar) particle, the two charged leptons from the W decays are correlated and preferentially emitted in the same spatial direction, which is not the case for quark-initiated W pair production.[3] In the case of a decay into a pair of Z bosons leading to a final state with four charged leptons, all the decay products are detected and precisely measured, allowing to reconstruct the Higgs boson mass with an accuracy of about 1% and to apply a more sophisticated spin-parity analysis. With W and Z boson pairs one thus has powerful selection criteria for the Higgs search. Unfortunately, the decay modes to pairs of Z bosons giving four leptons in the final state have very small branching fractions (reducing the accessible decay rate by a factor of 230), and are of no practical use at the Tevatron. These modes will, however, become essential for the subsequent Higgs boson searches at the LHC, where the production cross-sections are more favourable.

[3]It might be interesting for the reader to consult inset 11.2 dealing with the spin-parity of a particle.

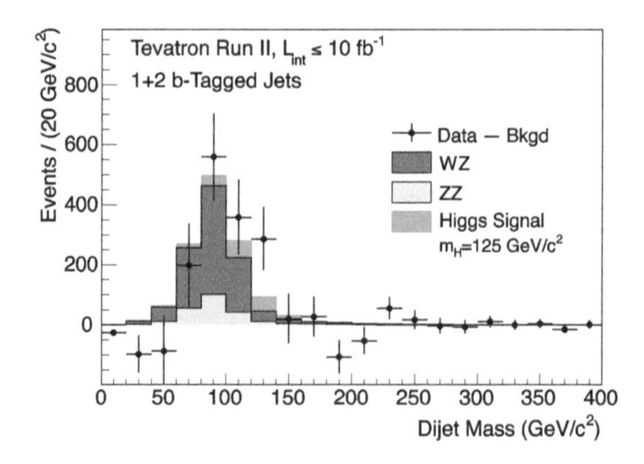

Fig. 11.8. Invariant mass distribution for two b-quark jets as seen by the combined CDF and DØ detectors at Fermilab, in the search for a VH, $H \to b\bar{b}$ signal ($V = W$ or Z). The points represent the distribution obtained from the data recorded by the experiments, after subtraction of all background components except for WZ and ZZ continuum production. Their contributions are shown as stacked histograms in red and yellow, respectively. The expected contribution from a 125 GeV Higgs boson, broadly accumulating around its mass, is shown by the green histogram. The black points show the data with error bars. *Image: CDF and DØ Collaborations, Phys. Rev. D 88, 052014 (2013).*

At a mass of 150 GeV, about six thousand Higgs boson events would have been produced at each of the two Tevatron interaction points for an integrated luminosity of 10 fb^{-1}, the total amount of data recorded by CDF and DØ by the time the complex was shut-down in 2011. Instead, had the mass been 500 GeV, only few tens of Higgs bosons would have been produced due to the fast decline of the gluon density in the proton for increasing momentum fraction, which is necessary to produce a heavier Higgs boson. In practice, taking into account the expected levels of backgrounds, a Standard Model alike Higgs boson of mass exceeding 200 GeV would have been beyond the reach of the Tevatron.

These are the main elements concerning the search for the Higgs boson at the Tevatron. In reality things are much harder still. A large number of final states was investigated, as it was impossible in view of the available statistics to expect observation in a single decay channel alone. Figure 11.8, taken from the final combined Tevatron analysis in 2012,[4] illustrates the difficulties well: it shows how small the expected signal for the production of

[4]This paper was released after the Higgs boson discovery by ATLAS and CMS at the LHC that fixed the mass of the Higgs boson at 125 GeV.

a WH or ZH pair via Higgsstrahlung and the decay $H \to b\bar{b}$ was, although it represented the most sensitive channel for the search of a Higgs boson with 125 GeV mass at the Tevatron.

In July 2010 at a large International Conference on High Energy Physics (ICHEP) held that year in Paris, both experiments DØ and CDF announced to have analysed almost 6 fb^{-1} of data each, and, in the absence of a Higgs signal, they were able to exclude the mass range from 158 to 175 GeV. Furthermore, the Tevatron data confirmed the LEP exclusion region at lower masses, excluding the mass range from 100 to 109 GeV.[5] These announcements came as a fright to many LHC physicists, as it meant that, had the Higgs boson a mass in these ranges, the CDF and DØ experiments could have announced with the remaining data still to be analysed, if not its discovery, at least a strong hint (see inset 11.1). And this just few months after the experiments at CERN started taking data, and very successfully so, but the amount of data collected was still more than a thousand times smaller than those accumulated at the Tevatron! While first tens, then hundreds, and finally thousands of physicists have worked at CERN with utmost dedication since the Aachen Conference of 1990, there was a danger of being overtaken by the Tevatron just when crossing the arrival line!

11.3. Higgs boson search at the LHC

Thanks to its larger energy and higher instantaneous luminosity, Higgs boson production should be much more abundant at the LHC than at the Tevatron, whatever its mass. As an example, if the Higgs boson had a mass of 150 GeV, it would be produced seventy times more often at a 14 TeV LHC than at the Tevatron (figure 10.1) assuming the same luminosity. Moreover, the design LHC instantaneous luminosity exceeds by a factor 25 the highest values achieved at the Tevatron, bringing the factor to 1700. At LHC collision energy of 7 or 8 TeV, the cross-section is still larger by a factor of 20 to 25. Furthermore, ATLAS and CMS were designed to explore a mass range up to about 1 TeV, well beyond the 200 GeV reach of the Tevatron. It should also be realised that the modern ATLAS and CMS detectors had much better performance compared to CDF and DØ whose design and construction were ten to twenty years older.

[5] To achieve this sensibility, the two Tevatron experiments put in common their searches in such a way that the result looks as coming from a single experiment but having twice the statistics of either experiment.

Inset 11.1: Statistics: evidence and discovery

What particle physicists call *evidence* and *discovery* corresponds to a precise definition. It is perfectly possible that local statistical fluctuations, for example in the invariant mass distribution of a background sample, give rise to a small peak mocking up a signal. Call n_0 the number of events expected in a background mass distribution in a specific mass interval. Due to the intrinsic randomness of the data, the number of observed events n_{obs} will differ from n_0, which is called a *fluctuation* of a random variable. If n_0 is sufficiently large, it can be assumed that these fluctuations follow a Gaussian statistical distribution in case of repeated identical experiments. The *standard deviation* of such a distribution is usually denoted by σ. By definition, 68% of these fluctuations, i.e. of differences between n_{obs} and n_0 in repeated experiments, are smaller than 1σ.

A fluctuation by more than one standard deviation has a probability of 32%. It is thus a rather common incident and nothing can be concluded from such an observation. In case where only an excess of events (as opposed to a deficit) is counted as a signal, the probability of a fluctuation by more than one standard deviation is halved to 16%, which is still fairly large. Should one, however, observe in a mass distribution an excess of events at a level of three standard deviations (or more), the probability that it is due to a background fluctuation is only 0.1% (or less). In such a case it is said to have a hint or *evidence* for a signal. If the observed peak exceeds 5σ, the probability for a background fluctuation is smaller than one in 3.5 million, in which case one announces a *discovery.*

If the deviation between background hypothesis and data depends purely on statistical grounds of a random variable, the requirement of reaching 5σ before claiming an observation seems overly conservative. However, this convention takes into account that most searches and measurements in physics are also subject to systematic uncertainties, whose statistical interpretation is questionable. The 5σ threshold covers these imponderables and avoids that physicists have to retract a discovery claim.

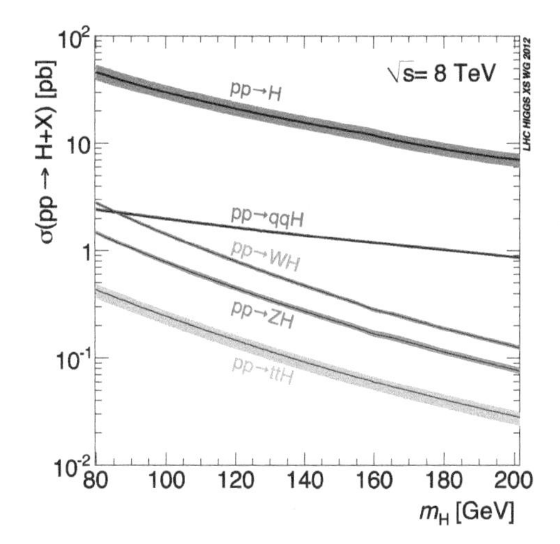

Fig. 11.9. Cross-section for Higgs boson production at the LHC with a collision energy of 8 TeV as a function of Higgs boson mass. In dark blue for the gluon-fusion mechanism, red for vector-boson fusion, green (orange) for WH (ZH) Higgsstrahlung, and light blue for top–antitop associated production. The width of the lines represents the uncertainty in the theoretical calculation in 2012. The vertical axis is in a logarithmic scale. *Image: LHC Higgs Cross Section Working Group.*

Figure 11.9 shows the mass dependence of the cross-sections for the various Higgs boson production mechanisms at an LHC collision energy of 8 TeV. As for the Tevatron, the gluon–gluon fusion channel dominates and the LHC can be thought of a *gluon–gluon collider*. The most abundant production channel is thus gluon fusion (figure 11.4(a)). Higgsstrahlung is still a very interesting mode, especially at low mass, but its production rates declines very rapidly with the Higgs boson mass. The vector boson fusion mechanism (figure 11.4(b)), unimportant at the Tevatron, is suppressed only by an order of magnitude at the LHC compared to gluon fusion. This is an interesting production mode as, besides the Higgs boson, the final state also contains two forward emitted jets and little additional activity as the two initial quarks do not interact strongly. This provides additional selection criteria allowing an efficient suppression of backgrounds. Finally, due to the high mass in the final state, the least abundant production mode is for a Higgs boson in association with a top–antitop pair (figure 11.4(d)).

Let us remember the times when the Higgs boson was not yet discovered and its mass was unknown. To understand the search strategies of ATLAS and CMS it is necessary to combine the production cross-sections (figure 11.9) with the Higgs decay probabilities (figure 11.6). The result is

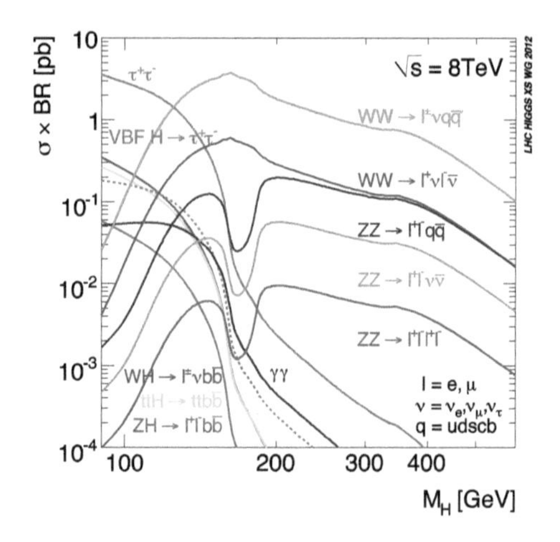

Fig. 11.10. Product of the Higgs boson production cross-section times its decay probability in a given final state as a function of mass at 8 TeV proton–proton collision energy. This figure is the result of combining the production cross-sections shown in figure 11.9 with the decay probabilities for the various channels in figure 11.6 and the decay probabilities of Ws and Zs into pairs of leptons or quarks. The channels allowing the detection of all final state particles, and thus full reconstruction of the Higgs boson mass, $H \to \gamma\gamma$ in grey and $H \to ZZ^{(*)} \to \ell^+\ell^-\ell'^+\ell'^-$ in light blue, have cross-sections of 50 and 3 fb, respectively, for a Higgs boson mass of 125 GeV. This means, for example, that for the four-charged-lepton final state with a luminosity of 10 fb^{-1} one expects to observe about ten events, taking into account detection efficiencies. The coordinate axes are logarithmic. *Image: LHC Higgs Cross Section Working Group.*

plotted in figure 11.10. At low mass one would think that the most interesting channels are those leading to $\tau^+\tau^-$ and $b\bar{b}$. However these two channels are plagued by very large backgrounds. The $b\bar{b}$ channel, for example, is less favourable than at the Tevatron, as the production rate for the background processes is much more enhanced than that of the signal. Even when considering the Higgsstrahlung or $t\bar{t}H$ production modes, these channels remain very challenging at the LHC.

Despite the small number of events, it is nonetheless thanks to the decays into two photons and into two Z bosons with subsequent decay to four charged leptons that the search for a low mass Higgs boson was ultimately successful. Let us discuss the specificities of these two channels. The mass domain that can be explored through $H \to \gamma\gamma$ is relatively limited, from 110 GeV to, at maximum, 150 GeV beyond which the branching fraction becomes too small. As the photon is massless, it does not couple directly to the Higgs boson. Instead, the decay proceeds via a loop of intermediate heavy quarks or a W boson (see top diagrams in figure 11.5), two

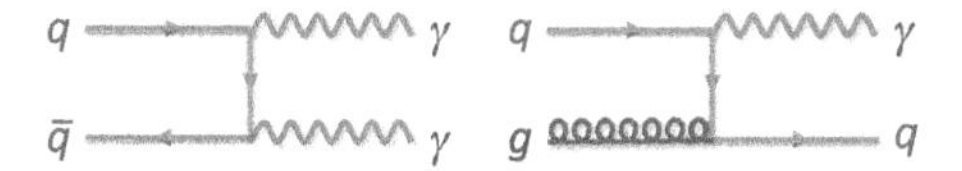

Fig. 11.11. Two production mechanisms giving rise to backgrounds in the search for a Higgs boson decaying into two photons. Left: example of a Feynman diagram leading to the direct production of two photons. This is a so-called *irreducible* background, as the final state consists of two photons exactly as for the signal. Right: example of diagram with production of a photon plus a quark jet in final state. With a small probability of the order of 10^{-4} this jet may hadronise into an almost single high-energy π^0, which decays into two nearby photons that may appear as a single photon. Such a process can give rise to di-photon candidates with large invariant mass representing a *reducible* source of background.

processes which interfere destructively. The interference further reduces the already small decay probability to just 0.2%, a small but non-zero value!

What about the backgrounds? A pair of photons can be produced directly through interactions of quarks or gluons from the incoming protons, without production of a Higgs boson or of whatever intermediate resonance (figure 11.11, left). A photon-jet pair event may also be misreconstructed as a di-photon event if the jet is misidentified as a photon (figure 11.11, right). These sources of photon pairs sum up to about one hundred times more events than the expected Higgs boson signal in the mass range considered. However, by measuring precisely the photon energies and directions it is possible to compute the invariant mass of the pair. If the two photons are produced directly or a jet is misidentified as a photon, the invariant mass has no specific value but follows a smoothly falling shape. On the contrary, if the two photons originate from the decay of a Higgs boson, a peak at the Higgs mass should be observed in the two-photon invariant mass distribution, provided the energy resolution of the electromagnetic calorimeter is sufficiently good. It was this objective of having a mass resolution as good as 1%, that is 1.2 GeV for a Higgs boson mass of 120 GeV, that determined (among other criteria) the design of the ATLAS and CMS electromagnetic calorimeters.

The *golden channel*, the one that should allow the detection of the Higgs boson whatever its mass, is the decay into two Z bosons, where each Z further decays into an e^+e^- or $\mu^+\mu^-$ pair.[6] Only for Higgs masses below 120 GeV the decay rate falls-off rapidly (figure 11.6) so that this

[6]There are three possible final states: $e^+e^-e^+e^-$, $e^+e^-\mu^+\mu^-$, and $\mu^+\mu^-\mu^+\mu^-$. This channel is usually called the four-lepton channel and is often denoted as $H \to ZZ^{(*)} \to \ell^+\ell^-\ell'^+\ell'^-$ (with $\ell, \ell' = e, \mu$) or simply as $H \to 4\ell$.

channel looses its importance. Thanks to the excellent performance of the electromagnetic calorimeters and central tracking detectors, the resolution of the reconstructed four-lepton invariant mass lies between 1 and 2 GeV. By selecting four-lepton final states in which at least one of the two lepton pairs[7] has an invariant mass close to Z boson, background is almost totally eliminated. A small residual contribution still exists and can have several origins.

- Direct production of $ZZ^{(*)}$ pairs (figure 11.7) produces a comparable number of events under the expected Higgs signal, but it has a continuous mass distribution, whilst the Higgs signal will produce a peak at the Higgs boson mass.
- The production of a single Z in association with a $b\bar{b}$ or $c\bar{c}$ quark pair (figure 11.12) can give rise to a four-lepton final state when the decaying bottom or charmed mesons decay into sufficiently isolated charged leptons.
- Finally, the production of a $t\bar{t}$ pair in which each top decays into a W plus a bottom quark may also leads to a four-lepton final state. This background however turns out to be negligible after applying all selection requirements.

As one may perceive from the above lines, the heart of an experimental physicist's analysis consists in reducing backgrounds, notably by applying appropriate selection criteria on photons and charged leptons to efficiently reject those that are produced from hadron decays. Once the selection is done, what remains to do is to estimate reliably the residual background contamination.

Higgs boson decays into two W bosons, where each W further decays into a charged lepton (e or μ) and corresponding neutrino, also represent a powerful channel covering a large mass domain. It is about one hundred times more abundant than $H \to ZZ^{(*)} \to 4\ell$ (figure 11.10), and relatively easy to distinguish from the backgrounds thanks to the presence of two charged leptons and missing transverse energy from the

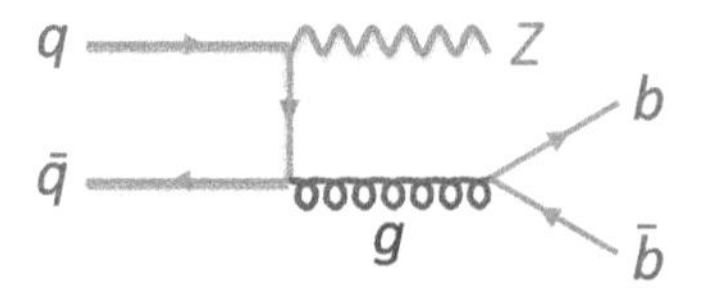

Fig. 11.12. Feynman diagram for production of a Z boson in association with a pair of bottom quarks, a potential background in the search for the decay $H \to ZZ^{(*)} \to \ell^+\ell^-\ell'^+\ell'^-$.

[7] For a Higgs boson mass smaller than twice the Z mass, one of the two Z bosons is produced off its mass shell (denoted Z^*) and the two leptons from its decay can have an invariant mass much smaller than the Z mass of 91 GeV.

neutrinos, although not as easy and purely as the $ZZ^{(*)}$ mode. However, the presence of two undetected neutrinos implies a poor mass resolution, of the order of 20%, to be compared with the 1–2% in the $\gamma\gamma$ and $ZZ^{(*)} \to 4\ell$ decay modes.

The searches for the Higgs boson by both ATLAS and CMS in the years 2011 and 2012 focused essentially on these three channels, $H \to \gamma\gamma$ (figure 11.13), $H \to ZZ^{(*)} \to \ell^+\ell^-\ell'^+\ell'^-$ (figure 11.14) and $H \to WW^{(*)} \to l\nu\ell'\nu'$, although other decay modes were investigated, too.

11.4. Here, at last!

On July 4th, 2012, CERN's main auditorium is packed. Numerous are those who had slept in the corridors, so as to be able early in the morning to sneak in this room with (only) 400 seats, built at a time well before the gigantic LHC collaborations. No rock star was expected that day in Geneva. CERN had just announced an exceptional seminar on the Higgs boson search during which the two spokespersons, Fabiola Gianotti for ATLAS and Joe Incandela for CMS, were going to present their team's latest results.[8] They were based on the totality of data recorded in 2011 at a collision energy of 7 TeV (about 5 fb^{-1} per experiment) and of the data collected up to June 2012 at 8 TeV (about 6 fb^{-1}).

There was in fact a certain degree of excitement in the particle physics community already since December 2011, when both experiments had presented results hinting that something might be going on at a mass around 125 GeV. The statistical level of these early results was not yet what could be termed evidence, but it was troubling and promising, as it was seen by both ATLAS and CMS, and even by the Tevatron experiments that by that time had analysed the totality of their data (albeit there was little information on the mass). Furthermore, through these studies the allowed mass range for the Higgs boson had considerably shrunk: either it had to be heavier than 550 GeV, or it had to be in a narrow mass range between 120 and 130 GeV.

To avoid any influence, even unconsciously, by these early results, it was decided by both ATLAS and CMS to perform a *blind* analysis. The mass region around 125 GeV had to be excluded from all preparatory studies defining the exact analysis and the step-by-step selection procedure, with

[8]The presentations of this memorable event are available at: indico.cern.ch/event/197461.

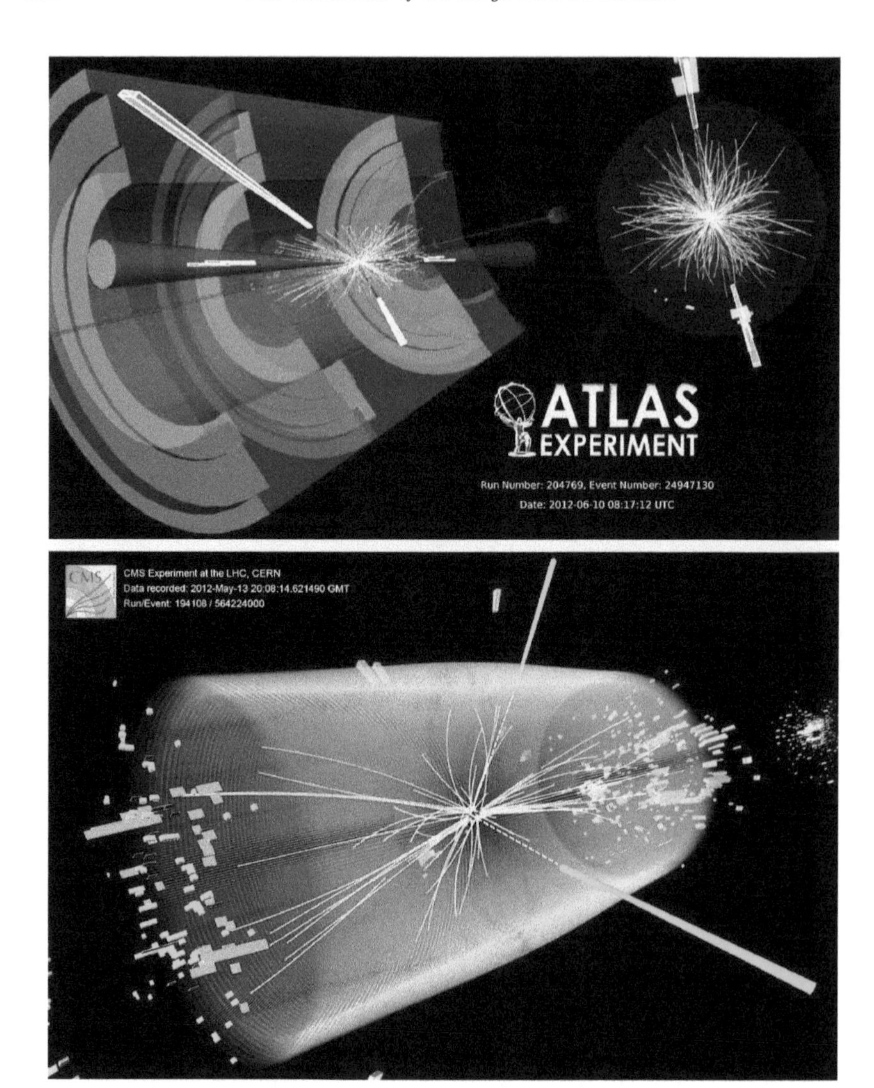

Fig. 11.13. Top: ATLAS candidate event for a Higgs boson decaying to two photons, with the photon energy deposits in the electromagnetic calorimeter represented by two yellow trapezohedrons of sizes proportional to the energy. The invariant mass of the two photons amounts to 125.9 GeV. The two cones in brown colour correspond to two high-energy jets of particles produced close to the beam direction. The event topology corresponds therefore to Higgs boson production through W or Z fusion (figure 11.4(b)). Bottom: candidate event for Higgs boson decaying to two photons as seen in the CMS detector. The energy deposits in the crystal electromagnetic calorimeter are shown in green. The invariant mass of the two photons is 125.0 GeV. The orange tracks correspond to charged hadrons emerging from the same and simultaneously occurring proton–proton collisions. *Images: ATLAS and CMS Collaborations, CERN.*

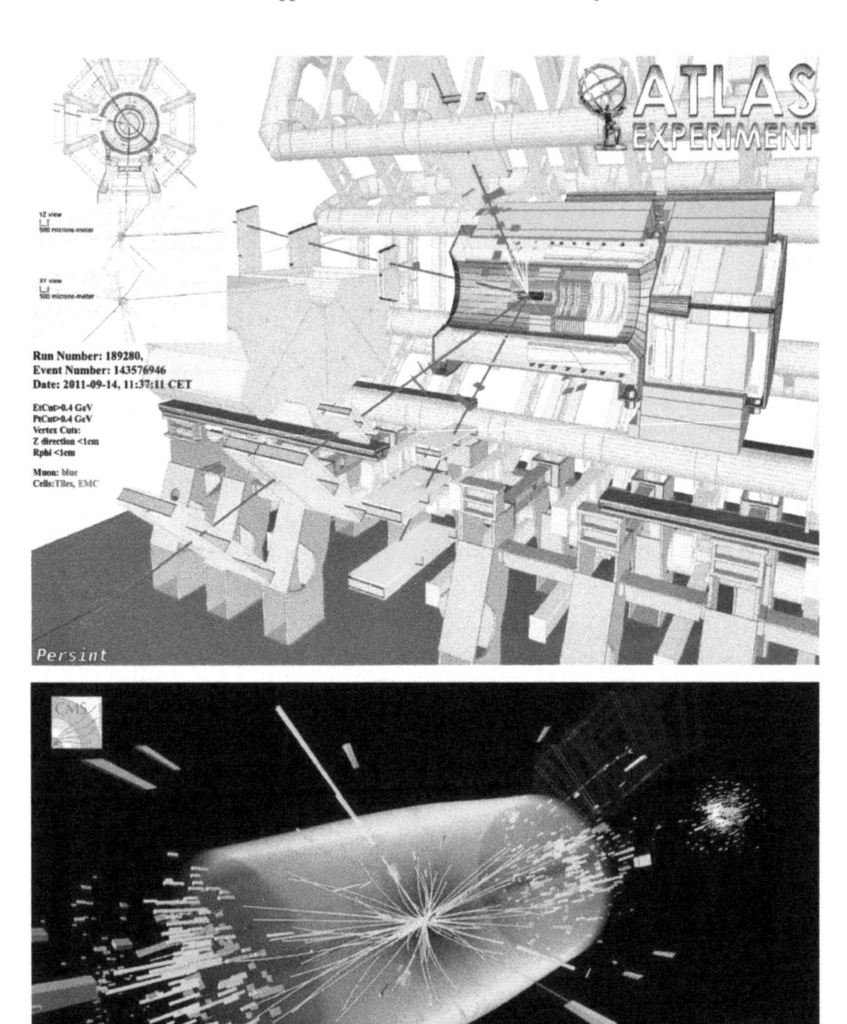

Fig. 11.14. Top: a candidate event for a Higgs boson decaying to two Z bosons seen by ATLAS. Each Z boson decays into a muon pair (blue lines). Bottom: $H \to ZZ^{(*)}$ candidate in CMS. One of the Z bosons decays to a pair of muons detected in the muon system (red lines), the other decays to an electron–positron pair, with energy deposits in the electromagnetic calorimeter shown in green. *Images: ATLAS and CMS Collaborations, CERN.*

Le Matin

CERN

Le boson
de Higgs
fêté comme
une rock-star

Fig. 11.15. Front-page of the Geneva newspaper *Le Matin* on July 5th, 2012.

each step being validated and approved. In both experiments, the signal region was thus *unblinded* only at the very last moment, when the full analysis chain was ready, and only to the members of each collaboration. Only the two spokespersons, the physics coordinators, and the CERN directorate had knowledge of all results prior to the July 4th presentation.

On that July 4th in the CERN auditorium, besides the many well-known and the many more anonymous physicists, there were also the five CERN directors general (DG), during whose mandates the LHC was built, with the present DG, Rolf Heuer, in the role of the master of ceremonies. Theorists at the origin of the electroweak symmetry breaking mechanism, François Englert,[9] Peter Higgs, Tom Kibble, and Carl Hagen were also present in the audience (figure 11.16). The session at CERN was broadcast in particle physics laboratories worldwide, and also in Melbourne, Australia, where the 2012 edition of the International Conference on High Energy Physics was about to start. One after the other, the spokespersons of CMS and ATLAS gave detailed reports of their collaboration's analyses and announced the discovery, greeted by a thunder of applause (figure 11.17), the planet over!

The repercussions of the discovery went well beyond the world of physics (figure 11.18). The television program CERN-TV on YouTube, also transmitting the event, registered fourteen thousand viewings during July 4th, 2012 alone. The video *What is the Higgs boson?* by CERN theorist John Ellis was viewed one hundred and sixteen

Fig. 11.16. François Englert and Peter Higgs after the presentations of the results by ATLAS and CMS announcing the discovery.

[9]Robert Brout, co-signatory with François Englert of the original theoretical paper, passed away one year earlier.

Fig. 11.17. Left: enthusiastic audience on July 4th in the CERN main auditorium at the announcement of the discovery. Right: Joe Incandela and Fabiola Gianotti, spokespersons of CMS and ATLAS, respectively, after the presentation of results of the two teams.

Fig. 11.18. A panel at CERN collecting press titles after the July 4th announcement.

thousand times on that day, and thousands of messages on the Higgs boson discovery circulated via Twitter.

What did the physicists really see? The discovery presented on July 4th was based on observations in essentially two channels, $H \to \gamma\gamma$ and $H \to ZZ^* \to \ell^+\ell^-\ell'^+\ell'^-$. Figure 11.19 shows the distributions of the invariant mass of photon pairs measured by ATLAS and CMS. An excess of events over the smoothly falling background is seen around a mass of 125 GeV. In both experiments, the excess is evaluated to be slightly above the 4σ level. Figure 11.20 shows the four-lepton mass distributions in the

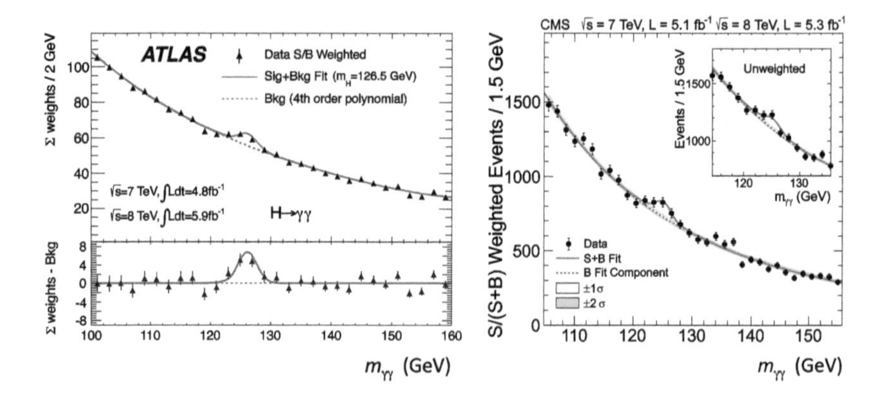

Fig. 11.19. Distributions of the two-photon invariant mass in the data samples used for the July 4th, 2012 Higgs boson discovery. The points with error bars represent the data. The large number of background events follow a smoothly falling shape. The deviation relative to this curve is assigned to the presence of a Higgs boson with mass 125 GeV. This excess, shown in the lower part of the ATLAS figure, corresponds to about 400 Higgs boson events. The data entries are weighted by their expected signal purity. *Images: ATLAS Collaboration, Phys. Lett. B 716, 1 (2012); CMS Collaboration, Phys. Lett. B 716, 30 (2012).*

$H \to ZZ^* \to \ell^+\ell^-\ell'^+\ell'^-$ search channel. The background here is much smaller than in the two-photon channel, but so is the signal. Nonetheless, the few events accumulating above background at around 125 GeV provided a signal excess above 3σ in each experiment. Remarkably, the excesses of events observed independently in both channels and experiments occurred at the same mass! When taking the $\gamma\gamma$ and four-lepton observations together, both ATLAS and CMS observed a combined signal at the 5σ level (figure 11.21) — a discovery!

One might be surprised by the modesty of the signals observed and shown in figures 11.19 and 11.20, especially in view of the huge human and financial investment represented by the LHC project. It must, however, be realised that this is only the very first observation, at the limit of the required significance, just as were the first W and Z bosons discovered at CERN in 1983 (figure 2.8). With the expected increase of the luminosity in the following years, the signals would bloom and eventually become almost as spectacular as that of the Z boson in figure 10.10. This evolution, well illustrated by figure 11.21, which demonstrates the growth in significance of the Higgs boson signal between December 2011, when the first hints appeared, and November 2012, when almost twice as many data were available than for the discovery. Figure 11.22 shows the two-photon mass spectrum as obtained by ATLAS (left) and the four-lepton one from CMS (right) by the end of 2012. Both peaks have become very significant.

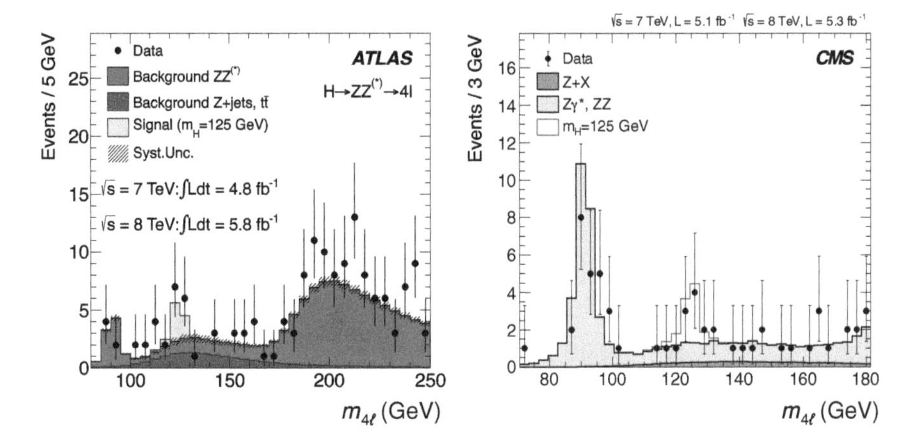

Fig. 11.20. Corresponding distributions of the four-lepton (electrons or muons) invariant mass, where the data are again represented by points with error bars. The histograms in red and purple for ATLAS (left), and in blue and green for CMS (right), are the expected background distributions, largely due to continuum electroweak $ZZ^{(*)}$ pair production. The peak at 90 GeV is due to the rare $Z \to \ell\ell'\ell'$ decay, expected in the Standard Model. The rise of the background distribution around 170 GeV corresponds to the transition from ZZ^* to the fully on-mass-shell ZZ. The excess of events around 125 GeV is assigned to Higgs boson production (in light blue or white the expected simulated signal distributions at that mass). *Images: ATLAS Collaboration, Phys. Lett. B 716, 1 (2012); CMS Collaboration, Phys. Lett. B 716, 30 (2012).*

11.5. Is it really *the* Higgs boson?

A new particle was discovered, but could one claim that it is related to the BEH mechanism responsible for electroweak symmetry breaking in the Standard Model? Is the boson the remaining degree of freedom originating from the complex scalar Higgs field after spontaneous symmetry breaking? We will see that with the dataset recorded in 2011 and 2012, the ATLAS and CMS physicists were able to learn a lot already about the newly discovered particle.

First of all, by combining the ATLAS and CMS measurements, its mass was measured to be $125.09 \pm 0.21 \pm 0.11$ GeV, that is with a two permil level precision![10]

[10] A measurement is always quoted with its uncertainty. Most often, one distinguishes between the statistical uncertainty that diminishes roughly as $1/\sqrt{L}$ with increasing integrated luminosity L, and the systematic uncertainty. The latter uncertainty quantifies the ignorance related to various ingredients entering the final result. In case of the Higgs mass measurement, it includes, for example, the uncertainty in the momentum and energy measurements for muons and electrons due to the calibration procedures employed in the different parts of the detector.

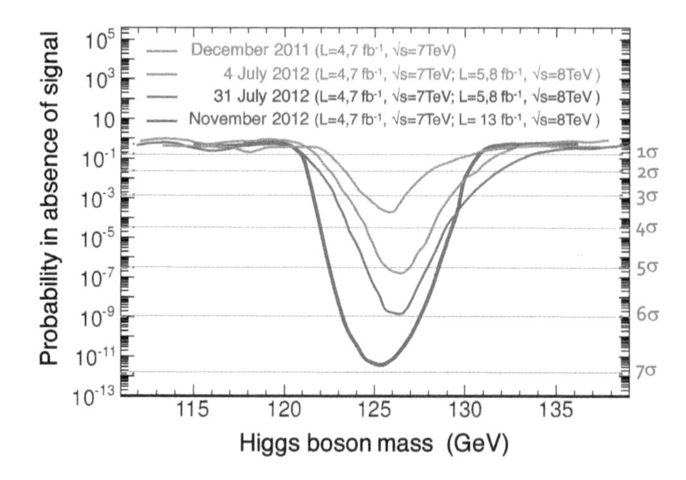

Fig. 11.21. Evolution with time of the Higgs boson signal in ATLAS, when combining results from the main Higgs decay channels $\gamma\gamma$, ZZ^*, WW^*. The left-hand vertical axis (ordinate) represents the probability for a background fluctuation to produce an excess at least as large as the one observed in the experimental data. The right-hand ordinate shows the corresponding standard deviations (see inset 11.1). This probability, also called *p-value*, is drawn versus the putative Higgs mass. The minimum of probability occurs where the Higgs boson manifests itself in the data. The width of a curve is governed by the mass resolution of the contributing channels. The four curves represent the four subsequent analyses shown at conferences or in publications. A hint for a Higgs signal around 125 GeV was already present in the 2011 data alone. The difference between the curves corresponding to the presentation of July 4th, 2012 (discovery announcement) and that of July 31st (publication) is largely due to the addition of the WW^* channel.

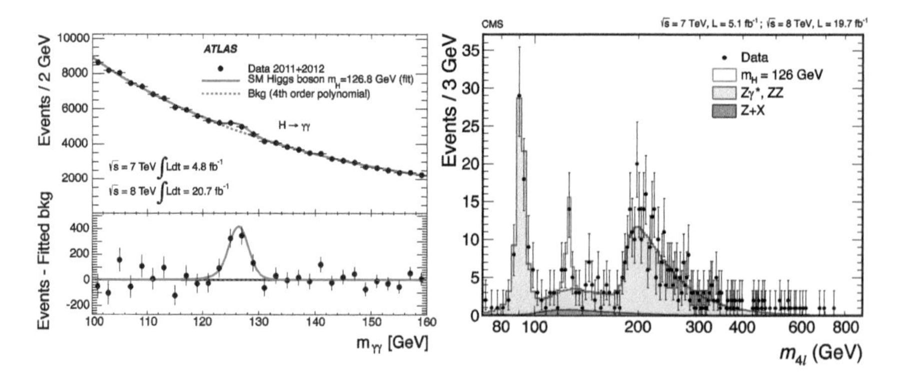

Fig. 11.22. Mass distributions for Higgs candidate events obtained from the analysis of the full 2011 and 2012 data samples. On the left, results for $H \to \gamma\gamma$ from ATLAS, and on the right CMS results for $H \to ZZ^* \to \ell^+\ell^-\ell'^+\ell'^-$. *Images: ATLAS Collaboration, Phys. Lett. B 726, 88 (2013); CMS Collaboration, Phys. Rev. D 89, 092007 (2014).*

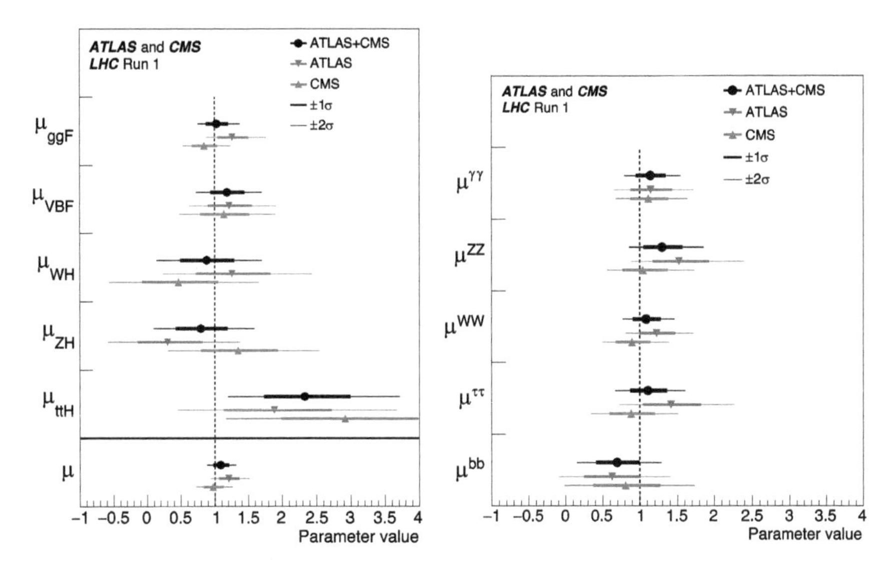

Fig. 11.23. Left: ratios μ_i of the measured production rates to the Standard Model predictions for production modes i, assuming the decay branching fractions correspond to the Standard Model ones. In the bottom part of the figure, the ratio μ corresponds to the total rate, including all production modes. Right: ratios μ^f of the measured decay branching fractions to the Standard Model predictions for final states f, assuming the production rates correspond to the Standard Model ones. These results were obtained using all the available data from 2011 and 2012, and combining the ATLAS and CMS measurements. If the observed particle is indeed the Standard Model Higgs boson, then these ratios should be equal to unity. This was indeed the case, within the (yet large) uncertainties of the measurements. *Images: ATLAS and CMS Collaborations, JHEP 08, 045 (2016).*

To further characterise the new particle, its spin and parity were also measured. If the spin turned out to be zero and the parity positive, this would be an additional indication in favour of the Higgs boson. The observation of the particle in the $\gamma\gamma$ decay mode already precludes a spin of one on the basis of spin-statistics properties of initial state and final state particles: two photons as decay products from a single spin-one particle are forbidden, while spin zero or two are allowed. It was thus clear that the particle had to be a boson. Would the spin turn out to be two, the observed particle could not be the Higgs boson, but, for example, a graviton. However, it is exceedingly difficult to think of a theory with a spin-two boson that features production and decay rates similar to those of the Standard Model Higgs boson. This would be quite a coincidence! Nonetheless, it was essential to check and ascertain that the particle observed was indeed a scalar. The angular distributions of decay products are affected by the particle's spin-parity, so that their measurement can provide the required information. The analysis of the 2012 data already answered this question to a large extent: the decay angular distributions clearly favour the spin-parity assignment 0^+, i.e. a scalar particle (see inset 11.2).

Inset 11.2: Measuring the Higgs boson's spin-parity

The parity transformation as defined in inset 1.12 on page 54 can be applied to a single particle, even if of spin zero. The properties of the associated quantum field under this mirror transformation allow one to define the parity of the particle, which is thus an intrinsic property. A particle of spin $J = 0$ and parity $P = +$ (or $-$) is called a *scalar* (*pseudoscalar*). In the Standard Model, the Higgs boson should be a scalar, whilst a particle such as the pion (a composite particle) is a pseudoscalar.

Spin and parity of a particle determine the angular distribution of its decay products. To be sure that the new particle observed at the LHC is indeed the Higgs boson, we must measure the momenta of its daughter particles. The figure underneath shows the experimental distribution obtained by CMS for the $H \to \gamma\gamma$ decays (*Eur. Phys. J. C 74, 3076 (2014)*). The distribution is for the angle θ^* between the photon emission direction and the Higgs line-of-flight, as seen from its own rest frame. This is compared with the distributions expected in case of a spin-parity $J^P = 0^+$ (red line) and two $J^P = 2^+$ hypotheses: assuming the spin-two particle is produced by gluon fusion (dashed blue histogram) or by quark–antiquark annihilation (green dotted histogram). These two options are rejected by the data with a probability of 94% and 85%, respectively.

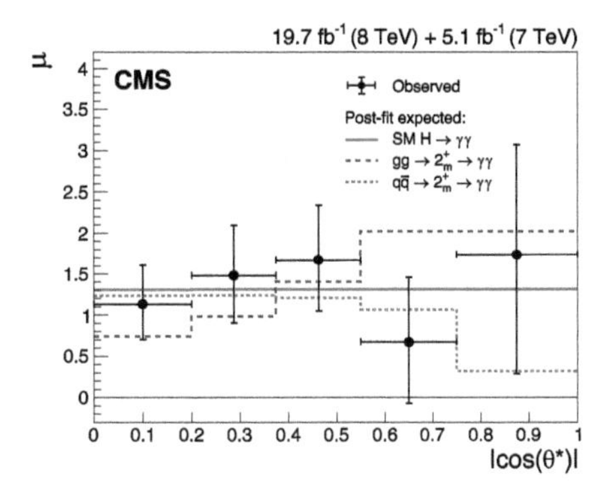

Similar analyses are performed with the $H \to ZZ^* \to \ell^+\ell^-\ell'^+\ell'^-$ and the $H \to WW^* \to \ell\nu\ell'\nu'$ channels. The context is more complicated, but also richer than in the $H \to \gamma\gamma$ case, because of the four-particle final states. For example, the figure below illustrates the five angles that can be measured in the four-lepton analysis.

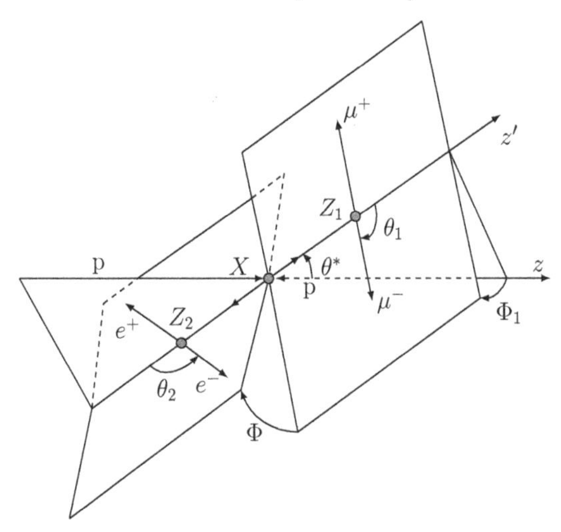

For every event in a mass window around the mass peak at 125 GeV, a likelihood ratio can be constructed from all the measured three-momenta in the event. The average value of this ratio over a sample of events depends on the spin-parity of the particle. The figure below compares the hypotheses of $J^P = 0^+$ and $J^P = 0^-$ to the data. For each one of these spin-parity hypotheses a large number of pseudo-experiments was simulated to derive the expected distributions of the likelihood ratio: in yellow for 0^+ and in blue for 0^-. The value of the likelihood ratio measured with the CMS data is indicated by the arrow. It is located in the central region of the 0^+ distribution, but far outside the 0^- distribution, which is rejected with a probability of 99.8% (*Phys. Rev. D 92, 012004 (2015)*).

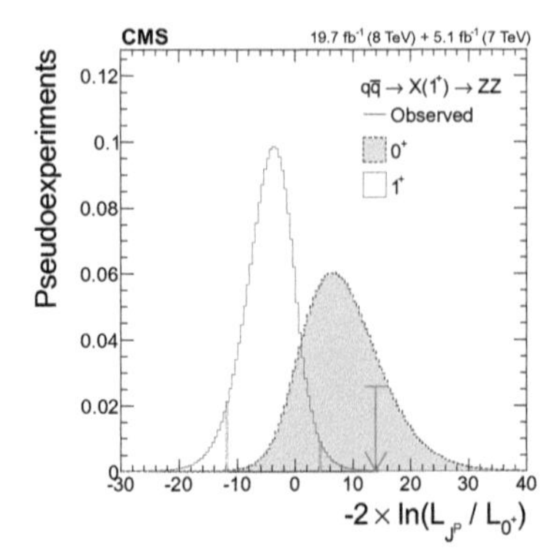

The figure below summarises the analyses performed by ATLAS for a number of spin-parity hypotheses including different production models (*Eur. Phys. J. C 75, 476 (2015)*). The expected values of the test statistics are shown in blue for Standard Model and in red for the alternative models. The measurements are shown in black and agree well (and better) with the Standard Model assumptions, whatever alternative model is tested.

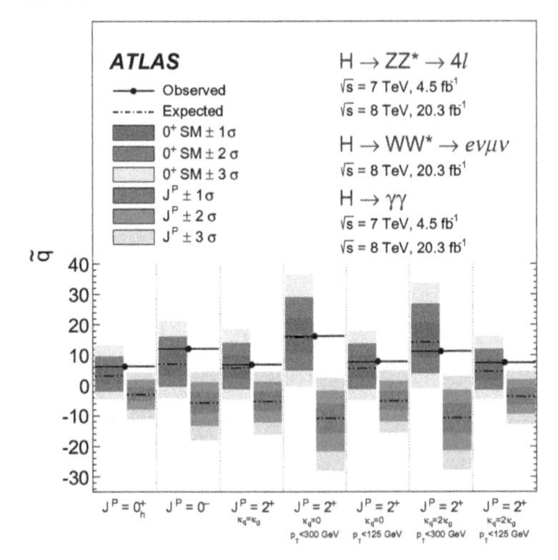

Another important and comprehensive test consisted in verifying that the observed particle is produced and decays according to the Standard Model predictions, that is, to test its couplings to all known particles. Having measured its mass, one could derive production rates and decay branching fractions from the calculations shown in figures 11.9 and 11.6, respectively. Figure 11.23 shows combined ATLAS and CMS Run-1 results comparing observed production rates (left) and decay branching fractions (right) with theoretical predictions. All essentially agree within the measurement uncertainties. The experimental observations after Run 1 also allowed to assert that the new particle couples to gauge bosons, W and Z, and to τ leptons. The decay (and thus coupling) to b quarks could not yet be firmly established at that time. Since the gluon-fusion process proceeds largely through the top loop (figure 11.4(a)), one also had indirect evidence that the Higgs boson couples to the top quark.

It is interesting to note that studying the Higgs boson is also a way to look for new particles into which the Higgs boson might decay. For example, since dark matter is massive, the Higgs boson could directly couple to it. If light enough (less than half the Higgs boson mass), the Higgs boson would decay into a pair of dark matter particles. Thanks to the small Higgs boson width of only 4 MeV, even small couplings to dark matter particles could have an observable effect on the measured branching fractions. The dark matter particles from the Higgs boson decay would traverse the detector unnoticed as their interaction with matter is expected to be extremely weak. The Higgs boson thus decays invisibly. Such events could nevertheless be experimentally detected by looking at associated Higgs boson production with other particles like forward jets (vector boson fusion), W or Z bosons, or — with more data — top–antitop quark pairs.

From the measurement of the production rates and branching fractions it was possible to deduce the coupling strengths of the Higgs boson to other particles, as shown on the (text book) left-hand plot of figure 11.24. It exhibits a key prediction of the theory, namely proportionality between the coupling to the Higgs field and the particle mass.[11] In contrast to the gauge interactions, the Higgs "force" is thus indeed non-universal!

[11]The couplings between Higgs boson and Standard Model particles depend linearly on the mass of the fermions but quadratically on the mass of the bosons. This is why the figure actually shows the couplings to fermions and the square root of the couplings to bosons.

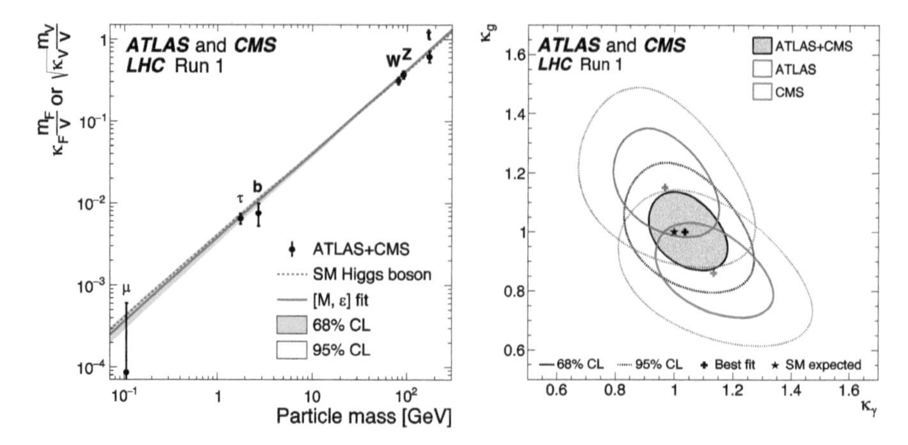

Fig. 11.24. Left: combined ATLAS and CMS Run-1 measurement of the coupling strengths of the Higgs boson to vector bosons (W and Z) and to fermions (b, t, τ). According to the Standard Model, this coupling should be proportional to particle mass (or the mass squared for vector bosons). This is, within yet large experimental uncertainties, well in agreement with experiment. Right: Higgs boson to gluon versus photon couplings normalised to the Standard Model predictions. The ATLAS and CMS measurements are shown separately, as well as their combination (grey area). It is in agreement with $\kappa_g = \kappa_\gamma = 1$ as predicted by the Standard Model. *Images: ATLAS and CMS Collaborations, JHEP 08, 045 (2016).*

It was also possible to measure the effective couplings of the new particle to gluons and the photon, which according to the Standard Model occur only through quantum loops involving massive virtual particles. The coupling to the gluon is dominated by a top loop (figure 11.4(a)), while that to the photon involves W as well as top loops (figure 11.5). Comparing these couplings to the Standard Model predictions was a powerful probe for the existence of heavy particles that could contribute to these loops, such as a fourth generation of fermions for example. The results, shown on the right-hand panel of figure 11.24, were found to be well compatible with the Standard Model. The measurement allowed to exclude a heavy fourth quark generation with Standard Model like couplings, whose presence would have increased the gluon fusion cross-section by a factor of nine.

Spin and parity, production rates and branching fractions, couplings: all measurements performed by ATLAS and CMS with the Run-1 data pointed to a Standard Model like Higgs boson. This was a giant step in the understanding of the role of spontaneous symmetry breaking in the potential of a new scalar field as the mechanism behind the mass of elementary particles. The celebration could start.

11.6. The Nobel Prize

The ultimate reward, which no scientist, whoever she or he might be, can admit to be a goal or even a motivation for a scientific career, the Nobel Prize,[12] is a rare distinction for a major discovery in a scientific domain. Usually it is the coronation of an exceptional scientific curriculum, requiring a high level of creativity, scientific rigour, determination, and sometimes also a bit of luck. The electroweak symmetry breaking mechanism proposed by Brout, Englert and Higgs in 1964 represents without doubts a major advance in physics, opening a new horizon for an entire generation of physicists. Forty-eight years were required to confirm this theoretical intuition, an exceptionally long time in the history of physics! After the July 4th, 2012 discovery and the quickly succeeding results confirming the nature of the new particle, experts and well-informed observers had no doubts that the Swedish Royal Academy of Sciences would attribute the 2013 Nobel Prize in physics to the authors of these achievements.

Tuesday, October 8th, 2013 at 10:45 am, Stockholm time. The countdown on the web page of the Nobel Committee for Physics indicated only few seconds left before announcing the new laureates. Physicists from the ATLAS and CMS collaborations at CERN and throughout the world had their eyes fixed on the screens where the announcement would be broadcast.

Then, what is rather rare in such occasions, the announcement of a first delay by half an hour was made, followed by several further delays, possibly due to heated discussions within the committee. Finally, at 11:45 am, the permanent secretary of the Swedish Royal Academy of Sciences solemnly announced the attribution of the Nobel Prize in Physics 2013 to François Englert and Peter Higgs (figure 11.25). The joy of physicists exploded. It was a historic day for the laureates but also for CERN and all those who contributed to the confirmation of this remarkable theoretical intuition (figures 11.26 and 11.27).

However, a question was looming in the minds of many: should CERN as the hub of this planetary undertaking, which mastered the scientific and technological challenges represented by the discovery of the Higgs boson, and thus of the underlying mechanism, not have been among the laureates?

[12]Excerpt from www.nobelprize.org: *"Physics was the prize area which Alfred Nobel mentioned first in his will from 1895. At the end of the nineteenth century, many people considered physics as the foremost of the sciences, and perhaps Nobel saw it this way as well. His own research was also closely tied to physics."*

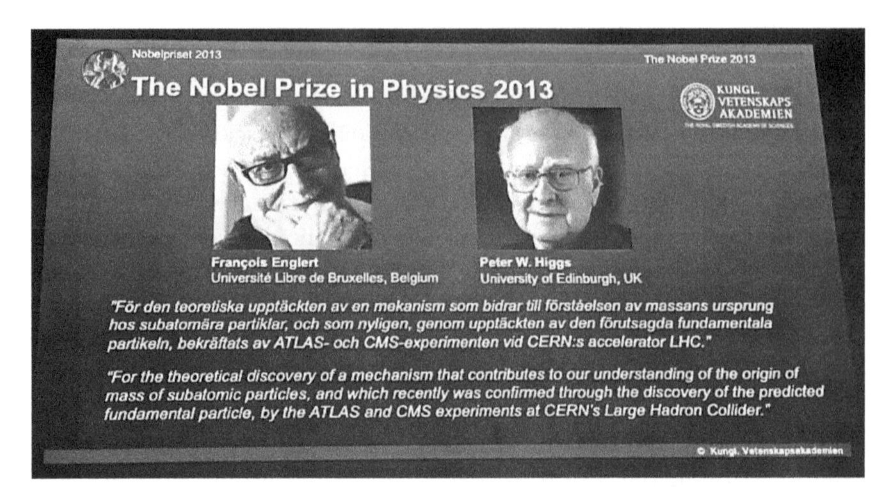

Fig. 11.25. Text published by the Swedish Royal Academy of Sciences in the morning of October 8th, 2013 on the attribution of the Nobel Prize in Physics to François Englert and Peter Higgs: *"For the theoretical discovery of a mechanism that contributes to our understanding of the origin of mass of subatomic particles, and which recently was confirmed through the discovery of the predicted fundamental particle, by the ATLAS and CMS experiments at CERN's Large Hadron Collider"*.

The members of the committee had made their choice. Did they not dare to make the step and attribute the Nobel Prize to an organisation? As science grows global and progress relies more and more on the force of collaboration, this question will surely reappear again on the Nobel Committee's agenda.

Fig. 11.26. Physicists at CERN rejoicing after the announcement that the Nobel Prize in Physics 2013 was awarded to François Englert and Peter Higgs. *Image: CERN.*

11.7. Extending and deepening the Higgs-boson studies

The Higgs boson will remain at the core of the ATLAS and CMS research programs throughout the lifetime of the LHC and its successor the High-Luminosity LHC (HL-LHC). With the events recorded during the

Fig. 11.27. Visit of Peter Higgs at the CMS detector in 2008. *Image: M. Brice, CERN.*

second data-taking period (Run 2) between 2015 and 2018 at a collision energy of 13 TeV, more and more detailed studies of the Higgs boson as well as searches for rare processes could be performed. The higher LHC energy increased the Higgs boson production cross-section by a factor of 2.3 compared to 8 TeV. Run 2 also saw a factor of almost six larger integrated luminosity compared to Run 1, so that the collaborations overall had about fourteen times more Higgs bosons available for their studies than after Run 1.

Figure 11.28 shows the di-photon and the four-lepton mass spectra, which are now familiar to the reader, as they look in the full Run-2 dataset. The $H \to \gamma\gamma$ signal at 125 GeV stands now clearly above the backgrounds. In the (rare) $H \to ZZ^* \to l^+l^-l'^+l'^-$ decay mode the peak now contains more than 200 events in each experiment. What a progress compared to the first observation in 2012 (figures 11.19 and 11.20)! The increased event yields allowed the collaborations to perform first studies of spectra (differential cross-sections) of variables sensitive to the Higgs boson production mechanisms. Examples of such spectra measured in the four-lepton channel by ATLAS are shown in figure 11.29: the Higgs boson transverse momentum on the left and the number of accompanying jets on the right. These

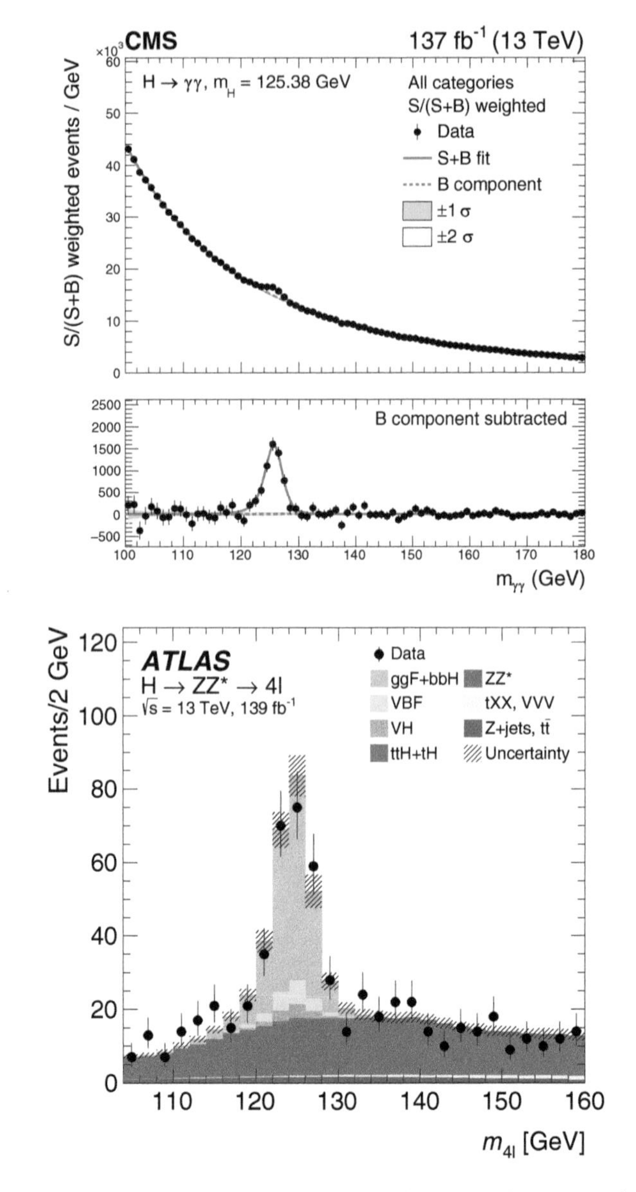

Fig. 11.28. Invariant mass distributions in the $\gamma\gamma$ final state from CMS (top) and the four-lepton final state from ATLAS (bottom), obtained using the 13 TeV data collected in Run 2 between 2015 and 2018. The colours under the Higgs boson peak in the lower plot indicate the contributions from the different production processes. *Images: CMS Collaboration, JHEP 07, 027 (2021); ATLAS Collaboration, Eur. Phys. J. C 80, 957 (2020).*

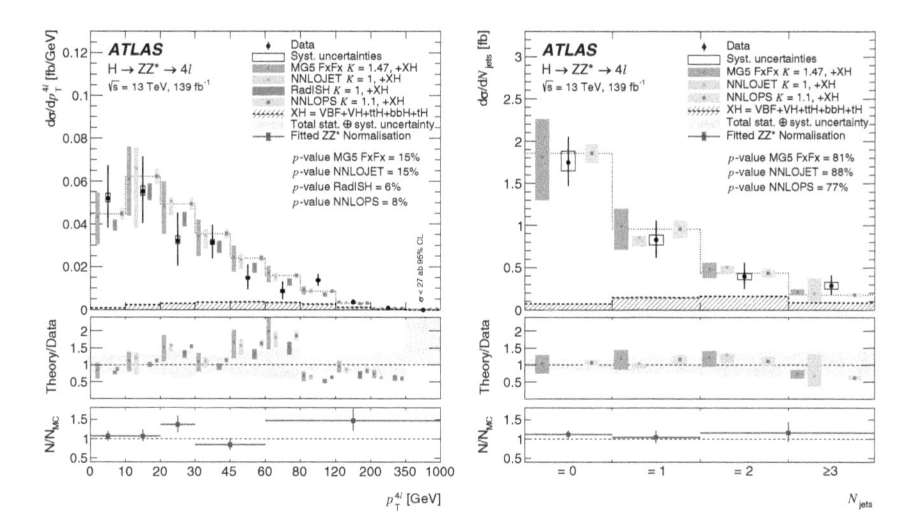

Fig. 11.29. Differential cross-section measurements in the $H \to ZZ^* \to l^+l^-l'^+l'^-$ channel versus the Higgs boson transverse momentum (left) and the number of jets produced in the events (right). The dots with error bars represent the data, and the coloured bars different theoretical predictions. Data and predictions are found in agreement. *Images: ATLAS Collaboration, Eur. Phys. J. C 80, 941 (2020).*

measurements test our understanding of the QCD aspects that come into play in the Higgs boson production.

The data accumulated during Run 1 were not sufficient to conclude on the (most abundant) $H \to b\bar{b}$ decay nor on the top–antitop associated production (ttH, figure 11.4(c)), which would allow to directly establish the coupling between top quark and Higgs boson. Also the $H \to \tau\tau$ decay was only observed through the combination of the ATLAS and CMS Run-2 data. Firmly establishing these processes was therefore among the most important goals of the LHC Run 2.

The search for the $H \to b\bar{b}$ decay at the LHC proceeded similarly as at the Tevatron (section 11.2) via the association with a W or a Z boson (Higgsstrahlung) allowing to trigger the events and to suppress the otherwise overwhelming background. The remaining (large) background contributions were determined simultaneously with the signal from a multiparameter fit to the selected data sample. Multivariate methods were applied to better distinguish signal from background events (see inset 10.1 on page 295). The left-hand panel of figure 11.30 shows the reconstructed $b\bar{b}$ mass obtained after subtracting all background events, except those belonging to VZ ($V = Z, W$) production. The VH, $H \to b\bar{b}$ signal shoulder (in red) is well visible on the right of the slightly larger

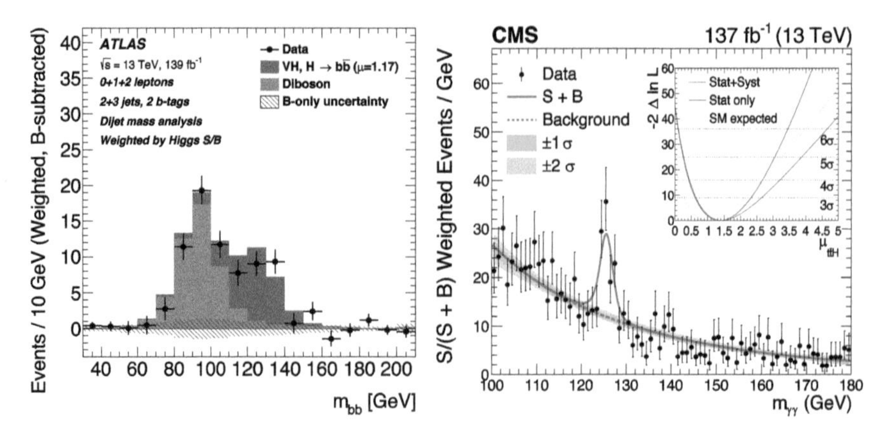

Fig. 11.30. Left: distribution of the $b\bar{b}$ invariant mass for events selected in the search for the $H \to b\bar{b}$ decay mode in data recorded by ATLAS during Run 2. The distribution has been corrected for all background contributions except VZ ($V = W, Z$). The data are shown by the dots with error bars. The colours indicate the $Z \to b\bar{b}$ (grey) and $H \to b\bar{b}$ (red) expected contributions, respectively. Right: di-photon invariant mass in the search for Higgs boson production associated with a top–antitop pair in the data recorded by CMS during Run 2. A clear peak at the Higgs boson mass indicating the presence of $H \to \gamma\gamma$ decays is visible over the smoothly falling background. The statistical significance is 6.6σ. *Images: ATLAS Collaboration, Phys. Lett. B 786, 59 (2018); CMS Collaboration, Phys. Rev. Lett. 125, 061801 (2020).*

VZ, $Z \to b\bar{b}$ background contribution. This figure is equivalent to the one obtained by the Tevatron experiments (figure 11.8), however, the $H \to b\bar{b}$ signal reported by ATLAS is much more visible with a significance at the 6.7σ level, thus providing a clear ground for observation.[13] Both CMS and ATLAS were also able to independently observe the $H \to \tau\tau$ decay using the full Run-2 datasets. Two spectacular examples of Higgs boson decays as seen in ATLAS and CMS are given in figure 11.31.

The second triumph of the LHC Run 2, the observation of top–antitop production in association with a Higgs boson ($t\bar{t}H$), occurred in parallel to the $H \to b\bar{b}$ observation in 2018. Since $t\bar{t}H$ production has a very small rate, about 1% of all produced Higgs bosons, it was necessary to add as many decay channels as possible, $H \to b\bar{b}$, $\tau\tau$, WW^*, $\gamma\gamma$ and even the tiny $H \to ZZ^* \to 4\ell$ mode, to first observe it with a partial Run-2 dataset. The analyses identify the presence of two top quarks via their decays to a b jet and a W boson, which further decays to a lepton and a neutrino or

[13]Both ATLAS and CMS announced the observation of the $H \to b\bar{b}$ decay in summer 2018 using a smaller dataset.

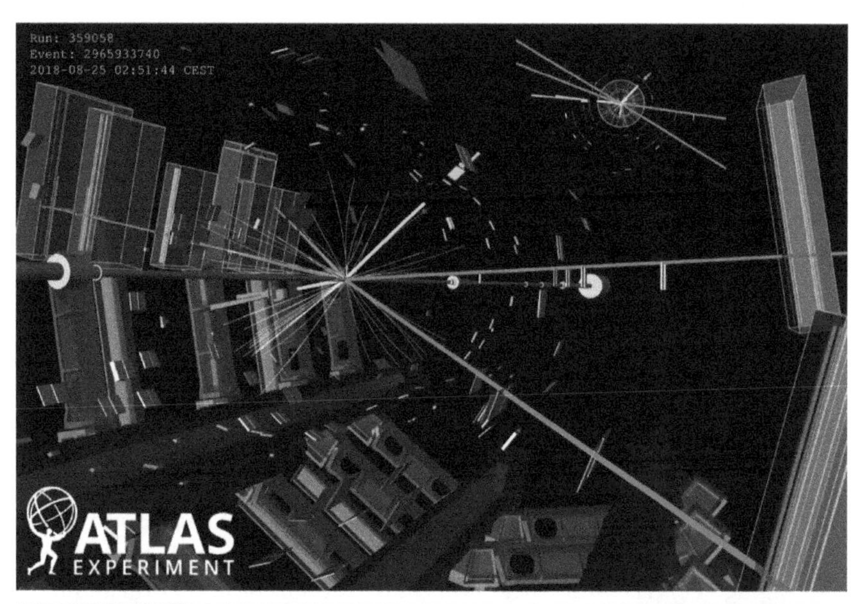

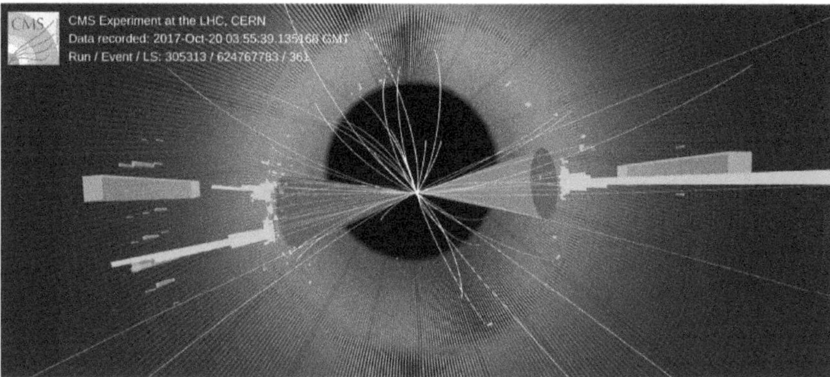

Fig. 11.31. Displays of Higgs boson candidate events. On top a six-lepton ZH candidate measured by ATLAS, where the Higgs boson decayed into two electrons (shown in green, with tracks followed by energy deposits in the calorimeter) and two muons (red tracks penetrating the full detector and being identified in the muon spectrometer), and the Z decayed into two muons. This event has a very high probability to show an actual Higgs boson production. The bottom display from CMS shows a boosted Higgs boson candidate decaying into two b quarks which hadronise into nearby jets (in the display to the left where two jets are visible among the tracks and calorimeter deposits) and recoiling against a high-energy jet (to the right with a single structure). The event is a striking example of the granularity of the LHC detectors, which is exploited in many high-mass new physics searches. *Images: ATLAS and CMS Collaborations.*

two jets. Once the full data were accumulated, ttH production could be firmly observed on grounds of the most sensitive $H \to \gamma\gamma$ channel alone, as can be seen on the right-hand panel of figure 11.30 by CMS. That channel not only allowed ATLAS and CMS to directly establish the top–Higgs boson coupling (found in agreement with the Standard Model expectation), but to also study the charge-parity structure of the top–Higgs coupling.

The collaborations had now established the Higgs boson couplings to the weak bosons as well as to the fermions of the third generation (τ-lepton, bottom and top quarks). This success and the large dataset titillated the ambition of the experimenters to look to the next rarer decay channels $H \to \mu\mu$ and $H \to Z\gamma$. The former mode tests the Higgs mechanism for second-generation fermions, while the latter decay occurs, akin to $H \to \gamma\gamma$, via quantum loops. While the expected branching fraction of 0.02% for $H \to \mu\mu$ is almost twice that of $H \to 4\ell$, the (irreducible) background from the $pp \to Z/\gamma^* \to \mu\mu$ process in the di-muon channel is profuse, about a thousand times larger than the signal, rendering this search very challenging. For $H \to Z\gamma$, the branching fraction of 0.01% (including Z decays to an electron or muon pair) is similar to the four-lepton case, but $Z\gamma$ has larger irreducible background. Both analyses proceed similarly to $H \to \gamma\gamma$ via a search for a Higgs boson peak at 125 GeV over a smooth background that is estimated from the data sidebands. A key aspect of these searches is the mass resolution, where the almost twice larger solenoid field and the larger precision tracker of CMS represents an advantage over ATLAS in the di-muon channel, leading to an about $\sqrt{2}$ better expected sensitivity.

Figure 11.32 shows the results of these investigations for the di-muon channel (left, a preliminary result from CMS) and $Z\gamma$ (right, from ATLAS). A small excess is seen on the left panel. The full statistical analysis results in an observed statistical significance in CMS (ATLAS) of 3.0σ (2.0σ) for $H \to \mu\mu$, and 2.2σ from ATLAS for $H \to Z\gamma$. All measured signals were found somewhat above, but in agreement with the Standard Model predictions. The combined significance of ATLAS and CMS allows to safely state evidence for the $H \to \mu\mu$ decay, as so often, much earlier than originally anticipated. Its observation and precise measurement will require the data of Run 3 (2022 through 2024) and the HL-LHC (scheduled to start in 2027).

Another fundamental measurement that became possible with the large Run-2 data samples is that of the natural width, Γ_H, of the Higgs boson

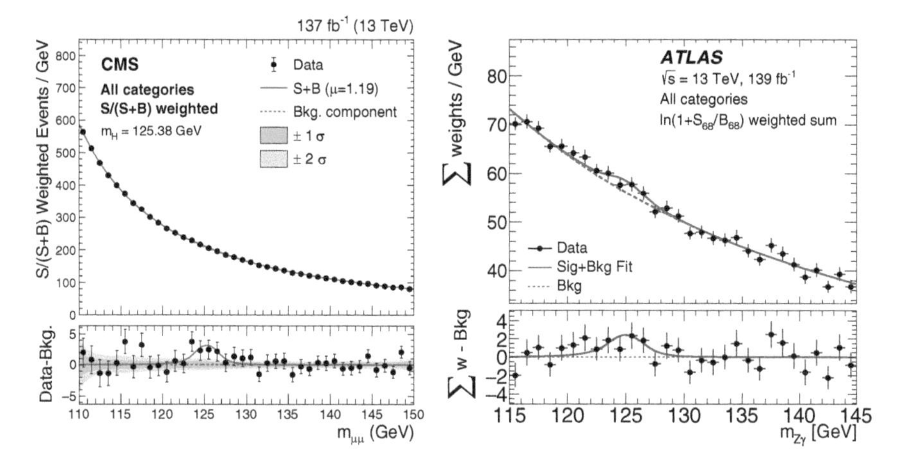

Fig. 11.32. Di-muon (left, CMS) and $Z\gamma$ (right, ATLAS) invariant mass distributions, weighted by purity (an estimation of the probability for each event to be of signal type), in the search for Higgs boson decays to these final states. The lower panels show the difference between data and smooth background fits. Overlaid are the background-only and signal plus background fits. *Images: CMS Collaboration, JHEP 01, 148 (2021); ATLAS Collaboration, Phys. Lett. B 809, 135754(2020).*

(see inset 2.1 on page 73 on the natural or total width of a particle), which is expected to be about 4 MeV in the Standard Model. A larger value of Γ_H would reveal the existence of yet undetected decay channels beyond those expected in the Standard Model, for instance decays to invisible particles (see section 13.7). The direct measurement of Γ_H in decay channels like $H \to ZZ^* \to 4\ell$ or $H \to \gamma\gamma$ through the shape of the observed signal peak is limited by the experimental resolution and provides only a very loose upper bound of about 1 GeV. More promising is to exploit the interference of the Higgs boson resonance with the continuum background in, for example, the $ZZ^{(*)}$ channel.[14] The small but non-zero Higgs boson width leads to a contribution to the $ZZ^{(*)}$ production cross-section at masses well beyond the Higgs boson peak. This so-called Higgs off-shell contribution originates from the far-away tail of the Higgs boson Breit–Wigner distribution. As illustrated on figure 11.33, this contribution interferes (negatively) with the $ZZ^{(*)}$ continuum production even at masses well above $2m_Z$, where

[14]Since quantum interference requires identical initial and final states, only the small $gg \to ZZ^{(*)}$ contribution of the continuum background interferes with the corresponding gluon-fusion Higgs production $gg \to H \to ZZ^{(*)}$. The dominant $qq \to ZZ^{(*)}$ process does not participate in the interference.

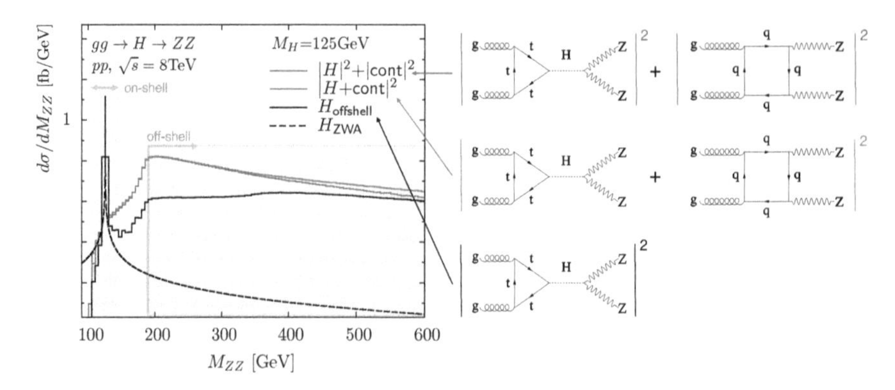

Fig. 11.33. The left-hand panel shows the predicted gluon–gluon fusion production cross-section (logarithmic scale) of Z boson pairs as a function of the $ZZ^{(*)}$ invariant mass. The Feynman diagrams corresponding to the production via an intermediate Higgs boson (on-shell or off-shell for high mass), and via the box diagram, which is dominant well above the Higgs mass, are shown on the right-hand side. The blue (respectively red) curve shows the expected cross-section with (without) the proper inclusion of the destructive interference effect. The black curve shows the Higgs boson contribution only, while the dashed curve shows the cross-section one would obtain with a quasi-zero-width approximation. *Image: N. Kauer and G. Passarino, JHEP 08, 116 (2012).*

both Z bosons become on-shell. By comparing the minute deficit of events in this Higgs off-shell region, which does not depend on Γ_H, with the Γ_H dependent event yield in the main Higgs boson peak,[15] one can deduce a constraint on Γ_H. In 2019, an analysis by CMS of the four-lepton final state, not yet using the full Run-2 dataset, led to the estimate $\Gamma_H = 3.2^{+2.8}_{-2.2}$ MeV. It should be noted that this measurement depends on the assumption that no unknown physics affects the high-mass spectrum, nor the on-shell Higgs boson production. More sensitive direct searches for invisible Higgs boson decays are performed by the LHC experiments by studying events with significant production of missing transverse energy, as we shall see in section 13.7.

The data at 13 TeV collision energy collected during Run 2 confirmed with increasing precision and breadth the Standard Model nature of the Higgs boson (see figure 11.34 and inset 11.3, where a more systematic approach to probe effects of possible new physics in the Higgs boson sector via the concept of *effective field theories* is introduced). Is there really only

[15]The on-shell $H \to 4\ell$ yield is proportional to $1/\Gamma_H$. Therefore, the ratio of measured off-shell to on-shell event yields normalised to their Standard Model predictions determines $\Gamma_H/\Gamma_H^{\text{SM}}$.

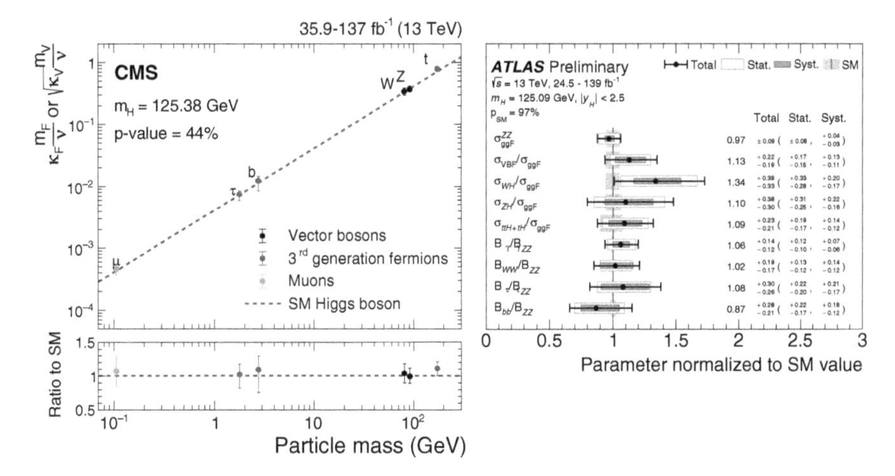

Fig. 11.34. Status of the Higgs boson production and decay rate measurements after the LHC Run 2. On the left panel the same plot as figure 11.24, indicating by different colours the couplings to bosons and fermions. It may be useful to recall that, albeit the observed coupling pattern reproduces perfectly the predicted coupling of the Higgs boson to mass, it does not *explain* the mass hierarchy among the matter particles. Yes, the top quark is heavier than the charm quark because the top quark has a stronger coupling to the condensed Higgs field. Why is that? Why is the top quark three hundred thousand times heavier than the electron? It remains a mystery. The right panel shows similar information as in figure 11.23 after a combination of various channels. Both figures illustrate the large gain in precision over the Run-1 results. *Images: CMS Collaboration, JHEP 01, 148 (2021); ATLAS Collaboration, ATLAS-CONF-2020-027 (2020).*

one, single, unique elementary Higgs boson, or could there be more? Does the Higgs boson couple to sectors possibly hidden from our sight? Answering these questions requires direct searches for additional scalar particles, which might indicate an extension of the symmetry breaking mechanism, as well as much more precise measurements of the Higgs boson couplings to known and invisible particles as will allow the forthcoming LHC Run 3 and, in particular, the upgraded HL-LHC described in chapter 15.

Another crucial sector of Higgs boson physics is the determination of the Higgs potential via the measurement of Higgs boson self-coupling. This is possible when two Higgs bosons are produced in the final state through the trilinar vertex $H \to HH$.[16] Di-Higgs production is an extremely rare occurrence, roughly a factor thousand less frequent than single Higgs boson

[16] It is also possible to produce two Higgs bosons without invoking Higgs self-coupling. This process interferes destructively with the $H \to HH$ one, leading to a further suppression of di-Higgs production.

production. The two Higgs bosons can be identified through decays to $bb+\gamma\gamma$, $bb+\tau\tau$, and $bb+bb$. The best limits on HH production by 2021 are about a factor of four larger than the predicted Standard Model cross-section. Di-Higgs production and the measurement of trilinear Higgs boson coupling is one of the main objectives of the HL-LHC discussed in chapter 15.

The reason why the clever, but also conceptually simple BEH mechanism works so seemingly well remains a mystery for many physicists. The scalar sector is also directly connected with profound questions: naturalness, vacuum stability and energy, flavour. The Higgs boson discovery allows us to directly study this sector, requiring a broad experimental programme that will extend over decades. This new field of research is still at its inception.

Inset 11.3: Probing new physics through effective field theories

For a large class of new physics scenarios beyond the Standard Model (BSM), processes occurring at energies much below the mass scale of the new particles can be parameterised by an *effective field theory* (EFT), where the Standard Model (SM) Lagrangian is supplemented by new terms, or operators,[a] which have the same field content and the same local gauge symmetry as the SM. For the sake of generality, one should consider all operators built out of the SM fields, possibly including derivatives, that fulfil the following constraints: they must be a Lorentz scalar with the same energy dimension as the Lagrangian and they must be invariant under the local SM symmetry group.

Let us do some power counting. In natural units ($\hbar = c = 1$), energy and mass have dimension 1, by definition. Length has dimension -1 (as $dx \propto E^{-1}$). The action $S = \int \mathcal{L} \cdot d^4x$ must be dimensionless (actually of dimension $\hbar$), therefore the Lagrangian $\mathcal{L}$ has dimension 4. Derivatives have dimension 1. This set is sufficient to determine the dimensions of the particle fields. Since kinetic terms in a Lagrangian must have dimension 4, we can derive from the term $(\partial\phi)^2$ (cf. the Lagrangian of inset 1.8 on page 38) that a bosonic field ϕ must have dimension 1. Similarly, the term $\overline{\psi}\partial\psi$ lets us conclude that a fermionic spinor field ψ must have dimension 3/2 (see inset 1.5 on page 24, considering that the γ matrices are dimensionless). We further note that mass terms in the Lagrangian, ϕ^2 and $\overline{\psi}\psi$, have dimensions 2 and 3, respectively. Simple interaction terms of the type ϕ^3 or $\overline{\psi}\gamma^\mu\psi A_\mu$ have dimensions 3 and 4, respectively.

An early prototype of an EFT is Fermi's "four-fermions" theory which was used before the advent of the SM to describe weak processes like β or muon decays at energies much lower than the mass of the mediator particle, the W boson, which was not known at the time. The figure below illustrates via Feynman diagrams how the tree-level β-decay process ($n \to p\, e^- \bar{\nu}_e$) can be parameterised at low energy by an "effective" interaction term involving only the fermionic fields of the incoming and outgoing particles, while integrating out the W propagator. The probability amplitude (matrix element) of the process is deduced from the operators indicated in the diagrams.

The operator on the left, a Lorentz scalar where the sum over the Lorentz indices μ and ν is implicit, is calculated using the SM Feynman rules applied to the diagram: fields ψ_f of the incoming and outgoing fermions, vertex terms with dimensionless weak coupling constant g, and the so-called *propagator* term associated with the internal virtual particle, the W, with four-momentum q. At low energy ($q \ll m_W$), the propagator term can be reduced to $1/m_W^2$. This procedure of integrating out the virtual heavy particle leads to the simplified operator on the right, which corresponds to an interaction term with a coupling constant $G_F = g^2/(\sqrt{2}m_W^2)$ of dimension -2 (the Fermi constant, see inset 1.10 on page 45) and a four-fermion operator of dimension 6 built only out of the known fields prior to the formulation of the full SM.[b]

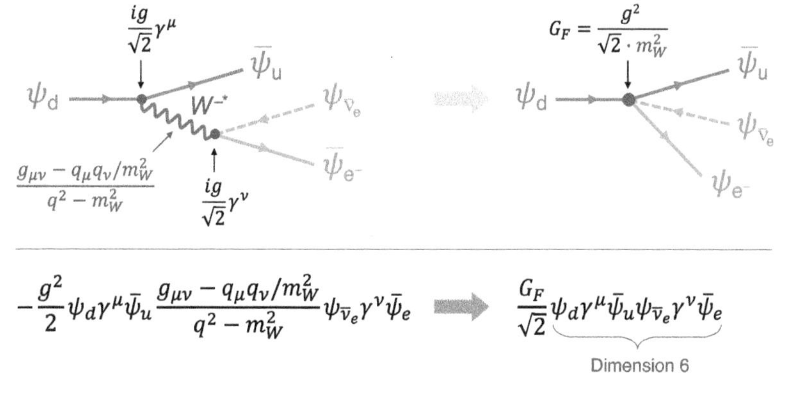

$$-\frac{g^2}{2}\psi_d\gamma^\mu\bar{\psi}_u\frac{g_{\mu\nu} - q_\mu q_\nu/m_W^2}{q^2 - m_W^2}\psi_{\bar{\nu}_e}\gamma^\nu\bar{\psi}_e \;\Longrightarrow\; \frac{G_F}{\sqrt{2}}\underbrace{\psi_d\gamma^\mu\bar{\psi}_u\psi_{\bar{\nu}_e}\gamma^\nu\bar{\psi}_e}_{\text{Dimension 6}}$$

Following to the same prescription, the EFT Lagrangian for BSM physics takes the form

$$\mathcal{L}_{\text{EFT}} = \mathcal{L}_{\text{SM}} + \sum_i \frac{c_i^{(5)}}{\Lambda} \mathcal{O}_i^{(5)} + \sum_i \frac{c_i^{(6)}}{\Lambda^2} \mathcal{O}_i^{(6)}$$

$$+ \sum_i \frac{c_i^{(7)}}{\Lambda^3} \mathcal{O}_i^{(7)} + \sum_i \frac{c_i^{(8)}}{\Lambda^4} \mathcal{O}_i^{(8)} + \cdots ,$$

where the $\mathcal{O}_i^{(D)}$ are SU(3) $\times$ SU(2) $\times$ U(1) (see insets 1.10 on page 45 and 1.13 on page 63) invariant operators of dimension D, and the parameters $c_i^{(D)}$ are called *Wilson coefficients*, which encapsulate the parts of the interactions that can be computed via perturbative methods. Operators of dimension 5 violate the lepton number[c] L, while dimension-7 operators violate $B - L$ (baryon number minus lepton number). In practice, mainly dimension-6 operators are considered, and sometimes also dimension-8 operators although they are suppressed by a factor $1/\Lambda^4$. This EFT framework is intended to parameterise observable effects of BSM physics where new particles with mass of order Λ are much heavier than the SM ones and heavier than the energy scale directly probed by the experiment. In the example of the Fermi interaction, we had a dimension-6 operator with $\Lambda \sim m_W$ and a Wilson coefficient of order unity. Inset 1.14 on page 67 lists the full set of dimension-4 operators of the SM Lagrangian after electroweak symmetry breaking.

A complete set of non-redundant BSM dimension-6 operators comprises 59 terms plus hermitian conjugates. Including flavour dependencies, there are several hundreds of dimensionless free parameters $\bar{c}_i = c_i^{(6)} v^2/\Lambda^2$ related to the Wilson coefficients[d] $c_i^{(6)}$. For instance, the presence of a heavy W' boson would manifest itself with a dimension-6 operator similar to the Fermi term. In practice, one considers only a subset of operators and free parameters for the study of a process with well-defined initial and final states. For instance, the table below shows dimension-6 operators involved in the analysis of Higgs boson production and decay.[e]

Wilson coefficient	Operator	Wilson coefficient	Operator
$c_{H\Box}$	$(H^\dagger H)\Box(H^\dagger H)$	c_{uG}	$(\bar{q}_p\sigma^{\mu\nu}T^A u_r)\widetilde{H}\,G^A_{\mu\nu}$
c_{HDD}	$\left(H^\dagger D^\mu H\right)^* \left(H^\dagger D_\mu H\right)$	c_{uW}	$(\bar{q}_p\sigma^{\mu\nu}u_r)\tau^I\widetilde{H}\,W^I_{\mu\nu}$
c_{HG}	$H^\dagger H\,G^A_{\mu\nu}G^{A\mu\nu}$	c_{uB}	$(\bar{q}_p\sigma^{\mu\nu}u_r)\widetilde{H}\,B_{\mu\nu}$
c_{HB}	$H^\dagger H\,B_{\mu\nu}B^{\mu\nu}$	c'_{ll}	$(\bar{l}_p\gamma_\mu l_t)(\bar{l}_r\gamma^\mu l_s)$
c_{HW}	$H^\dagger H\,W^I_{\mu\nu}W^{I\mu\nu}$	$c^{(1)}_{qq}$	$(\bar{q}_p\gamma_\mu q_t)(\bar{q}_r\gamma^\mu q_s)$
c_{HWB}	$H^\dagger\tau^I H\,W^I_{\mu\nu}B^{\mu\nu}$	$c^{(3)}_{qq}$	$(\bar{q}_p\gamma_\mu\tau^I q_r)(\bar{q}_s\gamma^\mu\tau^I q_t)$
c_{eH}	$(H^\dagger H)(\bar{l}_p e_r H)$	c_{qq}	$(\bar{q}_p\gamma_\mu q_t)(\bar{q}_r\gamma^\mu q_s)$
c_{uH}	$(H^\dagger H)(\bar{q}_p u_r\widetilde{H})$	$c^{(31)}_{qq}$	$(\bar{q}_p\gamma_\mu\tau^I q_t)(\bar{q}_r\gamma^\mu\tau^I q_s)$
c_{dH}	$(H^\dagger H)(\bar{q}_p d_r\widetilde{H})$	c_{uu}	$(\bar{u}_p\gamma_\mu u_r)(\bar{u}_s\gamma^\mu u_t)$
$c^{(1)}_{Hl}$	$(H^\dagger i\overleftrightarrow{D}_\mu H)(\bar{l}_p\gamma^\mu l_r)$	$c^{(1)}_{uu}$	$(\bar{u}_p\gamma_\mu u_t)(\bar{u}_r\gamma^\mu u_s)$
$c^{(3)}_{Hl}$	$(H^\dagger i\overleftrightarrow{D}^I_\mu H)(\bar{l}_p\tau^I\gamma^\mu l_r)$	$c^{(1)}_{qu}$	$(\bar{q}_p\gamma_\mu q_t)(\bar{u}_r\gamma^\mu u_s)$
c_{He}	$(H^\dagger i\overleftrightarrow{D}_\mu H)(\bar{e}_p\gamma^\mu e_r)$	$c^{(8)}_{ud}$	$(\bar{u}_p\gamma_\mu T^A u_r)(\bar{d}_s\gamma^\mu T^A d_t)$
$c^{(1)}_{Hq}$	$(H^\dagger i\overleftrightarrow{D}_\mu H)(\bar{q}_p\gamma^\mu q_r)$	$c^{(8)}_{qu}$	$(\bar{q}_p\gamma_\mu T^A q_r)(\bar{u}_s\gamma^\mu T^A u_t)$
$c^{(3)}_{Hq}$	$(H^\dagger i\overleftrightarrow{D}^I_\mu H)(\bar{q}_p\tau^I\gamma^\mu q_r)$	$c^{(8)}_{qd}$	$(\bar{q}_p\gamma_\mu T^A q_r)(\bar{d}_s\gamma^\mu T^A d_t)$
c_{Hu}	$(H^\dagger i\overleftrightarrow{D}_\mu H)(\bar{u}_p\gamma^\mu u_r)$	c_W	$\epsilon^{IJK}W^{I\nu}_\mu W^{J\rho}_\nu W^{K\mu}_\rho$
c_{Hd}	$(H^\dagger i\overleftrightarrow{D}_\mu H)(\bar{d}_p\gamma^\mu d_r)$	c_G	$f^{ABC}G^{A\nu}_\mu G^{B\rho}_\nu G^{C\mu}_\rho$

Source: ATLAS Collaboration, ATLAS-CONF-2020-053 (2020).

Example diagrams for three operators are shown in the following table.

Coefficient	Operator	Example process
c_{uG}	$(\bar{q}_p\sigma^{\mu\nu}T^A u_r)\widetilde{H}\,G^A_{\mu\nu}$	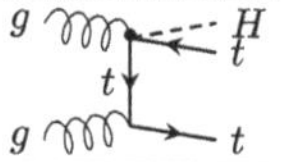
c_{uW}	$(\bar{q}_p\sigma^{\mu\nu}u_r)\tau^I\widetilde{H}\,W^I_{\mu\nu}$	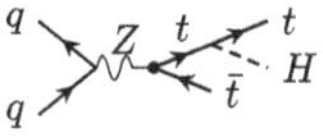
c_{uB}	$(\bar{q}_p\sigma^{\mu\nu}u_r)\widetilde{H}\,B_{\mu\nu}$	

By combining Higgs-boson measurements, ATLAS constrained Wilson coefficients with the results shown in the following figure. The presence of new physics would manifest itself by one or several coefficients differing from zero (their value in the Standard Model), which is not observed within the present experimental uncertainties.

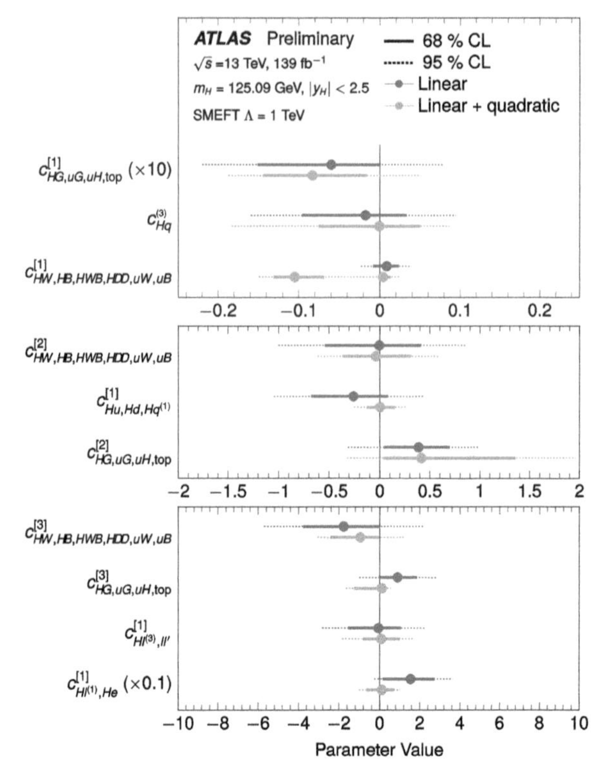

Source: ATLAS Collaboration, ATLAS-CONF-2020-053 (2020).

The figure summarises experimental constraints on the parameters $c_i^{(6)}$ assuming a new physics scale of $\Lambda = 1$ TeV and two different flavours of the EFT model in blue, using only Λ^{-2} terms, and orange, using Λ^{-2} and Λ^{-4} terms. The ranges shown correspond to 68% (solid) and 95% (dashed) confidence level intervals.

A major motivation for the use of this framework is the possibility to re-interpret the constraints on the EFT parameters as constraints on masses and couplings of new particles in concrete BSM scenarios. In other words, the translation of experimental data into a theoretical framework has to be done only once in the EFT context, rather than for each BSM model separately. The EFT framework is also a convenient tool to combine results from different searches and measurements.

---

[a]In quantum field theory (cf. inset 1.3 and section 1.9), the fields are operators acting in the space of states, denoted *Fock space*, describing the particle content of the Universe, which in practice is the process under study.

[b]Although $G_F = (\sqrt{2}v^2)^{-1}$ (inset 1.10) is proportional to the reciprocal of $\Lambda^2 \sim m_W^2$, it does not *predict* the W-boson mass.

[c]A dimension-5 operator built out of Majorana neutrinos (which are their own antiparticles, see section 3.5) and the Higgs field gives rise to a Majorana mass term which violates the lepton number conservation.

[d]Additional assumptions such as *maximum flavour conservation* can greatly reduce these numbers.

[e]Definitions of some operators appearing in the table are well beyond the scope of this book.

Chapter 12

Testing the Standard Model

Since the LHC Run 1 and even more since Run 2 (figure 12.1), the Standard Model is a complete and self-consistent theory. It represents the legacy of 20th century particle physics, unifying quantum theory and special relativity and describing all laboratory data. The discovery of the Higgs boson is a triumph for the imagination and rigour of the scientific endeavour. It is also a triumph for the greatest experimental undertaking ever, at the frontier of accelerator and detector technologies, global data sharing, analysis and collaboration. The Higgs mass of 125 GeV is in agreement with the prediction from the global electroweak fit (cf. figure 11.2) and it lies marginally within the requirement for vacuum stability (see inset 12.2 below). The Higgs discovery does thus not come with a strict requirement for new physics below the Planck scale, were it not for all the reasons outlined in chapter 3. We have now two beautiful and extremely precise theories. On one hand the Standard Model describing electroweak and strong interactions (though not their unification), predicting, for instance, the anomalous magnetic moment of the electron to a relative precision of 10^{-10} in agreement with

Fig. 12.1. The LHC is a Standard Model particle factory. The 140 fb^{-1} Run-2 integrated luminosity corresponds to the production of 7.8 million Higgs bosons, 275 million top quarks, 8 billion Z bosons, 26 billion W bosons, and roughly 160 trillion b hadrons.

experiment. On the other hand, there is general relativity, the theory of gravitation. It has been tested to an accuracy of order 10^{-5} (Cassini probe).[1] However, the Standard Model and general relativity do not work in regimes where both are important, that is at very small scales.

As we have seen in the case of the Higgs boson studies and throughout the history of particle physics, new physics may be revealed by precisely studying properties of known particles and processes. We will thus devote this chapter to some highlights of the so-called "Standard Model physics" at the LHC, which explores known phenomena but, contrary to what the name suggests, thereby also probes new physics.

With increasing scattering momentum transfer, Standard Model processes at the LHC can be categorised as follows.

- **Soft QCD**: study of particle spectra. The transverse momenta are typically smaller than a few GeV. More than 99.999% of the proton–proton collisions belongs to this type. Measurements of soft QCD processes serve to understand and improve the modelling of phenomena that cannot be predicted from first principles. Among these are the multiplication (*showering*) of partons, the formation of hadrons (*hadronisation*), or collective phenomena such as long-range correlations among the produced particles. An important practical application of these studies is the accurate simulation of pile-up interactions.

- **Hard QCD**: study of jets. Typical jet transverse momenta are greater than tens of GeV up to several TeV. A fraction of approximately 10^{-5} of the collisions belongs to this category. The measurements probe QCD calculations, the evolution of the strong coupling strength with energy, the determination of parton distribution functions, the study of parton showers, etc.

- **Hard QCD and electroweak processes**: W, Z, H, top-quark production and decays to stable identified particles. Typical transverse momenta greater than tens of GeV are involved. A fraction of 10^{-6} and less of the collisions belongs to this category. The measurements

[1] In 1997, a joint NASA and ESA mission called *Cassini-Huygens* was launched to explore the planet Saturn and its largest moon, Titan. The physicists B. Bertotti, L. Iess and P. Tortora published in 2004 a measurement of the frequency shift of radio photons to and from the Cassini spacecraft as they passed near the Sun. The observed delay of the photons was found in agreement with the prediction from general relativity within a (record) relative precision of 2.3×10^{-5}.

probe precise higher order QCD calculations, parton distribution functions, electroweak physics, etc.

Many elements of the Standard Model, besides the Higgs boson, require further consolidation, control and improved precision, both in the strong-interaction (QCD) and electroweak sectors. They hold fundamental value, for instance in the precise determination of parameters of Nature and for a better understanding of scattering dynamics, and are critical to fully exploit the search potential of the LHC. Indeed, the detailed knowledge of rates and properties of Standard Model background processes is crucial to identify many new physics processes. There is another important spin-off: the study of Standard Model processes at hadron colliders is typically more complex than the search for new physics signatures and throughout the years has been a driver of experimental and theoretical innovation.

12.1. Production rates of particles at the LHC

Among the first studies, the ATLAS and CMS collaborations have undertaken the measurement of production cross-sections of all possible final states containing a given set of Standard Model particles (quarks, gauge bosons and, later, Higgs boson[2]). Exemplary for the 13 TeV data, these measurements are summarised in figure 12.2. From left to right, the final state complexity and the sum of masses of the produced particles increases with a corresponding decrease of the cross-sections. For Z-boson production (among others), the cross-sections are also measured for final states with a given number of accompanying jets (mostly from gluon radiation). The experimental measurements are compared with theoretical predictions in the Standard Model. These predictions for a given final state F depend on two ingredients: (*i*) the parton–parton cross-sections at given parton momentum fractions x_1 and x_2 of the colliding protons to produce F, and (*ii*) the probability of a given parton to have that momentum fraction which depends on the parton distribution functions (see inset 10.4 on page 306). The full prediction is obtained from the sum over all possible $x_{1,2}$, which is computed by numerically solving a convolution integral. The parton distribution functions that describe the proton content in terms of quark and gluon densities are obtained from experimental

[2]Leptons are not considered here, as in proton–proton collisions they are always produced via the decay of a heavy quark or a gauge or Higgs boson.

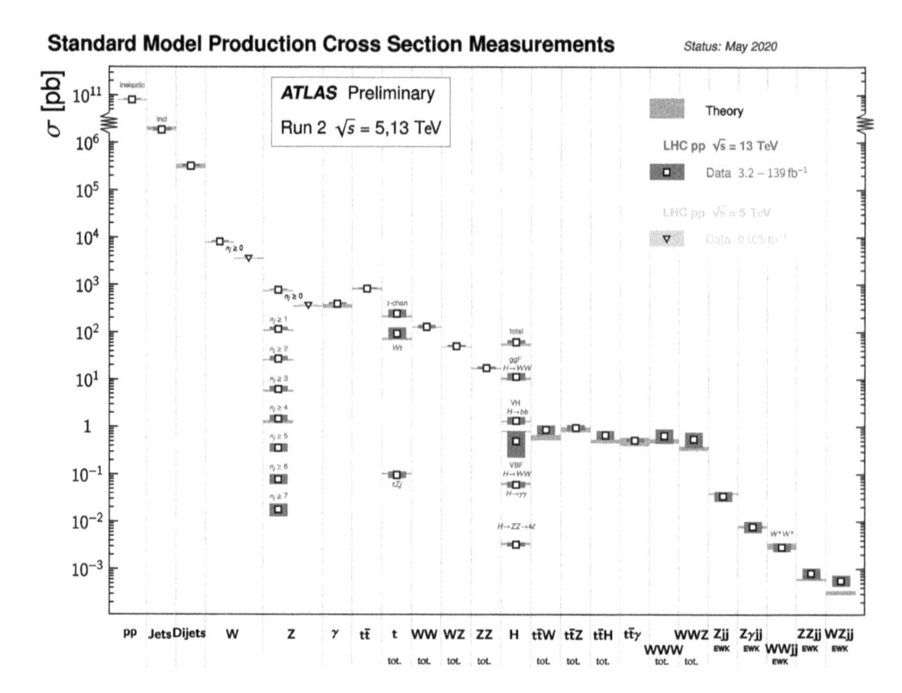

Fig. 12.2. Compilation of cross-section measurements of Standard Model processes at the LHC with 13 TeV proton–proton collision energy (for W and Z production also results at 5 TeV are shown). From left to right, the cross-sections decrease with the increasing complexity of the final state. The grey horizontal bands show the corresponding theoretical predictions. Agreement between measurement and theory is observed over fourteen orders of magnitude in cross-section. This figure is on its own a manifestation of the variety of the physics channels studied at the LHC and of the Standard Model's predictive power. *Image: ATLAS Collaboration, ATL-PHYS-PUB-2020-010 (2020), figure modified by authors.*

data of, in particular, electron–proton scattering processes. As witnessed in figure 12.2, agreement between measurements and predictions is observed over fourteen orders of magnitude in cross-section, representing an impressive quantitative test of the Standard Model.

The parton distribution functions also govern the dependence of the production cross-sections on the centre-of-mass energy, which is shown for the example of the weak gauge bosons in figure 12.3. While the parton–parton cross-section decreases as $1/\hat{s}$ with the parton–parton collision energy squared $\hat{s} = x_1 x_2 s$ (s is the proton–proton collision energy squared, cf. figure 10.8), the steep rise of the parton densities (in particular for gluons)

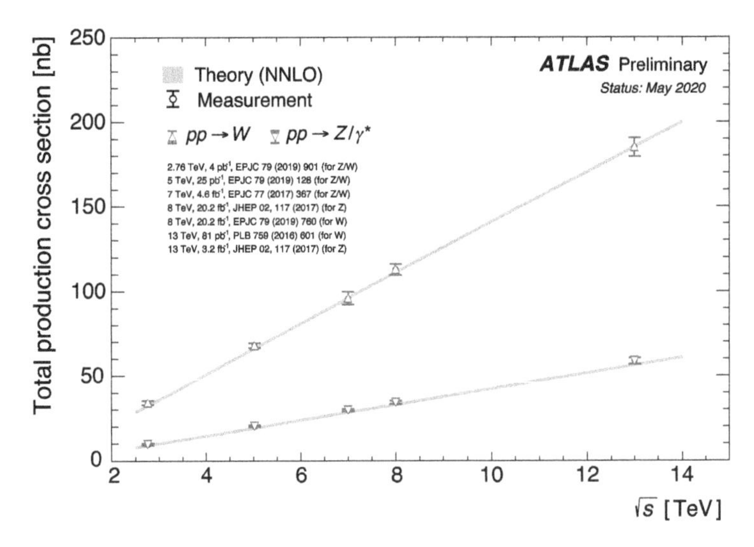

Fig. 12.3. Measured and predicted cross-sections for inclusive W (blue) and Z-boson (red) production versus proton–proton centre-of-mass energy. The increase in cross section with collision energy is a manifestation of the parton distribution functions in the proton. The theory prediction uses next-to-next-to leading order (NNLO) calculations in the strong coupling constant, which means that it includes second-order quark and gluon quantum loops (that is, quantum loops within quantum loops). *Image: ATLAS Collaboration, ATL-PHYS-PUB-2020-010 (2020).*

at low $x_i \sim 0.006$ required to produce a weak boson ($m^2_{W,Z} \sim \hat{s}$) explains the energy-dependent pattern displayed in figure 12.3.[3]

In addition to the total production cross-section for a given final state, the number of events may be large enough to measure the production rate as a function of kinematic variables, for instance the transverse momentum of one or of a combination of final state particles, or the invariant mass of a pair of gauge bosons or heavy quarks, leading to so-called *differential cross-sections.* As an example, figure 12.4 shows the double differential cross-section of jet production versus transverse momentum p_T and rapidity y (see inset 10.5 on page 307). Note that for better visibility of the various rapidity bins the cross-sections have been multiplied by the constant factors indicated in the figure. The data are given by the dots (with uncertainty bars too small to be visible) and the QCD predictions including

[3]The inclusive W boson cross-section is about 3.4 times larger than that of Z-boson production because of their different couplings to the initial state quarks. A factor of two is due to the electric charge of the W. Another factor $1/0.77^2$ stems from the different neutral current couplings. Recall also that the $W \to \ell\nu_\ell$ branching ratio is with its 11% about 3.3 times that of $Z \to \ell\ell$.

one sub-level of quantum corrections by the red lines. Also here, impressive agreement with the data is observed over many orders of magnitude in cross-section.

Figure 12.5 shows the differential cross-section for the simultaneous production of two pairs of charged leptons $pp \to \ell^+\ell^-\ell'^+\ell'^-$ ($= 4\ell$), with $\ell, \ell' = e, \mu$, as a function of the four-lepton invariant mass $m_{4\ell}$. This measurement is experimentally clean and exhibits a rich contribution of different physics processes. It is dominated around the Z mass ($m_{4\ell} \sim m_Z \sim 92$ GeV) by Drell–Yan Z-boson production and decay to four charged leptons.[4] At $m_{4\ell} \sim 125$ GeV one notices the $H \to ZZ^* \to 4\ell$ peak. So-called *continuum* $pp \to ZZ^* \to 4\ell$ production via quark–quark and (much smaller) gluon–gluon scattering rises slowly with $m_{4\ell}$ and suddenly bumps up at $m_{4\ell} \sim 2m_Z \simeq 184$ GeV, when both Z bosons are produced on their mass shell (yellow line in figure 12.5). Also shown is a small contribution from top–antitop production in association with one or two vector bosons which can also produce four charged leptons albeit with additional missing transverse energy. The data are in good agreement with the Standard Model prediction over the whole mass spectrum.

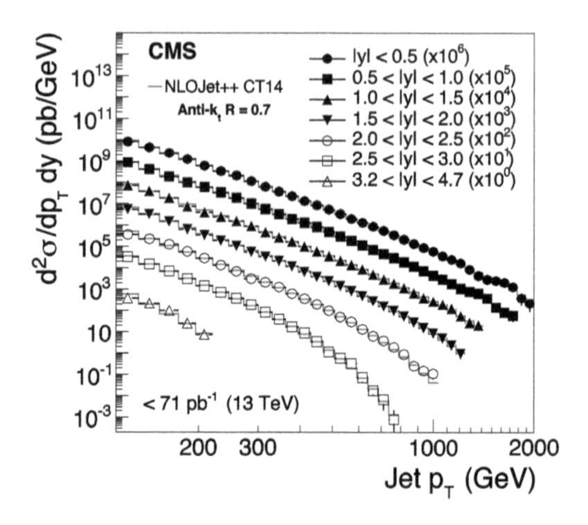

Fig. 12.4. Double differential jet cross-section versus jet transverse momentum and rapidity. *Image: CMS Collaboration, Eur. Phys. J. C 76, 451 (2016), figure modified by authors.*

12.2. Multi-boson production and vector boson scattering

The Brout–Englert–Higgs mechanism providing the W and Z bosons with mass allows for the existence of longitudinal spin states of these bosons — usually denoted by W_L and Z_L — besides the transverse states, which are

[4]The rare $Z \to \ell^+\ell^-\ell'^+\ell'^-$ process is predicted by quantum electrodynamics via so-called *inner bremsstrahlung* of a virtual photon by one of the charged leptons. Its branching fraction is measured to be $(4.4 \pm 0.3) \times 10^{-6}$ in agreement with the prediction.

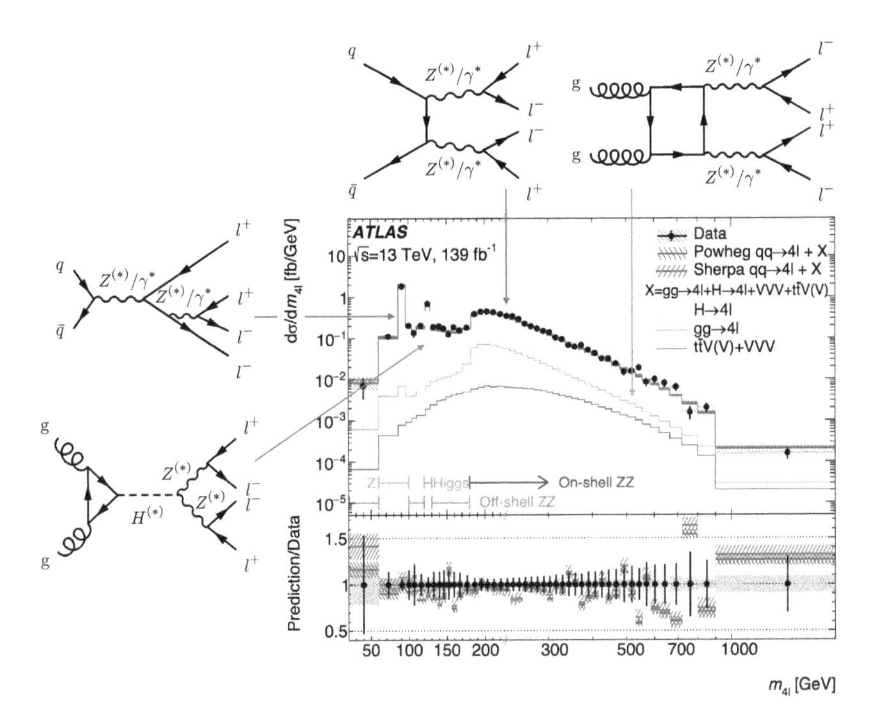

Fig. 12.5. Observed (data points with error bars) and predicted (histograms) differential cross-section of the four-lepton invariant mass by ATLAS. The lower panel shows the ratio of prediction to data, where the error bars are statistical and the grey band indicates the systematic uncertainties. The blue and red histograms and uncertainty bands correspond to two different predictions. The coloured histograms in the top panel illustrate the various processes contributing to the final state, which are depicted by the Feynman graphs to the left and the top of the figure. See text for more discussion. There is one data event with $m_{4\ell} = 2.1$ TeV, while 0.4 events are expected from simulation for $m_{4\ell} > 2$ TeV. *Image: ATLAS Collaboration, JHEP 07, 005 (2021), figure modified by authors.*

the only ones allowed for massless vector bosons.[5] It turns out that, were it only for Z boson and photon exchange, the scattering amplitude among

[5]A massive particle of spin 1, as is the case for the W and Z bosons, can have three quantum states of spin projections along its flight direction, with values $+1$, 0 and -1. However, a *real* massless spin-1 particle, as the photon, can have only the two states, $+1$ and -1, which — as the photon always propagates with the speed of light — we shall call helicity states. In the classical Maxwellian description of an electromagnetic wave, the electric and magnetic fields are transverse to the direction of propagation in vacuum: this is the origin of the denomination of the photon polarisation states as transverse. More exactly, in the classical terminology, the helicity state $+1$ corresponds to right circular polarisation of light and the -1 state to the left circular polarisation. Note

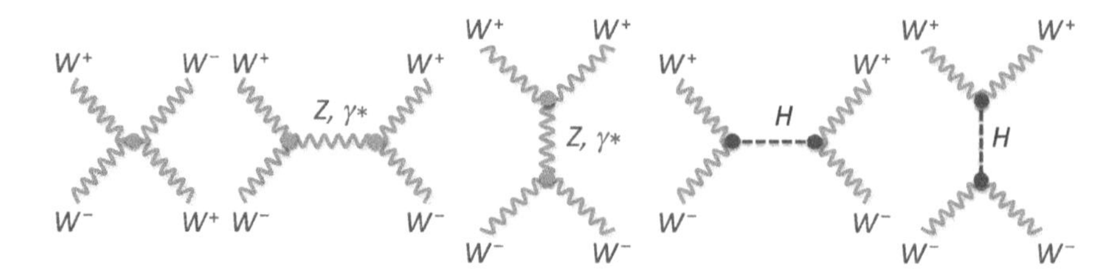

Fig. 12.6. Feynman diagrams for the interaction involving two incoming and two outgoing W bosons. The interaction proceeds either through direct coupling (W-boson scattering) or the exchange of a Z boson or a photon (diagrams on the left). In the Standard Model this scattering also proceeds through the exchange of a Higgs boson (diagrams on the right). The first diagram on the left corresponds to a *quartic* gauge boson coupling. The second and third diagrams involve the *triple* gauge boson couplings WWZ and $WW\gamma$. The first, third and fifth diagrams can also propagate the interaction among W bosons with the same charge.

longitudinally polarised massive weak bosons, such as $W_LW_L \to W_LW_L$, rises with the square of the energy at which the scattering process occurs (invariant mass of the outgoing weak bosons).[6] Above the TeV energy scale the quantum probability for the processes would exceed 100%, which is unphysical. This is called *violation of unitarity* and usually signals that the theory is incomplete, a situation we already met in section 2.4. Historically this realisation signalled the incompleteness of the Standard Model and was an argument in favour of the existence of the Higgs boson. Indeed, the Higgs boson acts as "moderator" to unitarise high-energy longitudinal vector boson scattering if its mass is not too large (below a TeV), which is certainly the case. Figure 12.6 illustrates the Feynman diagrams which come into play here. For longitudinally polarised W bosons, the three diagrams on the left violate unitarity at large energy, which is restored (to all orders in quantum loop) by the two right-hand diagrams. The diagrams are inter-connected through gauge symmetry.

What would happen if the elementary Higgs boson did not exist? Either new, possibly composite scalar states would replace the Higgs boson in the right-hand diagrams of figure 12.6, or the large scattering amplitude would lead to some new form of strong interaction giving rise to the formation of resonances in the W_LW_L invariant mass spectrum.[7] An experimental study

that virtual photons can go off their mass shell and thus acquire unphysical longitudinal polarisation.

[6] A similar unphysical rise would occur in $Z_LZ_L \to Z_LZ_L$ scattering.

[7] Such resonances could occur in a *Technicolour* extension of the Standard Model, a replication of QCD at high energy, which we have briefly introduced in section 4.4. In

of high-energy $W_L W_L$ scattering, or vector boson scattering in general, is thus intimately related to probing the mechanism of electroweak symmetry breaking in the Standard Model.

But how is it possible to distinguish the rare vector scattering process $WW \to WW$ from the much more abundant continuum production of W boson pairs via quark–antiquark fusion $q\bar{q} \to W^+W^-$? Fortunately, there exists a mechanism at the LHC that allows to isolate vector boson scattering from other processes, which is very similar to the vector boson fusion we discussed in relation to Higgs boson production in the previous chapter and in figure 11.4. The corresponding diagram is shown in figure 12.7.

It involves two energetic jets and two weak bosons in the final state. The two jets are emitted in opposite forward directions so they form a large invariant mass. That di-jet mass is a formidable selection variable to discriminate vector boson scattering against the strong-interaction background production of two jets associated with two weak bosons.

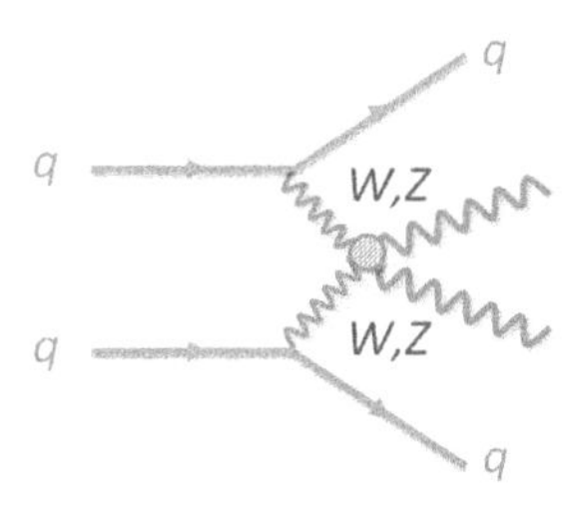

Fig. 12.7. Scattering of two W or Z bosons off each other, where each boson was radiated by a quark from an incoming proton. The dashed circle includes all Standard Model processes, like the ones displayed in figure 12.6 for the case of two W bosons, and possible new physics processes. The two outgoing quarks lead to two energetic jets in opposite forward directions and large di-jet invariant mass.

The process in figure 12.7 has another benefit that, depending on the charge of the two radiating quarks, it allows to produce two W bosons with the same charge, which distinguishes it from conventional strong-interaction di-jet plus associated W-pair production and thus greatly reduces backgrounds. The same-charge requirement restricts the analysis to the (from the left) first, third and the fifth diagrams of figure 12.6. Since leptonic decays of the W bosons are considered, the final state corresponds to two forward jets, two leptons (electron or muon) of the same charge, and missing transverse energy from the two neutrinos. Applying such an event selection and plotting the invariant mass of the two jets gives the distribution shown on the left panel of figure 12.8 for 13 TeV

this theory, the longitudinal W and Z bosons are condensates of so-called technifermions and no elementary scalar particles exist. With the emerging measurements from the Higgs boson sector, in agreement with the Standard Model, Technicolour is disfavoured.

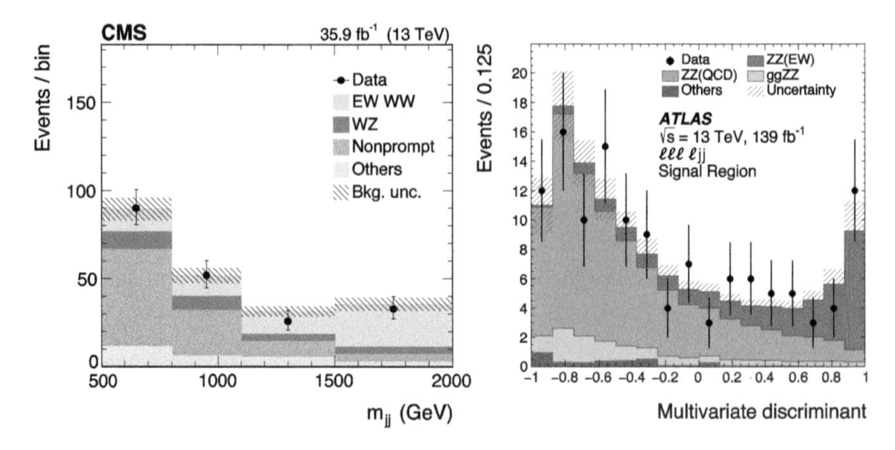

Fig. 12.8. Left: di-jet invariant mass in events selected to search for electroweak production (among which vector boson scattering) of two same-charge $W(\to \ell\nu_\ell)$ bosons and two jets. The dots show the data with statistical error bars, the orange histogram the expected contribution from electroweak production, and the other colours show background contributions. The non-prompt background originates mostly from leptonic decays of heavy quarks, hadrons misidentified as leptons, and electrons from photon conversions. The hatched area indicates the uncertainty on the background contribution. Right: output of a multivariate discriminant trained to enhance signal in a search for electroweak production of two $Z(\to \ell^+\ell^-, \nu\bar{\nu})$ bosons and two jets. Also here, the dots show the data with statistical error bars, the red histogram the expected electroweak contribution, in green the contribution to the same final state via strong interactions, and in dark grey other backgrounds. Both analyses lead to the observation of the respective processes with more than 5σ statistical significance. *Image: CMS Collaboration, Phys. Rev. Lett. 120, 081801 (2018); ATLAS Collaboration, arXiv:2004.10612 (2020); figures modified by authors.*

proton–proton collision data taken by CMS in 2016. The clear excess of the data over background at large mass provides an observation of electroweak $W^\pm W^\pm jj$ production (indicated by the orange colour). Why electroweak production and not vector boson scattering? In effect, vector boson scattering is not the only process contributing to the $W^\pm W^\pm jj$ final state. For example, the same final state can be obtained by inserting a Z boson in-between the incoming W's, just as for the third diagram of figure 12.6. The sum of all diagrams, which includes very large cancellations among them, corresponds to a gauge invariant representation of a process. The theory becomes incomplete if one diagram is extracted without the others. The right-hand panel in figure 12.8 shows the corresponding observation of electroweak $ZZjj$ production from ATLAS, where one Z boson decays to a pair of electrons or muons, and the other decays to a pair of charged leptons or neutrinos. The cross-section for this process is lower than that of electroweak $W^{\pm}W^{\pm}jj$ production, requiring the full Run-2 dataset to pin it down. The signal is indicated by the red histogram of a multivariate discriminator distribution separating the electroweak $ZZjj$ production from strong-interaction backgrounds giving the same final state.

Although these results represent an important milestone, the test of restoring unitarity via the Higgs boson requires the measurement of high-energy longitudinally polarised W boson scattering, which is about a factor of ten smaller than the inclusive signal observed so far. The W polarisation states can only be measured on a statistical basis using the event kinematics. CMS has estimated the fraction of W_L using multivariate techniques to increase the statistical sensitivity. However, due to the yet insufficient number of events, only an upper limit of about 1.2 fb on this cross-section could established, which is about a factor four above the expected value. This study requires the full luminosity to be delivered by the High-Luminosity LHC (HL-LHC) to be conclusive, as we will discuss in chapter 15.

Vector boson scattering as depicted in the leftmost diagram of figure 12.6 involves the *quartic* (or *quadruple*) coupling among four gauge bosons. Other diagrams in figure 12.6 involve *triple* gauge bosons. New physics may enter these processes and alter the measured rates, leading to *anomalous* triple or quartic gauge boson couplings. For example, the Standard Model forbids triple gauge couplings among purely neutral bosons (Z or photon). If such a process is detected it would be a signal of new physics. The search for anomalous triple gauge boson couplings started already at LEP and the Tevatron. Anomalous quartic gauge boson couplings require higher energy and luminosity and can only be probed at the LHC. Such couplings would spoil the gauge invariance of the electroweak $pp \to VVjj$ production and require a high-energy Standard Model extension to restore it.

Another process to study triple and quartic gauge boson interactions is to look for the very rare tri-boson production involving (among others) the diagrams shown in figure 12.9. While the predicted inclusive cross-sections are not that small, for example 505 fb and 350 fb for WWW and WWZ production, respectively, decreasing with the number of Z bosons, the requirement to use leptonic W and Z decays to distinguish them from overwhelming backgrounds reduces the accessible rates to a few femtobarns or less in the fully leptonic channels. Categories from two leptons with the same charge and missing transverse energy for the triple W boson channel and up to six charged leptons for the triple Z boson channel are considered by the analysis. Figure 12.10 shows a five-lepton event in the CMS detector

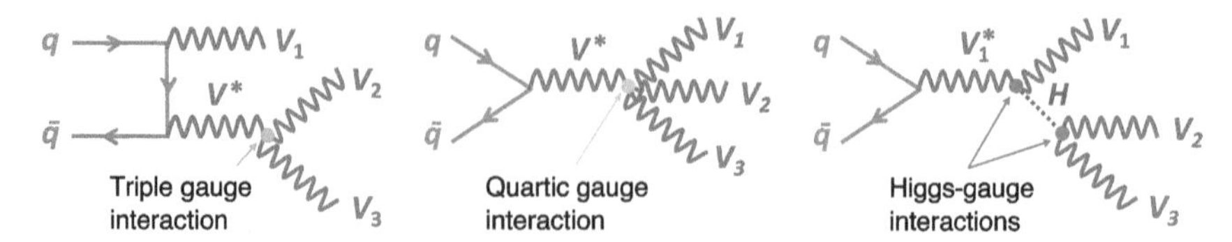

Fig. 12.9. Diagrams contributing to tri-boson production at the LHC. The letters $V_{1,2,3}$ stand for either W or Z bosons. Other processes, not involving gauge or Higgs boson interactions, are not shown.

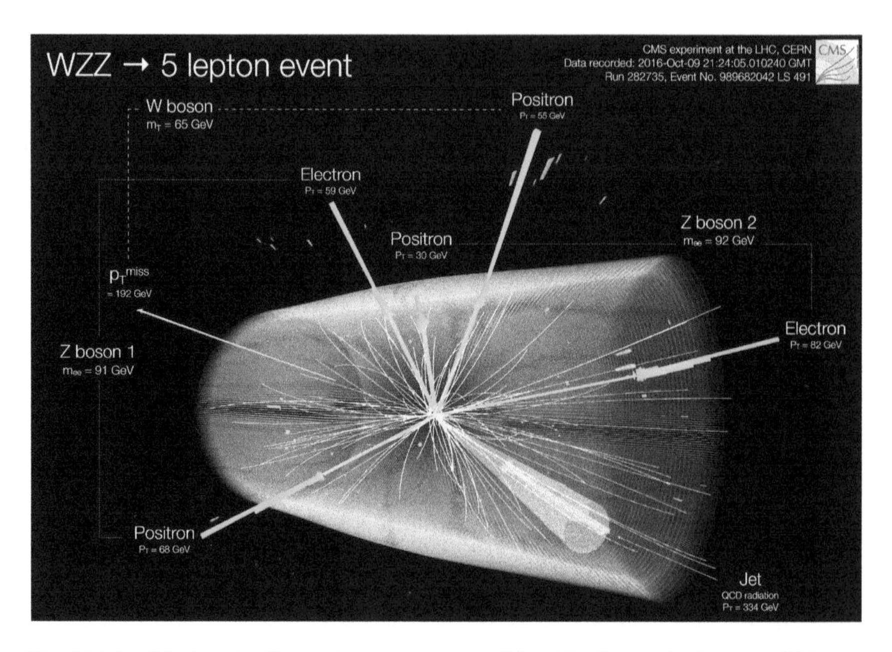

Fig. 12.10. Display of a five-lepton event compatible with the production of a W boson and two Z bosons in the CMS detector. The green tracks and energy depositions in the calorimeter correspond to identified electrons and positrons. Two electron–positron pairs have invariant masses consistent with a Z boson. The pink arrow indicates the direction of the missing transverse energy from the W-boson candidate. *Image: CMS Collaboration, Phys. Rev. Lett. 125, 151802 (2020).*

as a candidate for WZZ production. Results for the four individual modes and their combination are given in figure 12.11. The combination has a statistical significance of 5.7σ and demonstrates the observation of tri-boson production. The individual WWW and WWZ channels exhibit more than 3σ statistical significance. All cross-sections are compatible with the Standard Model predictions within yet large experimental uncertainties.

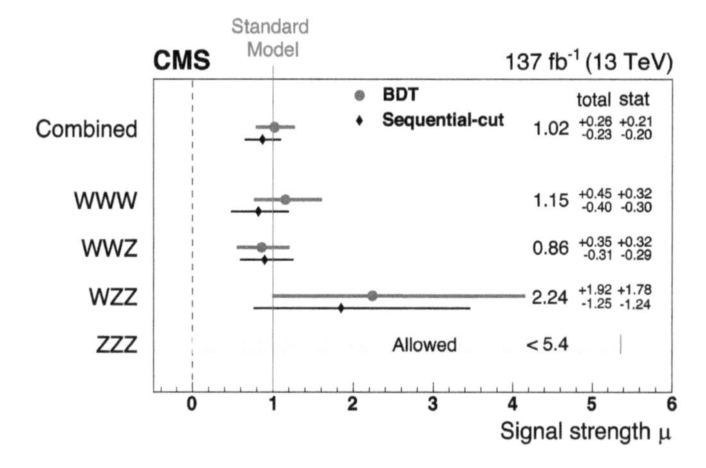

Fig. 12.11. CMS measurement of the signal strength defined by the ratio of measured-to-predicted production cross-sections for the various three-gauge-boson final states. The results are given for two different analysis methods, one using a multivariate classification method in form of a boosted decision tree (BDT, see inset 10.1 on page 295) and one using a traditional set of fixed requirements ("cuts") on topological and kinematic variables. *Image: CMS Collaboration, Phys. Rev. Lett. 125, 151802 (2020), figure modified by authors.*

12.3. Study of the top quark

The top quark is a centrepiece of the LHC scientific program. Despite its high mass of 173 GeV, which makes it as heavy as a gold atom made of 79 protons and 118 neutrons, the top quark is to the best of our current knowledge an elementary particle. Its large mass makes it a sensitive probe of any new physics phenomenon that couples to mass as is the case for the Higgs boson. The top quark gives the largest contributions to the radiative corrections affecting the Higgs boson mass (cf. section 3.7), which needs to be addressed by any new physics model attempting to alleviate the naturalness problem (see following chapter on new physics searches). The top quark decays extremely fast, within half a yoctosecond (that is $\tau_t \simeq 0.5 \times 10^{-24}$ seconds,[8] giving it a natural width of $\Gamma_t = 1/\tau_t \simeq 1.5$ GeV) via the weak interaction almost exclusively into

[8]The top quark decays more than twelve orders of magnitude faster than the next heaviest quark, the b quark (actually a b hadron such as a B^+ meson, as the b quark forms a hadron before decaying via the weak interaction), with mean lifetime $\tau_b \simeq 1.5 \times 10^{-12}$ seconds. The reason for this is that the top quark decays via an on-mass-shell W boson, while the relatively light b quark ($m_b \simeq 4.2$ GeV) decays via an off-mass-shell W boson. In terms of math, we find approximately $\tau_t^{-1} = \Gamma_t \approx G_F m_t^3/36$ compared to

a W boson and a bottom quark. This time is smaller than the typical hadronisation time of $1/\Lambda_{\rm QCD} \sim 3 \times 10^{-24}$ s. The top quark decays within 0.1-femtometre flight length, a length scale where QCD is asymptotically free (the typical hadron size is one femtometre). Top physics at the LHC thus allows to explore the interactions of a bare quark at energies up to and beyond the TeV scale. Weak and strong interaction effects can be reliably predicted using the traditional approaches of perturbation theory with small contributions from hard-to-control hadronic effects. This allows to accurately interpret the data and extract fundamental physical parameters, and to look for deviations from the Standard Model that would indicate new physics. Furthermore, a precise knowledge of top production is an indispensable prerequisite to many new physics searches as top production is a very important background which must be known and controlled very precisely.

12.3.1. *Search and discovery at the Tevatron*

The hunt for the top quark set in after discovery of the bottom quark in 1977. The τ-lepton was discovered two years before, which already hinted at the existence of a third family of elementary fermions. The discovery of the top quark in 1995 by the CDF and DØ experiments was thus not a complete surprise, but it certainly ended up much heavier than initially expected. It is precisely this feature that made it escape observation at previous colliders. Before the Tevatron no machine in the world had sufficient collision energy to produce top quarks.[9] Immediately after the start of LEP in 1989 it became clear that the Z boson did not decay into a top–antitop pair, thus its mass had to exceed $m_Z/2 \simeq 46$ GeV. Subsequently, precision measurements of Z and W-boson properties, whose Standard Model predictions

---

$\tau_b^{-1} = \Gamma_b \approx G_F^2 m_b^5 |V_{cb}|^2/850$, where $G_F \simeq 1.17 \times 10^{-5}$ GeV^{-2} is the Fermi constant and $|V_{cb}| \simeq 0.042$ is a CKM matrix element (cf. section 1.8),

[9]The top quark was intensively searched for at the CERN Sp$\bar{\rm p}$S collider. After a false alarm of a 40 GeV top in 1985, the final result of UA1 using a combination of the muon and di-muon plus jets channels was a lower mass limit of 60 GeV at 95% confidence level, while UA2 with the electron plus jets channel obtained $m_t > 69$ GeV. The observation of neutral meson oscillation in the $B_d^0 \overline{B}_d^0$ system by the ARGUS experiment at the e^+e^- collider DORIS at DESY (Hamburg) in June 1987 allowed to infer that top mass should be larger than 50 GeV. Namely, a B_d^0 meson can transform itself into a $\overline{B}_d^0$ and vice versa through a double W-boson and top-quark exchange in form of a *box diagram*. The frequency of the oscillation, measured by ARGUS, depends directly on the top mass and allows to constrain it.

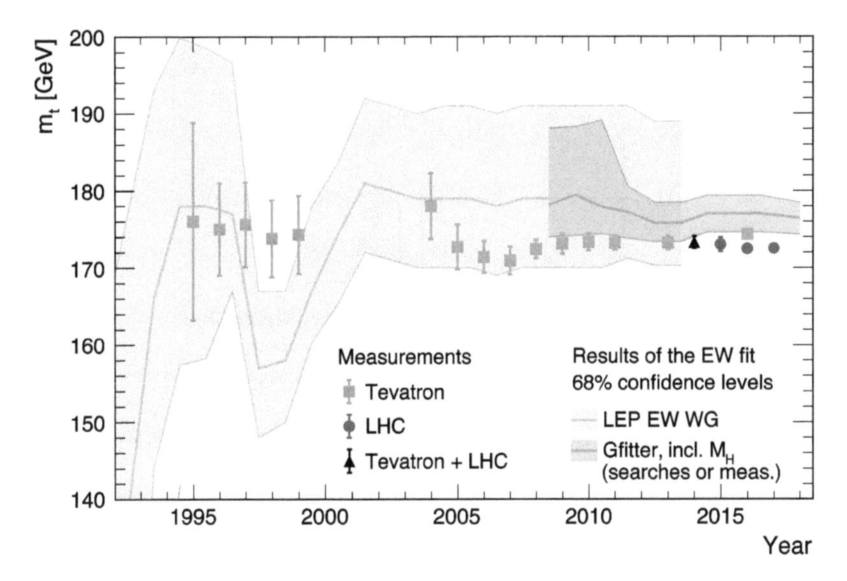

Fig. 12.12. Indirect determinations (blue area) and direct measurements of the top-quark mass (orange, red and black points with error bars) as a function of years. The green area corresponds to indirect determinations taking into account experimental constraints on the Higgs-boson mass. That constraint significantly improved after the Higgs boson discovery in 2012. At the 1995 discovery of the top quark by CDF and DØ, direct and indirect determinations of m_t were in agreement presenting a soaring confirmation of the Standard Model. *Image: Gfitter group, Eur. Phys. J. C 78, 675 (2018), figure modified by authors.*

exhibit a quadratic dependence on the top quark through quantum corrections, allowed to constrain the top-quark mass (figure 2.16). Just before the discovery at the Tevatron, the top mass was that way predicted to be $178 \pm 11\,^{+18}_{-19}$ GeV, where the central value and first uncertainty were for a fixed assumed Higgs-boson mass of $m_H = 300$ GeV and the second uncertainty corresponded to the variation 60 GeV $< m_H <$ 1000 GeV (the Standard Model predictions of the observables depend logarithmically on m_H). Figure 12.12 shows the predictions and the direct measurements of the top-quark mass versus year. The indirect LEP results represented useful guides for the Tevatron experimentalists. The first direct m_t measurement was found in excellent agreement with the indirect determination thereby providing a crucial consistency test of the electroweak Standard Model. After the top quark discovery, the fit to electroweak precision data changed gears and targeted constraints on the Higgs-boson mass (cf. section 11.1 and figure 11.2).

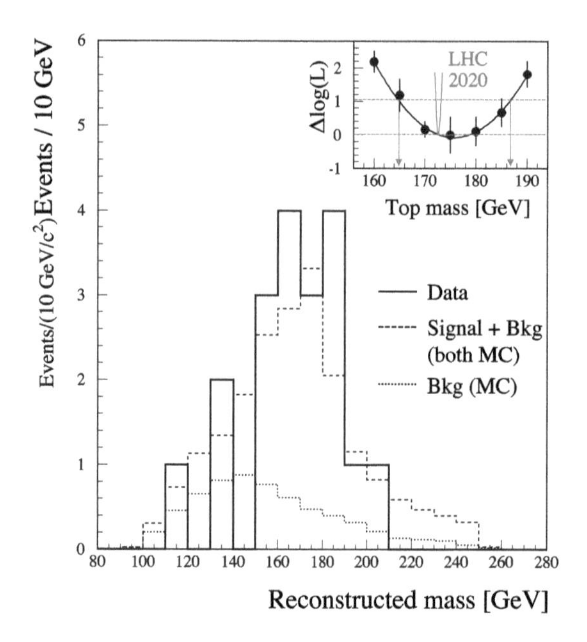

Fig. 12.13. Reconstructed mass distribution in lepton (electron or muon) plus four jets events among which at least one is tagged to contain a b hadron. The solid line shows 20 selected events. The dotted line indicates the expected background, and the dashed line the expected signal plus background distribution (in agreement with the measured distribution taking into account statistical uncertainties). The upper right inset shows the variation of a χ^2-like test statistic as a function of the assumed top-quark mass. The downward arrows indicate the 1σ range with respect to the minimum at 176 GeV. The red parabola indicates for comparison today's best LHC measurement (here ATLAS). *Image: CDF Collaboration, Phys. Rev. Lett. 74, 2626 (1995), figure modified by authors.*

In March 1995, the collaborations operating and exploiting the CDF and DØ experiments at Fermilab's Tevatron proton–antiproton collider announced the observation of the top quark with a mass of $m_t = 176 \pm 8 \pm 10$ GeV for CDF and $m_t = 199\,^{+19}_{-21} \pm 22$ GeV for DØ, where the first uncertainties are of statistical origin and the second represent systematic uncertainties in the measurement processes. The signal significance was 4.8σ for CDF and 4.6σ for DØ. Figure 12.13 shows a distribution of the reconstructed mass measured by CDF for top-quark candidate events with one electron or muon and four jets assigned to the process $p\bar{p} \to t(\to W^+b)\bar{t}(\to W^-\bar{b})$, where one W boson decays as $W \to \ell\nu_\ell$ and the other as $W^- \to q\bar{q}'$, and where the quarks hadronise to jets. Among those jets, the b quarks form a b hadron which can be experimentally tagged. The compatibility of these observations for both cross-section and mass with the

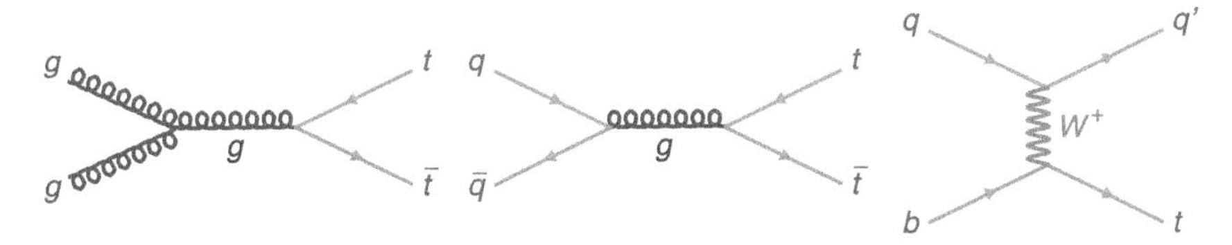

Fig. 12.14. Feynman graphs for top-quark pair production through strong interaction (left and middle) and electroweak single top-quark production (right). Whereas at the LHC, with its higher relative gluon density due to the high collision energy and the valence quark composition in the proton, the gluon fusion process dominates with about 90% over quark fusion (10%), the situation is opposite (15% versus 85%) at the Tevatron $p\bar{p}$ collider. The cross-section of inclusive electroweak production is between 2.2 (Tevatron) and 2.8 (LHC) times smaller than that of strong production.

Standard Model predictions was a formidable success. CDF and DØ confirmed and significantly refined these results using data taken up to 2011, when the Tevatron was closed. The combination of the CDF and DØ measurements including all available data resulted in $m_t = 174.3 \pm 0.7$ GeV, which is twenty times more precise than the first measurements from 1995. The top quark is thus the quark whose mass is most precisely known, down to a few-permil level, thanks to the properties mentioned in the beginning of this section.[10]

The top quark is mostly produced through strong interactions as a top–antitop quark pair via gluon and quark fusion (cf. left two diagrams in figure 12.14) giving rise to relatively clean events in particular if both W bosons decay leptonically. Although in more than one-third of the cases a top quark can also be singly produced via electroweak interactions,[11] it took until 2009 for that process to be observed for the first time at the Tevatron. The reason is the large backgrounds from top–antitop production and W boson plus jets processes, requiring highly optimised analyses based on multivariate methods to detect a significant signal.

[10]We have seen in section 11.5 that the coupling between the top quark and the Higgs boson was established and found to have the expected size within the about 10% experimental precision of the combined fit. While it is clear that the large top-quark Yukawa coupling has important consequences and makes the top quark unique in the Standard Model, one may wonder whether the very fact that it is close to unity has a reason or is an accident of Nature: $g_t = \sqrt{2}m_t/v = (2\sqrt{2}G_F)^{-1} \approx 1$, with $v = 246.22$ GeV^2 and the Fermi constant $G_F = 1.166 \times 10^{-5}$ GeV^{-2}. The answer to this is unknown at present, but we assume most physicist would concede it to be a numerical accident.

[11]This process can either occur via virtual W-boson exchange (this is the dominant single top quark production process both at the Tevatron and the LHC), via quark–antiquark annihilation into a virtual W boson and transformation into a top and bottom quark pair, or via gluon–b-quark fusion into a top quark and a W boson.

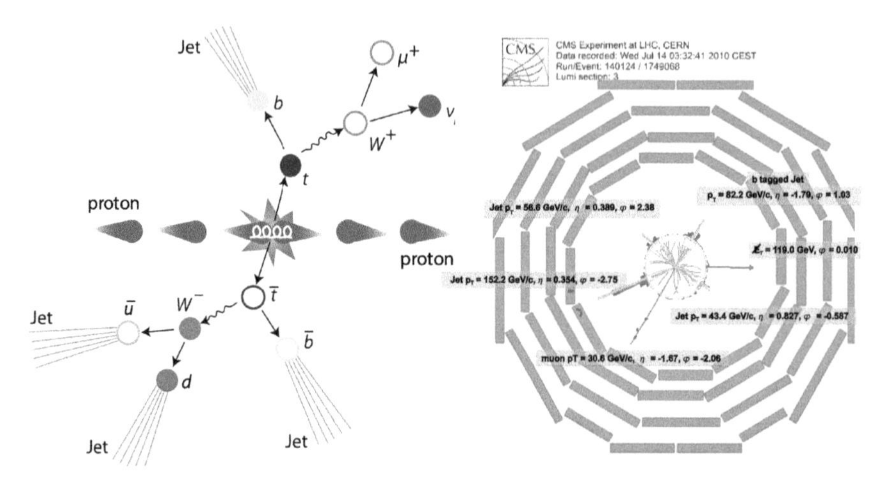

Fig. 12.15. On the left-hand side, schematic representation of top pair production through two-gluon fusion, followed by decays into the "one-lepton channel". Here, one of the two W bosons (lower hemisphere) decays into a quark–antiquark pair producing two jets in the detector, while the other W decays into an antimuon and a muon-neutrino. Each top quark also decays into a b quark, which forms a jet of hadrons among which a b hadron that lives about 1.5 picoseconds before decaying into lighter hadrons with a slightly displaced vertex that can be reconstructed in the tracking detector, allowing the b jet to be *tagged*. The presence of one or two undetected neutrinos reveals itself through missing transverse energy ($E_T^{\rm miss}$). The single-lepton mode has the advantage of a more complete reconstruction as only one neutrino is produced, but it suffers from larger background from more abundant W plus jets production than the di-lepton mode. The right-hand picture displays an example of a one-lepton top-quark candidate event measured with the CMS detector. *Images: Wikipedia (figure modified by authors); CMS Collaboration, CMS PAS TOP-10-004 (2010).*

12.3.2. *Top-quark physics at the LHC*

Top–antitop production at the LHC (see figure 12.15 for an example event) is dominated by gluon–gluon scattering in the initial state. At 13 TeV proton–proton collision energy, the cross-section of 830 pb (300 pb for electroweak single top-quark production) exceeds that at the Tevatron by a factor of 120, justifying the often employed predicate "top factory" for the LHC. Between Run 1 (8 TeV collision energy in 2012) and Run 2 (13 TeV) of the LHC, the top–antitop production cross-section increased by a factor 3.3. During Run 2, at peak luminosity of 2×10^{34} cm^{-2}s^{-1}, the LHC produced 17 top–antitop pairs per second in each ATLAS and CMS! A total of 275 million (115 million) top quarks (top–antitop pairs) were produced in the 140 fb^{-1} dataset of Run 2. Figure 12.16 shows the measured and predicted top–antitop cross-section as a function of centre-of-mass energy

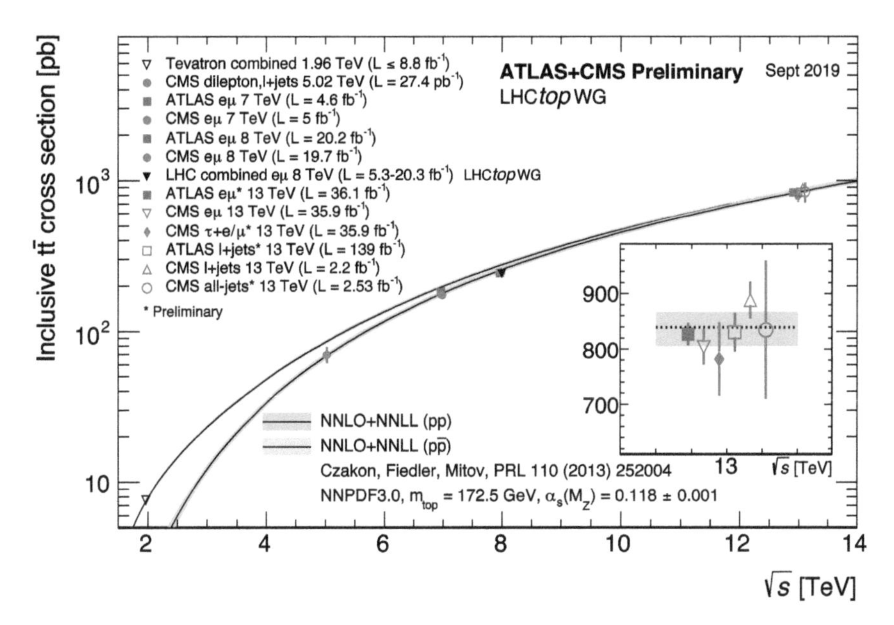

Fig. 12.16. Measurements of the inclusive top–antitop production cross-section as a function of centre-of-mass energy in proton–antiproton collisions at 2 TeV from the Tevatron and in proton–proton collisions at 5, 7, 8 and 13 TeV at the LHC. The cyan band corresponds to the Standard Model prediction of the cross-section for $p\bar{p}$ collisions and the green band that for pp collisions. The size of the bands indicates the theoretical uncertainties. At lower energy, the former cross-section is larger due to the higher valence quark than valence antiquark density in the proton. With increasing energy, gluon–gluon fusion becomes the dominant production mechanism, which is independent of the quark structure. This is also the reason why very high-energy hadron colliders do not need to collide protons with antiprotons, which are difficult to produce. *Image: LHC Top Working Group, 2019.*

at the LHC and the Tevatron (note the logarithmic ordinate scale). Good agreement over two orders of magnitude in cross-section is observed within the experimental uncertainties reaching the remarkable precision of 2.4% (ATLAS, di-lepton channel) and 5% theoretical uncertainty (at 13 TeV). Any new physics that would modify this production rate by (significantly) more than these uncertainties could thus be discerned through such measurements. For example, supersymmetric top-squark pair production (cf. section 13.3) could lead to a top–antitop pair final state with additional missing energy which may remain undetected, thus adding to the measured $t\bar{t}$ cross-section. This is, however, not seen in the data.

With so many events produced and recorded the experimentalists could learn much about the top quark, which is by now the best-known quark flavour (even if it did not yet reveal all its secrets). For example, what is its electric charge? The observation of the decay $t \to bW^+$ suggests that it should indeed be $+2/3$ as predicted. However as it is observed through $t\bar{t}$ pair production, $t\bar{t} \to b\bar{b}W^+W^-$, a value of $-4/3$ with $t_{-4/3} \to bW^-$ and $\bar{t}_{-4/3} \to \bar{b}W^+$ could just as well be compatible with electric-charge conservation. This uncertainty can be removed either by measuring the charge of the W boson and b jet individually, or through the study of $t\bar{t}$ production in association with a photon. According to the Standard Model, the ratio $\sigma(t\bar{t}\gamma)/\sigma(t\bar{t})$ is directly proportional to the square of the top quark's electric charge for a photon radiated from one of the final state top-quark legs. This measurement, however, requires a large data sample so that the first appraisal of the top-quark charge was obtained via determining the W and b charges using 7 TeV Run-1 data from ATLAS. Whereas the W charge is easily reconstructed from the charge of the lepton it decays to, the b-quark charge is estimated by computing a weighted sum of the charges of all particles associated to the b jet. The resulting b-jet charge times W-boson charge distribution is very broad (cf. figure 12.17), albeit sufficient statistics allows to quite precisely determine its mean value. The measurement resulted in a charge q_t of $0.64 \pm 0.02 \pm 0.08$, where the first uncertainty is statistical and the second systematic (dominated by theoretical uncertainties related to the top-quark modelling). The measured charge is in good agreement with the Standard Model value of $+2/3$, and certainly excludes $q_t = -4/3$.

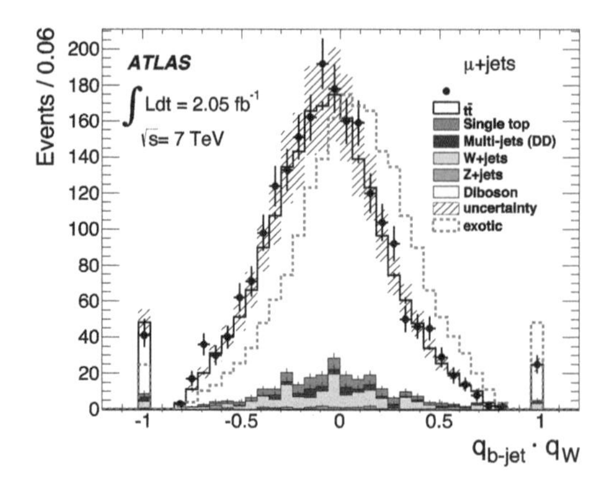

Fig. 12.17. Reconstructed product of the b-jet charge times the $W \to \ell\nu_\ell$-boson charge. The mean value of the data (dots with error bars) exhibits a shift that is in agreement with the Monte Carlo simulation that uses a top-quark charge of $+2/3$. Another simulation with an exotic charge of $-4/3$ is shown in red. It is incompatible with the data. The coloured histograms are non-$t\bar{t}$ backgrounds. *ATLAS Collaboration, JHEP 11, 031 (2013), figure modified by authors.*

Similarly, the decay $t \to bW^+$ with angular momentum conservation suggests that the top quark should be a spin-1/2 matter particle as predicted. A direct proof requires the study of angular distributions of and correlations among the final state particles. Such studies have been undertaken both at the Tevatron and the LHC with results comforting the Standard Model. The interpretation of these measurements is complex as the top quarks are reconstructed through decay products where at least one of the W bosons decays to a charged lepton and a neutrino which leads to missing energy. When the other W boson decays to two quarks, the final state of $t\bar{t}$ production will have one charged lepton, missing transverse energy, and four jets, two of which are b jets. The assignment of these jets to the initial top quarks goes via a reconstruction of their combined invariant masses, but this is a statistical process, which sometimes fails. The comparison with the Standard Model proceeds in general through a Monte Carlo simulation which includes to the best possible the theory of top-quark production and decay as well as a simulation of the detector and the full reconstruction chain. Uncertainties in this simulation process lead to systematic effects in the measurement (cf. section 10.1.4), the dominant of which stem from the modelling of the production and decay processes. All these studies indicate, so far, that the top quark is as Standard Model like as it can be.

The large number of top quarks produced at LHC allows the ATLAS and CMS physicists to not only measure inclusive cross-sections, but to also determine differential cross-sections in all kinds of kinematic variables. These measurements are crucial probes of top-quark modelling and exhibit sensitivity to new physics effects (such as the above-mentioned top-squark production). Figure 12.18 shows a measurement of the differential cross-section of the top-quark transverse momentum in top–antitop events with two leptons by CMS. An experimental precision of about 5% per measurement bin is achieved, which is sufficient to show that the theoretical calculations predict too hard spectra, one of which is clearly incompatible with the data. A reason for the mismodelling could lie in the parton density functions used or in the modelling of soft QCD effects (such as the radiation of additional partons, the so-called *parton shower*) or of the formation of hadrons in a jet (*hadronisation*), which cannot be calculated from first principles.

Just as at the Tevatron, given its pivotal importance for electroweak and Higgs-boson physics, large efforts are invested by the LHC experimenters to further improve the precision on the top-quark mass. There are two ways to measure the top-quark mass at the LHC: (i) via kinematic reconstruction from the measurement of the invariant mass of the top-quark decay products, or (ii) via a precise top-antitop cross-section measurement and

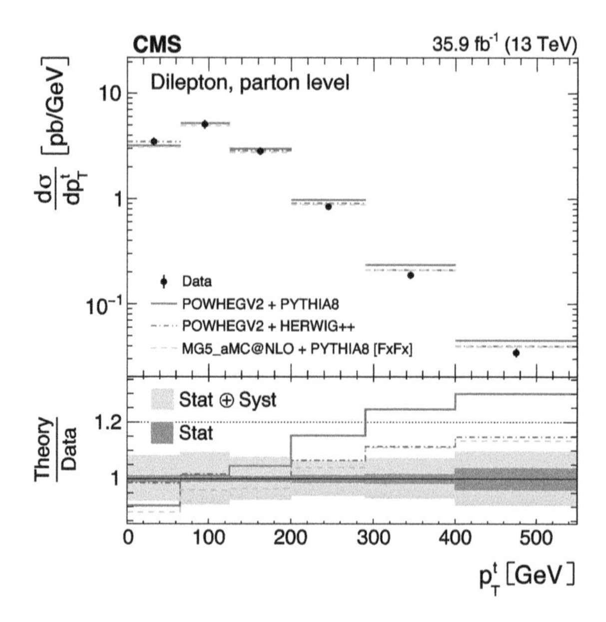

Fig. 12.18. Measurement of the differential cross-section of top–antitop production as a function of the top-quark transverse momentum (p_T). The data correspond to the markers whereas the lines correspond to various predictions obtained from Monte Carlo generators. The first generator given in the label (e.g., POWHEGV2) is used for the calculation of the hard top–antitop production including possible additional hard jet radiation, and the second generator (e.g., PYTHIA8) is used for the simulation of additional jet activity from parton showers and their hadronisation (see inset 10.4 on page 306). The lower panel shows the ratio of the predictions to the data. The orange band indicates the total uncertainty of the measurement. None of the predictions fully reproduces the data. *Image: CMS Collaboration, JHEP 02, 149 (2019).*

exploiting the dependence of the prediction on the top-quark mass. The latter method is less precise and depends on the assumption that no new physics contributes to the measured cross-section. On the positive side, it draws in a cleaner definition of the top-quark mass parameter (see inset 12.1 below).

The most precise kinematic measurement is obtained from the reconstruction of the top decay fragments in the one-lepton $t\bar{t}$ channel where one top quark decays fully hadronically, $t \to bW^+$ followed by $W \to q\bar{q}'$, whilst the other decays in a semileptonic mode ($t \to b + W \to b + \ell\nu_\ell$). The event is cleanly tagged by the semileptonic top-quark decay, and the mass of the hadronically decaying top quark is reconstructed from the invariant mass of the three jets generated by the three quarks b, q and $\bar{q}'$. Figure 12.19 shows a corresponding three-jet mass distribution measured by CMS, with

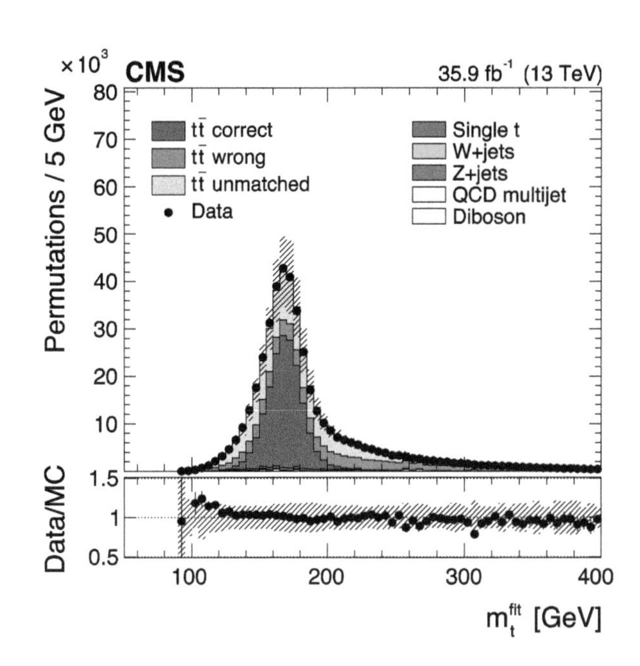

Fig. 12.19. Distribution of the three-jet invariant mass resulting from a fit to decay products of $t \to b + W \to b + 2$ jets in selected one-lepton $t\bar{t}$ events. Since the event contains at least four jets among which are two b jets, there are two possible three-jet combinations and two solutions for the neutrino momentum reconstruction, leading to a total of four permutations per selected event. The fraction of correct assignments (permutations) is increased by adding a kinematic requirement, which has been applied in this figure. The points correspond to the experimental data, the coloured histograms to predictions obtained through Monte Carlo simulation. The colours indicate properly matched, wrongly matched, and unmatched permutations. *Image: CMS Collaboration, Eur. Phys. J. C 78, 891 (2018), figure modified by authors.*

a well-recognisable top peak around 170 GeV. The measured distribution is then compared with properly calibrated Monte Carlo simulation and the m_t parameter in the simulation adjusted to best reproduce the data. The result from that analysis, $m_t = 172.25 \pm 0.08 \pm 0.62$ GeV, is like all top-mass measurements at the LHC dominated by systematic uncertainties mostly related to the theoretical physics modelling and to the knowledge of the jet energy calibration. At the end of Run 2, the kinematic top-quark mass measurements of ATLAS and CMS have each achieved a combined uncertainty of about 0.5 GeV and are in good agreement. The results are about 2 GeV lower than the combined measurement from the Tevatron, exhibiting a tension that needs to be understood before combining the two. The list of recent measurements by ATLAS and CMS is summarised in figure 12.20.

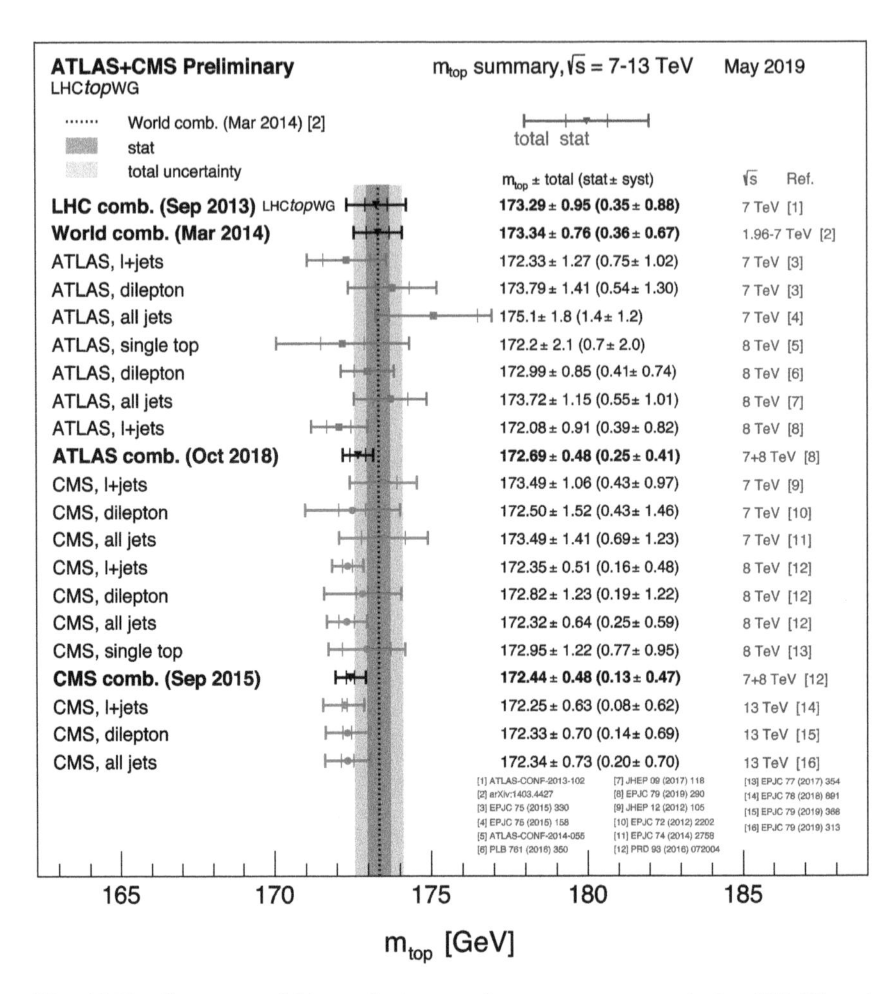

Fig. 12.20. Summary of kinematic top-quark mass measurements by ATLAS and CMS up to 2019, using different types of processes and final states. The ATLAS and CMS combinations of Run-1 measurements, $m_t = 172.69 \pm 0.48$ GeV and $m_t = 172.44 \pm 0.48$ GeV exhibit agreement and comparable precision. *Image: LHC Top Working Group, 2019*

Further progress in precision requires improved calibrations and theoretical modelling.

Another key parameter is the top quark's natural width, Γ_t, which is related to the particle's lifetime and decay modes (see inset 2.1 on page 73). It is driven by the fast weak-interaction decay to bW. Other decays, possibly resulting from new physics, might alter the width, so that it is important to precisely measure it. In the Standard Model, theoretical calculations predict a value for the decay width of 1.32 GeV for a top-quark mass of 172.5 GeV. Due to the limited resolution of about 20 GeV in the reconstructed three-jet mass, it is difficult to measure the small contribution from the natural width rendering a precise measurement very challenging. ATLAS has performed an analysis using the full Run-2 dataset and employing instead of the one-lepton $t\bar{t}$ mode a channel where both top quarks decay semileptonically. The measurement of the invariant mass of the leptons and b-jets resulted in a top-quark width of $\Gamma_t = 1.9 \pm 0.5$ GeV, in agreement with the Standard Model.

Inset 12.1: Top-quark mass: what do we really measure?

Contrary to the masses of leptons, quark masses are not physical observables. This is due to the confinement property of QCD, which prevents quarks to evolve as free particles (see inset 1.13 on page 63), and, in the case of the top quark, to its very short lifetime. Akin to the coupling constant α_S, quark masses are simply parameters of the QCD Lagrangian. In the Standard Model, they are related to the Yukawa coupling strengths to the Higgs field. In quantum field theory, quark masses are thus prone to quantum corrections like those illustrated in the following QCD Feynman diagrams, which are used to calculate the quantum correction to the bare mass parameter (see section 3.7) at a given order in perturbation theory (the so-called quark *self-energy*).

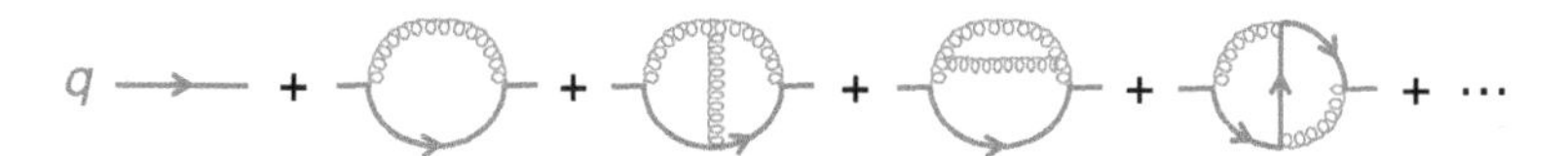

These diagrams with loops are divergent (infinite) and, as usual, a renormalisation procedure (see again inset 1.13) is required to obtain finite values for the physical observables. Two renormalisation schemes are often employed in this case. The first one is the $\overline{\text{MS}}$ scheme, which leads to a mass that varies with the energy scale of the process under study, similar to the variation of the coupling constants (see section 1.10). The second scheme is the on-shell scheme in which the mass parameter represents a scale independent *pole mass*, which can be identified with the physical mass in the limit of an infinite lifetime (stable particle). The pole mass is the one used in many calculations such as for the consistency test of the Standard Model (see section 12.5) or for evaluating the stability of the Universe (inset 12.2). The relation between $\overline{\text{MS}}$ mass and pole mass is known to high precision in QCD perturbation theory as a function of α_S. For example, the top-quark pole mass is about 10 GeV larger than the corresponding $\overline{\text{MS}}$ mass at the energy scale equal to the top-quark mass.

The question, not entirely solved at the moment, is whether the kinematically measured top-quark mass at the LHC and previously the Tevatron corresponds to a pole mass and, if not, how much it differs from it and how large is the associated uncertainty. The analyses adjust the top-quark mass parameter incorporated in the Monte Carlo simulation to best reproduce the reconstructed invariant mass distribution of the top decay products in data. This mass parameter depends on the modelling of the top-quark pair production, their decays, as well as the fragmentation and hadronisation of the jets formed, and the amount of additional jets in the event. The ambiguity between this "Monte Carlo mass" used in the measurement process and the theoretical pole mass is being intensively discussed among experts. Since this difference is due to non-perturbative effects, it may be that it is already accounted for in corresponding uncertainties applied by the experimentalists. The additional systematic uncertainty is estimated between zero (already included), over a few hundred MeV (typical size of non-perturbative hadronic corrections), up to a GeV as promoted by some theorists. Since these numbers are of similar order of magnitude as the present experimental uncertainty, it needs to be addressed to allow further progress in the precision of the top-quark mass.

Exploiting other m_t-dependent processes like the $t\bar{t}$ pair production cross-section provides a well-defined mass definition, but its uncertainty presently exceeds 1 GeV and is thus not competitive with the kinematic method. With the advent of a new high energy e^+e^- collider (see chapter 15), capable of measuring the variation of the production cross-section around threshold centre-of-mass energy for $t\bar{t}$ pair production (around 350 GeV), the top-quark mass could be measured with a precision of 50 MeV or better.

12.4. Precision measurements of the *W*-boson mass

The discovery of the W and Z bosons in the early eighties at the Sp$\bar{\text{p}}$S collider at CERN (see section 2.3) has led to the first measurements of the

W mass, with a final uncertainty of about 350 MeV. Once the experiments at the LEP e^+e^- collider (section 2.4) in the 1990s provided very precise measurements of the Z-boson mass — ultimately reaching a 2.1 MeV (i.e. 0.02 permil) uncertainty,[12] it became clear that huge efforts had to be undertaken to bring the W mass measurement precision in a similar ballpark, eventually by using the Z boson as a standard candle to calibrate the lepton energy measurement in the detectors.[13]

While LEP was running above 180 GeV centre-of-mass energy (LEP-2 between 1996 and 2000), about forty thousand W^+W^- pairs were recorded (figure 12.21) and a 33 MeV uncertainty (about half-half statistical and systematic) on the W-boson mass was reached from a combination of measurements of the W^+W^- threshold cross-section, the leptonic decay spectrum of the W boson, and via the kinematic reconstruction from the decay products. Despite a more difficult experimental environment, the Tevatron experiments CDF and DØ made use of the large numbers of W and Z bosons produced in the proton–antiproton collisions — a few million with decays to an electron or a muon — to improve the precision down to 16 MeV. The main limitation in these measurements was the systematic uncertainty from the lepton energy calibration, which is itself statistically limited by the number of leptonically decaying Z bosons (about a factor ten less than the number of

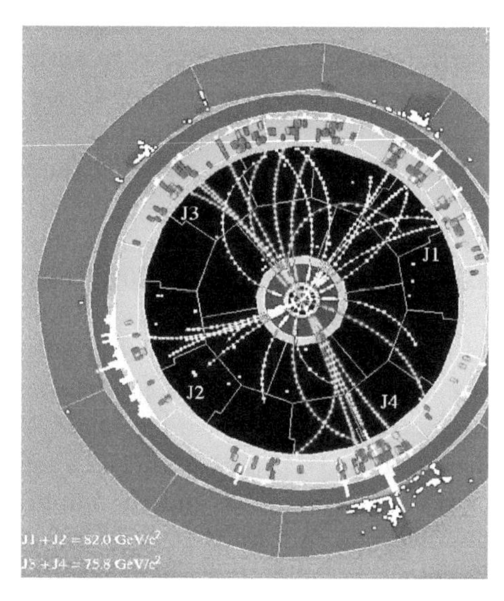

Fig. 12.21. Transverse view of an $e^+e^- \to W^+W^-$ event measured by the ALEPH detector at LEP. Each W boson decays to a pair of jets (indicated by the same colour) whose invariant mass is compatible with that of the W boson. Ten thousand W^+W^- events were recorded by each LEP experiment. *Image: ALEPH Collaboration, CERN.*

[12]This tremendous precision was achieved thanks to the very precise calibration of the LEP beam energies, not from the absolute calibration of the lepton energy measurements in the LEP detectors.

[13]The 350 MeV uncertainty mentioned above was reached by the UA2 experiment precisely using as a calibration the very accurate Z mass measured at LEP.

corresponding W bosons). The refinement of the Tevatron data analysis is still ongoing and further improvements are expected.

Although the LHC benefits from an even larger number of W and Z bosons and new generation detectors, it suffers from two majors drawbacks compared to the Tevatron: a larger pile-up interaction rate, which affects the quality of the missing transverse energy measurement needed to compute the W transverse momentum, and a symmetric proton–proton initial state so that the W and Z are produced via the annihilation of a valence quark from one proton and a sea antiquark from the other proton.[14] As a consequence, more sea antiquarks, such as strange, charm, and even bottom, contribute to the production and do so differently between the W and the Z bosons, thus complicating the use of the Z boson as a calibration channel.

In defiance of these difficulties and after a painstaking effort, ATLAS released towards the end of 2015 the first measurement of the W mass at the LHC, using the data collected at 7 TeV at the start of Run 1 (hence with reduced luminosity but also pile-up). The measurement is based on the simultaneous exploitation of the lepton transverse momentum and the W transverse mass measurements,[15] and uses both the muon and electron decays of the W boson. Figure 12.22 gives an example of the large number of W bosons available and the impressive agreement with the simulated distributions. A total of 14 million reconstructed W bosons and 1.8 million Z bosons as calibration channel were used in the measurement. ATLAS found $m_W = 80\,370 \pm 7 \pm 11 \pm 14$ MeV, where the first uncertainty is statistical, the second due to experimental systematic uncertainties, and the third from theoretical modelling effects. The total uncertainty amounts to 19 MeV. The result is compatible with the previous measurements and has a competitive precision with the Tevatron. It is also compatible with the indirect prediction of the W-boson mass from the global electroweak fit, discussed in the next section. A summary of W mass measurements is given in figure 12.23.

[14]At the Tevatron and $Sp\bar{p}S$ colliders, thanks to their proton–antiproton initial state, the bulk of the W and Z bosons were produced from the annihilation of a u or d valence quark from the proton and a corresponding valence antiquark of the antiproton — see inset 2.4 on page 88.

[15]See inset 2.3 on page 86 for a discussion of invariant and transverse masses. The transverse mass is calculated from the visible decay products as follows: $m_T = \sqrt{2p_T^{\ell}E_T^{\text{miss}}(1 - \Delta\phi)}$, where $\Delta\phi$ is the difference in azimuth between the lepton and the missing transverse energy directions.

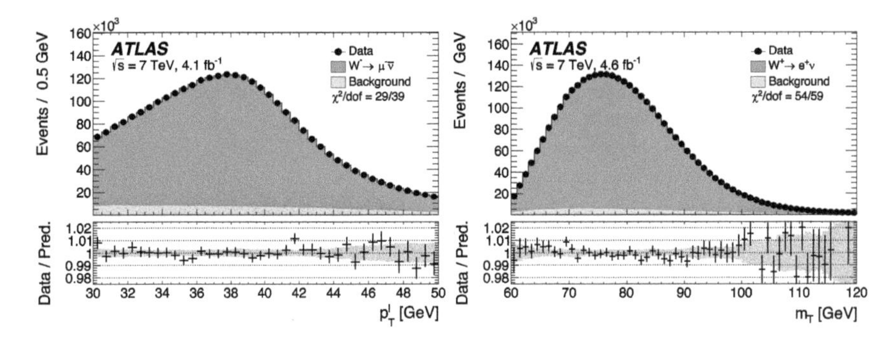

Fig. 12.22. Measurement of the W mass by ATLAS. The two figures show the distributions of lepton transverse momentum for $W^- \to \mu^- \bar{\nu}_\mu$ candidates (left panel) and of the transverse mass for $W^+ \to e^+ \nu_e$ candidates (right). The data are compared to the Monte Carlo simulation including signal and estimated background contributions. Meticulous detector calibration and physics-modelling corrections are applied to the simulated events. The lower panels show the corresponding ratios of data to prediction with statistical uncertainties given as error bars and systematic uncertainties as grey bands. The test statistic χ^2 divided by the number of degrees of freedom (equal to the number of histogram bins minus one) on each figure indicates the compatibility between data and simulation. A ratio of about one is expected for good statistical compatibility. *Images: ATLAS Collaboration, Eur. Phys. J. C 78, 110 (2018).*

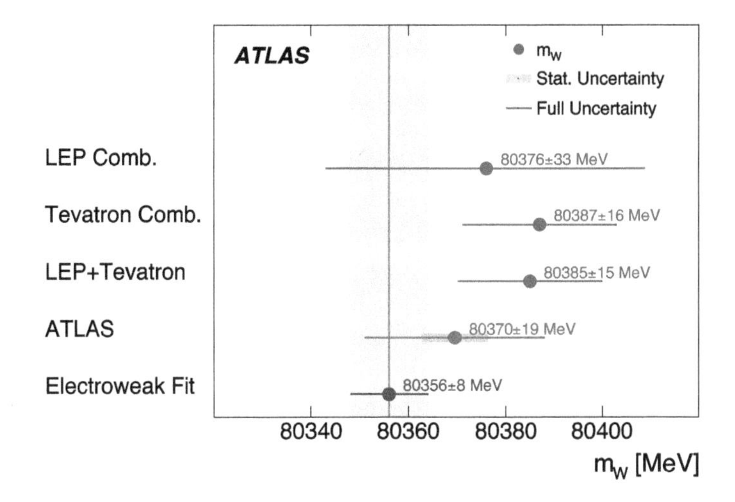

Fig. 12.23. The ATLAS W-boson mass measurement from 2015 compared to the LEP and Tevatron combinations (and their combination), and to the Standard Model prediction from the global electroweak fit using measurements of the top-quark and Higgs-boson masses as well as other electroweak observables (see discussion in section 12.5). *Image: ATLAS Collaboration, Eur. Phys. J. C 78, 110 (2018).*

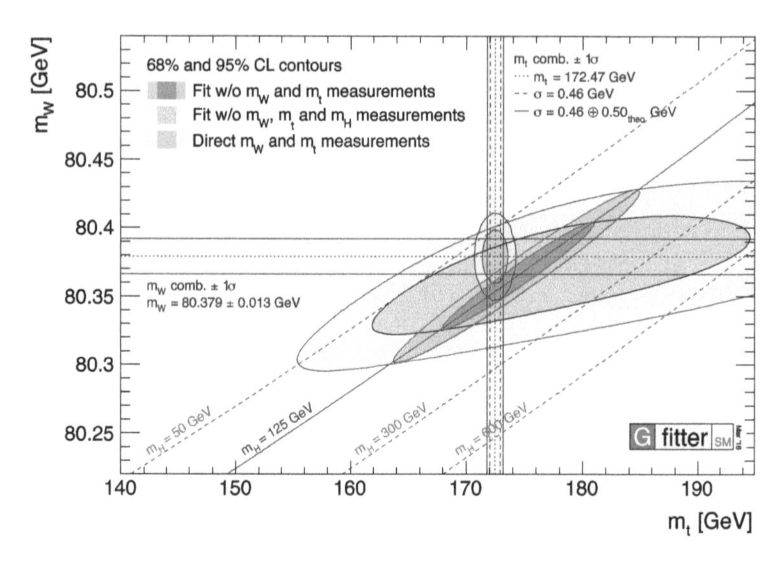

Fig. 12.24. Correlations between the masses of the W boson, the top quark, and the Higgs boson in the Standard Model. The contours in grey indicate the predicted values of W-boson and top-quark masses as indirectly deduced from quantum corrections as those shown in figure 2.16. The constraint is improved to the blue region when including the Higgs-boson mass measured by ATLAS and CMS. The green bands and green ellipse indicate the direct m_W and m_t measurements. The diagonal lines indicate the central contours of the prediction for different Higgs-boson masses. Values larger than 160 GeV would have been incompatible with the direct m_W and m_t measurements (cf. figure 11.2). The agreement between indirect measurements, resulting from theoretical predictions, and direct measurements is remarkable and represents a powerful consistency test of the Standard Model. *Image: Gfitter Group, Eur. Phys. J. C 78, 675 (2018).*

As the LHC operates at increasing luminosity and pile-up, which significantly complicates the W mass measurement, dedicated data-taking periods with reduced instantaneous luminosity and a pile-up of about two interactions per bunch crossing have been performed during Run 2. Although the integrated luminosity and thus the number of W and Z bosons recorded is much smaller compared to the standard runs, a few million W bosons are sufficient to improve the mass measurement precision down to, possibly, 10 MeV, if the modelling uncertainties can be controlled.

12.5. Standard Model self-consistency test

The importance of a precise measurement of top-quark, W and Higgs boson masses in view of testing the internal consistency of the electroweak sector of the Standard Model is well illustrated in figure 12.24. It shows, in the plane of the W-boson mass versus the top-quark mass, the present agreement among the direct measurements (for which combinations of all available experimental results were made) and the prediction from the Standard Model. That prediction is computed without using the direct m_W and m_t measurements to allow for an independent comparison. The difference between the grey and blue ellipses is the addition of the Higgs-boson mass measurement, which dramatically improves the precision of the fit. Direct measurements and prediction from the fit are in good agreement. The fit would suggest a slightly smaller m_W value, as also seen in figure 12.23. With the future data from the LHC it might be possible to reduce the present experimental uncertainties by a factor of two. It may even be possible, via an accurate measurement of the Z polarisation, to increase the precision on the weak mixing angle, $\sin^2\theta_W$, parameterising the weak coupling constant, over that achieved at LEP,[16] making the Standard Model test yet more incisive. Although this perspective justifies every experimental effort at the LHC, the real revolution for this physics would come with a future electron–positron collider as discussed in section 15.2.1.

Inset 12.2: Higgs boson, top quark, their masses and the stability of the vacuum

The measured Higgs-boson and top-quark masses have the intriguing consequence that the vacuum of the Standard Model is not stable would one extrapolate the Standard Model all the way to the Planck scale without injecting new physics. This peculiar situation is illustrated by the figures below. They show the regions of stability (green), instability (red), and metastability (yellow) of the Standard Model vacuum as a function of the Higgs-boson mass on the abscissa and the top-quark mass along the ordinate. Large values of the Higgs-boson mass correspond to the region of non-perturbativity where field-theory calculations based on Feynman diagrams result in amplitudes violating unitarity. The figure on the right zooms into the small rectangle on the left-hand picture in the vicinity of the measured mass values.

[16]A preliminary effective weak mixing angle measurement from ATLAS in 2018 gave a result with an uncertainty competitive with Tevatron measurements, but yet a factor of two less precise than the combination of all LEP and SLD measurements at the Z-boson pole.

One notes that the experimentally favoured region lies in the narrow metastability band. What does this mean?

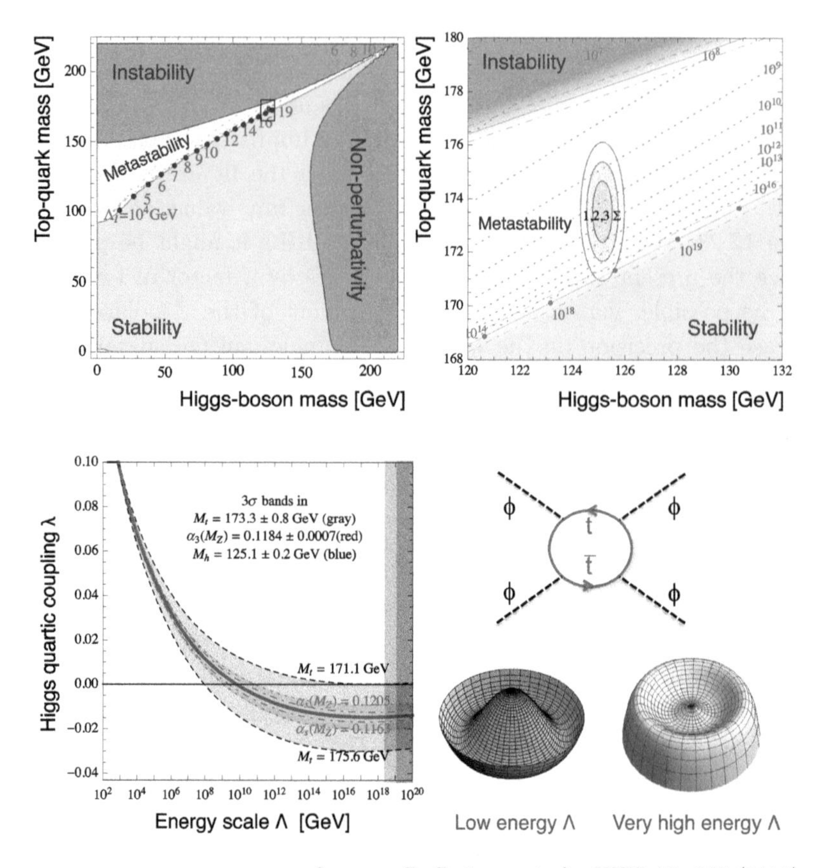

Images: D. Buttazzo et al., JHEP 12, 089 (2013).

Let us recall that the vacuum is defined as the minimum of the Higgs potential, $V(\phi) = \mu^2|\phi|^2 + \lambda|\phi|^4$, where ϕ is the Higgs field, $\mu^2 = -m_H^2/2$ the Higgs mass term, and $\lambda \simeq 0.13$ the Higgs quartic coupling. This potential has its minimum at $\sqrt{2} \cdot |\phi| = v = \sqrt{-\mu^2/\lambda}$ and takes the shape of a Mexican hat for $\mu^2 < 0$ and $\lambda > 0$. As for all couplings, λ evolves with the energy scale Λ, mostly due to the coupling of the Higgs boson to the top quark shown on the lower right-hand graph above. Beyond about $\Lambda \sim 10^{10}$ GeV — the exact value depends on the top-quark mass (see lower left figure above) — $\lambda(\Lambda)$ turns negative and the Higgs potential takes the shape of a doggy's pot in the complex ϕ plane (figure on the lower right above). The potential $V(\phi)$ acquires a second, infinitely deep (negative) minimum as the value of $|\phi|$ goes to infinity. Only (yet unknown) quantum gravity effects appearing at these extremely high energies might prevent such an infinite downward slide.

What is the relevance of this ancillary deep (and diverging) minimum at large scale Λ, and what are the characteristics and consequences of the physics at an extremely high-energy scale to our present-day cooled-down world? In quantum field theory all energy scales, however large, are accessible through virtual processes, even those for which λ becomes negative. The transition probability through quantum tunnelling between our vacuum — in the local minimum of the Higgs potential — and the infinitely deep high-energy vacuum is, albeit very small, non-zero. Such a transition could lead to a local formation of a high-energy vacuum bubble, which would spread out with the velocity of light until it would engulf the entire Universe. If this probability, integrated over all space–time (taking into account the present age of the Universe, about 13.8 billion years), turns out to be less than one, it would mean that our Universe has survived until now all quantum fluctuations. This is what is understood by a *metastable vacuum.*

Physicists are questioning whether there is a fundamental reason behind the apparent location of the Universe in the narrow metastability band rather than in the large lawn of stability. Could a more precise measurement of the top-quark mass give further insight? Could one establish a link between the Standard Model and quantum gravity at the Planck scale? Some theorists conjecture that the evolution of the potential close to the Planck scale could be at the origin of cosmological inflation. All this is of course highly speculative as we do not know whether the Higgs potential evolves unshaken by new particles or forces up to the Planck scale. We may even doubt that this is the case.

12.6. Forward physics: the TOTEM, ATLAS-ALFA/AFP, CT-PPS, and LHCf experiments

An early measurement performed at any new hadronic collider inaugurating a new collision energy is that of the elastic and inelastic proton–proton interaction cross-sections, whose sum is referred to as the total cross-section. The measurement of the inelastic proton–proton cross-section at the LHC is also of practical importance for the hard-collision physics, as it determines the rate of events in the crossing of the proton bunches, that is, the pile-up rate. One distinguishes two main types of inelastic collisions: *diffractive* and *non-diffractive* processes. Diffraction includes single-diffractive dissociation, double-diffractive dissociation, and central diffractive production (see inset 12.3 below). Diffractive collisions are characterised by small scattering angles and by preserving the structure of at least one of the two incident particles (except in case of double diffractive dissociation), and no exchange of colour or, more generally, of additive quantum numbers among the incoming particles. On the contrary, non-diffractive inelastic collisions alter the proton structure and involve colour exchange among the colliding partons. All hard collision types leading to the production of heavy particles or high-transverse momentum objects described so far in this book belong to this category. The elastic cross-section amounts to about one-quarter of the total cross-section. In forward physics studies, it is the measurement of this component that allows to precisely determine the total cross-section as we will see in the following.

Inset 12.3: Total proton–proton cross-section, diffraction and related phenomenology

Proton collisions can be classified as elastic (el), diffractive, and non-diffractive (ND) (to which belong the hard scattering interactions). Elastic and diffractive collisions are characterised by the exchange of quanta with the quantum numbers of the vacuum, i.e. without exchange of colour or any additive quantum number.

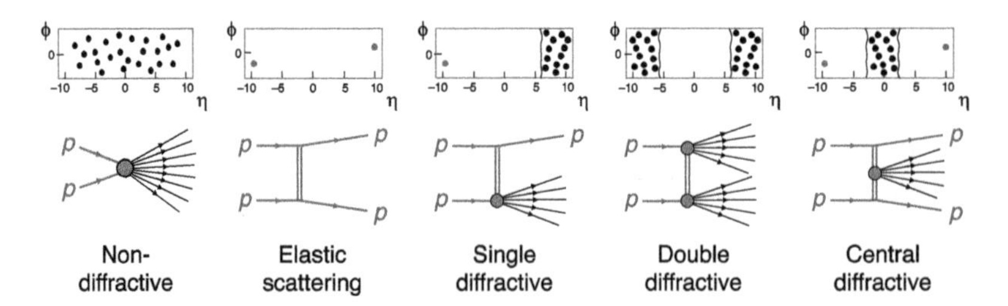

Diffractive collisions can be further subdivided into single diffractive dissociation (SD) in which one of the two incident protons is dislocated and the other remains intact, double diffractive dissociation (DD) where both protons are dislocated, and central diffractive production (CD) where the two incident projectiles remain unscathed, but a cluster of particles is produced centrally (see figure above).[a] Diffractive processes are characterised by large rapidity gaps $\Delta\eta$ (see inset 10.5 on page 307) as illustrated by the η–ϕ distributions of the produced particles (essentially hadrons) sketched on top of the diagrams in the figure.

Diffractive phenomena are long-range, large-impact-parameter QCD processes. They cannot be described in a perturbation theory approach and are usually modelled by means of a specific phenomenology (initially based on the so-called *Regge Theory*, developed in the early 1960s before the advent of QCD). The (quasi or effective) particle exchanged between the colliding protons is called *pomeron*,[b] drawn as double vertical lines in the above diagrams. It can be thought of as a colourless collection of (at least two) gluons.[c]

The total proton–proton cross-section can be decomposed as follows:

$$\sigma_{\text{tot}} = \sigma_{\text{el}} + \sigma_{\text{inel}}, \text{ with } \sigma_{\text{inel}} = \sigma_{\text{SD}} + \sigma_{\text{DD}} + \sigma_{\text{CD}} + \sigma_{\text{ND}}.$$

At 13 TeV collision energy at the LHC, the elastic cross-section, $\sigma_{\text{el}} \sim 30$ mb, is responsible for about 27% of σ_{tot} with the inelastic cross-section, $\sigma_{\text{inel}} \sim 80$ mb, contributing to the rest (cf. figure 12.29). Diffractive contributions cannot be easily measured directly as most of the produced particles escape the detector in the very forward region. They can be inferred from rapidity gap studies and theoretical considerations, leading to the contributions illustrated in the pie chart on the right.

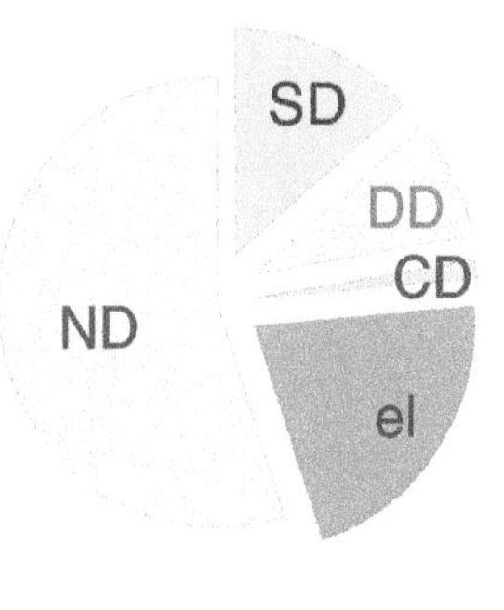

The total cross-section, σ_{tot}, and the elastic differential cross-sections are related through a quantum mechanical principle based on unitarity — the *optical theorem*[d] — connecting the total interaction cross-section σ_{tot} and the imaginary part of the forward elastic scattering amplitude $A(t = 0)$, where $t = |p_{\text{incident}} - p_{\text{scattered}}|^2$ is the momentum transfer squared. In natural units ($\hbar = c = 1$), it reads

$$\sigma_{\text{tot}} = \frac{4\pi}{|\mathbf{p}|} \cdot \text{Im}A(0)\,, \tag{12.1}$$

where $|\mathbf{p}|$ is the magnitude of the three-momentum of the incident (and scattered) particles.

The rise of the total cross-section with the proton–proton collision energy is constrained by a logarithmic bound, the Froissart–Martin bound introduced in 1961 and 1966, respectively, by the French physicists Marcel Froissart and André Martin.[e] It is rooted in the mathematical properties of quantum mechanical amplitudes for finite-range interactions and by the Pomeranchuk theorem, which states that the high-energy limit of the interaction cross-section of a particle hitting a target is the same as that of its antiparticle hitting the same target.

The optical theorem allows the LHC physicists to measure σ_{tot}. To see this, we can start from the relation

$$|A(t)|^2 = \frac{d\sigma_{\text{el}}}{d\Omega} = \frac{|\mathbf{p}|^2}{\pi} \cdot \frac{d\sigma_{\text{el}}}{dt},$$

where we have used on the right-hand side the approximation $d\Omega = \int \sin\theta d\theta d\phi = 2\pi \sin\theta d\theta = \pi d\theta^2 \approx (\pi/|\mathbf{p}|^2) \cdot dt$, with $\theta^2 = t/|\mathbf{p}|^2$ for a small scattering angle θ. By squaring equation (12.1) and using $|A(0)|^2 = \text{Im}^2 A(0)(1+\rho^2)$ with $\rho = \text{Re}A(0)/\text{Im}A(0)$, one finds

$$\sigma_{\text{tot}}^2 = \frac{16\pi}{1+\rho^2} \cdot \left.\frac{d\sigma_{\text{el}}}{dt}\right|_{t=0}. \tag{12.2}$$

The measurement of σ_{el} as a function of t and extrapolation to $t = 0$ (zero-degree scattering) thus determines σ_{tot}^2.

The measurement of σ_{el} requires the knowledge of the luminosity L of the dataset used. Alternatively, using the relations $N_{\text{tot}} = N_{\text{el}} + N_{\text{inel}} = L \cdot \sigma_{\text{tot}}$, one finds from equation (12.2)

$$\sigma_{\text{tot}} = \frac{16\pi}{1+\rho^2} \cdot \frac{1}{N_{\text{el}} + N_{\text{inel}}} \left.\frac{dN_{\text{el}}}{dt}\right|_{t=0}. \tag{12.3}$$

which is independent of the luminosity. The number of elastic and inelastic events, N_{el} and N_{inel}, can be measured with dedicated counting devices of adequate geometrical coverage around the interaction point.

The quantity ρ in equation (12.3) is typically taken from theory, but can also be measured through the interference of electromagnetic and strong-interaction contributions to the elastic scattering at very-forward angles (t of order 0.001 GeV^2). This is a difficult but key measurement in diffractive and elastic scattering experiments as its centre-of-mass energy dependence, $\rho(s)$, together with $\sigma_{\text{tot}}(s)$, informs scattering models involving Pomeranchuk-theorem-violating processes, such as the putative exchange of a pomeron counterpart dubbed *odderon* with *odd* charge parity (the pomeron has even charge parity). The value of ρ being of order 0.10–0.15, its effect on the σ_{tot} measurement is however small.

The elastic forward scattering cross-section is determined by extrapolating to zero momentum transfer the elastic-scattering differential cross-section

$$\frac{d\sigma_{\text{el}}}{dt} = \left.\frac{d\sigma_{\text{el}}}{dt}\right|_{t=0} \cdot e^{-B|t|}, \tag{12.4}$$

where the parameter B is fitted to data and typically of order 20 GeV^{-2}. Depending on the specific collider optics chosen, the differential cross-section $d\sigma_{\text{el}}/dt$ may be measured between $t \sim 0.01$ GeV^2 and $t \sim 0.1$ GeV^2. The inelastic cross-section σ_{inel} is obtained by subtracting σ_{el} (integrated over the full t range) from σ_{tot}.

---

[a]This list is not exhaustive, as combinations of single diffraction or double diffraction with central diffraction can also occur, as well as multi-pomeron exchange (more than two) extensions of the central diffraction case. But these contributions are marginal, although subject to theoretical and experimental studies.

[b]Named after the Soviet theoretical physicist Isaak Yakovlevich Pomeranchuk (1913–1965).

[c]For elastic scattering, a non-negligible contribution comes from electromagnetic interaction, which interferes with the strong-interaction one.

[d]The optical theorem is a direct manifestation of the principle of unitarity (conservation of probability) in scattering reactions, cf. section 1.3.

[e]The Froissart–Martin bound states that the total cross-section cannot rise faster than $C \cdot \ln^2(s/s_0)$, where s is the collision energy squared, and s_0 and C are unknown constants.

The inelastic cross-section can be measured by counting the number of proton–proton collision events, N_{inel}, with a minimum of final-state activity in the detector. By independently measuring the integrated luminosity L of the data sample with the help of a van-der-Meer calibration scan (cf. inset 10.2 on page 297), one can derive the inelastic cross-section via $\sigma_{\text{inel}} = N_{\text{inel}}/(A \cdot L)$, where A is the acceptance of the event selection for inelastic scattering events taken from simulation. Since the calculation of soft scattering events relies on phenomenological models rather than precise QCD calculations, the theoretical uncertainty in A is large and dominates the measurement. A measurement by ATLAS at 13 TeV centre-of-mass energy found $\sigma_{\text{inel}} = 78.1 \pm 0.6 \pm 1.3 \pm 2.6$ mb, where the first uncertainty is experimental, the second due to the integrated luminosity, and the third accounts for the acceptance correction. The total uncertainty is thus 3.7%.

A more precise value of σ_{inel} can be obtained from the measurement of the *elastic* forward scattering rate via the optical theorem as explained in inset 12.3. Such a measurement, however, requires specific instruments to measure small-angle proton scattering very close to the beam. They are located at distances of a few hundred metres from the interaction point. Their detecting elements are either special wire chambers, scintillating fibres, or silicon strip or pixel detectors located in thin-walled retractable cylindrical structures called *Roman Pots* (figures 12.25 and 12.26) that can be inserted into the beam pipe to within millimetres from the circulating beams without interfering significantly with the machine operations (in particular without breaking the ultra-high vacuum of 10^{-10} to 10^{-11} mbar in the beam pipe).[17] These tracking stations are complemented by time-of-flight measuring devices with 50-picosecond resolution or better to assign protons

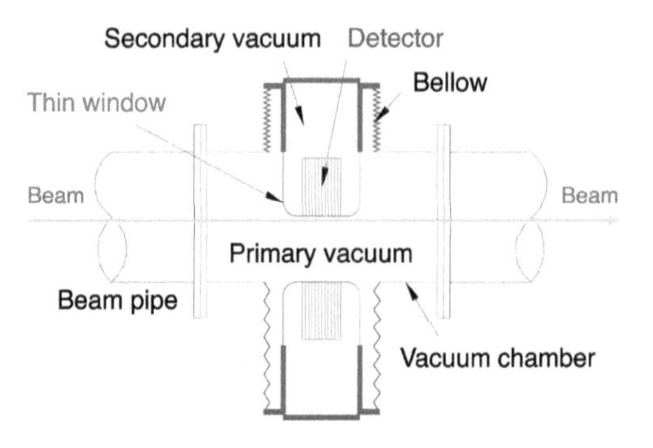

Fig. 12.25. Illustration of the principle of a Roman Pot detector. See description in text.

[17]The Roman Pot experimental technique was introduced at the ISR for the detection of forward protons from elastic and diffractive scattering. Detectors are placed inside a secondary vacuum vessel, called pot, and moved into the primary vacuum of the machine through vacuum bellows (cf. figure 12.25). This technique physically separates the detectors from the primary vacuum in the beam pipe and preserves it against emission of gas molecules by the detector elements. Roman Pots (though not yet called that way) allowed physicists at the ISR in 1973 to measure the total proton–proton interaction cross-section via elastic forward scattering between 23 and 53 GeV. The cross-section was found to rise from 39 to 43 mb, after passing through a shallow minimum at around 15 GeV, measured at lower-energy experiments. It was the first experimental proof of the logarithmic (actually, $\ln^2(s)$) rise of the high-energy cross-section with centre-of-mass

to an interaction vertex along the longitudinal coordinate of the luminous region covered by the crossing beams. Such forward detectors are installed on either side of ATLAS and CMS.

Fig. 12.26. Photo of the ATLAS-ALFA Roman Pot close to the ATLAS detector in the LHC tunnel. The Roman Pot is installed on one of the two parallel beam pipes, where the beam is leaving the interaction point, to measure protons scattered at small-angle. *Image: Ronaldus Suykerbuyk/CERN.*

Next to CMS is the independent TOTEM experiment, an international collaboration of about 100 scientists, 16 institutes and 8 participating countries. TOTEM was considered from the outset of the LHC physics program. Subsequently all LHC experiments considered and performed forward physics, diffractive physics and total cross-section measurements. For example, next to ATLAS are the ALFA and AFP detector systems, which are fully integrated in the ATLAS experiment.

ALFA consists of Roman Pot scintillating fibre tracking stations placed at distances of 237 and 241 metres along the beam on either side of the ATLAS interaction point. AFP hosts four silicon-pixel tracking units located 205 and 217 metres upstream and downstream of ATLAS. Figure 12.27 shows the location in the LHC tunnel of the TOTEM experiment with its Roman Pot stations at distances of 213 and 220 metres on either side of CMS. Each station consists of three Roman Pots making a total of twelve Roman Pots. Additional Roman Pots are installed at ± 147 metres from the CMS interaction point, but these are not used for elastic cross-section measurements. The scattered protons are transported through a number of LHC machine elements before they reach the high-resolution position-measuring detectors in the Roman Pots.

Not only dedicated forward detectors are needed to measure elastic scattering, also the colliding beam optics must be specially adapted to achieve an as small as possible momentum transfer t, as required by equation (12.3) in inset 12.3. To maximise the sensitivity of the position

energy squared s, in agreement with studies of cosmic rays and a prediction by Werner Heisenberg from 1953.

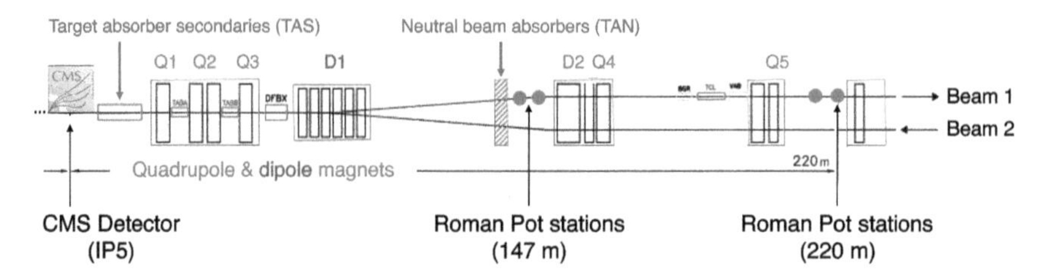

Fig. 12.27. Location of one arm of the TOTEM detectors in the LHC tunnel. Two arms are positioned symmetrically on both sides of the interaction point at the centre of CMS (Interaction Point 5, IP5). The relevant LHC interaction region elements, focusing quadrupole magnets (Qi), separation dipoles (Di), and other LHC machine optic elements through which protons scattered at small angles pass before reaching the Roman Pots (RP147 and RP220), are indicated. *TOTEM Collaboration, CERN-LHCC-2004-002 (2004), figure modified by authors.*

measurement in the Roman Pots to the scattering angle θ^* (the momentum transfer is given by $t = -(\theta^* p)^2$, with p the beam momentum) and at the same time minimise its dependence on the vertex position, special optics are designed to achieve minimum beam divergence at the interaction point. The beam settings used correspond to a focusing length β^* (see inset 6.2 on page 178) of ninety metres up to two kilometres,[18] a low-luminosity configuration (up to 10^{30} cm^{-2}s^{-1}) and no pile-up. This is to be compared with the value of $\beta^* = 25$–80 centimetres used for high-luminosity running during Run 2 in ATLAS and CMS.

The measurement of σ_{tot} via equations (12.3) and (12.4) requires the simultaneous measurement of the inelastic and the elastic rates, and the extrapolation of the elastic rate into the concealed region down to zero four-momentum transfer squared t. In TOTEM, in order to measure N_{inel}, charged particles from inelastic collisions are detected by means of two ancillary tracking telescopes that are incorporated into the CMS detector itself, just in front of and just behind the very-forward hadron calorimeter, covering together a rapidity range from 3 to 7. Two such detector sets are symmetrically placed on each side of the interaction point at the centre of CMS.[19] The measurement of the inelastic scattering rate is not needed if the luminosity of the dataset is known (cf. equation (12.2)).

[18]The beam divergence is proportional to $1/\sqrt{\beta^*}$.
[19]They are used only for the forward-physics dedicated runs and are removed before initiating high-luminosity physics runs to protect the detectors against radiation.

Using a dedicated run performed in 2015 at 13 TeV centre-of-mass energy and $\beta^* = 90$ metres, TOTEM measured $\sigma_{\text{el}} = (31.0 \pm 1.7)$ mb, $\sigma_{\text{inel}} = (79.5 \pm 1.8)$ mb, and a total cross-section $\sigma_{\text{tot}} = (110.6 \pm 3.4)$ mb, where a value $\rho = 0.1$ was assumed in all cases. Using a $\beta^* = 90$ metres run at 8 TeV centre-of-mass energy and an independent luminosity measurement, ATLAS-ALFA obtained $\sigma_{\text{tot}} = 95.35 \pm 1.4$ mb, which is so far the most precise total cross-section measurement at the LHC.

Figure 12.28 shows the final measurement and fit performed by ATLAS-ALFA and figure 12.29 a compilation of proton–(anti)proton cross-section measurements versus centre-of-mass energy. It includes older measurements performed at much lower energy at the ISR (cf. footnote 17 and section 6.2) and results derived from cosmic-ray measurements at much higher energies up to 100 TeV. Such energies could also be probed with a future hadron collider such as the FCC-hh project discussed in chapter 15.

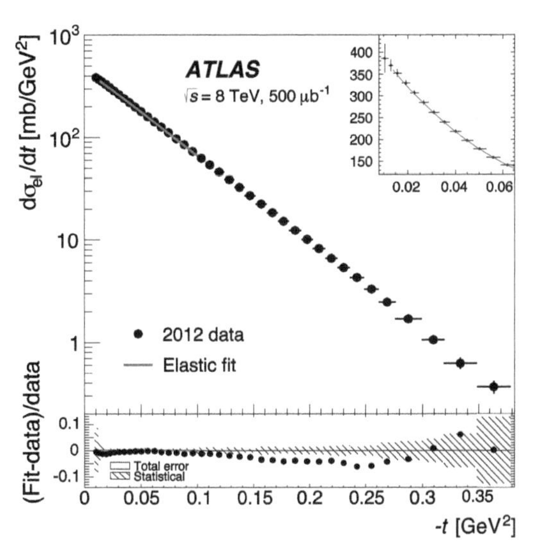

Fig. 12.28. Differential elastic cross-section measured by ATLAS-ALFA (solid dots) using 8 TeV proton–proton collision data taken with $\beta^* = 90$ m beam optics. The red curve shows the fit of the function (12.4) to the data with σ_{tot} and B as free parameter. The upper right insert shows a zoom into small $|t|$. The lower panel shows the relative difference between fit and data together with the experimental uncertainties. *ATLAS Collaboration, Phys. Lett. B 761, 158 (2016), figure modified by authors.*

There is a particular class of diffractive phenomena, called *central diffractive processes* $pp \to ppX$ (see inset 12.3), whereby both incident beam protons remain intact, while in the central part — well separated from the projectiles — the two emitted pomerons (or quasi-real photons for QED interactions, see section 12.7) interact and fuse into a final state X. The striking signature of these processes is the presence of two quasi-elastic very-forward-going isolated protons and the centrally produced cluster of particles X. ATLAS-ALFA/AFP as well as a joint effort from the CMS and TOTEM collaborations, the CMS-TOTEM Precision Proton Spectrometer

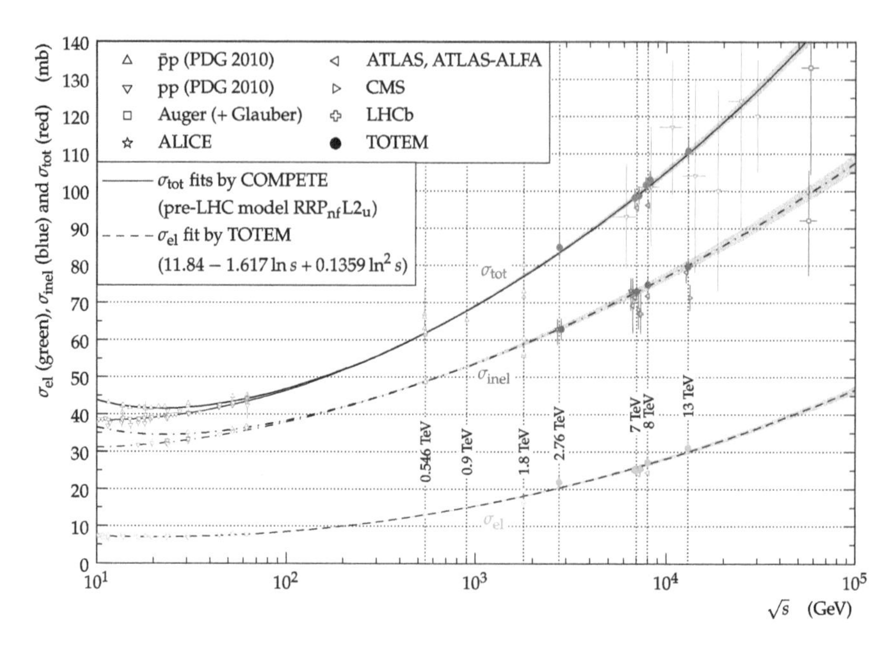

Fig. 12.29. Total proton–proton and proton–antiproton interaction cross-section measurements versus centre-of-mass collision energy $\sqrt{s}$ from 30 GeV (ISR) to 13 TeV (LHC) and cosmic-ray experiments (Pierre Auger Observatory); proton–antiproton collider measurements are from CERN colliders and the Tevatron. The curves are phenomenological models allowing for the maximal rate of rise of the cross-section with increasing energy as $\ln^2(s)$ (Froissart–Martin bound, see inset 12.3). The proton–antiproton total cross-sections at around 0.6 and 2 TeV centre-of-mass energies are in qualitative agreement with expectations of asymptotically same values for particle and antiparticle cross sections (Pomeranchuk theorem, see inset 12.3). The figure also shows measurements of the corresponding elastic and inelastic cross-sections up to 13 TeV. *Image: TOTEM Collaboration, Eur. Phys. J. C 79, 103 (2019).*

(CT-PPS),[20] study these diffractive processes. The forward-proton precision measurements of the protons (for position and timing) are provided by the Roman Pots and are complemented by the ATLAS or CMS trackers and calorimeters to measure X. The main physics motivations are the search for new resonances decaying to di-lepton, di-jet, $t\bar{t}$ or di-boson final states as well as the study of anomalous WW or ZZ quartic couplings. High luminosity is required for these studies. The CT-PPS team has recorded

[20]For high-luminosity studies as those described in section 12.7, CMS and TOTEM combine their data and formed in 2016 the CMS-TOTEM precision proton spectrometer (CT-PPS) with major involvement from CERN, Brazil, Portugal, Poland.

an integrated luminosity of 15 fb^{-1} in 2016, 40 fb^{-1} in 2017 and more than 55 fb^{-1} in 2018.

We should mention here one additional forward experiment, LHCf, a calorimetric detector dedicated to the measurement of neutral particles (such as neutrons, photons from π^0 or η decays to di-photons, and also neutral kaons) with pseudo-rapidity larger than 8.4. LHCf is placed next to the neutral beam absorbers (TAN, cf. figure 12.27) at 141 metres on either side of ATLAS after the dipole magnets that bend the two proton beams and any other charged particles. LHCf's measurements are beneficial to improve Monte Carlo generators used to simulate cosmic-ray shower developments in the atmosphere in the 10^{15} to 10^{17} eV incident energy range.[21]

12.7. The LHC as a photon collider

A particularly interesting type of events is produced in ultra-peripheral[22] elastic collisions of protons or heavy nuclei of charge Z by the extremely high Coulomb fields which can polarise the vacuum and lead to an effective photon–photon scattering process (figure 12.30). The coherent action from the large number of protons in a lead nucleus ($Z = 82$) gives rise to an electromagnetic field reaching up to 10^{25} volts per metre, which is

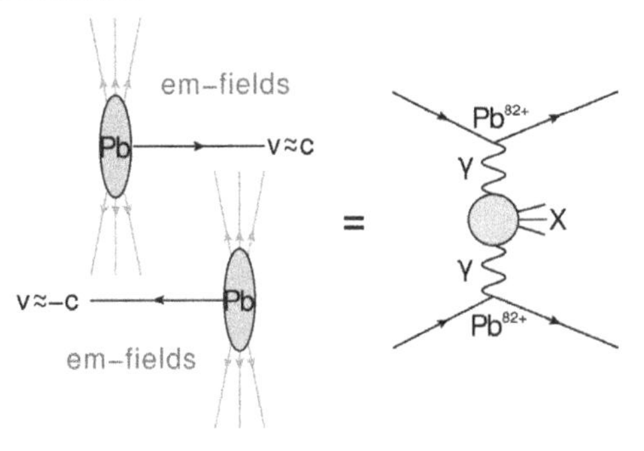

Fig. 12.30. Illustration of an ultra-peripheral collision of two lead ions. The electromagnetic interaction between the ions can be described as an exchange of photons that can couple to form a given final state X, for example two photons (light-by-light scattering) or two weak gauge bosons. *ATLAS Collaboration, Nature Physics 13, 852 (2017).*

[21]This is equivalent to a 10 TeV proton–proton centre-of-mass LHC collision energy, expressed in electron-volts, for a projectile hitting a fixed target, usually a nucleus in the Earth's atmosphere. However, cosmic-ray energies reach as high as 10^{21} eV. At such high energies, the cosmic-ray flux seems to be composed of a mix of protons and nuclei up to iron.

[22]Ultra-peripheral means that the impact parameter of the collision (transverse distance between the protons or nuclei centres) is significantly larger than the radius of the nuclei.

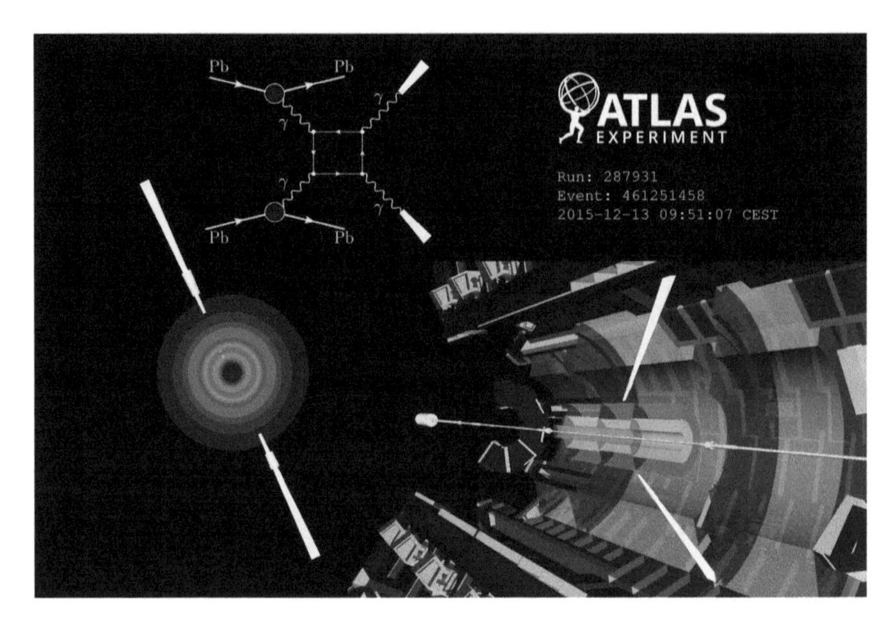

Fig. 12.31. Display of an ultra-peripheral lead–lead collision event measured in ATLAS. In yellow are shown energy deposits of two photons in the electromagnetic calorimeter on opposite transverse sides of the detector. There is no other activity in the detector, which is the clean signature of light-by-light scattering. The Feynman diagram of the process is shown in the upper part of the display. It is suppressed by the electromagnetic coupling constant to the fourth power, $\alpha^4 = 3 \times 10^{-9}$, which makes light-by-light scattering rare and difficult to observe. *ATLAS Collaboration, Nature Physics 13, 852 (2017).*

much larger than the Schwinger limit above which QED effects become important.[23]

For a given value of the impact parameter of the collision, the photon flux from one nucleus is proportional to Z^2. The cross-section of the so-called *light-by-light scattering*, in which the two quasi-real photons interact and produce another pair of outgoing photons emitted in some angle with respect to the beam axis is thus proportional to $Z^4 \sim 5 \times 10^7$. Lead–lead nuclei collisions are therefore particularly suited for the study of this phenomenon, despite the much lower instantaneous luminosity and

[23]The Schwinger limit is denoted after the American physicist and architect of QED Julian Schwinger (1918–1994). It corresponds to the critical electric field strength, E_c, where, in contradiction with the linear classical Maxwell differential equations, non-linear effects due to e^+e^- pair production set in. The Schwinger limit is given by $E_c = m_e^2c^3/(e\hbar) \simeq (0.511\text{ MeV})^2/(e\hbar c) = 0.511^2 \times 10^6\text{ V}/(197 \times 10^{-15})\text{ m} \simeq 1.3 \times 10^{18}\text{ V/m}$.

lower nucleon–nucleon energy compared to proton–proton collisions.[24] At the LHC, as heavy-ion collisions are pile-up free and the ions remain intact in this process, events with only two energetic photons and no other particle detected are a very distinct signature of light-by-light scattering. Such a spectacular and beautiful event, recorded in the ATLAS detector, is shown in figure 12.31. Light-by-light scattering occurs at very low transverse momentum, so that the signal photons peak sharply back-to-back in the transverse plane, that is, they have an acoplanarity $A_\phi = (1 - \Delta\phi_{\gamma\gamma}/\pi) \sim 0$. The A_ϕ distribution measured by ATLAS in the large 2018 lead–lead collision dataset is shown in figure 12.32. A clear, yet small signal excess over background is seen at small acoplanarity leading to a first direct observation of light-by-light scattering with more then 8σ statistical significance. This four force-carrier interaction (two incoming and two outgoing photons) is an integral part of electroweak theory and, at the same time, can be sensitive to deviations from the Standard Model predictions due to unaccounted-for new physics (such as, for example, contributions from axion-like particles).

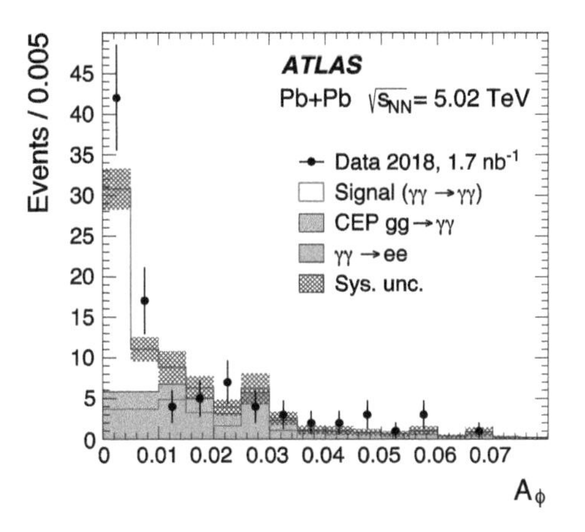

Fig. 12.32. Measured acoplanarity A_ϕ of selected photon pairs (dots with statistical error bars), compared to the expected light-by-light scattering signal (red line) and background contributions (blue and grey shaded areas). The hatched band indicates the uncertainty on the expectation. *ATLAS Collaboration, Phys. Rev. Lett. 123, 052001 (2019), figure modified by authors.*

Photon scattering can also lead to heavy final states, like the production of a W^+W^- pair. That process involves triple and quartic boson couplings as those discussed in section 12.2 above. The advantage here is that it is very pure, with reducible background contributions from strong interactions only. To study this highly suppressed $\gamma\gamma \to W^+W^-$ process, the higher energy and much larger luminosity of ultra-peripheral proton–proton collisions is required. Events are selected

[24] Even the ultra-intense laser experiments are not yet powerful enough to directly probe light-by-light scattering.

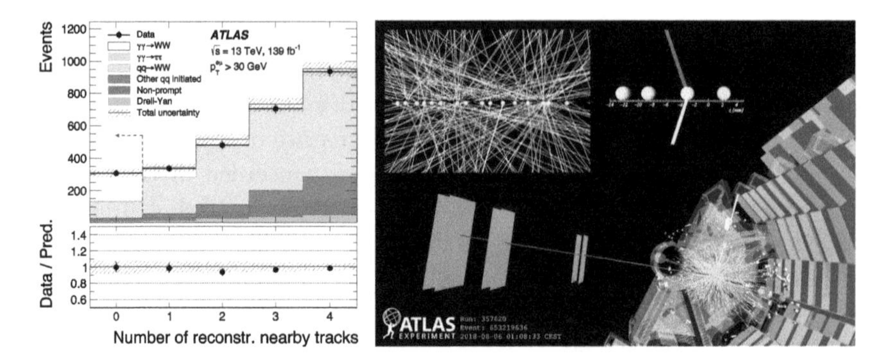

Fig. 12.33. Left: number of reconstructed (additional) tracks in the vicinity of the electron and muon tracks for events selected in an ATLAS search for $\gamma\gamma \rightarrow W^+W^-$ production. The coloured histograms show background contributions, with the largest originating from the $qq \rightarrow W^+W^-$ hard scattering process, and the black empty histogram is the expected signal. The black markers show the data with statistical error bars. The lower panel shows the ratio of data to signal plus background predictions. The hatched band indicates the total uncertainty on the prediction. A clear signal is seen at zero nearby tracks. An example signal candidate event is displayed on the right side. The event is consistent with the production of a pair of W bosons decaying into a muon (red line through the muon chambers) and an electron (yellow blocks corresponding to energy depositions in the calorimeter) and neutrinos (not detected). The many particles from pile-up interactions reconstructed in the inner detector are shown in orange. The zoom into the electron–muon vertex shows that no other particle originates from the proton–proton interaction corresponding to the WW production as expected from a photon–photon collision. *ATLAS Collaboration, Phys. Lett. B 816, 136190 (2021), figure modified by authors.*

where one W-boson candidate decays into an electron and a neutrino, and the other into a muon and neutrino. The primary challenge of this measurement is the background from W^+W^- production from the proton's constituents, which is hundreds of times more abundant than the signal from photon–photon interactions. To suppress this background, only collisions where no other charged particle is measured in the vicinity of the electron and the muon from the W decays are selected (see figure 12.33). Such additional tracks would occur from the underlying parton collisions in a hard proton–proton scattering background event. In 2020, ATLAS observed for the first time two W bosons produced from light-by-light scattering with a statistical significance of 8.4σ.

These measurements required the clean environment of ultra-peripheral lead–lead collisions or a precise vertex reconstruction in proton–proton collision events to enrich the selected event sample in photon collisions. An elegant way to very effectively suppress backgrounds in these events is to reconstruct the intact, but deviated protons in the forward Roman Pot detectors (AFP for ATLAS and TOTEM for CMS). Indeed, the momentum transfer between initial and final state in elastic collisions can be calculated from the measured deflection angles of the protons and compared to the corresponding measurement from the final state particles. A matching case corresponds to signal, while background events with coincidental forward proton signals will usually fail to match. Both ATLAS and CMS/TOTEM have published proof-of-principle measurements of single forward-proton tagged $\gamma\gamma \to \ell^+\ell^-$ production,[25] where $\ell = e, \mu$.

12.8. Precision

High-precision measurements are key for a better understanding of particle interactions. Many results are dominated by uncertainties in the theoretical calculations or the event generators, or the measurement precision and modelling accuracy of the parton distribution functions of the proton. The experimental collaborations at the LHC and elsewhere perform measurements to reduce these uncertainties and help theorists improve the modelling of particle reactions in order to push forward precision tests of the Standard Model. A lot is yet to be learned. We may cite in this context William Thomson Kelvin, from a speech held to the British Association for the Advancement of Science in 1871.

"Accurate and minute measurement seems to the non-scientific imagination a less lofty and dignified work than looking for something new. But [many of] the grandest discoveries of science have been but the rewards of accurate measurement and patient long-continued labour in the minute sifting of numerical results."

Fig. 12.34. Lord Kelvin (1824–1907). *Wikipedia.*

[25] A more correct way of writing the observed reaction is $pp \to p(\gamma\gamma \to \ell^+\ell^-)p^{(*)}$, where the asterisk indicates that one of the protons (not tagged) may have dissociated following electromagnetic excitation.

Chapter 13

The Quest for New Physics

The astounding success of the Standard Model to describe the phenomena of fundamental physics should not satisfy us and draw us away from its problems and shortcomings we summarised in chapter 3. The LHC collaborations therefore explore every possible avenue for hints of new physics — as never before physics experiments had a comparable potential. Both collaborations, ATLAS and CMS, have a deep and broad search program trying to reveal clues to the cardinal questions of modern physics. We cannot review the complete program here, and thus restrict the discussion to examples with choices influenced by current theoretical prejudice leading us to favour some models rather than others, or simply because specific models give rise to predictions more easily verifiable at LHC energies. It is of course quite possible, even plausible, that Nature surprises us and new physics will appear in a way different than what is predicted (figure 13.1). Some of the models proposed could turn out to be obsolete by the time the LHC reaches new heights in terms of data rates at the dawn of the high-luminosity phase of the LHC, the HL-LHC, in 2027.

13.1. Searches for new physics with jets

The study of jets, which are the manifestation of confined quarks and gluons in the final state, is one of the key activities of the ATLAS and CMS collaborations, first of all as a testing ground of QCD (see, for example, figure and discussion in inset 10.4 on page 306). However, invariant mass spectra of events with two or more jets could also reveal the presence of new physics, such as substructure of quarks leading to excited quark states (figure 13.2), quark–quark, quark–gluon, or gluon–gluon resonances, for example originating from dark matter mediators as discussed in section 13.7 below.

Fig. 13.1. About the difficulty of choosing a theory to initiate a new analysis. In face of what Nature may have in reserve for us, we must remain open to all possibilities theorists have not yet thought about. *Image: H. Murayama, Lepton-Photon Conference 2003.*

At 13 TeV collision energy, jets with the largest transverse momentum reaching up to 3 TeV are spectacular objects. An example of such a di-jet event measured by ATLAS is shown on figure 13.3. The experimentalists look for bumps in the mass spectra over a smoothly falling background. Technically, this is done by fitting the measured shape with a flexible, polynomial function, which properly describes the background, but does not allow to "swallow" a possible signal peak. The comparison of the fit with the

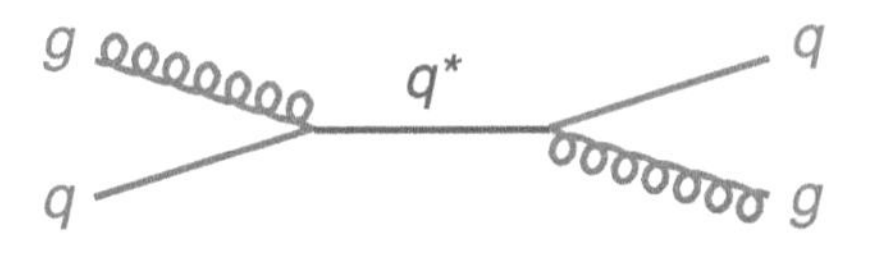

Fig. 13.2. Feynman graph of a hypothetical excited quark (q^*) production process via quark–gluon fusion. This is analogous to excited atoms when a photon (light) hits an atom, electrons orbiting the nucleus are energised and pushed into higher, metastable orbits. After some time, an electron will radiate a photon and fall down to its original orbit. The excited atom has relaxed to its ground state (de-excitation). Similarly, if quarks were composite particles, i.e. bound states of smaller objects, one could excite a quark by injecting energy, and then observe it relax back into its ground state via emission of a gluon.

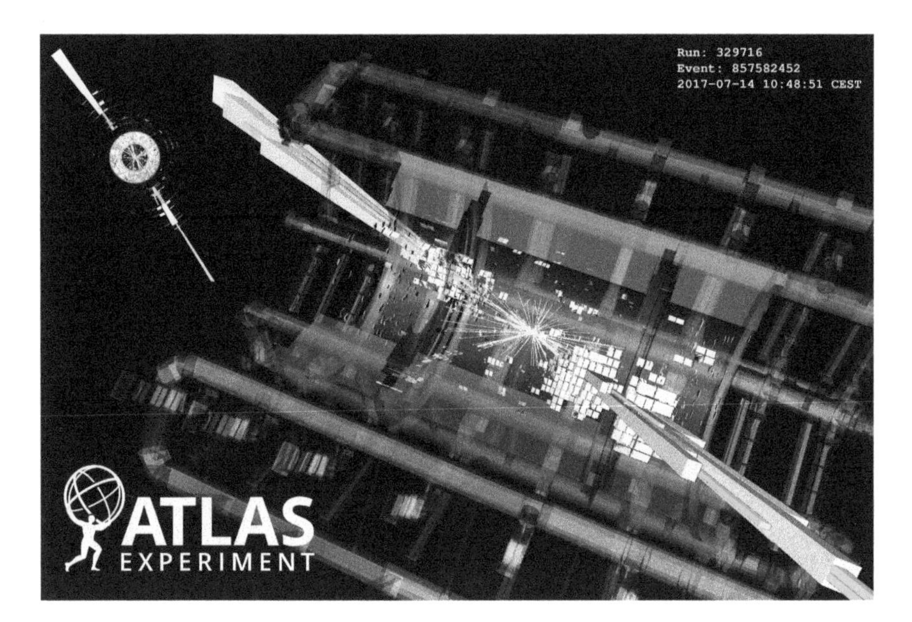

Fig. 13.3. Display of a di-jet event with an invariant mass of 9.5 TeV in the ATLAS detector. Each jet has about 3 TeV transverse momentum. *Image: ATLAS Collaboration, Phys. Lett. B 796, 68 (2019).*

data could then reveal a bump. Figure 13.4 shows examples of such searches: on the left for relatively central, generic di-jet events, and on the right for b-jet events where both jets were tagged to originating from b quarks. Owing to the large cross-section of strong-interaction jet production, the amount of events in each of the bins of the histograms is huge, requiring a highly faithful fit to avoid a fake signal claim. No excess or bump was seen in any of these searches by ATLAS and CMS,[1] allowing the collaborations to set constraints on possible new physics phenomena. For example, excited quarks with masses below 6.7 TeV are excluded at 95% confidence level.[2]

[1]The probability values (p-values) in figure 13.4 are quoted for the most discrepant regions detected by a statistical "bump hunter" algorithm. They take into account the *look-elsewhere effect* or *trials factor* discussed in inset 13.1 and are perfectly in agreement with the statistical fluctuations expected from a background-only distribution.

[2]How can one derive from the non-observation of a signal a lower limit on the mass of a hypothetical particle? In the above search, we have discussed the production of an excited quark (figure 13.2). Its production cross-section through strong interaction is predicted by QCD as a function of its mass. The lighter the excited quark, the larger its cross-section. The non-observation of an excess at high mass allows one to exclude

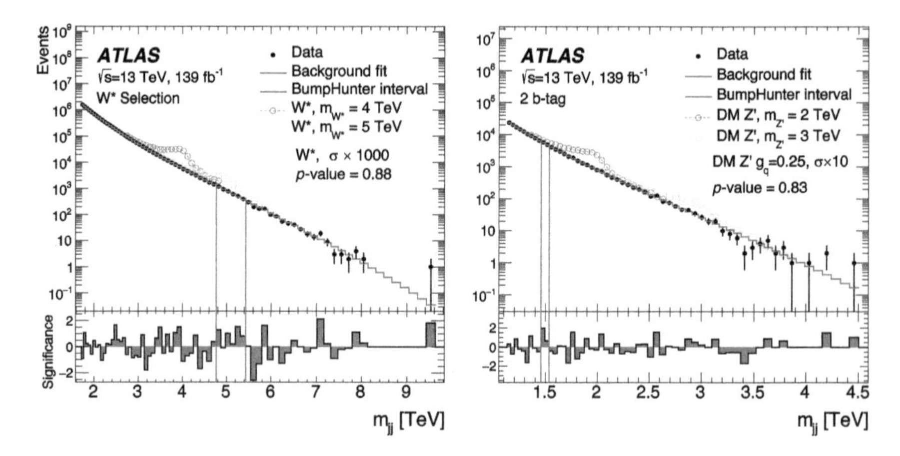

Fig. 13.4. Invariant mass spectra for pairs of jets (left) and b-jets (right) measured using the complete 13 TeV Run-2 data. The red histograms indicate the result of a background fit to the data. The open circles illustrate hypothetical resonances of, respectively, 4 and 5 TeV (left panel) and 2 and 3 TeV (right panel). The narrow bottom panels show the deviation (in terms of their significance) between the data and the background fit. The blue vertical lines indicate the largest discrepancy seen, which is non-significant in both cases. The highest-mass event in the left panel is displayed in figure 13.3. *Image: ATLAS Collaboration, JHEP 03, 145 (2020).*

13.2. Di-lepton resonances — mediators of grand unification?

In section 4.2 we discussed the possibility of unifying electroweak and strong interactions through models based on an extended gauge group incorporating the gauge group of the Standard Model $G_{SM} = SU(3) \times SU(2) \times U(1)$. The symmetry of the extended gauge group would break at some characteristic energy scale through a set of associated (extended) Higgs fields with appropriate potentials, analogous to electroweak symmetry breaking in the Standard Model. In the case of grand unified theories, the symmetry breaking can proceed in several steps at different energy scales. For example,

excited quarks lower than the heaviest mass to which the analysis is sensitive. This conclusion is however only true if the coupling between the proton (its quarks and gluons) and the hypothetical particle is known. If, for example, one searches for new forces mediated by a boson with unknown coupling strength to quarks or gluons, light resonances decaying to two jets may still exist, but with weak couplings to the proton. This motivates continuing searches in all mass ranges with increasing luminosity (see also the discussion in section 13.7).

the extended group E_6 could be broken as follows: $E_6 \rightarrow SO(10) \otimes U(1) \rightarrow SU(5) \otimes U(1) \otimes U(1) \rightarrow G_{SM} \otimes U(1)$.

Just as for the Standard Model, where the electroweak symmetry breaking through the vacuum expectation value of the Higgs field results in massive W and Z gauge bosons, the breakdown of the grand unified symmetry should give rise to massive gauge bosons. These are numerous, most of them very massive (see inset 4.2 on page 132), so that their direct production via particle collisions is beyond the reach of present-day accelerators. However, among the neutral gauge bosons associated with the splinter U(1) gauge groups, some could be light enough to be within the reach of the LHC. A possible discovery could provide information on the nature of grand unification and its symmetry breaking mechanism.

There are other theoretical models, for example associated with extra space dimensions introduced to alleviate the hierarchy problem (see sections 4.3 and 13.4), which feature massive, TeV-scale gauge bosons. Depending on the specificity of the models, these bosons can be of spin one, when associated to Kaluza–Klein excitations of the Standard Model W and Z bosons or to gauge symmetries in extra dimensions. In other models, they can be of spin two, when they correspond to excitations of the graviton.

These heavy gauge bosons would show up experimentally as heavier versions of the Z or W bosons, and are therefore generically called Z' and W' bosons. They may decay to pairs of leptons or quarks, or into Standard Model gauge bosons (for example $Z' \rightarrow WW$, $W' \rightarrow WZ$). Their couplings to fermions and bosons depend on the model considered. The experimentally most favourable cases are the *golden* decays into lepton pairs, $Z' \rightarrow e^+e^-$ and $Z' \rightarrow \mu^+\mu^-$, or $W' \rightarrow e\nu_e$ and $W' \rightarrow \mu\nu_\mu$. The dominant production mechanism at the LHC is shown in figure 13.5. Whether it is the Z' signal or the Standard Model background due to Z or virtual photon production (denoted *Drell–Yan* mechanism), the production mechanism is the same and leads to the same

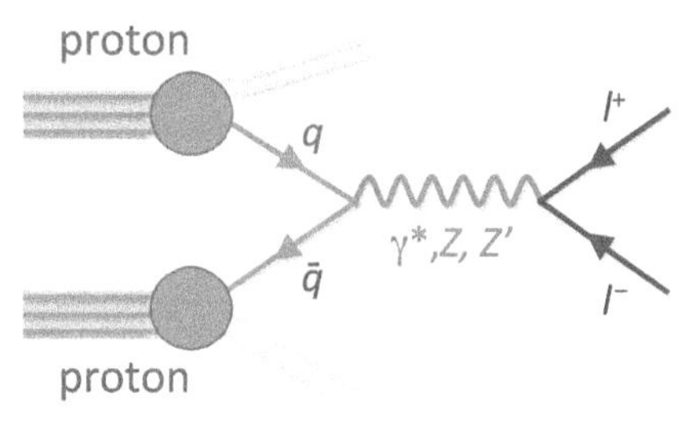

Fig. 13.5. Production of a lepton pair (e^+e^- or $\mu^+\mu^-$) in a proton–proton collision. A quark and an antiquark from the incident particles can annihilate via a massive virtual photon, a Z boson (which can also be virtual), or a possible hypothetical Z' boson. Subsequently they all decay into a lepton pair.

final state.[3] The Z' should thus appear as a peak in the two-lepton invariant mass distribution, just as was the case for the Z boson (figure 9.8). Since the neutrino is not directly measured by the detector, the $W' \to \ell\nu_\ell$ ($\ell = e, \mu$) signal can only be reconstructed in an incomplete way using the transverse balance of the event, allowing to estimate the transverse momentum of the neutrino. The thus calculated *transverse mass* has a broader distribution than the invariant mass, but a sharp Jacobian shoulder on the high-transverse mass side. Since high-energy electrons are more precisely measured in the electromagnetic calorimeters than high-momentum muons in the tracking detectors (even when exploiting both the inner tracker and the muon spectrometer), the di-electron channel is generally more sensitive than the di-muon one when looking for a mass peak. However the di-muon channel is more appropriate to measure decay angular distribution or forward–backward asymmetries allowing to differentiate the various Z' models or between spin 1 and spin 2 (graviton).

Both ATLAS and CMS have looked for a peak in the neutral and charged di-lepton mass spectra in the data recorded during Run 1 (8 TeV) and Run 2 at 13 TeV. Figure 13.6 shows these spectra for the electron cases from ATLAS for the full Run-2 dataset, compared to expectations for Standard Model backgrounds obtained through Monte Carlo simulations.[4] Putative Z' (left) and W' (right) signal resonances are also displayed. Some of the expected backgrounds, such as the one due to top–antitop production (shown in red), are said to be *reducible*, when selection on event kinematics and event topology can suppress them without reducing too much the signal efficiency. Other backgrounds, such as the non-resonant production of lepton pairs due to the Drell–Yan mechanism, are said to be *irreducible*,

[3]Identical initial and final states would lead to interference between the Z' signal and the Z/γ^* Drell–Yan process, the study of which would provide further insight into the signal, and offer experimental sensitivity even if the Z' is too heavy to be produced on-mass-shell at the LHC.

[4]One notices the difference with respect to the di-jet resonance search discussed in the previous section. There, the background was determined from a phenomenological fit assuming a smoothly falling spectrum. The same approach was also performed by ATLAS in the di-lepton resonance search for the numerical results reported here. The advantage of the fit approach is that it does not depend on simulations. Its shortcoming is that it is less sensitive to small deviations (rather than a bump) in the spectrum due to new physics. It is also quite tricky to prove that a fit function is flexible enough to fully describe the data.

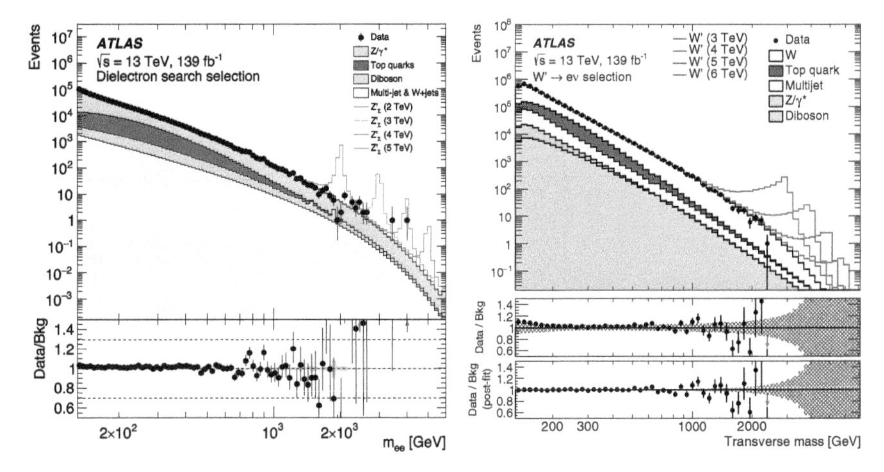

Fig. 13.6. Measured mass spectra in the electron–positron (left) and electron plus electron-neutrino (right) channels in the search for new heavy gauge bosons. The spectra use the full Run-2 dataset taken by ATLAS. Note that the scales on both the abscissa and ordinate are logarithmic. The experimental data are given as black points with statistical error bars. They are compared with background expectations for Standard Model processes obtained by simulations. These backgrounds are dominated by pair-production through a virtual photon or a Z boson, in fact a quantum mixture of these two processes, called the Drell–Yan mechanism. Also shown are hypothetical Z' (left) and W' (right) signal resonance, which are excluded by the data. *Image: ATLAS Collaboration, Phys. Lett. B 796, 68 (2019).*

as their topology and kinematics are the same as for the signal. Although some very high mass di-lepton events, as that shown in figure 13.7, were observed in the detectors, no significant excess or bump was found.

A statistical analysis of the data allows to determine, for every hypothetical value of the possible Z' (or W') mass, an upper limit on the product $\sigma \cdot B$ of the production cross-section times decay branching ratio into two leptons (figure 13.8). This statistical limit is independent of the underlying theoretical model, except for the assumption of the natural width of the Z'. As any specific model makes a prediction of $\sigma \cdot B$ as a function of the Z' mass, one can derive a lower mass limit on the Z'. Usually the collaborations use a reference model to benchmark the power of a search. Here it is the sequential Standard Model, denoted $Z'_{\rm SSM}$, where one assumes that the Z' has the same couplings to fermions as the Standard Model Z boson, except for the difference in the boson masses. In this particular framework, the negative results in the Run-2 dataset require that the $Z'_{\rm SSM}$ mass must

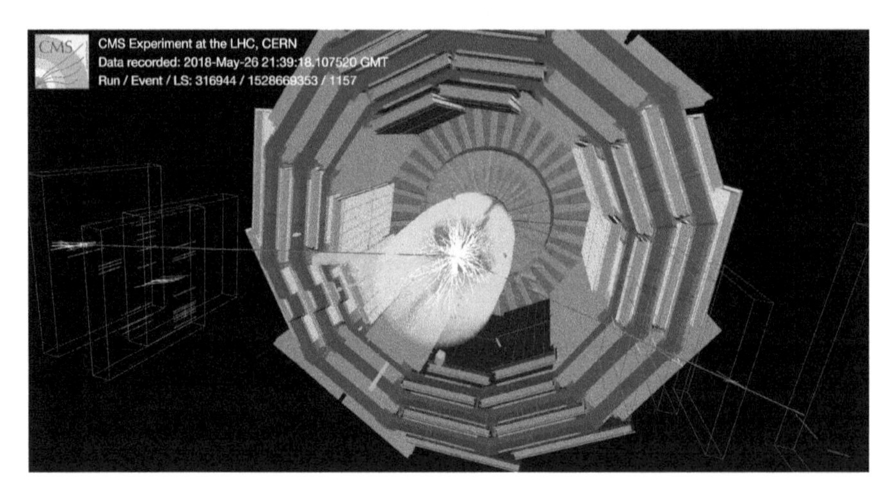

Fig. 13.7. Proton–proton collision event with the highest mass (3.1 TeV) muon pair (red lines) observed during Run 2 by CMS. Both muons go through the muon drift tubes in the detector central region with a transverse momentum of around 1.5 TeV each. The tiny curvature of the muon tracks used to measure their momentum is not visible by eye. They appear as straight lines. *Image: CMS Collaboration, 2018.*

be larger than 5.2 TeV to have escaped detection up to now (see the crossing between the dotted line representing the theoretical prediction of the $Z'_{\rm SSM}$ model and experimental limit in figure 13.8). Similarly, a lower limit of 6.0 TeV was set on a $W'_{\rm SSM}$ boson. More interesting than the somewhat simplistic sequential Standard Model is an interpretation in terms of grand unified models based on the E_6 group. Such a model is denoted by Z'_ψ in figure 13.8. As it predicts smaller couplings than the sequential Standard Model, the lower limit derived is only 4.6 TeV.

Grand unification models are not the only ones predicting heavy Z'-like bosons. Some theories, attempting to explain electroweak symmetry breaking in a less *ad hoc* way than the BEH mechanism of the Standard Model, also predict the existence of Z' bosons in the TeV mass range. In such models, particles playing the role of Higgs bosons can be bound states of massive fermions (composite Higgs models), or even of a top and antitop pair, or pairs of still more massive new particles. The Z' bosons expected in such theories can have couplings to Standard Model fermions very different from the Standard Model Z boson. In some models, they couple preferentially to top quarks so that they are searched for in channels like $Z' \to t\bar{t}$. Searches for such resonances excluded (narrow) Z' bosons below 3.9 TeV. In comparison with so heavy Z' bosons, the top quarks

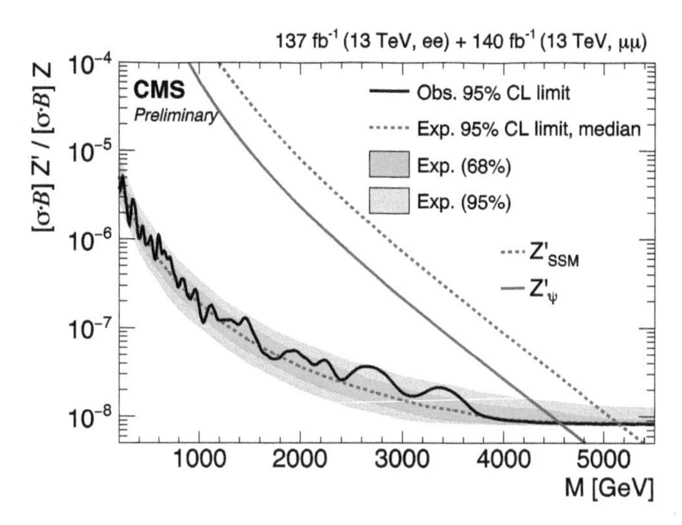

Fig. 13.8. Experimental limits on the mass of a heavy neutral Z' boson as derived by CMS. The vertical axis shows the product $\sigma \cdot B$ of the Z' production cross-section times its decay branching fraction to a pair of leptons versus the Z' mass (M). On the basis of invariant mass spectra, such as those shown in figure 13.6, it is possible to determine a 95% confidence level upper limit for each value of the Z' mass. These upper limits observed in the data are shown by the solid black curve. The dotted line shows the corresponding expected curve one would obtain if the data were spot on the expected (simulated) distributions. The green and yellow bands indicate the 1σ and 2σ variations of the expected curve. The observed curve fluctuates within these bands. The presence of a signal would show up as a significantly less stringent observed limit than the expected one. The curves labelled $Z'_{\rm SSM}$ and Z'_{ψ} correspond to different theoretical models for Z' production, each predicting a relation between $\sigma \cdot B$ and the Z' mass. The intersection of these curves with the solid black line determines the observed lower limit on the Z' mass allowed in the corresponding model. *Image: CMS Collaboration, CMS-PAS-EXO-19-019 (2019).*

appear light so that they receive a strong Lorentz boost (see inset 1.5 on page 24) merging the top-quark decay products (if fully hadronic, i.e. $t \to W(\to qq')b$, where the quarks further fragment and hadronise into jets) in a single "fat" jet (or, more scientifically, "large-radius" jet), which exhibits a three-prong substructure. Sophisticated experimental methods have been developed to analyse jet substructure and reconstruct boosted top quarks, but also hadronically decaying boosted weak and even Higgs bosons. Figure 13.9 shows an example of a heavy Z' candidate decaying into two hadronically decaying top quarks. The invariant mass formed by the two top-quark candidates amounts to 4.8 TeV, so that the top quarks decay in a highly collimated way forming three sub-jets each.

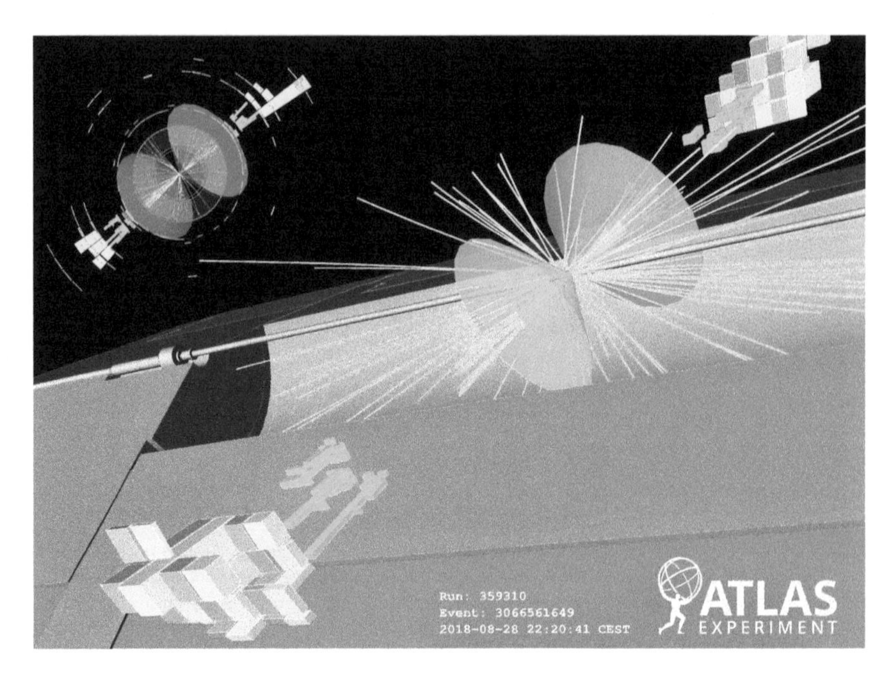

Fig. 13.9. Display of a $Z' \to t\bar{t}$ candidate measured by ATLAS where both top quarks decay hadronically into three quarks each ($t \to W(\to qq')b$), which fragment and hadronise into jets. Due to the large mass difference between the Z' and the top quark, the top quarks receive a strong Lorentz boost so that their decay products (the three jets) are merged in a single "fat" jet. The excellent granularity of the calorimeter and sophisticated jet substructure reconstruction techniques allow the experimenters to reconstruct the three jets. In this particularly beautiful example, showing an event with 4.8 TeV top–antitop mass, the three sub-jets can be seen with the naked eye. *Image: ATLAS Collaboration, JHEP 10, 61 (2020).*

Inset 13.1: The statistical "look-elsewhere effect" at work

Preliminary analyses of the first $3\,\mathrm{fb}^{-1}$ of 13 TeV data taken in 2015 by ATLAS and CMS revealed an excess of events at around 750 GeV diphoton invariant mass in the search for a heavy resonance in $pp \to \gamma\gamma$ production. The excess was seen by both collaborations, albeit weaker in CMS. The results were first presented by the collaborations in December 2015, provoking an eruption of phenomenological activity. About 500 articles were released after the presentations until the end of 2016, but most during the first months of 2016. In spring 2016, reanalyses of the 2015 data were published by ATLAS and CMS confirming the December results. The following figures show the measured spectra of both experiments.

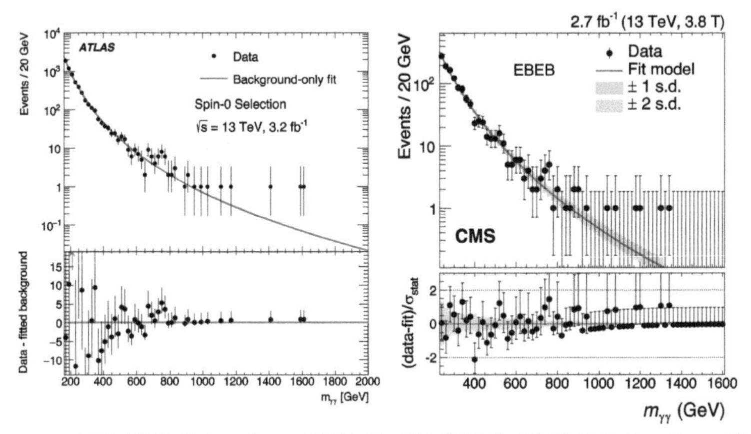

Images: ATLAS Collaboration, JHEP 09, 001 (2016); CMS Collaboration, Phys. Rev. Lett. 117, 051802 (2016).

CMS also included a small dataset taken without magnetic field for which a dedicated reconstruction was developed. The photons are tightly identified and isolated and have a typical purity of 94%. The background modelling used empirical functions that were fit to the full invariant mass spectra. ATLAS observed the lowest background-only p-value (see figure 11.20) at around 750 GeV for a resonance with a large natural width of about 45 GeV (6% with respect to the mass). The, respectively, local and global significance was found to be 3.9σ and 2.1σ.

The global significance was derived by running background-only pseudo-experiments, modelled according to the fit to data, and by evaluating for each experiment the mass and width that leads to the largest excess, that is, the lowest p-value. One then counts the fraction of experiments with a p-value lower than that observed in data. This procedure corrects the local p-value for the *trials factor* (also, more colloquially, called the "look-elsewhere effect"). Indeed, the local p-value corresponds to a non-normalised probability that does not have a well-defined interpretation. Only the global p-value defines a proper probability and is thus the correct reference value. CMS also found its lowest p-value at around 750 GeV with, however, a narrow width. Combining 8 TeV and 13 TeV data, a global (local) significance of 1.6σ (3.4σ) was seen.

The first data taken in 2016, about five times more abundant than those taken in 2015, did not reproduce the excess in neither experiment. The excesses in the 2015 data were thus the result of a statistical fluctuation which, given the global significance, is not that unlikely to occur. In this interpretation one should also take into consideration that the *actual* trials factor is larger than the global factor quoted for these analyses as there are many signatures probed by the experiments. This truly global significance of a local excess is hard or impossible to estimate in a thorough manner, but the additional trials factor should be kept in mind. In that respect, having a second experiment with a similar non-significant excess does not remove the trials factor if the results from both experiments are retained. Removing the 2015 data and looking solely at 750 GeV in the 2016 data does, however, properly remove any trials factor.

Why this buzz in spite of the relatively modest global significance of the bumps in the 2015 data? There are several reasons: theorists remembered the Higgs boson discovery, which started with relatively small excesses in December 2011, becoming significant by summer 2012. The situation here was, however, quite different. Firstly, the Higgs boson was indirectly constrained to be in a mass region between 114 GeV and about 135 GeV, while there was no such constraint on a di-photon resonance, so the trials factor is much larger in the latter case. Secondly, the Higgs boson signals seen were in agreement with expectation, while again there was no expectation in the di-photon case, which further increases the trials factor of the di-photon excess. The experience with the 750 GeV di-photon excess falls under the statistically ill-defined, but scientifically important category "extraordinary claims require extraordinary evidence", popularised in 1980 by the American astronomer Carl Sagan.

13.3. Where is supersymmetry?

We saw in section 4.1 that supersymmetry, which attributes to every Standard Model particle a partner particle that differs by half a unit in spin, could solve (or alleviate) several of the problems of the Standard Model, among which the hierarchy and dark-matter problems. In particular the hierarchy problem motivates supersymmetry at the TeV scale that is in reach of the LHC. The BEH mechanism and the existence of a Higgs boson, is a precondition for supersymmetry, but even in the simplest supersymmetric version of the theory, a single Higgs boson (originating from a single complex Higgs doublet field) is not sufficient. In fact, as discussed in section 4.1, at least two Higgs field doublets (both acquiring a vacuum expectation value), leading to five physical Higgs bosons, are required, out of which three are neutral and denoted h, H, and A, and two are charged: H^+, H^-. All five Higgs bosons have spin zero; h, H and $H^{\pm}$ are scalars (as the Standard Model Higgs boson), and A is a pseudoscalar (it changes sign under a parity inversion). While two among the neutral Higgs bosons can be very massive, the lightest scalar, defined to be h, can be very much Standard Model alike in terms of production and decay properties. If the A boson is much heavier than the Z boson, one reaches a situation that is called the *decoupling limit*: the light h has essentially Standard Model properties, and the masses of the other Higgs bosons (A, H, H^+, H^-) are very similar. Contrary to the Standard Model, where the Higgs mass is essentially a free parameter, the mass of the lightest Higgs boson in supersymmetry is bound from above! In lowest order perturbation theory, which means excluding Feynman diagrams with loops involving virtual particles, the h boson should be lighter than the Z boson. This was, however, experimentally excluded by LEP in the 1990s. Was that already the death of supersymmetry? It turned out that, when including radiative corrections due to virtual particles, the upper limit on the h-boson mass increases to about 135 GeV. The discovery of "a" Higgs boson at 125 GeV was thus comforting, bolstering supersymmetry enthusiasts as it could be interpreted as the h, respecting the upper mass bound, even if no other massive neutral or charged scalar boson was yet observed. However, the significant radiative corrections required to move the Higgs boson mass to 125 GeV puts the supersymmetry spectrum at the TeV scale, making it harder to find at the LHC.

13.3.1. *A challenging search*

The production rate of a supersymmetric particle, a *sparticle*, depends on its mass and the strength of its coupling to the proton. The heavier it is, the smaller its production cross-section. However, searches at previous colliders tell us that sparticles, if they exist at all, should be more massive than their Standard Model partners, otherwise they should have been found already. *Squarks* (denoted by the symbol $\tilde{q}$) and *gluinos* ($\tilde{g}$), the, respectively, spin-zero and spin-one-half supersymmetric partners of quarks and gluons, should have the largest production cross-sections of all sparticles, as they are subject to strong interactions. Most interesting among the squarks are those of the third generation, in particular the top squark[5] (or *stop*), which compensate the effects of the virtual top quark in the loops contributing to the Higgs boson mass (figure 4.3). Its mass should not much exceed a TeV for the cancellation to remain *natural*, that is, to avoid too much fine-tuning between positive top-quark and negative top-squark contributions to the Higgs boson mass.[6] It might even be that only the squarks of the third generation, the stop and sbottom, are light enough to be produced with noticeable rates at the LHC, whilst the squarks of the first and second generations are beyond reach. Another conceivable scenario could be that all squarks and gluinos are too heavy to be produced at the LHC, and only the gauginos and the higgsinos, the partners of the Standard Model gauge bosons W, Z, and photon, and of the Higgs bosons (see figure 4.2), respectively, and possibly the sleptons, partners of the leptons, are light enough to be produced at the LHC.

Let us consider a few examples of experimental signatures and corresponding searches, where, if not mentioned otherwise, the conservation of R-parity is assumed (cf. section 4.1). Figure 13.10 shows the simplest among the Feynman diagrams for the production of squarks and gluinos at the LHC. Only processes leading to hadronic final states with two, three or four jets are shown here. In such cases, the experimental signature will essentially depend on the presence of high-momentum jets and the large *missing energy* carried away by the two escaping neutralinos, $\tilde{\chi}_1^0$, the stable lightest supersymmetric particle (LSP). As the $\tilde{\chi}_1^0$ carries neither colour nor electric charge, it does not measurably interact with the detector material.

---

[5]Each Standard Model quark, being a fermion and having two spin states, needs two scalar supersymmetric squark partners, denoted squark-left ($\tilde{q}_L$) and squark-right ($\tilde{q}_R$). These states have the same quantum numbers and can therefore mix to form light and heavy squark states, $\tilde{q}_1$ and $\tilde{q}_2$, which are different superpositions (linear combinations) of the original $\tilde{q}_L$ and $\tilde{q}_R$ states. The $\tilde{q}_1$ and $\tilde{q}_2$ are the physical states which are (or would be) created in the proton–proton collision.

[6]The top quark gives the largest contribution to the radiative corrections of the Higgs mass, δm_H^2, in presence of a high new physics scale Λ. The stop quark cancels this contribution, but if the stop is heavier than the top residual logarithmic contributions $\delta m_H^2 \propto \ln\left(\Lambda^2/m_{\tilde{t}}^2\right)$ remain.

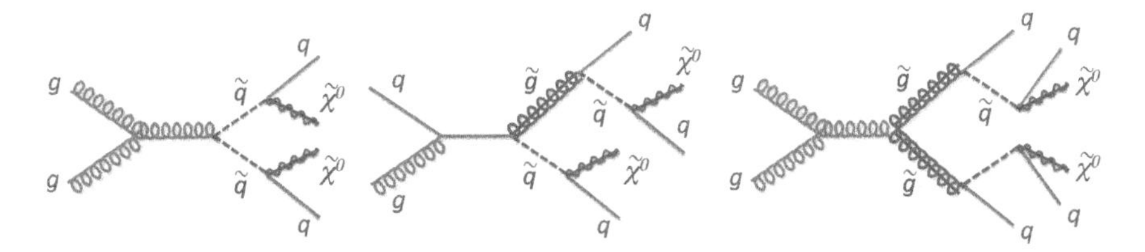

Fig. 13.10. Feynman diagrams for squark and gluino production leading to final states with two, three or four jets. The initial-state quarks and gluons — on the left-hand side of each diagram — are constituents of the incoming colliding-beam protons. Note the systematic appearance in the final state of two $\tilde{\chi}_1^0$, the lightest supersymmetric particle (LSP), escaping direct detection, as it does not interact with detector material. They can, however, be detected indirectly through the measured missing energy (or energy imbalance) in the event.

We remind the reader that R-parity conservation warrants that no single supersymmetric particle (which has negative R-parity) can be produced, as the initial state, the colliding protons, has positive R-parity. Supersymmetric particles can thus only be produced in pairs. Similarly, a supersymmetric particle can only decay into another supersymmetric particle (plus a Standard Model particle) so that the LSP is necessarily stable, as there is no lighter supersymmetric particle to decay to. If the LSP is only weakly interacting with the Standard Model fields, it is a dark-matter candidate.

Figure 13.11 sketches the case of a more complex supersymmetry process, with a cascade of sequential supersymmetric particle decays giving rise to a specific signature made of five quark jets, two high transverse momentum leptons from the decays of heavy $\tilde{\chi}_2^0$ and $\tilde{\ell}_R$, and significant missing energy due to the two lightest neutralinos. Such event configurations were among the first ones investigated by ATLAS and CMS, without however giving indications of a signal. Figure 13.12 shows a fully hadronic supersymmetric candidate event, which, even if it has the features of supersymmetry, can be perfectly interpreted as a Standard Model background process. Only the accumulation of such events above the expected background would give a statistical hint for something new. This is different from the bump searches discussed in the previous sections. There, the background could be determined from the side-bands assuming a smooth distribution, and the signal be distinguished from the background by its mere form. In most supersymmetry searches this is not possible because the final state is under-constrained due to the unmeasured lightest neutralinos. Therefore, supersymmetry searches are often among the most challenging new physics analyses at the LHC as they require a precise estimate of

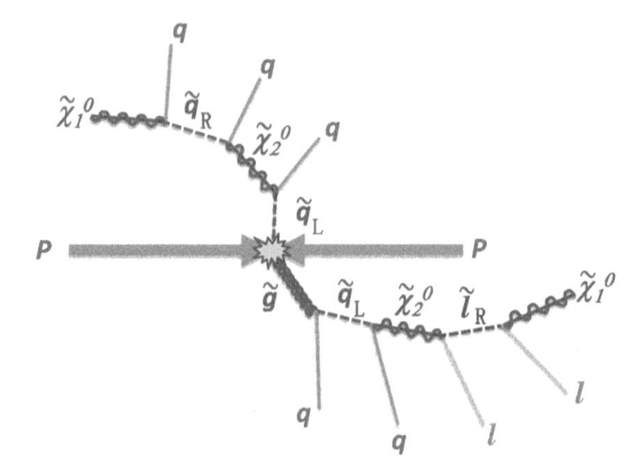

Fig. 13.11. Schematic representation of a proton–proton collision producing a squark–gluino pair, followed by a cascade of decays including also two types of neutralinos, the heavier $\tilde{\chi}_2^0$ and lightest (LSP) $\tilde{\chi}_1^0$ (see figure 4.2). The gluino decays into a quark–squark pair. The squark then decays into a quark and a $\tilde{\chi}_2^0$, which further decays into a lepton and a slepton ($\tilde{\ell}_R$). The slepton finally decays into a lepton and a $\tilde{\chi}_1^0$, which is stable as R-parity conservation is assumed, and thus escapes the detector leaving missing energy. The whole decay cascade proceeds instantly, so that all final state particles originate from the primary proton–proton interaction vertex.

the expected abundance of background events in often extreme kinematic regimes. Background could, for example, consist of top-quark pair events, where each top decays to bW and the W boson further decays either to two jets or to a lepton plus a neutrino. As for the lightest neutralino, the neutrino escapes the detector unnoticed, only visible through the missing energy it leaves (we encountered it already in section 2.3 when discussing the discovery of the W boson through events with a light lepton and a neutrino, identified through the presence of missing energy). The experimental signature is thus similar to squark or gluino pair production. Another important Standard Model background present in such searches is Z-boson production with associated jets, where the Z decays into two neutrinos (corresponding to 20% of all Z-boson decays).

At this stage we should recall how missing energy is measured (more details are given in inset 7.5 on page 213). Due to the unknown energy the colliding partons in each proton have (each parton carries only a fraction of the proton's momentum), the centre-of-mass of the parton–parton collision has an unknown Lorentz boost along the beam direction. Since, moreover, collision debris are scattered along the beam pipe outside the acceptance of the detector, it is not possible to determine the full missing energy of

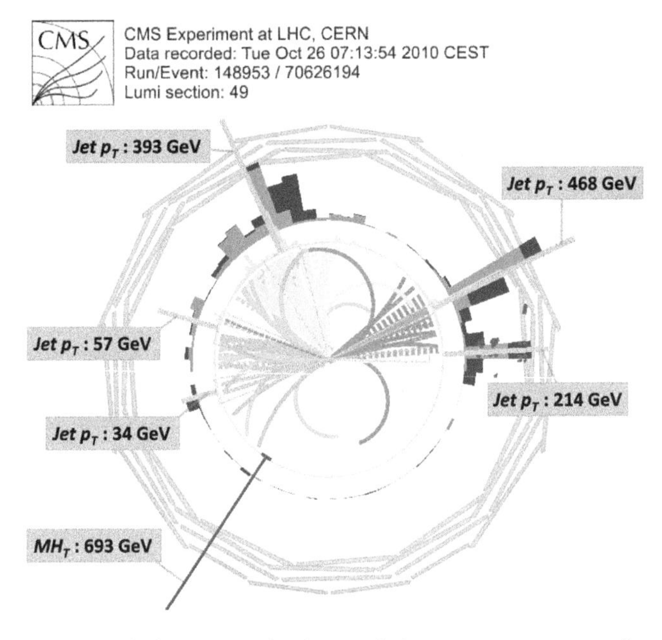

Fig. 13.12. Event with five jets and a large missing transverse energy (here denoted by MH_T) observed in the CMS detector. Events of such topology and kinematic properties fulfil all criteria of supersymmetric candidates, but remain nonetheless perfectly explainable in terms of certain Standard Model background processes. *Image: CMS Collaboration, 2010.*

the event from the three-dimensional vectorial sum of all measured final-state particle momenta. However, fortunately, in direction transverse to the beam direction, the two-dimensional vectorial sum of all momenta should indeed be zero, as the colliding protons possess no transverse momentum.[7] Any transverse imbalance of energy can therefore be assigned to one or more escaping particles. So far the theory. In reality, measurement resolution effects, in particular in presence of high-momentum jets, can lead to mis-measurement of missing transverse energy, and thus produce events that resemble supersymmetric signal events. This is another source of background that the experimentalists need to consider and estimate.

[7]Because of a small (couple of hundred microradians or 0.01 degrees) crossing angle between the colliding proton beams, introduced to prevent encounters in the region where the two beams share the same vacuum chamber, the centre-of-mass of the colliding system has a small transverse Lorentz boost. It has, however, a negligible impact on the measured missing transverse energy.

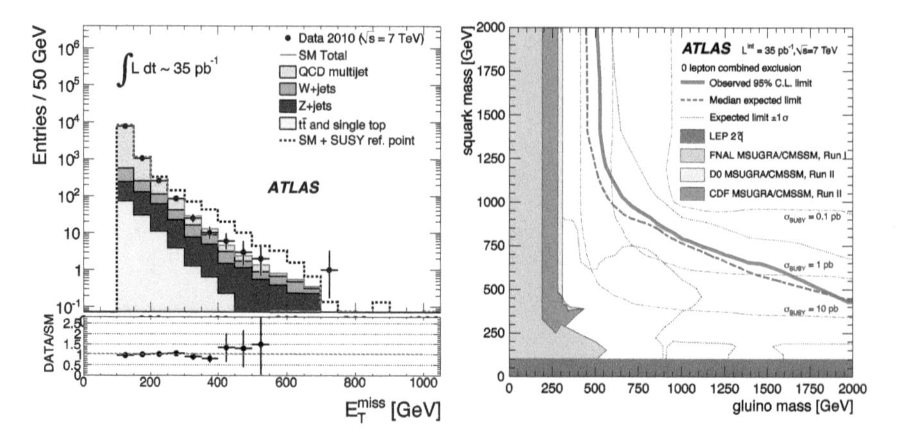

Fig. 13.13. Results from one of the first ATLAS searches for supersymmetric squark and gluino production (see figure 13.10) using only the small dataset of $35\,\mathrm{pb}^{-1}$ accumulated in 2010. The left panel shows the distribution of missing transverse energy. The coloured regions indicate the contributions from the various Standard Model background processes, where "QCD multijet" is due to mis-measured events. The other backgrounds, $W+\mathrm{jets}$, $Z+\mathrm{jets}$, and $t\bar{t}$ exhibit genuine missing energy due to decays involving neutrinos. The dotted line shows the expected contributions for a specific supersymmetric model. The data, given by the dots with statistical error bars, are in agreement with the expected background, allowing to exclude the specific supersymmetric model shown. The right-hand panel shows limits on the squark-mass versus gluino-mass plane assuming a massless lightest neutralino. The coloured bands indicate limits derived at the Tevatron proton–antiproton and LEP electron–positron colliders. Owing to the large collision energy of the LHC, the tiny 2010 dataset was sufficient to significantly extend the reach of this search. *Images: ATLAS Collaboration, Phys. Lett. B 701, 186 (2011).*

13.3.2. *Squarks and gluinos*

Let us look at the outcome of some specific searches. Figure 13.13 shows on the left-hand panel the distribution of missing transverse energy in an early ATLAS search from 2011 using only $35\,\mathrm{pb}^{-1}$ of 7 TeV proton–proton collision data. The data are in agreement with the Standard Model background prediction, allowing ATLAS to set (at the time) stringent limits on the squark and gluino masses as shown in the right-hand panel. Squarks and gluinos of equal mass are excluded below 780 GeV with 95% confidence level. ATLAS and CMS continued these searches throughout the years, increasing further their sensitivity by using much larger datasets at 13 TeV collision energy, as well as more sophisticated analysis techniques — without, however, uncovering any sign of squark or gluino production.

The currently best limits exclude squarks with mass up to about 2.0 TeV, and gluinos up to 2.3 TeV.[8]

The 2010 results from the LHC already held a strong message. The science magazine Nature titled *"Beautiful theory collides with smashing particle data — latest results from the LHC are casting doubt on the theory of supersymmetry"*. Theorists who had spent their entire careers on the study of supersymmetry, and who had placed great hopes in the LHC, up to considering supersymmetry to be "around the corner", had to face the new reality. *"We're painting supersymmetry into a corner"*, said an ATLAS physicist. That corner has become ever more hidden since the LHC produces data.

13.3.3. *Stop and sbottom*

With the increasing luminosity it became possible in 2012 to search for the holy grail of supersymmetry, the top squark, $\tilde{t}$. If gluinos are too heavy to be produced in significant quantities and squark mixing between left and right states makes third generation squarks lighter than the first and second generation squarks (see footnote 5), stop and sbottom squarks could be the first discovered sparticles, despite their low cross-sections and large backgrounds.[9] In the simplest process, each produced stop decays to a top quark and the lightest neutralino as drawn in the diagram of figure 13.14. Depending on the available phase space, that is, the mass difference between stop and neutralino, the

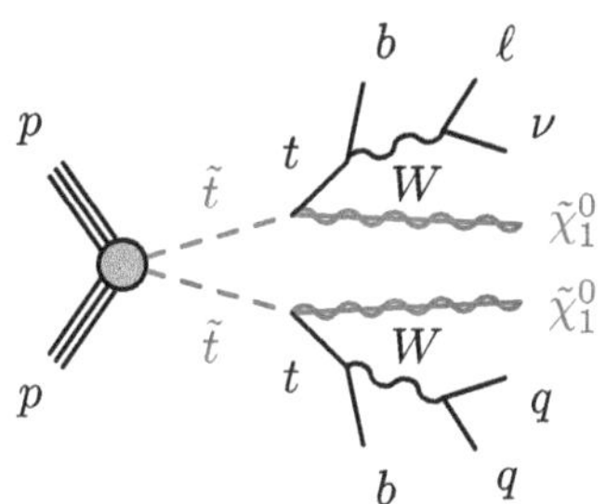

Fig. 13.14. Feynman graph of stop pair production and decay to top quark and lightest neutralino $\tilde{\chi}_1^0$. The top quarks further decay each into a W boson and a b jet. In this representation, one W boson decays leptonically and the other hadronically. The final state thus counts one reconstructed charged lepton, two jets and two b jets. If both W bosons decay leptonically the event has two charged leptons and two b jets, and if both W bosons decay hadronically there is no charged lepton but four jets and two b jets. All events have missing energy from the two stable lightest neutralinos.

[8]These limits are to be taken with a grain of salt: different supersymmetry model assumptions may significantly weaken the experimental constraints. For the purpose of clarity and simplicity, the experiments employ so-called *simplified models*, for which often 100% branching fractions to the studied final states are assumed, while a full supersymmetric implementation, if realised in Nature, would likely feature more complex processes.

[9]For equal mass, the stop pair production cross-section is approximately seven times smaller than that of top pair production.

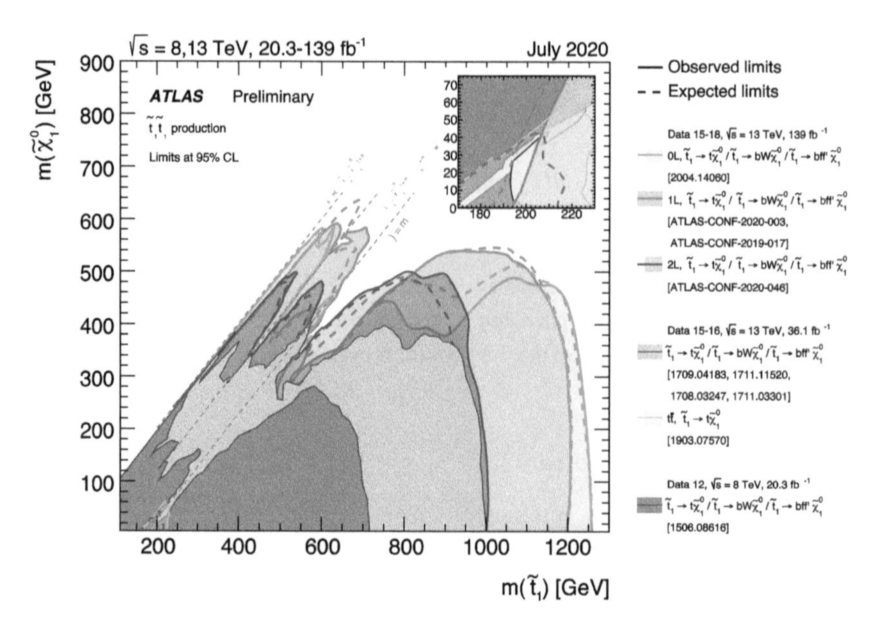

Fig. 13.15. Summary of ATLAS searches for top squarks as of summer 2020. The figure shows a complex set of exclusion limits in the LSP ($\tilde{\chi}_1^0$) mass versus stop mass plane. The colours indicate different analyses according to the stop and top-quark decays. At large stop mass and low LSP mass, the two-body decay $\tilde{t} \to t\tilde{\chi}_1^0$ is kinematically possible, which allows a better separation of top-quark backgrounds. When the LSP mass approaches that of the stop, the stop decays through a virtual top (three-body decay $\tilde{t} \to Wb\tilde{\chi}_1^0$) or even a virtual W boson (four-body decay $\tilde{t} \to \ell\nu b\tilde{\chi}_1^0$), which requires more involved analyses as the final state particles have lower momenta and are thus more difficult to trigger (online selection) and to separate from large Standard Model backgrounds. One notices the large increase in the excluded parameter space since the Run-1 (8 TeV) searches shown in dark grey. *Image: ATLAS Collaboration, 2020.*

top quark is either on-mass-shell (for a mass difference larger than the top-quark mass) or it is off-mass-shell (virtual). If the mass difference is even lower than the W-boson mass, both top quark and W boson are virtual. This has important consequences for the searches as small mass differences with virtual intermediate particles lead to lower final state momenta and missing energy. Such events are harder to distinguish from the background, which is dominated by Standard Model top-quark pair production. None of the stop searches performed by ATLAS and CMS have yet revealed a significant signal. Figure 13.15 shows the resulting exclusion limits in the $\tilde{\chi}_1^0$ versus stop mass plane as obtained by ATLAS. The various colours indicate the different analyses used to cover all final states (see figure 13.14) and kinematic scenarios. Top squarks up to 1.2 TeV are excluded, although the limits are becoming less stringent for a heavier neutralino. The exclusion limits are similar for pair production of bottom squarks. These results challenge the naturalness paradigm underlying TeV-scale supersymmetry.

13.3.4. *Gauginos and sleptons*

Perhaps all squarks and gluinos are beyond reach of the current LHC sensitivity and electroweak gauginos (neutralinos and charginos) and/or sleptons are the lightest sparticles. Gauginos and sleptons have much lower production cross-sections than strongly produced sparticles and thus require large datasets to be probed. Figure 13.16 shows a diagram for the associated production of the lightest chargino and next-to-lightest neutralino, and their subsequent decay into W and Z bosons plus lightest neutralinos, respectively. It can be identified via the occurrence of three charged leptons (two from the Z and one from the W decay) together with missing energy from the lightest neutralinos. The Standard Model backgrounds stem from WZ di-boson production, which can only be distinguished from the signal through the additional missing energy.

Figure 13.17 shows exclusion contours and cross-section limits obtained by CMS for the process depicted in figure 13.16, for varying $\tilde{\chi}_1^0$ and $\tilde{\chi}_1^{\pm}/\tilde{\chi}_2^0$ masses. Sleptons are assumed to be too heavy for the $\tilde{\chi}_1^{\pm}$ or $\tilde{\chi}_2^0$ to decay to them. To simplify the presentation, the two gauginos are assumed to have the same mass. The negative searches lead to the exclusion of gauginos up to 750 GeV and for the lightest neutralinos up to 350 GeV. The parameter region where all gauginos have similar masses $m_{\tilde{\chi}_1^0} \approx m_{\tilde{\chi}_2^0} \approx m_{\tilde{\chi}_1^{\pm}}$ is called *compressed*. Similar to the stop case, the searches become less sensitive in that region as the signal spectra are softer and resemble more the backgrounds. Larger datasets are needed to probe compressed spectra, which are not as theoretically contrived as it may seem. If the three lightest gaugino states are mostly of higgsino type, their masses are expected to be very similar. Gaugino decays via sleptons are a favourable case due to the larger leptonic rate than in weak boson decays. The exclusion limits in such cases can reach up to 1150 GeV. In most of the models considered by

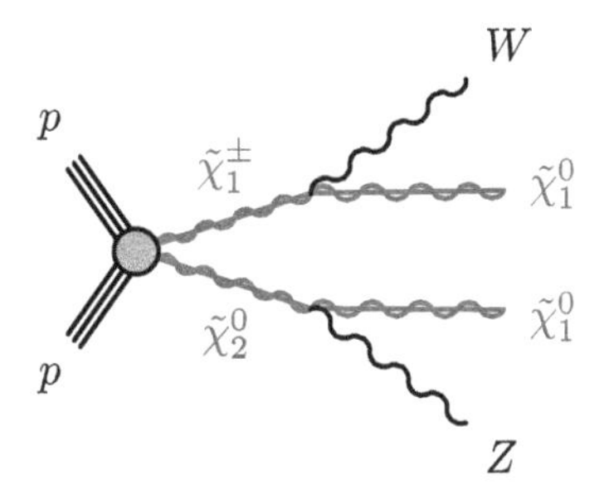

Fig. 13.16. Feynman graph of pair production of electroweak gauginos, here the lightest chargino ($\tilde{\chi}_1^{\pm}$) and the next-to-lightest neutralino ($\tilde{\chi}_2^0$). If sleptons are heavier than the gauginos, they decay via emission of a W boson and Z boson, respectively, and can be identified through their leptonic decays.

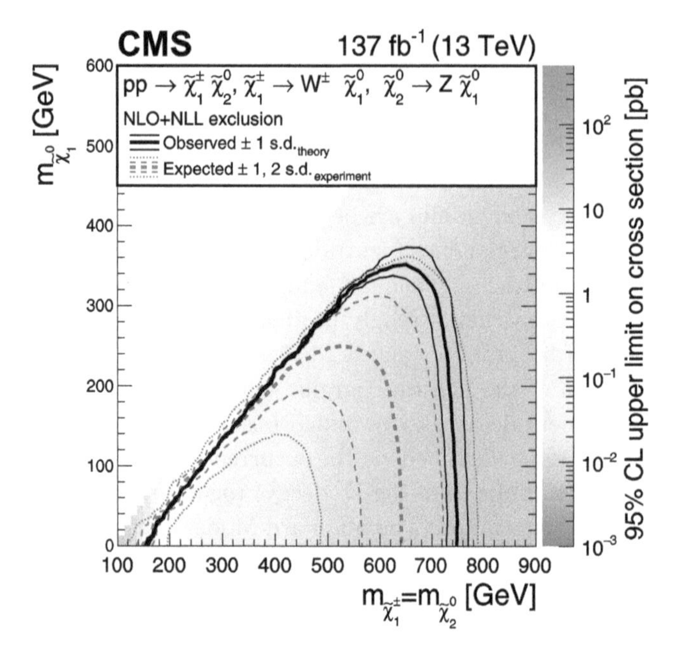

Fig. 13.17. Results of a CMS search for gaugino production using the full Run-2 dataset (see figure 13.16 for the corresponding process graph). The figure shows exclusion limits in the LSP ($\tilde{\chi}_1^0$) mass versus $\tilde{\chi}_1^\pm$ mass plane, assuming $\tilde{\chi}_1^\pm$ and $\tilde{\chi}_2^0$ to be mass degenerate. The solid black line gives the observed limit, while the red dashed line shows the expected limit for the background-only assumption. The observed limits exclude a slightly larger mass region than expected owing to an under-fluctuation in the data compared to the background expectation. The colours indicate the upper limits on the allowed cross-sections: production rates larger than these would have led to a signal in the data that would be in contradiction with the observation. The values of these upper limits increase towards the diagonal where the $\tilde{\chi}_1^0$ mass approaches that of the $\tilde{\chi}_1^\pm$ and $\tilde{\chi}_2^0$ and the separation from background processes becomes more difficult. *Image: CMS Collaboration, JHEP 04, 123 (2021).*

ATLAS and CMS, chargino pair production has a lower cross-section than the associated $\tilde{\chi}_1^+\tilde{\chi}_2^0$ production shown in figure 13.16. The cross-section depends on the mixing properties of the states: neutralinos can be *bino*, *wino* or *higgsino* like; charginos wino or higgsino like, depending on the dominant contribution.[10]

[10]There are a total of eight spin-half partners of the electroweak gauge and Higgs bosons: the neutral bino (superpartner of the U(1) gauge field), the winos, which are a charged pair and a neutral particle (superpartners of the W bosons of the SU(2) gauge fields), and the higgsinos, which are two neutral particles and a charged pair (superpartners of the Higgs field's degrees of freedom). The bino, winos and higgsinos mix to form four

ATLAS and CMS have also looked for slepton pair production, whose simplest processes are $pp \to \tilde{\ell}^+\tilde{\ell}^- \to \ell^+\ell^-\tilde{\chi}_1^0\tilde{\chi}_1^0$. The final state consists of two leptons with the same flavour, opposite charge, and missing energy. In the selectron (electron slepton) and smuon cases, the exclusion limits reach 700 GeV, while for the stau the limits are much less constraining (less than 400 GeV) as the final state τ leptons are unstable and their reconstruction suffers from reduced efficiency and larger backgrounds.

13.3.5. *The Higgs sector*

Yet another way to look for supersymmetry is to establish the existence of several Higgs bosons. We have seen above and in section 4.1 that the minimal supersymmetric extension of the Standard Model requires two Higgs field doublets giving rise to five physical Higgs bosons: h, H, A, H^+, H^-, where the observed Standard Model like Higgs boson is identified with h, and the four remaining Higgs bosons are more massive. Depending on the supersymmetry models considered, the general properties of the Higgs bosons can be predicted allowing the experimenters to develop dedicated search strategies. As for the Standard Model Higgs boson, these additional Higgs bosons couple to the mass of the particles,

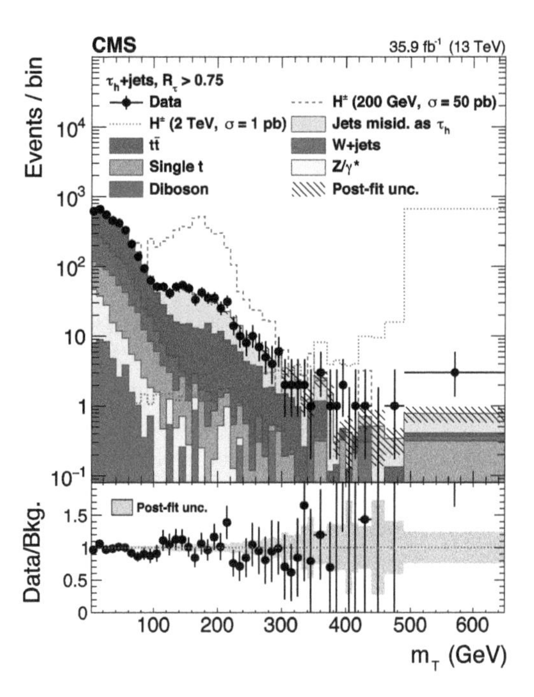

Fig. 13.18. Distribution of the transverse mass of $\tau\nu_\tau$ final states measured by CMS to search for a charged Higgs boson. Dots with error bars are the data, the coloured histograms give the various Standard Model background contributions, and the dotted lines illustrate two different $H^\pm$ mass hypotheses (both excluded by the data). The lower panel shows the ratio of data over predicted total background. *Image: CMS Collaboration, JHEP 07, 142 (2019).*

---

charged states called charginos ($\tilde{\chi}_i^\pm$, $i = 1, 2$) and four neutral states denoted neutralinos ($\tilde{\chi}_i^0$, $i = 1, 2, 3, 4$). Their indices i are ordered according to the increasing mass of the $\tilde{\chi}_i$ state. See the discussion in section 4.1 and the summary table in figure 4.2.

and are therefore searched for through decays to third generation quarks or leptons, and heavy weak bosons.

Figure 13.18 shows the distribution of the transverse mass of $\tau\nu_\tau$ final states measured by CMS in 13 TeV data to search for $H^\pm \to \tau^\pm\nu_\tau$. Overlaid are putative signal shapes, which are excluded by the data. The main background in this search is due to top-quark pair production and, in the hadronic τ-lepton decay channel, misidentified jets. For heavy $H^\pm$ the decay $H^\pm \to tb$ takes over from that to $\tau^\pm\nu_\tau$. Depending on the parameters, charged Higgs boson production may have low cross-section, so that — at present — no large mass ranges can be excluded.

The heavy neutral H boson is searched for through decays into weak W and Z-boson pairs, covering all masses up to the TeV scale. This search is quite similar to the Standard Model one before its discovery. Due to CP invariance, the pseudoscalar CP-odd A boson does not decay to W or Z pairs. Instead it is searched for via its decays to a τ-lepton pair for a light A and to a top-quark pair for a heavy A beyond the threshold of twice the top-quark mass.

Towards the end of Run 2, none of these searches have led to a significant signal. The inherent flexibility of supersymmetry and its Higgs sector allows it to survive the negative results so far so that it remains important to continue these searches. Also the precise measurement of the h boson's properties will give important clues about the degree of decoupling of the supersymmetric Higgs sector. However, even with the full data sample of the HL-LHC it will be difficult to fully refute supersymmetry via exclusion of its extended Higgs boson sector.

13.3.6. *Supersymmetry endgame?*

What conclusion can be drawn from the so far fruitless searches (figure 13.19) for TeV scale supersymmetry? If sparticles exist, can we assert that they are too heavy to be produced at the LHC? Are there gaps in the search mesh that would allow supersymmetry production to escape the sharp eyes of the LHC experimenters? Did ATLAS and CMS physicists really cover all possible scenarios?

Despite the naturalness argument requiring some of the sparticles to be at or below the TeV scale to moderate the Higgs sector, it is still well possible that supersymmetry lies just above the LHC mass reach. Naturalness is more a qualitative argument than a quantitative one. Is 1% fine-tuning really so much worse than 10%? If sparticles are too heavy to be directly

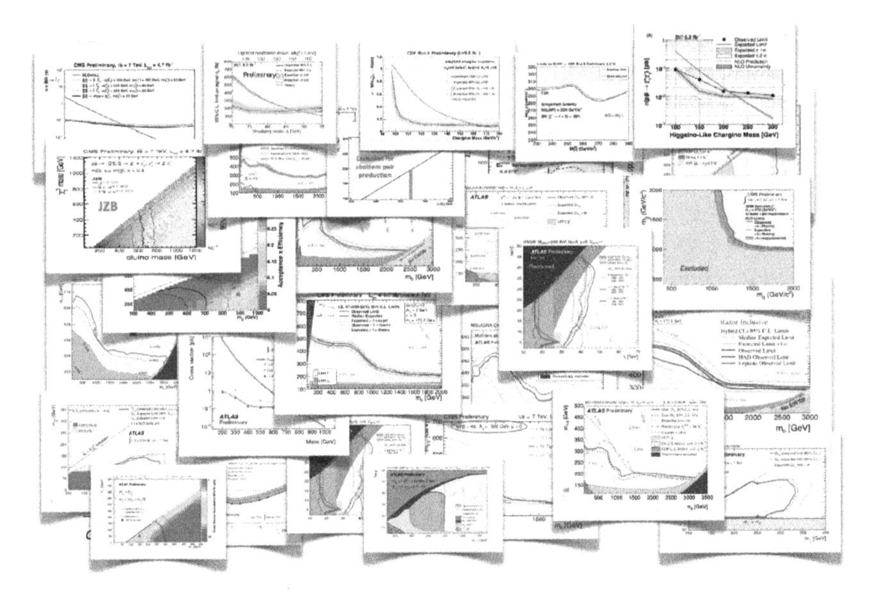

Fig. 13.19. A sample of the many early searches for supersymmetry performed at the LHC and the Tevatron. All these studies strive to cover as many conceivable scenarios as possible. At the end of LHC Run 2, however, no positive signal of supersymmetry was found.

produced at the LHC, the only hope of the experimentalists, until a more powerful accelerator is available, is to see their effects in virtual processes, by studying precision observables or rare processes. We will revisit this possibility when discussing flavour physics in chapter 14.

If sparticles with sufficiently low masses to be produced at the LHC exist, how could they have escaped our attention? Theorists and experimentalists work on such "stealth supersymmetry" scenarios, trying to investigate all possible options how supersymmetry could have camouflaged. One possibility, already mentioned above is that the spectrum of sparticles masses is compressed. The resulting signatures are soft and hard to disentangle from Standard Model backgrounds. Smart analysis techniques, whose discussion lies beyond the scope of this book, have been developed to address such cases. The exclusion limits on third generation squark and gaugino production are far less constraining than those on first and second-generation squarks and gluinos. More data and another increase in collision energy will help improve these.

It might also be that R-parity is non-conserved, which would allow sparticles to decay into Standard Model particles only, possibly removing one of the attractive features of supersymmetry,[11] a stable dark-matter candidate. R-parity conservation is arbitrarily imposed and not enforced by any known symmetry.

However, uncontrolled broken R-parity quickly leads to proton decay (for example $p \to \pi^0 e^+$), which is (so far) not observed in Nature. Sparticle decays via R-parity violating interactions change dramatically the signature of supersymmetric events since the lightest neutralino may decay to leptons or quarks (see figure 13.20) or a mixture of both. Such events will not feature large missing energy, but could lead to many-leptons final states or resonance production and thus resemble non-supersymmetry new physics signatures. Over the course of Run 2 ATLAS and CMS have developed a broad programme of searches covering such scenarios. Some parameter settings can lead to the occurrence of heavy long-lived particles giving rise to interesting signatures in the particle detectors. We will discuss such scenarios in section 13.8.

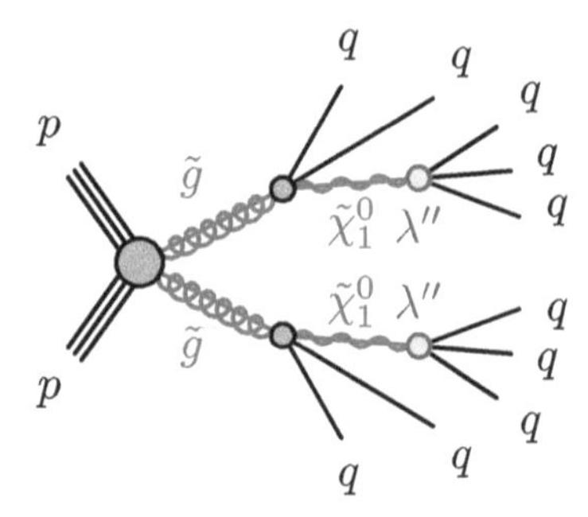

Fig. 13.20. Feynman graph for gluino pair production with decays to four jets and two lightest neutralinos. Due to the non-zero coupling λ'', breaking R-parity, each neutralino decays into three jets, leaving the event without missing energy. Depending on the value of λ'', the decay can occur promptly or delayed.

Fig. 13.21. The experimental hall of NA62 at CERN for the study of the extremely rare $K^+ \to \pi^+ \nu\bar{\nu}$ decay.

It is also possible to look for effects from virtual supersymmetric particles contributing to quantum corrections. This type of study could establish evidence for sparticles that are too massive to be produced and directly observed at the LHC. Many of the observables sensitive to high-mass quantum corrections stem from the charged-lepton and quark-flavour sector of the Standard Model, a field dubbed *flavour physics* that we will discuss in chapter 14. Examples could be the modification of a suppressed decay

[11]Theorists have found ways to maintain viable dark-matter candidates in R-parity violating supersymmetry.

rate in the Standard Model such as that of a B meson decaying to two muons (section 14.1.2), charged-lepton flavour violating decays such as $Z \to \ell\ell'$, $H \to \ell\ell'$ (where $\ell \neq \ell'$), $\tau \to \mu\gamma$, $\mu \to e\gamma$, or the modification in the mixing and CP properties of neutral B and K mesons. Many of these processes are studied at the LHC. Other observables, such as precision measurements of the muon magnetic moment or the extremely rare kaon decays $K^+ \to \pi^+\nu\overline{\nu}$ and $K^0_L \to \pi^0\nu\overline{\nu}$ with predicted branching fractions as small as 9×10^{-11} and 3×10^{-11}, respectively, are studied with dedicated experiments.[12] With the exception of the muon magnetic moment, where a longstanding discrepancy between measurements and the Standard Model prediction subsists,[13] and some non-significant anomalies linked to lepton universality in B-meson decays (discussed in section 14.1.4), all other results are in agreement with prediction. Because of the sensitivity of these measurements to heavy virtual particles, the success of the Standard Model flavour physics sector represents since long a source of discomfort for supersymmetry and other beyond the Standard Model physics, as are the anomalies seen a source of excitement among particle physics.

13.4. Strong gravity in extra space dimensions

In section 4.3 we have introduced the theoretical hypothesis of a bulk, made of the three traditional space dimensions plus compactified extra spatial dimensions. The Standard Model particles would live on the three-dimensional *brane*, while gravity extends through the entire bulk, where it is strong. This set-up offers a solution to the problem of the enormous difference (hierarchy) between the scale of electroweak interactions and that of gravity, and the instability that comes with their interdependence. The gravitons in the extra dimensions form *towers* of *Kaluza–Klein excitations.*[14]

[12]The currently ongoing NA62 experiment (figure 13.21) at CERN studies $K^+ \to \pi^+\nu\overline{\nu}$ decays. Protons of 400 GeV extracted from the SPS hit a stationary beryllium target producing about one billion secondary particles per second of which about 6% are K^+. Their decays are studied in a dedicated 270-metre long decay and measurement facility with as primary challenge the suppression of overwhelming backgrounds. In 2020 NA62 reported first evidence of 3.5σ for the $K^+ \to \pi^+\nu\overline{\nu}$ decay. The measured branching fraction is in agreement with the Standard Model prediction.

[13]In 2021, the first result using only a fraction of the available data of a renewed Muon g-2 experiment at Fermilab (USA) confirmed the BNL measurement. The combined Fermilab and BNL experimental precision reaches 0.35 parts per million and the discrepancy with the Standard Model prediction rises to 4.2σ.

[14]Would the Standard Model fields propagate into the large extra dimensions, they would associate Kaluza–Klein towers that should have been observed already.

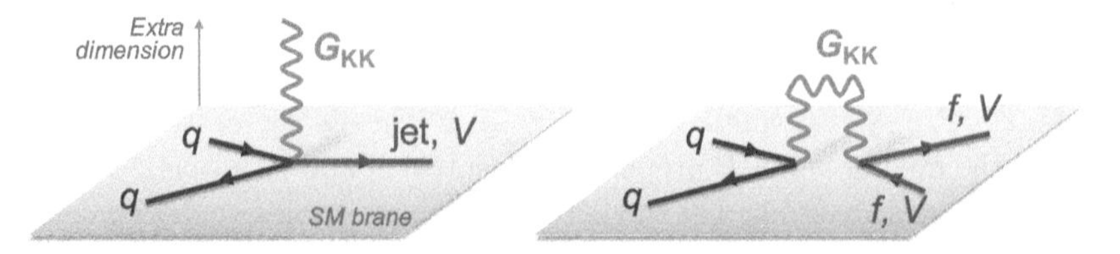

Fig. 13.22. Quark–antiquark scattering (gluon–gluon scattering would be equivalent) in presence of extra spatial dimensions. The plane in blue represents the three (space) plus one (time) dimensional brane of our world to which the Standard Model fields are confined and thus also the colliding protons and the final-state particles measured in the detectors. In the reaction on the left-hand side of the picture, a Kaluza–Klein graviton tower G is emitted in the extra dimensions, which appears as missing energy for an observer confined to the blue brane. The graviton emission is associated with a quark or a gluon, giving rise to a single jet in the detector, or a vector boson V (photon or Z). The picture on the right-hand side sketches a case where Kaluza–Klein gravitons participate only as virtual particles in the production of a fermion or vector boson pair. Their contribution is responsible for an increase in the production cross-section.

In case of large extra dimensions of compact radius R reaching micrometre scale (the ADD scenario, see inset 4.4 on page 139), the small coupling of the graviton to the Standard Model particles is compensated at large enough energy by the colossal number of accessible Kaluza–Klein states that is summed over. Indeed, the mass difference of the Kaluza–Klein states in a tower is given by the (small) energy scale, $R^{-1} = M_f(M_f/\overline{M}_P)^{2/n} \sim (M_f/\mathrm{TeV})^{(n+2)/2} \cdot 10^{(12n-31)/n}\,\mathrm{eV}$,[15] of the (large) extra dimension.[16] The most direct manifestation of large extra dimensions would then be the presence of a large number of Kaluza–Klein gravitons, G_{KK}. The produced gravitons in the process $pp \to \mathrm{jet} + G_{\mathrm{KK}}$ do not interact in the detector, so that the typical signature is a *mono-jet*, that is, a high-energy jet recoiling against missing energy (see left-hand illustration in figure 13.22).

---

[15]Here, $\overline{M}_P$ is the reduced Planck mass given by $\sqrt{\hbar c/(8\pi G_N)} \simeq 2.4 \times 10^{18}\,\mathrm{GeV}$, M_f is the new fundamental mass scale of gravity and n the number of extra dimensions (see inset 4.4). The relation for R^{-1} yields mass splittings for $M_f = 1\,\mathrm{TeV}$ and $n = 2, 4, 6$ of 0.3 meV, 20 keV, and 7 MeV, respectively.

[16]There exist strong astrophysical constraints on the ADD model: (i) the requirement that Kaluza–Klein gravitons do not carry away more than half of the energy emitted by the supernova SN1987A gives the bounds $M_f > 14\,(1.6)\,\mathrm{TeV}$ for $n = 2\,(3)$; (ii) Kaluza–Klein gravitons produced by all supernovae in the Universe lead to a diffuse γ-ray background generated by graviton decays into photons; measurements by the EGRET satellite imply $M_f > 38\,(4.1)\,\mathrm{TeV}$ for $n = 2\,(3)$; (iii) limits on γ-rays from neutron-star sources on graviton halos emitting photons imply $M_f > 200\,(16)\,\mathrm{TeV}$ for $n = 2\,(3)$; (iv) the decay products of the gravitons forming the halo can hit the surface of the neutron star, providing a heat source; the low measured luminosities of some pulsars imply $M_f > 750\,(35)\,\mathrm{TeV}$ for $n = 2\,(3)$.

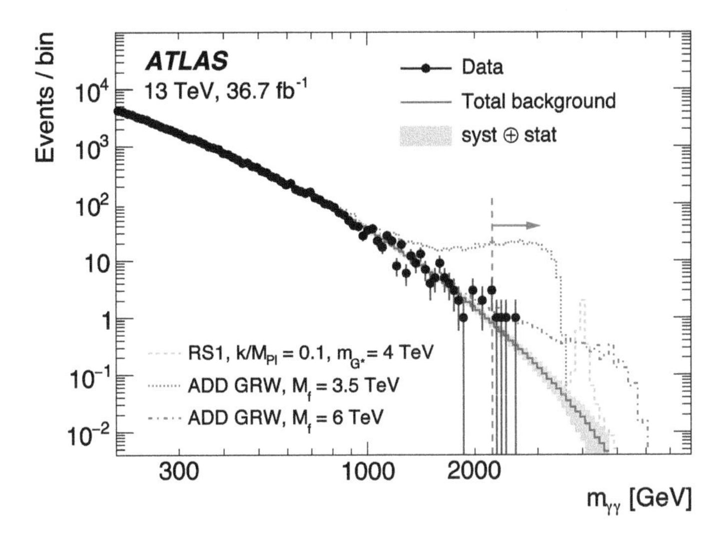

Fig. 13.23. Invariant mass spectrum for photon pairs in data recorded in 2015 and 2016 by ATLAS. The blue histogram is the Standard Model prediction including the experimental background due to jets mis-identified as photons. The red histograms are the predictions of ADD-type models with fundamental gravity energy scales taken to be $M_f = 3.5\,\text{TeV}$ and $M_f = 6\,\text{TeV}$, respectively. The green histogram is a prediction of a Randall–Sundrum-type model, for a specific choice of the model parameter (denoted $k/\overline{M}_P$) and the mass of the excited graviton. In this model, a resonance peak is expected at the mass of the first excited level of the graviton. The chosen parameter values correspond in both models to the limits of sensitivity established in this study by counting the number of events above the mass indicated by the red arrow. *Image: ATLAS Collaboration, Phys. Lett. B 775, 105 (2017), figure modified by authors.*

Alternatively, the extra dimensions may be detected via virtual Kaluza–Klein graviton exchange. It forms a set of finely spaced resonances, which manifests itself as a non-resonant deviation from the expected Standard Model shape at high invariant mass of lepton or photon pair events (see right-hand illustration in figure 13.22). An example for an ATLAS search for such a phenomenon in di-photon events is shown in figure 13.23. The data points are the measured di-photon invariant mass. The blue histogram is the Standard Model prediction including a small contribution from mis-identified photons which is estimated from data. The orange shape indicates the uncertainty in the prediction. This is the same type of background as the one affecting the search for and measurement of the Higgs boson decaying into two photons. The dashed and dotted red curves indicate the expected shapes from putative graviton exchange in the ADD scenario using the high-energy cut-off prescription proposed by Gian Giudice, Riccardo Rattazzi and James Wells (GRW). These scenarios are

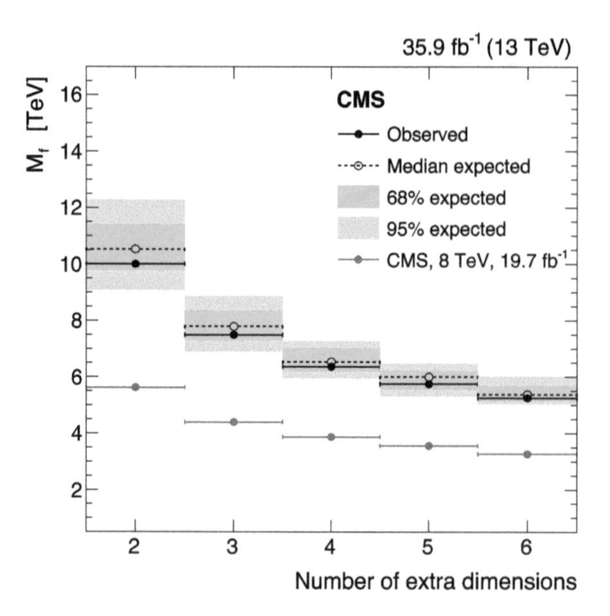

Fig. 13.24. Lower limits at 95% confidence level on the fundamental Planck scale M_f as a function of the number of extra spatial dimensions as obtained by CMS in an analysis of mono-jet events recoiling against missing energy. The solid black lines give the limits obtained from the data. The dashed black line shows the corresponding expected limits assuming absence of extra spatial dimensions. The yellow and green bands show the 1σ and 2σ variations of the expected limits, which are compatible with the observed limits in all cases. The blue lines show the less-constraining results obtained with Run-1 data. With increasing number of dimensions, the partonic cross-section of graviton production (see left sketch of figure 13.22) decreases, therefore the reduced limits on M_f. *Image: CMS Collaboration, Phys. Rev. D 97, 092005 (2018), figure modified by authors.*

excluded by the data. Figure 13.24 shows upper limits obtained by CMS using mono-jet events on the fundamental Planck scale M_f versus the number of extra spatial dimensions n. As the partonic jet plus Kaluza–Klein graviton cross-section scales with $M_f^{-(n+2)}$, the limits on M_f decrease with increasing number of extra dimensions.

There is a modification of the ADD model with extremely small extra dimensions — of the order of the Compton length, see inset 3.2 on page 117 — in which Standard Model gauge fields can propagate into the bulk and then form Kaluza–Klein excitations with mass interval of the order of one TeV and thus visible at the LHC. However, if the mass interval were too large, such a scenario may escape the current LHC limits.

The above scenarios posited that the extra dimensions are flat (for example, their topology can be that of a torus) and that they factorise with

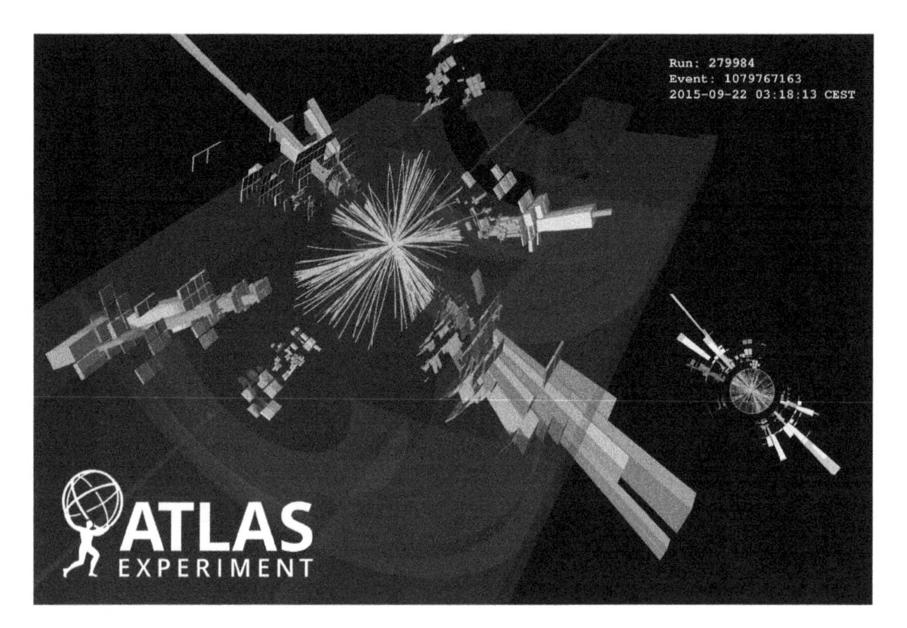

Fig. 13.25. Display of an event with six jets and a sum of transverse momenta of 6.4 TeV. This event was selected by an early ATLAS search for micro black hole production in multi-jet events using 13 TeV proton–proton collision data. *Image: ATLAS Collaboration, JHEP 03, 026 (2016).*

the usual four-dimensional space–time. Non-factorised geometries allow the generation of a large mass hierarchy even for small extra dimensions. The Randall–Sundrum model with one warped extra dimension (see inset 4.5 on page 141) implements such a non-factorisation by exponentially scaling ("warping") the four-dimensional space–time metric with the coordinate of the extra spatial dimension, into which only gravity extends. As a consequence, all mass scales in the bulk are of the order of the Planck scale, so there is no mass hierarchy. However, we on the four-dimensional Standard Model brane (also called the *TeV brane* in this model) see a red-shifted mass scale that is reduced by the exponential warp factor. On the other hand, the Kaluza–Klein gravitons-to-Standard-Model couplings are increased by the warp factor so that weak scale Kaluza–Klein gravitons with weak scale couplings should produce universal spin-2 resonances. This is a spectacular signature! The ATLAS di-photon analysis introduced above also searched for such type of phenomena. The green resonance curve in figure 13.23 shows a 4 TeV Randall–Sundrum graviton, which is just excluded by the data. If such a signal were discovered, in order to truly identify these resonances as spin-2 gravitons, one would need to demonstrate that it is indeed spin-2 through the analysis of angular distributions of the measured final states, and that the couplings are universal (as general relativity does not distinguish particles type or flavour) through the measurement of their decay branching fractions.

While these are exciting scenarios of new physics, we should be aware that their calculation (and Monte Carlo signal simulation) relies on phenomenological models describing the interaction of gravitons with Standard Model particles. The problem is that today there does not exist a complete theoretical scheme unifying gravity and the Standard Model. Thus these predictions must be taken with some caution. *Ad hoc* energy cut-off scenarios, like the GRW one above, are introduced to protect the effective theories against breakdown at large energy.

13.5. Black holes at the LHC?

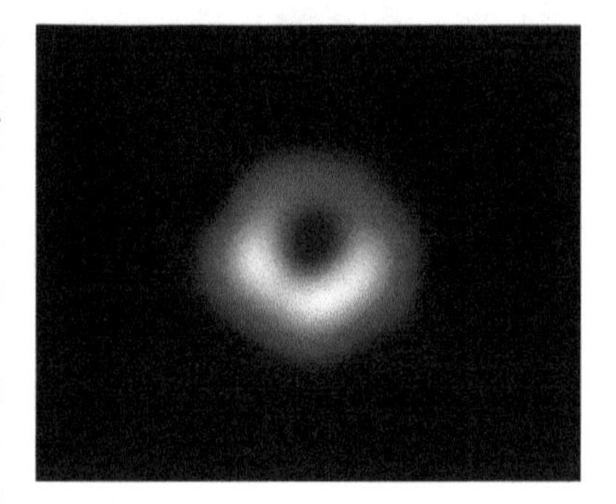

Fig. 13.26. Putative image of a supermassive black hole (6.5 billion times the mass of the Sun) at the core of the galaxy Messier 87 (a massive galaxy in the nearby Virgo galaxy cluster), captured in 2019 by the Event Horizon Telescope, a planet-scale array of eight ground-based radio telescopes. *Image: ESO, April 10th, 2019.*

Models with extra space dimensions can predict the production of *microscopic quantum black holes* at the LHC, if the fundamental energy scale of gravity, M_f, is smaller than the centre-of-mass energy of the colliding partonic constituents in proton–proton collisions. The LHC would then enter trans-Planck energy scales. An object of mass m becomes a black hole if it extends less than its Schwarzschild radius $r_s = 2G_N m/c^2$. For example, the Sun with mass 2.0×10^{30} kilograms, would become a black hole if it were compressed to a diameter of about six kilometres. The Earth would need to be compressed to less than two-centimetre diameter. At the other extreme, figure 13.26 is a recently obtained picture of a colossal black hole as seemingly present at the center of most (all?) galaxies. In extra $4+n$ spatial dimensions, the Schwarzschild radius would become $r_s(n) = 2G_N^{(4+n)} M_f/c^2$, and the cross-section for micro black hole production given by the geometric area would be $\sigma_{\rm BH} \sim \pi r_s^2(n)$. For M_f of a few TeV and not too many extra dimensions one would find a cross-section of order picobarns with 14 TeV proton–proton

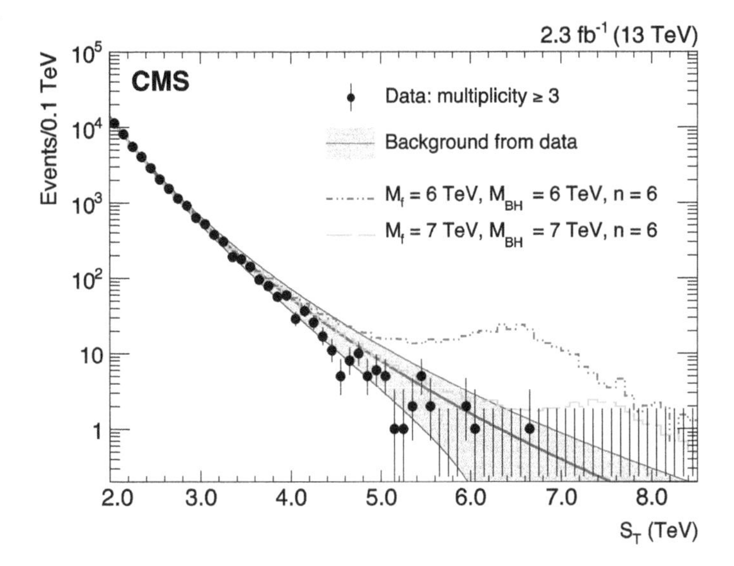

Fig. 13.27. Result of an early CMS search for quantum black hole production using 13 TeV collision data taken in 2015. The plot shows the distribution of the total transverse energy, S_T, for inclusive multiplicities of at least two objects (electrons, muons, photons, or jets). The observed data are shown by points with error bars, the solid blue lines along with the grey shaded band show the main background estimation along with the uncertainty band, obtained from a phenomenological fit. Also shown are distributions of putative quantum black holes. *Image: CMS Collaboration, Phys. Lett. B 774, 279 (2017), figure modified by authors.*

collisions, allowing to produce several microscopic black holes per hour at design luminosity.[17]

[17]The discovery of microscopic quantum black holes at the LHC would be of comparable importance, both scientifically and philosophically, to the discovery of relativity or of quantum mechanics one hundred years ago, as it would imply the existence of additional dimensions of space. It would also represent a major indirect support to string theory, even if there the scale of extra spatial dimensions is many orders of magnitude smaller than what can be probed at the LHC (a natural expectation would be 10^{-35} metres, the Planck's length). The existence of TeV-scale micro black holes would have as a consequence a complete modification of high-energy particle physics compared to its past half-century practice. Once the threshold for the formation of black holes is reached, there would be no sense in looking further for underlying substructure in quarks or leptons. Beyond this threshold, all collisions would just lead to the production of mini black holes. High-energy particle physics will become the study of their properties, with the aim of understanding the number of additional spatial dimensions, their size, their geometry, ways to curl-up, and thus study the nature of graviton-electroweak-strong interactions.

According to Stephen Hawking's thermodynamical theory of black holes, these micro black holes would evaporate in 10^{-26} seconds via *Hawking radiation*. The possible life cycle of a 10 TeV black hole created in an LHC collision may be described as follows[18]: (i) *Creation* (mass 10 TeV, time 0 second): the micro black hole is created in a proton–proton collision, it is asymmetric, may vibrate and rotate, and may be electrically charged; (ii) *"Baldness" phase* (mass 10–8 TeV, time 0–1 $\times$ 10^{-27} seconds): emission of gravitational and electromagnetic waves, and charged particles; the black hole is solely characterised by mass and angular momentum; (iii) *Slowing down* (mass 8–6 TeV, time 1–3 $\times$ 10^{-27} seconds): the black hole is not black but radiates by reducing its angular momentum; its form becomes spherical; (iv) *Schwarzschild phase* (mass 6–2 TeV, time 3–20 $\times$ 10^{-27} seconds): after loosing its angular momentum, the micro black hole evaporates its mass via Hawking radiation; (v) *Measurement phase* (mass 2–0 TeV, time 20–22 $\times$ 10^{-27} seconds): the black hole shrinks down to the new Planck mass (see inset 4.4 on page 139) and fully decays into all particles with probabilities according to their degrees of freedom.

The decay products can be any among the Standard Model particles, irrespective of their internal quantum numbers. Such decays would thus give rise to spectacular spherical events with a total transverse energy of several TeV and with comparable numbers of jets and leptons of all types, as well as of gauge bosons (species-democratic decays). These features would make such events rather distinct from the expected QCD background consisting essentially of jets. Figure 13.25 shows the display of an event with six high-momentum jets that was selected by an ATLAS search for quantum black holes using jets. Figure 13.27 shows a spectrum of the total transverse energy measured by CMS in inclusive events with at least two reconstructed particles. Such a selection was used for a very early analysis using 13 TeV proton–proton collisions taken in 2015 to search for strong gravity effects and the production of quantum black holes. The figure also shows two signal distributions for illustration.

Numerous searches for strong gravity and black hole production by ATLAS and CMS using the full Run-2 datasets did not reveal any sign of the existence of extra spatial dimensions.

[18]This cycle was described in an article by B.J. Carr and S.B. Giddings in the German science magazine *Spektrum der Wissenschaft* in 2005: http://www.spektrum.de/artikel/837198.

Inset 13.2: Is the LHC dangerous?

The possibility of producing micro black holes has generated some controversy regarding the dangers of the LHC before its start-up (see figure, the front page of the British newspaper *The Sun* on September 1st, 2008, a week before the start of the LHC operation). After four years of operation at 13 TeV proton collision centre-of-mass energy, it is fair to say that none of the prophesied/feared/apprehended disasters has occurred. We survived ... so far!

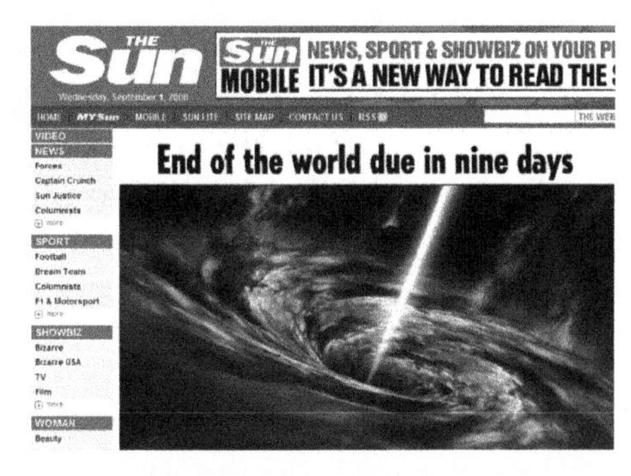

Could quantum micro black holes indeed be potentially dangerous and even swallow our planet Earth (see the artist's view below)? Should a sort of precautionary principle be invoked? The clear scientific answer here is: no! Hawking radiation will lead to so fast evaporation allowing the black hole to barely travel a fraction of a fermi (10^{-13} centimetres) before disintegrating. Yes, but what if this theory were incorrect?

Cosmic rays with energies sometimes well in excess of what the LHC is able to produce have been bombarding the Earth and its atmosphere for billions of years, interacting with the nuclei in the atmosphere. Recent measurements performed with large air-shower detectors show that the interaction rates are of the order of tens of events a day per square kilometre. If this rate is extrapolated to the surface of the Earth and integrated over just a few hundred million years, this number corresponds to several tens of years of LHC operation at its maximal luminosity. There is thus a priori no reason to worry, especially when considering the case of our Sun which has also been bombarded by cosmic rays for billions of years without being destroyed.

Is this comparison fully relevant? Micro black holes produced by the LHC could have a very small velocity, if the collision centre-of-mass is almost at rest in the laboratory frame. In fact, a fair fraction of them would remain trapped in the terrestrial magnetic field and if their lifetime were long, such black holes would have had ample time to accrete mass at the centre of the planet. On the contrary, black holes produced through interactions of cosmic rays with nuclei (practically at rest) in the atmosphere or the bowels of the Earth, would have very large velocities and would cross the Earth in a fraction of a second, not having enough time to accrete sufficient mass. Does this invalidate the conciliating argument relying on cosmic rays? Looking at it more closely, the argument holds true. Some of the micro black holes potentially produced by cosmic rays are electrically charged and thus must loose energy through electromagnetic interactions with terrestrial matter. Even the most pessimistic models predict that they stop within the Earth. Those that are neutral, however, do not remain so for a long time, as they interact with protons via strong gravity. These arguments become even stronger when considering high-density astrophysical objects, such as neutron stars, which, through cosmic ray bombardment, would have been transformed into stellar black holes long ago. Despite this, neutron stars identified as radio pulsars are still present here around us. More than two thousand neutron stars have been identified in the Milky Way (including the Magellanic Clouds), the closest known one being about 500 light years away from Earth.

Among the zoo of particles or exotic entities whose existence is predicted in a number of more-or-less speculative theories, and which have triggered similar fears as black holes, let us mention three: *bubbles of quantum mechanical vacuum*, *magnetic monopoles*, and *strangelets*. The appeasing arguments based on cosmic-ray interactions mentioned above are directly applicable to the first two types of objects, which we will thus dismiss. Strangelets are aggregates of *strange matter* with size of an atomic nucleus, that could potentially be produced in heavy-ion collisions. Strange matter is a hypothetical state of matter made of equal amounts of free u, d, and s quarks (recall that ordinary stable matter is solely made of u and d valence quarks, confined in protons and neutrons).

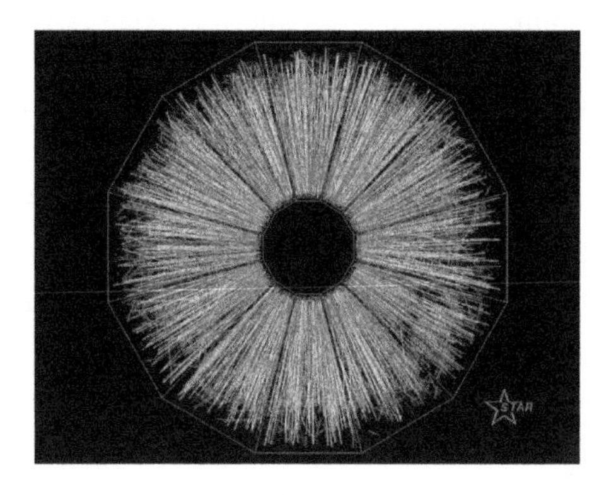

If strangelets were stable objects, which is not expected in the majority of theories predicting their existence, they could accrete ordinary matter, transform it gradually into strange matter and grow without limits. All strangelet theories, however, agree on the fact that the production rate for strangelets — provided they exist — should be very large already at heavy-ion collision energies well below the LHC, for example, at the RHIC collider at Brookhaven (see above the display of a gold-ion collision recorded with the STAR detector) whose exploitation started some twenty years ago. At the start-up of this machine in the year 2000, the same fears were expressed by the same prophets of doom, and we are still here, alive and well, after many years of RHIC operation and many months of operation with heavy ions at the LHC (see chapter 14).

More details on the discussion in this inset can be found in *J. Ellis et al., CERN-PH-TH/2008-136, arXiv:0806.3414 (2008).*

13.6. Higgs boson compositeness

We have discussed in section 4.4 the prospect of the Higgs boson as a composite particle, a bound state of a hypothetical new, high-scale strong interaction sector that is still to be discovered.

The first sign of Higgs boson compositeness to look for are deviations in the Higgs boson coupling properties from the Standard Model predictions. For most composite Higgs models, these deviations are expected to be of the order $\sqrt{1 - v^2/f^2}$ for both fermionic and bosonic Higgs boson couplings, where $v = 246\,\text{GeV}$ is the vacuum expectation value of the Higgs field and f is the mass (or symmetry breaking) scale of the new strong sector. Figure 13.28 shows the constraints on the joint bosonic and fermionic coupling modifiers κ_V and κ_F, respectively, obtained by ATLAS from the combination of available Run-2 Higgs boson coupling results. The numerical results are $\kappa_V = 1.03 \pm 0.03$ and $\kappa_F = 0.97 \pm 0.07$, with an impressive precision already and agreement with the Standard Model value of one. The results can be used to set a limit on the Higgs boson compositeness scale of $f > 1.0\,\text{TeV}$.

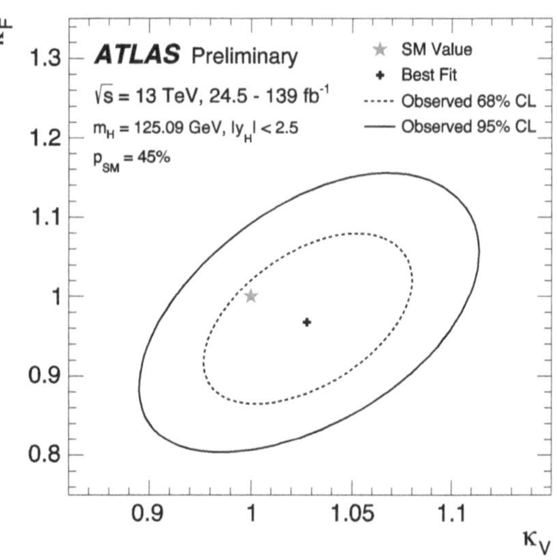

Fig. 13.28. Measured *coupling modifiers* defined as the ratio of the measured Higgs boson coupling strength to either gauge bosons, κ_V, or fermions, κ_F, to the expected Standard Model value. By definition, $\kappa_V = \kappa_F = 1$ if the measured Higgs boson couplings match their Standard Model predictions. *Image: ATLAS Collaboration, ATLAS-CONF-2020-027 (2020).*

Composite Higgs models predict the existence of fermionic resonances (for instance top and bottom quark partners) with masses close to the scale f, and bosonic resonances with masses slightly above. These resonances are expected to couple mainly to heavy Standard Model quarks. Figure 13.29 shows a typical diagram for the single production (through the electroweak force) and decay of a top or bottom partner (denoted T or B), together with the display of a simulated event in CMS. While the Wtb and Zbb coupling strengths are predicted by the Standard Model, the WTb and ZBb couplings, and thus the heavy-quark production cross-sections, depend on the particular new physics scenario assumed. Pair production via gluon fusion is also investigated by the experiments. Here, the production occurs through the strong interaction for which the cross-section is predicted. Depending on the T and B mass,

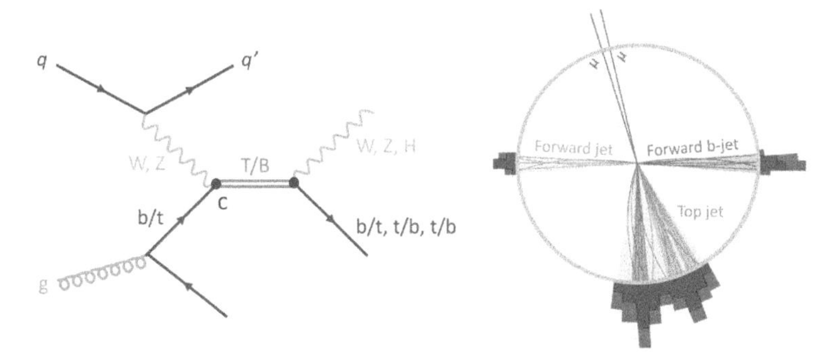

Fig. 13.29. Typical diagram for the single production and decay of a heavy top or bottom quark partner, T or B, with their possible decay channels (left). The cross-section depends on the model-dependent coupling c. The right-hand side shows the display of a simulated T production event with a Z boson decay into a muon pair.

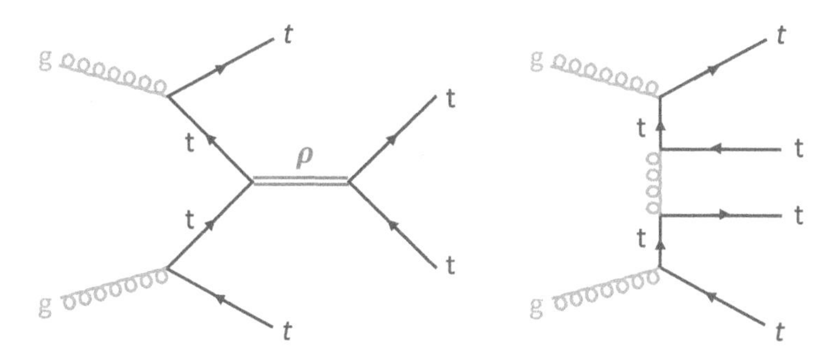

Fig. 13.30. The diagram on the left-hand side shows the production of two top quarks and two top antiquarks involving a neutral bosonic resonance ρ^0. The diagram on the right-hand side shows a Standard Model production mode of the same four top quark final state.

single or pair production may be the advantageous search avenue: while single production is expected to have a smaller coupling, pair production has a larger phase space penalty.

As many bosonic resonance models predict a large coupling to the top quark, a promising search channel could be four top-quark production (see diagram in figure 13.30). The Standard Model production cross-section is predicted to be very small, $\sigma_{t\bar{t}t\bar{t}}^{\mathrm{SM}} \simeq 12.0\,\mathrm{fb}$, as the final state is so massive, almost 700 GeV in rest masses, requiring very hard gluon–gluon scattering in the initial state. Experimentally, this is a very busy event topology with high jet and b-jet multiplicity, possibly leptons and missing transverse momentum. The cleanest decay channels require two same-sign leptons or three leptons to suppress the Standard Model background coming from top-quark pair production accompanied by radiative jets. The presence of two neutrinos, and hence of large missing energy, prevents from searching for a resonance peak in the two-top-quark invariant mass. Figure 13.31 shows the result of an ATLAS search for four-top-quark production yielding evidence of a signal with 4.3σ significance, and a measured cross-section of 24 ± 7 fb, which is slightly above, but in agreement with the Standard Model prediction.

We have presented here only a small selection of searches for new high-scale symmetries motivated by a solution to the gauge hierarchy problem. Alternative models for new heavy quark partners introduce, for example, vector-like quarks, which are hypothetical fermions that are strongly interacting and have left-handed and right-handed components with same colour and electroweak quantum numbers. This property allows vector-like quarks to have gauge invariant bare mass terms, thus not relying on the BEH mechanism. This is an advantageous property, as the measurement of the gluon-fusion Higgs-boson cross-section in agreement with the Standard Model prediction has excluded heavy fermions that couple to the Higgs boson via Yukawa terms. Vector-like quarks have similar properties as the heavy T and B top and bottom quark partners. They can be singly or pair produced and decay to bW, tZ or tH. Also exotic $X_{5/3} \to tW$ processes may exist.

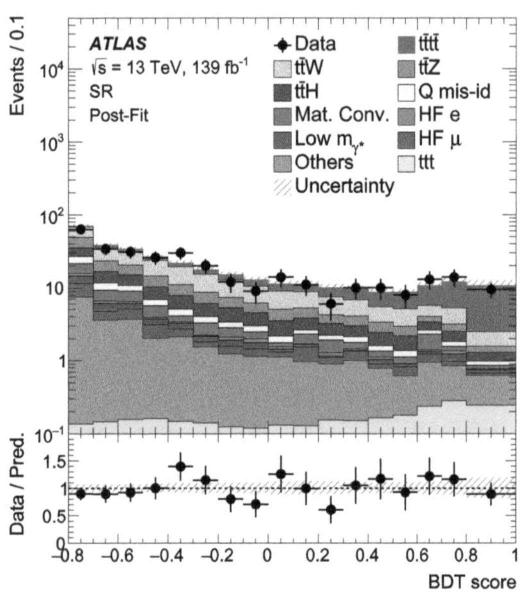

Fig. 13.31. Output of a multivariate boosted decision tree (BDT) algorithm (see inset 10.1 on page 295) in a search for four top quarks by ATLAS. The data are represented by the dots with error bars. The coloured shapes indicate the estimated signal (red) and background contributions. The signal accumulates at large BDT score. The lower panel shows the ratio of data to predicted signal plus background contributions. *Image: ATLAS Collaboration, Eur. Phys. J. C 80, 1085 (2020).*

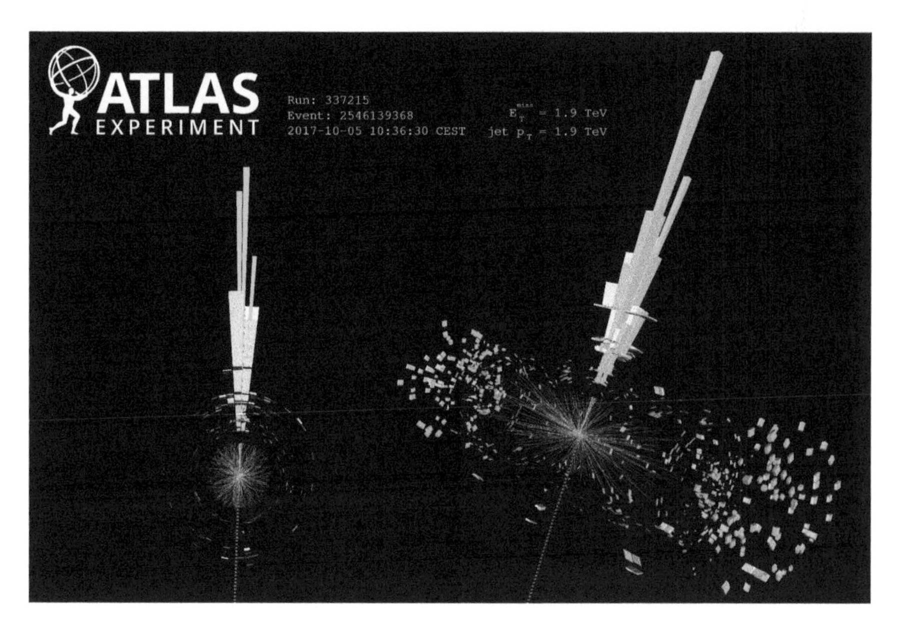

Fig. 13.32. A spectacular mono-jet event recorded by ATLAS with one high-transverse-momentum (1.9 TeV) jet recoiling against a similar amount of missing transverse energy. Such events are candidates for dark matter, but can also be produced in Standard Model processes, such as $pp \to Z(\to \nu\bar{\nu}) + \text{jet}$. *Image: ATLAS Collaboration, Phys. Rev. D 103, 112006 (2021).*

13.7. The search for dark matter at the LHC

Although all experimental evidence of dark matter (see section 3.1) relies on its gravitational interaction with ordinary matter, the LHC may be able to shed light on its nature if dark matter is made of particles. This aspiration is driven by the supersymmetry inspired WIMP model (the WIMP miracle introduced in section 4.1), according to which there are compelling reasons to believe that the coupling strength of the dark-matter particles to Standard Model particles is of the order of the weak interaction, provided that their masses are in the range 10 GeV to a few TeV. The lightest supersymmetric neutralino $\tilde{\chi}_1^0$ could be an obvious candidate for such a particle. In addition, present cosmological models aiming to explain the origin of the large structures of the Universe favour *cold* dark matter (non-relativistic, i.e. $v_{\rm DM} \ll c$), which excludes for instance standard (light) neutrinos. Plausible cold dark-matter models expect relatively heavy dark-matter particles, which could be produced in the proton–proton collisions

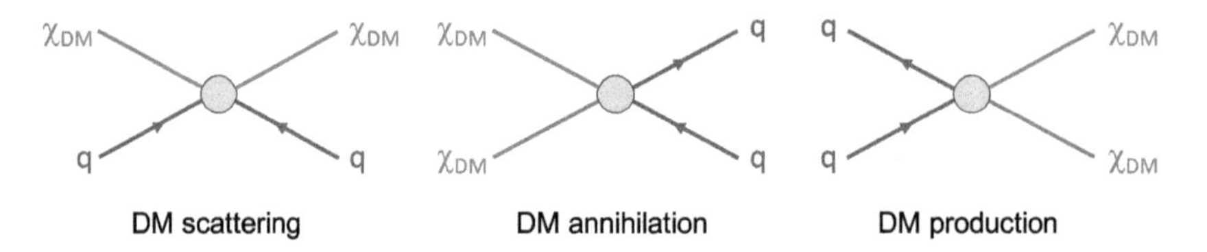

Fig. 13.33. Diagrams illustrating three types of interactions of dark matter with Standard Model particles (here represented, for example, by a quark). The green blobs represent the yet unknown interaction process. Depending on the nature of the dark matter particle (fermion or boson), it could be for instance the annihilation via a heavy boson mediator or the exchange of a supersymmetric squark.

at the LHC.[19] Therefore, the large interest of the ATLAS and CMS physicists to search for dark-matter production and help unravel the dark-matter puzzle. Another type of massive cold dark-matter candidate which could be produced at the LHC is a heavy sterile neutrino proposed in some extensions of the Standard Model.

Figure 13.33 depicts a threefold way through which dark-matter particles may interact with Standard Model particles. Each of the diagrams also represents a different detection strategy exploited by experiments underground (dark-matter scattering), in space (dark-matter annihilation), and using accelerators (dark-matter production).

The first category of experiments endeavours to detect the elastic scattering of omnipresent fossil (from Big-Bang times) dark-matter particles on nuclear matter, as sketched on the left-hand diagram of figure 13.33, with earth-bound detectors. The expected nuclear recoil energies being very small, in the keV to few tens of keV range at best, they are extremely difficult to detect due to thermal and other sources of background noise. To achieve clean conditions and protect from cosmic rays, these experiments are based in underground laboratories such as ancient mines with a kilometre or more overburden of rock. A first type of detectors (for instance, EDELWEIS, CDMS, now upgraded in SuperCDMS, and CREST) exploits

[19]A hypothetical particle very different from a typical WIMP but considered a viable dark matter candidates is the *axion*, which we have already briefly introduced in section 3.4 when discussing the strong CP problem. Even very light (micro-eV mass), cold axions could have been produced abundantly in the early Universe, where the exact mechanism depends on whether Peccei–Quinn symmetry breaks before or after cosmic inflation. These cold axions were never in thermal equilibrium with the rest of the Universe (which explains why they are cold despite their tiny mass) and could provide the missing dark matter. Searches for axion-like particles, albeit not the primary QCD axion candidates, are also performed at the LHC.

both the minute ionisation or scintillating signal together with the heat signals (phonons) that would be deposited in *bolometers*, i.e. cryogenic detectors made of several kilos of ultra-pure germanium or other scintillating crystals, kept at a very low temperature below 20 milli-kelvin. This type of detectors have the potential to provide the best limits in the 0.5–5 GeV WIMP mass range.

Fig. 13.34. Photo of the XENON1T experiment, housed in the 3.6-kilometre underground hall of the Gran Sasso laboratory in Italy. The cryostat is located inside the large water tank (left) next to a three-story building containing auxiliary systems. The heart of XENON1T is a liquid xenon time projection chambers with 3.2 tonnes ultra-pure liquid xenon of which 2 tonnes are employed as target material in the active volume. *Image: Roberto Corrieri and Patrick De Perio.*

A second detector type relies on the simultaneous detection of ionisation and scintillation signals that would be generated by the nuclear recoil in more than 100 kilograms of liquid xenon. In 2020, the leading experiments in this domain are XENON1T (figure 13.34), located in the Gran Sasso road tunnel close to Rome in Italy, LUX, which is operated in a mine in South Dakota, USA, and the PANDAX-II experiment operated in the Jinping underground laboratory, China. These experiments are most sensitive for WIMP masses between 5 and 1000 GeV. All observations until now can be explained by backgrounds due to neutron scattering from cosmic ray interactions, or residual radioactivity in the detector construction materials or the walls of the deep underground experimental halls. Figure 13.35 shows upper limits on the WIMP–nucleon scattering cross-section versus WIMP mass. These upper limits, which have significantly improved in recent years, reach the $10^{-46}\,\mathrm{cm}^2$ level over a large mass domain. This is two orders of magnitude smaller than the neutrino–nucleon cross-section for low-energy (anti)neutrinos produced in a nuclear reactor! The XENON and PANDAX experiments are being upgraded to several tons of sensitive material and new projects (LZ, DarkSide-20k, DARWIN) aim at pushing the sensitivity to new levels in order to eventually cover within the next decade the entire WIMP parameter space for masses above 5 GeV. Experiments with even higher sensitivity will hit the so-called *neutrino floor*, which is background from coherent neutrino–nucleon scattering

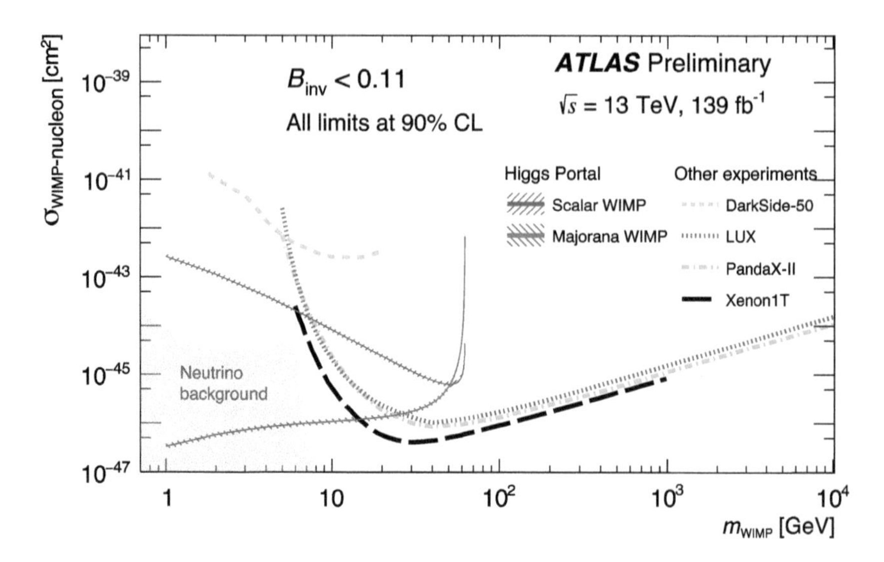

Fig. 13.35. Upper limits on the WIMP–nucleon elastic scattering cross-section versus the WIMP mass. The black, green, pink, and yellow dashed and dotted curves are obtained from underground experiments searching for direct signals of WIMP–nucleon scattering processes. They are most sensitive for WIMP masses in the range 5–1000 GeV. At lower mass, in particular below 1 GeV, the nuclear recoil energy becomes too small to be separated from background noise. The blue and red solid lines are limits derived from an ATLAS search for Higgs boson to invisible particles using the vector-boson fusion topology. The limits are valid to arbitrary low WIMP mass, but only up to half the Higgs-boson mass, above which the decay is kinematically forbidden. The cyan area indicates expected background from coherent neutrino scattering dominated at low WIMP mass by neutrinos originating from the solar carbon–nitrogen–oxygen (CNO cycle) fusion process $^3\text{He} + {}^4\text{He} = {}^7\text{Be} + \gamma$ with $^7\text{Be} + e^- \to {}^7\text{Li} + \nu_e$, and at intermediate mass by boron-8 neutrinos via $^7\text{Be} + p \to {}^8\text{B} + \gamma$ with the β decay $^8\text{B} \to {}^8\text{B}^* + e^+ + \nu_e$. Above 10 GeV WIMP mass the neutrino background is around $10^{-49}\,\text{cm}^2$, slowly increasing to $10^{-48}\,\text{cm}^2$ towards 1 TeV, dominated by neutrinos created in cosmic-ray showers in the atmosphere and from supernovae explosions. *Image: ATLAS Collaboration, ATLAS-CONF-2020-008 (2020), figure modified by authors.*

due to neutrinos from the Sun at low WIMP mass (below about 10 GeV) and from cosmic-ray showers above that mass. New technologies including directional information will be needed to further improve the search sensitivity in that domain.

The second way to look for dark matter consists in trying to detect the present-day annihilation products of fossil dark-matter particles left over after the freeze-out soon after the Big Bang, but still occasionally scattering against each other in regions of higher density, towards galactic centres

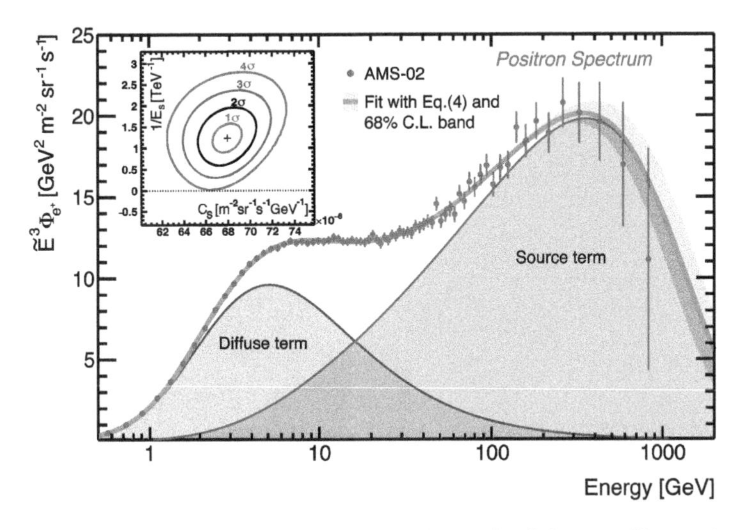

Fig. 13.36. Positron flux measured by the space-borne AMS detector (data points with error bars). The green curve depicts a phenomenological fit to the spectrum parameterising the sum of a low-energy diffuse term from the collision of cosmic rays with interstellar gas, and a high-energy source term dominated by a astrophysical sources. *Image: AMS Collaboration, Phys. Rev. Lett. 122, 041102 (2019).*

for example. The annihilation process might lead to lepton–antilepton or quark–antiquark final states (figure 13.33, centre). Annihilation into a photon pair would produce a monochromatic line in cosmic-ray spectra in the MeV to TeV range. This method is for example employed by the satellite-borne Fermi γ-ray Space Telescope and the ground-based telescopes of the HESS collaboration in Namibia, without finding positive signals until now. In case of annihilation into a lepton or quark pair, the signal could appear as an excess in the production of antiparticles in cosmic ray spectra, characterised by a sharp upper limit, a *Jacobian peak* shape, in the antiparticle spectrum with a sharp cut-off at the dark-matter particle mass.

This is the technique employed by the Alpha Magnetic Spectrometer (AMS) experiment mounted on the International Space Station orbiting the Earth at an altitude of about 350 kilometres. Many antiparticle spectra have been analysed by the AMS collaboration: positrons, antiprotons, even antihelium nuclei have been observed. Figure 13.36 shows the positron flux versus energy based on 1.9 million positrons measured by AMS. The observed spectrum can be fit by the sum of a low-energy diffuse cosmic-ray scattering contribution and a high-energy astrophysical source term.

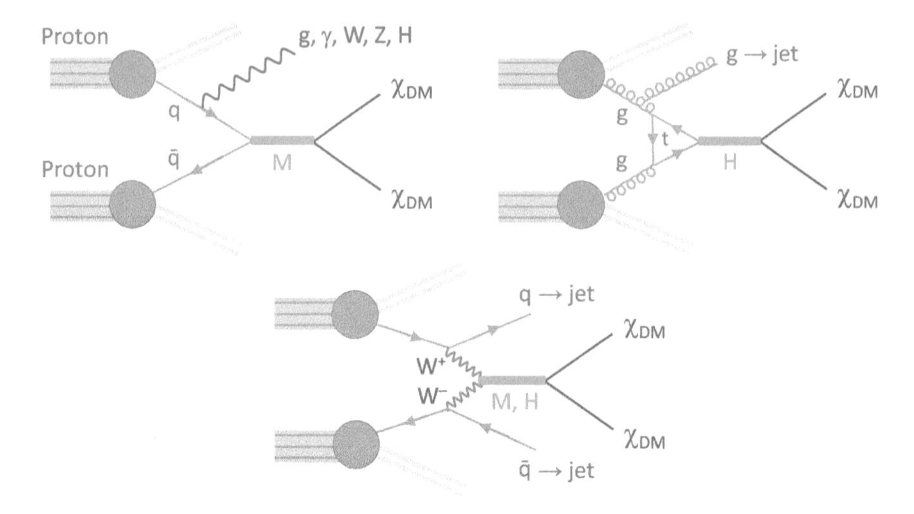

Fig. 13.37. Diagrams of processes used to search for dark-matter particles in proton–proton interactions at the LHC. The nature of the interaction between the dark-matter particles and the proton depends on the specific model considered (here illustrated by a heavy mediator M such as a Z' or Higgs boson). For the two processes on top, the emission of a high-energy boson in the initial state generates the missing energy due to the energy carried away by the undetected dark-matter particles. The experimental signature for such events is the presence of a jet, photon, W, Z, or Higgs boson balancing the missing energy. The bottom diagram refers to a vector boson fusion production mode with two opposite jets in the forward and backward hemispheres balancing the missing transverse energy.

It exhibits an excess of positrons above 25 GeV compared to naive power-law trends, and evidence for a cut-off at around 800 GeV positron energy. Further measurements and theoretical work are required to pin down the origin of the source contributions. Certainly, any interpretation in terms of heavy dark-matter annihilation is inconclusive at present.

The third way is the direct production of WIMP pairs in high-energy proton collisions (right-hand diagram in figure 13.33) provided dark matter couples to the constituents of the proton. The search for dark-matter particles is indeed among the highest priorities of the LHC research program. As dark matter is invisible, inclusive production of dark-matter particles is hard to detect and, in particular, to trigger on with the fast, few microseconds per event, online selection of the LHC experiments. ATLAS and CMS therefore use the associated production with Standard Model particles together with missing energy to hunt for dark-matter. Figure 13.37 shows the three main production mechanisms studied for this search.

The most striking and among the strongest signatures of the first process (upper left diagram in figure 13.37) is the mono-jet one. It consists of a high-momentum gluon jet radiated from the initial state, against which recoils the missing energy from the escaping WIMPs. A beautiful example of such an event measured by ATLAS is shown in figure 13.32. The main background in this search stems from $pp \to Z$ + jets production, where the Z-boson decays to two neutrinos. Its contribution can, fortunately, be directly measured with the data by using well identified Z-boson decays to electron and muon pairs. Figure 13.39 shows the measured missing transverse energy distribution in a mono-jet search by ATLAS using the full Run-2 dataset. The comparison with the precise Standard Model prediction does not show any significant hint for an excess in data that might originate from a dark-matter signal. The interpretation of this search in terms of limits on dark-matter production depends on the assumed coupling strength between the mediator M (cf. upper left diagram in figure 13.37) and the proton on one side, and that between M and the dark matter particles on the other side.

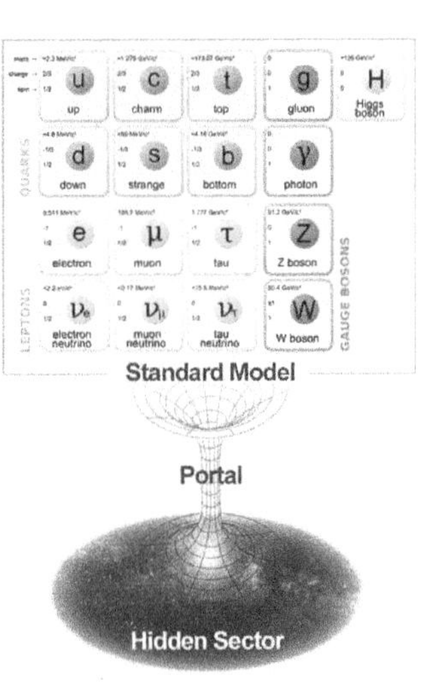

Fig. 13.38. Could a portal interaction link the Standard Model to a hidden or dark sector? *Illustration: Institute for Basic Science, Korea.*

Since dark-matter particles are sensitive to gravity and hence have mass, the Higgs boson could (should, given the WIMP miracle?) decay into them, provided it is kinematically possible, that is, $m_{\chi_{\mathrm{DM}}} < m_H/2 \approx 60\,\mathrm{GeV}$. The Higgs boson might be the mediator or *portal* to an entire *hidden* or *dark sector* (figure 13.38), which could have a richer structure than only χ_{DM} particles.[20] The most appropriate channel to look for *invisible*

[20]There might be other portal fields. For example, several experiments, including at the LHC, are searching for a dark photon. This is a hypothetical very light particle which shares some properties with the Standard Model photon. However, while photons couple to the electromagnetic charge, dark photons couple to the so-called dark charge, which might be carried by other dark-sector particles. A vector portal would allow the mixing between the photon and dark photon. An axion portal could connect the photon to the axion. There are only several possible portals identified so far, and each portal is a tool to search for dark-sector particles.

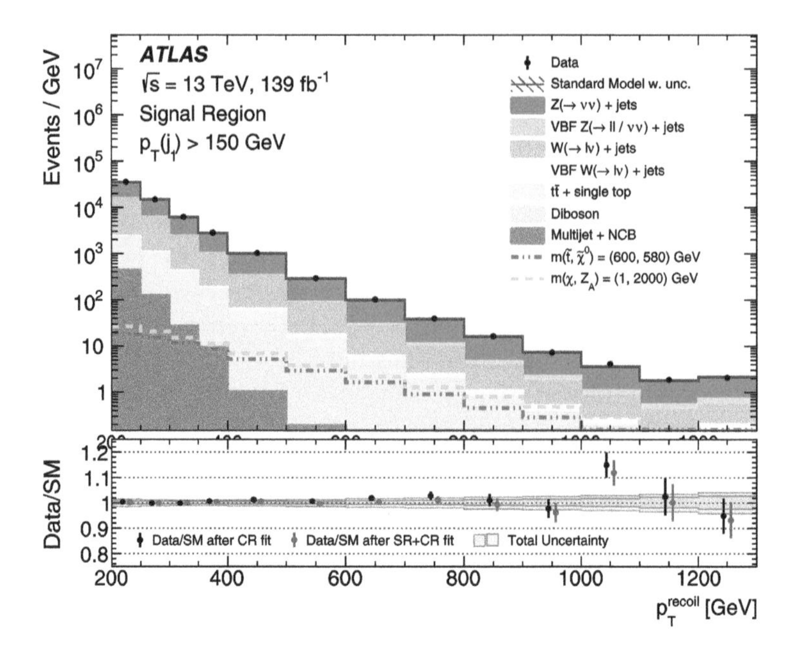

Fig. 13.39. Distribution of the recoil transverse momentum, quantifying the missing transverse energy, in events with a high-momentum jet measured by ATLAS. The dots with error bars represent the observed data and the coloured histograms the various background distributions. Also shown are putative signals assuming supersymmetry or a Z' mediator. The lower panel shows the ratio of data to Standard Model prediction, exhibiting the high precision obtained (and needed) in this challenging search. *Image: ATLAS Collaboration, Phys. Rev. D 103, 112006 (2021), figure modified by authors.*

Higgs boson decays is vector-boson fusion production (see lower diagram in figure 13.37). Also here, no sign of dark matter production is observed yet by neither ATLAS nor CMS. ATLAS has combined results from this analysis using the full Run-2 dataset with other Higgs boson measurements and obtained a 95% confidence level upper limit of 9% on the $H \to$ invisible branching fraction. The limits from the vector-boson fusion analysis, translated into upper bounds on the WIMP-nucleon elastic scattering cross-section, are shown in figure 13.35. They provide complementary information to direct dark-matter searches with underground experiments for low-mass dark matter.

In the above example, the Higgs boson acted as mediator between the colliding protons and the dark-matter particles. Yet, the Higgs boson certainly does nor exclusively decay to dark matter, nor would any other mediator produced at the LHC do so: the mere fact that it was produced proves that it must couple to the proton's constituents, quarks or gluons. The mediator particle (or force field) should thus also decay to quarks or gluons. Consequently, searches for a di-jet or di-lepton resonance as those discussed in section 13.1 are indirectly sensitive to dark-matter production by constraining the mediator. By definition, these searches have no dependence on the dark-matter mass, although the interpretation in terms of mediator and WIMP mass limits have some residual dependence due to phase space considerations.

The complementarity between dark matter and mediator searches is illustrated by the diagrams shown on top of figure 13.40. The left diagram shows the production of a WIMP pair via exchange of a vector or axial-vector Z' boson. The cross-section for this process depends on the couplings g_q and g_χ. To allow detection, the Z' production must be associated with a high-momentum Standard Model companion. Assuming a specific set of coupling strengths, the limits obtained from, e.g. the mono-jet analysis can be expressed in the WIMP mass versus mediator mass plane, as shown in the lower panel of figure 13.40 by CMS. By construction, only the region $m_\chi < m_{Z'}/2$ allowing the Z' to decay into a χ pair is kinematically accessible. The right-hand diagram shows the corresponding Z' production, but this time it decays into a Standard Model fermion pair, e.g. a quark and an antiquark. Here the limits depend on the assumed coupling strengths for g_q and g_f. One sees that, whether the coupling to χ or that to Standard Model fermions is stronger, the one or the other search will be more sensitive. The particular choice realised in the lower panel of figure 13.40 gives a clear advantage to the di-jet resonance search, in spite of a large coupling to dark mater (here $g_{\rm DM}$ stands for g_χ).

In closing this section, we should emphasise that, while the LHC can effectively search for the production of invisible particles, it cannot prove — in case of a positive signal — that these particles constitute the observed dark matter in the Universe as it does not have access to the signal's lifetime. Dark matter requires stable or very long-lived particles. While direct dark matter experiments do not have this handicap as they measure signals from fossil WIMPs, they will have difficulty to study the very off-mass-shell mediator process.

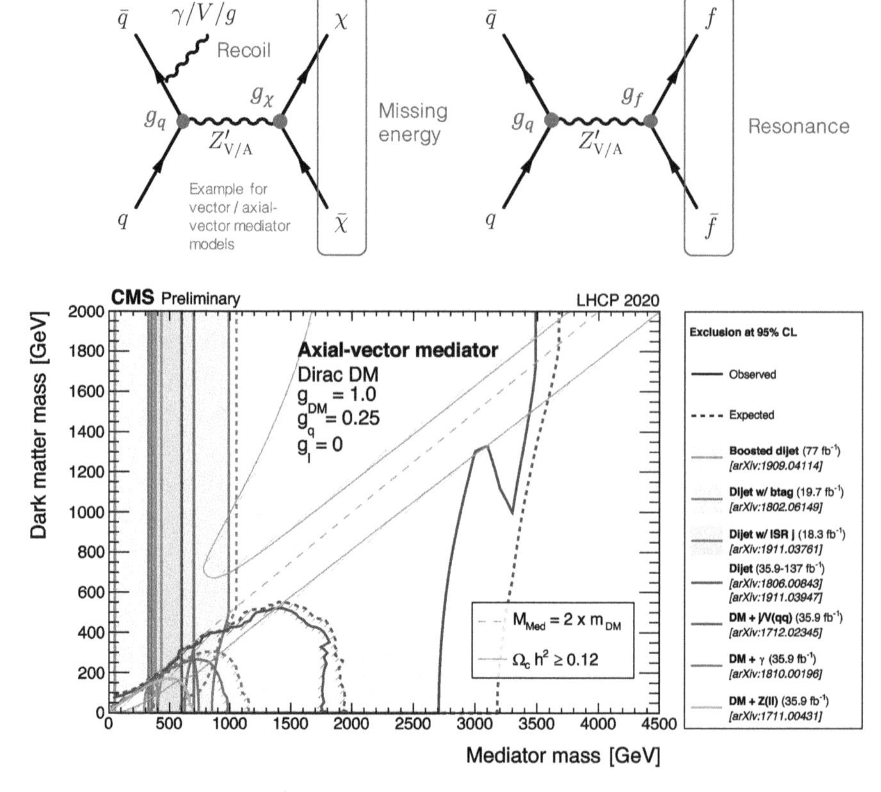

Fig. 13.40. The upper diagrams show the production of an axial-vector or vector Z' resonance in proton–proton collisions at the LHC. In the left-hand process, the Z' decays to a dark-matter pair and the production is accompanied by a hard gluon, photon or vector boson, which allows to experimentally detect it. In the right-hand process, the Z' decays to a fermion pair (for example two quarks giving rise to two jets) which forms a resonance and can be detected in a dedicated two-particle resonance search. The lower panel shows exclusion limits obtained by CMS in the dark matter versus the mediator mass plane. Results for both the upper left and upper right processes are shown. The limits strongly depend on the values assumed for the quark and dark matter couplings, g_g and g_χ. *Lower figure: CMS Collaboration, 2020, figure modified by authors.*

13.8. Long-lived massive particles

Long-lived particles are ubiquitous in the Standard Model (see figure 13.41). Among the matter particles only the electron and the neutrinos are considered stable. The quarks are confined in protons and neutrons. Both (a neutron if bound in a nucleus) are expected to have a finite lifetime even if very long (cf. section 4.2). Leptons or hadrons decaying via the weak interaction

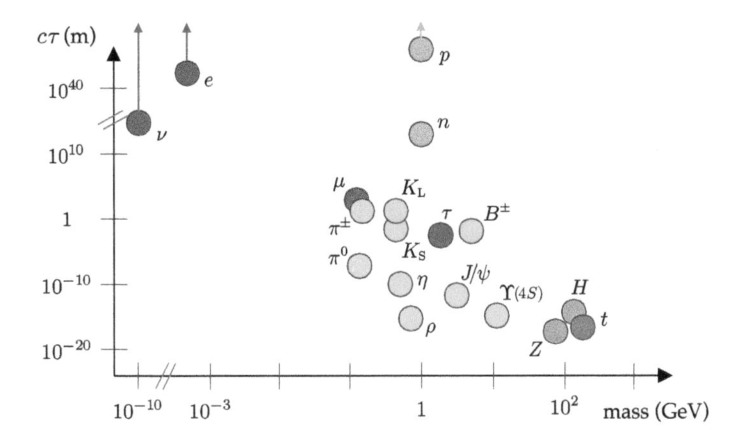

Fig. 13.41. Lifetime (expressed in path length $c\tau$, where one metre corresponds to a lifetime of 3.3×10^{-9} seconds) versus mass for Standard Model particles. In red and purple elementary fermions, in blue baryons, and in green other hadrons. *Image: Brian Shuve.*

such as the τ-lepton, neutral kaons, charm or bottom mesons have lifetimes that are measurable through their path length and displaced decay vertex in a particle detector. Charged pions, kaons and muons also decay via weak interactions but their lifetimes are so long that they appear as quasi-stable particles in most experiments (unless those specialised to study their decays).[21]

The Standard Model does, however, not feature heavy long-lived particles. The heaviest composite particle whose lifetime has been measured is the B_c^+ meson with a mass of 6.3 GeV. New massive long-lived particles are predicted in many new physics models. They can occur due to large virtuality when a decay can only occur via a very heavy virtual particle (as is the case for weakly decaying particles),[22] because of low couplings,[23] and

[21]For example, the decay width of a charged pion, which decays through a highly off-mass-shell W boson into a muon and a muon-neutrino, is proportional to $(m_\pi/m_W)^4 \sim 10^{-11}$. On the contrary, the neutral pion decays three hundred million times faster via the electromagnetic interaction into two photons.

[22]A branch of supersymmetry, called *split supersymmetry*, attempts to solve the lack of evidence for supersymmetry in precision measurements and rare processes by postulating light gauginos only, while the squarks and sleptons are very heavy with masses beyond 10 TeV. Split supersymmetry maintains dark matter and grand unification but gives up on naturalness. The large virtuality of the gluino decay via squarks can make the gluino long-lived.

[23]Low couplings may occur in some supersymmetric scenarios where the gravitino, the spin-3/2 partner of the spin-2 graviton, is the lightest supersymmetric particle, but also in supersymmetry with R-parity non-conservation.

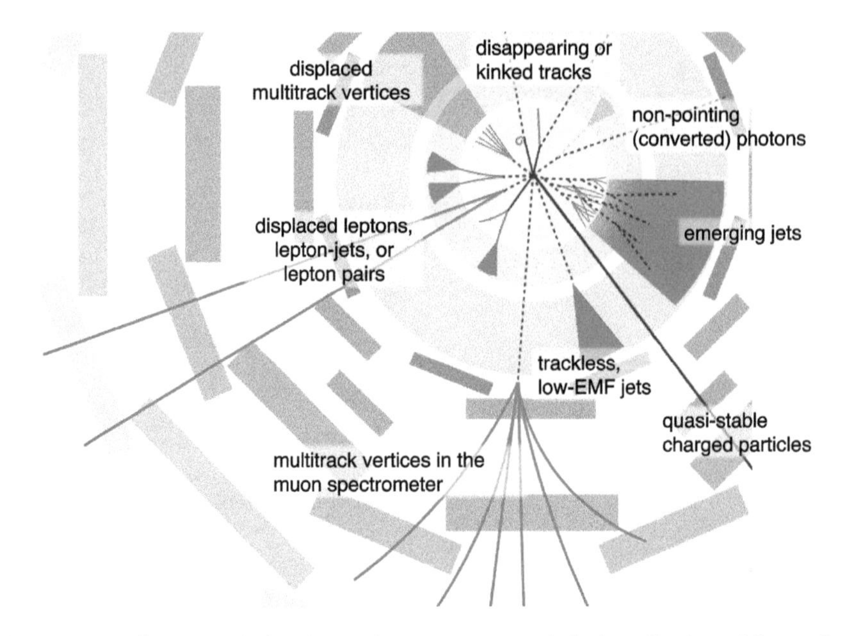

Fig. 13.42. Illustration indicating various ways to search for long-lived particles at the LHC. See text for a discussion. *Image: Heather Russel.*

due to mass degeneracy in a cascade decay.[24] The search for massive long-lived particles is a key element of the LHC new physics search programme.

Figure 13.42 illustrates the various signatures long-lived massive particles leave in a particle detector. The LHC experiments exploit measurements of specific ionisation loss in the tracking detectors, the time-of-flight in the calorimeters and muon systems, as well as the reconstruction of displaced vertices, and kinked or disappearing tracks. Looking for calorimeter deposits outside of the colliding proton bunches allows to detect very long-lived (up to several years!) strongly interacting massive particles that were stopped in the calorimeter layers. Some signatures need dedicated triggers, most require novel analysis strategies to estimate the signal efficiency and determine backgrounds from data.

Figure 13.43 shows lower exclusion limits on the gluino mass versus mean lifetime and flight length in the detector from ATLAS. Different searches cover the full lifetime spectrum excluding gluinos below a mass

[24]This may occur via a scale-suppressed colour triplet scalar from *unnaturalness* or anomaly-mediated supersymmetry breaking scenarios with a wino-like lightest chargino.

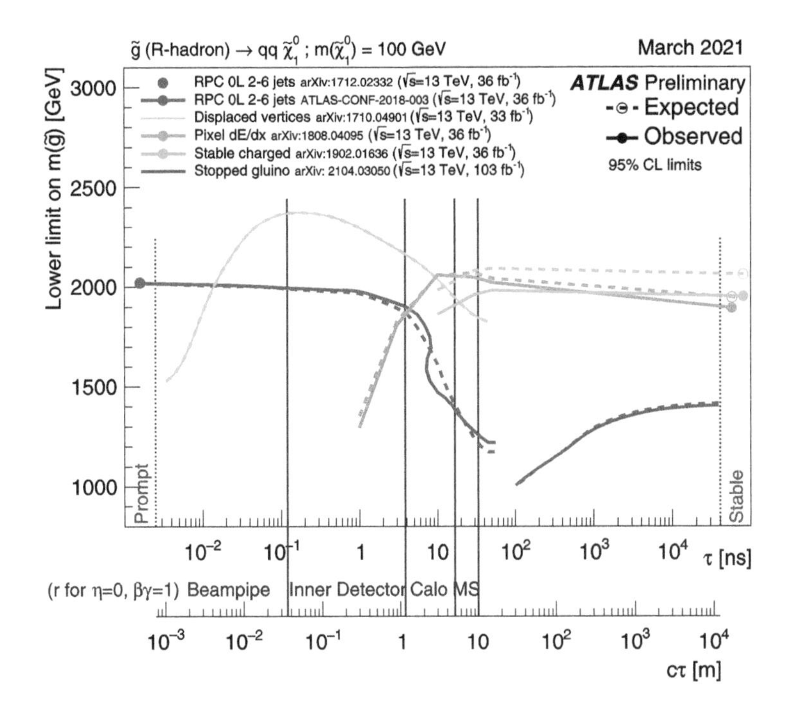

Fig. 13.43. Exclusion limits on the gluino mass versus mean lifetime and flight length in the detector. Varying search strategies cover the full lifetime spectrum. See text for a discussion. *Image: ATLAS Collaboration, ATL-PHYS-PUB-2021-019 (2021).*

of about 2 TeV. The limit from the prompt-gluino-decay analysis using the jets and missing energy signature is shown in red on the left. One observes that this analysis is not blind to scenarios with long-lived gluinos if their lifetime is short enough so that they still decay before the calorimeter,[25] after which the sensitivity rapidly declines. The displaced-vertex analysis (cyan curve) searches for heavy particles decaying in the inner tracking detector by reconstructing the tracks from the decay and fitting them to a common vertex. Such an analysis has low backgrounds, which explains the strong mass limit in figure 13.43. For longer lifetimes, analyses look for isolated tracks with large specific ionisation loss measured in the tracking detectors and/or slow time-of-flight (green and orange curves). In mauve is

[25] We may recall that particles decay according to an exponential law. Their survival probability is given by $P(t) = e^{-t/\tau}$, with τ the mean lifetime after which $1 - 1/e = 63\%$ of the particles have decayed. Even if the mean lifetime of a particle is relatively long, some will decay early (see also inset 2.1).

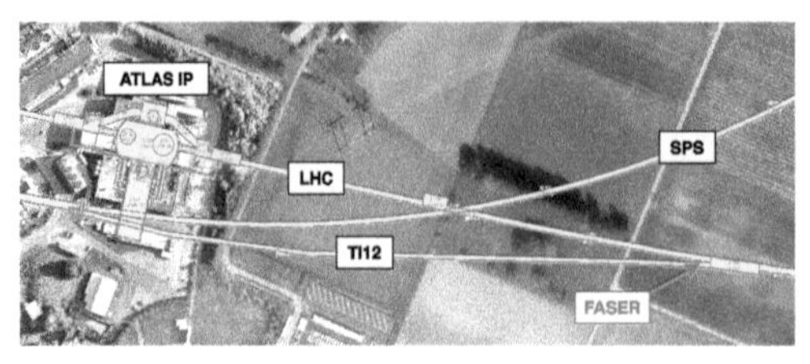

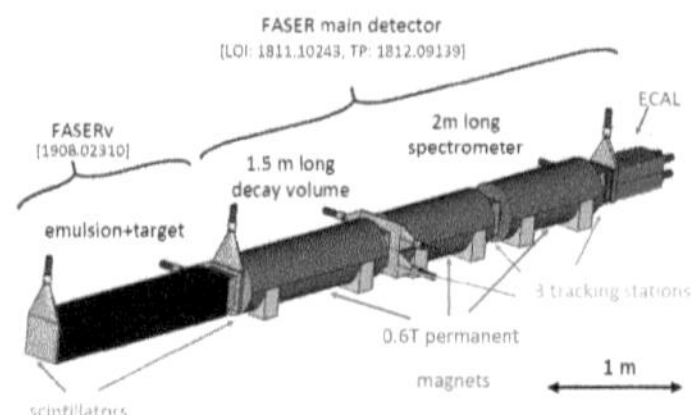

Fig. 13.44. Illustration of the location and design of the FASER experiment to search for light and weakly-interacting particles (such as dark photons). The particles would be produced in the ATLAS interaction point and travel about 480 metres through underground rock before decaying in the detector volume into, for example, an electron–positron pair. The decay particles are measured by silicon tracking layers immersed in a 0.5-tesla magnetic field, and a calorimeter at the far end of the detector. The experiment is being constructed during the LHC long shutdown 2019–2021 and will operate during Run 3. *Image: FASER Collaboration, arXiv:1812.09139 (2018), arXiv:2001.03073 (2020).*

shown the result from the "stopped gluino" search briefly discussed above. It extends to very long lifetimes, while still looking for the decay of the long-lived particle.

We did not cover all signatures sketched in figure 13.42. The search for long-lived particles is a vibrant field, which allows to use the particle detectors in unconventional and creative ways. The redundancy and complementarity of the ATLAS and CMS designs is thereby a particular asset. New experiments, such as FASER (see figure 13.44), searching for light and heavy neutral long-lived particles are being proposed or under construction around the LHC interaction points, further enriching the exploitation of this unique research facility.

13.9. Paradox

"*Where is everybody*" asked allegedly in 1950 the Italian-American physicist Enrico Fermi faced with the apparent contradiction between the conjectured prevalence of extraterrestrial civilisations and the lack of any evidence for it. This question became the famous *Fermi paradox*, which did not fall into oblivion, rather, it is today as relevant as it was then. Theorists have turned Fermi's interrogation to modern particle physics. Despite the strong reasons for the existence of new physics at the fundamental level, be it new TeV-scale particles or forces to alleviate the naturalness problem, signs of weakly interacting massive particles to explain dark matter, or an axion to solve the strong-CP problem, no evidence for such phenomena has been found yet. So: "where is everybody"?

We do not know the answer to this question. We might speculate about reasons for it, such as our searches just didn't yet reach the required sensitivity, or that (contrary to our experience) naturalness might not be a good criterion to understand fundamental physics and we'd have to accept that fine-tuning is ubiquitous in Nature. Our Universe, in all its beauty, is a specific disposition among innumerable other, more mundane choices, here for no other reason than its exceptional conditions that allow life to accrue. As we will develop in chapter 15, these deep and momentous questions can only be addressed by furthering experimental work at the LHC and beyond.

Chapter 14

LHCb and ALICE: The Physics of Flavour and of Hot & Dense Matter

From the very beginning of the LHC project it was planned to have, besides the two large general-purpose experiments, ATLAS and CMS, two more specialised detectors, one to study in detail aspects of heavy-ion physics and the quark–gluon phase of matter, and another one for the detailed study of beauty physics and related aspects of matter–antimatter asymmetry. The first study group converged rapidly towards a unique project, the ALICE experiment. For the second topic, three proposals were initially put forward. Two of these were fixed-target experiments, which means a target bombarded by the LHC proton beam. In the first option a gas jet would be injected in the LHC beam tube, the gas providing the target, whilst in the second option, the proton beam would be extracted from the machine with a curved crystal and the extracted beam would hit an external target. The third option was to operate the LHC in the proton–proton collision mode, as in ATLAS and CMS. The LHCb experiment incorporates the best of these ideas with an asymmetric detector, typical of fixed-target configuration experiments, but operating in a proton–proton collision mode. LHCb thus takes advantage of the much larger centre-of-mass energy available with head-on collisions,[1] providing an about two hundred times larger b-quark production cross-section than for a fixed target exposed to an LHC beam.

[1] As we saw in section 6.2.2, the centre-of-mass energy of a 6.5 TeV proton beam hitting a fixed target is $\sqrt{2E_{\text{beam}}m_p} \approx 0.11\,\text{TeV}$, compared to 13 TeV for head-on collisions.

14.1. Flavour physics and the LHCb experiment

So-called beauty physics[2] concerns the study of hadrons containing at least one b quark. In this chapter, only two types of studies are presented: first, the subtle differences between matter and antimatter, phenomenologically described by the concept of CP symmetry violation (see section 1.8), and second, the search for very rare B-meson decay modes. These two domains of study could both provide windows on physics beyond the Standard Model. Just as the radiative corrections to W mass and Z properties allowed to constrain top-quark and Higgs-boson masses before their discovery, it is possible that new particles, heavier than allowed by kinematics to be directly observed at the LHC, could manifest themselves virtually in the quantum loops, measurably modifying B-meson properties.

Let us consider the case of B_s^0 mesons for which the decay rate into two muons (figure 14.1) is extremely small, with a branching ratio expected to be a few 10^{-9} in the Standard Model. During the LHC Run 2, about 10^{11} B_s^0 mesons were produced within the acceptance of the LHCb detector for $9\,\mathrm{fb}^{-1}$ integrated luminosity at $13\,\mathrm{TeV}$. It is thus possible to measure this branching fraction and verify if it comes out as expected. Likewise, the existence of new particles could modify the CP symmetry violation properties.

Besides the fact that in a single year LHCb allows to record about one hundred times more b-hadrons than has been possible to produce in ten years of running of the B factories (section 2.5), the LHC allows, just as the Tevatron was capable of doing, but with a still larger quantity of data and more higher-performance detectors, to study the B_s^0–$\overline{B}_s^0$ system. With a production rate during Run 2 of about fifty thousand $b\bar{b}$ quark pairs and one million $c\bar{c}$ pairs each second, a major difficulty the LHCb experiment has to face is the online selection of events that indeed contain a b or c quark, without saturating the data acquisition system.

14.1.1. *The LHCb detector*

Due to the smallness of the b-quark mass compared to the beam energy, B hadrons at the LHC are produced in a particular kinematic configuration,

[2]Beauty or B physics is one area of the more general field of *flavour physics*, which primarily studies the properties of bottom, charm and strange mesons and baryons, but also those of charged leptons. Flavour physics is a very active sub-field of particle physics with its own community, experiments and conferences.

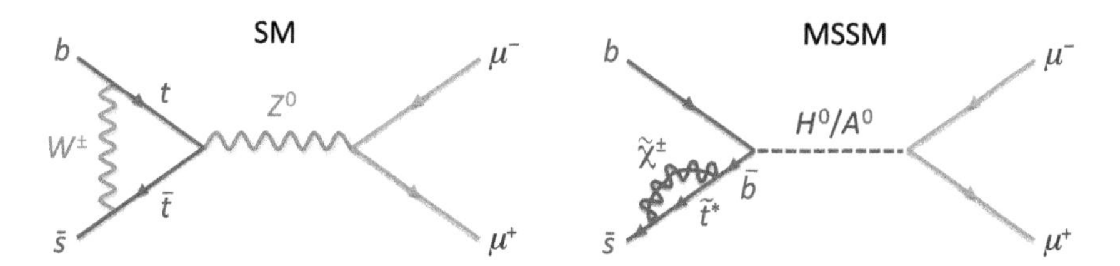

Fig. 14.1. On the left, one of the Feynman diagrams responsible for the decay $B_s^0 \to \mu^+\mu^-$ in the Standard Model, with a W in a quantum loop. The coupling between the W, the strange quark (s) and the top quark (t) is small, therefore the small value for the predicted decay rate. On the right, putative contributions to the B_s^0 decay rate in the framework of the minimal supersymmetric standard model (MSSM). The presence of additional heavy Higgs bosons, H^0, A^0, and of supersymmetric partners of gauge bosons ($\tilde{\chi}^\pm$), would offer other decay possibilities resulting in an increased branching fraction.

emitted preferentially at small angles relative to the beam direction, that is, in the forward direction. The LHCb detector (figure 14.2) has thus been designed very differently from ATLAS and CMS, which are optimised for central collisions producing heavy particles. The components of the LHCb detector are aligned along the beam axis over a length of about 25 metres, all on one side relative to the collision point and thus taking benefit of only half of the delivered luminosity. The detector systems are mostly traditional: the tracking detector combines silicon microstrips in the vicinity of the beam line with drift tubes of five-millimetre radius outwards; the electromagnetic and hadronic calorimeters are of sampling type, with lead plates for the first one and iron plates for the second, the shower detection being achieved through scintillator pads; the muon spectrometer is made of classical wire chambers and an original device of a GEM (Gas Electron Multiplier)[3] type, with an overall detector surface of about 450 square metres.

A distinctive characteristic of LHCb is its use of not one but two Cherenkov detectors (see inset 14.1), the RICH (Ring Imaging Cherenkov), allowing detection of the light-rings emitted by crossing particles and thus

[3] A GEM detector is yet another variant of a wire chamber. It consists of an electrically isolating plate sandwiched between two copper plates at an electric tension of a few hundred volts. This sandwich is pierced with about 10 to 30-micrometre holes spaced by typically 150 micrometres in a grid-like configuration, and the set-up is immersed in a gas. The ionisation electrons, liberated by the passage of a charged particle through the gas, give rise to an electron avalanche in the holes, as they would do in a usual wire chamber. This is a very fast detector, able to withstand a high particle rate.

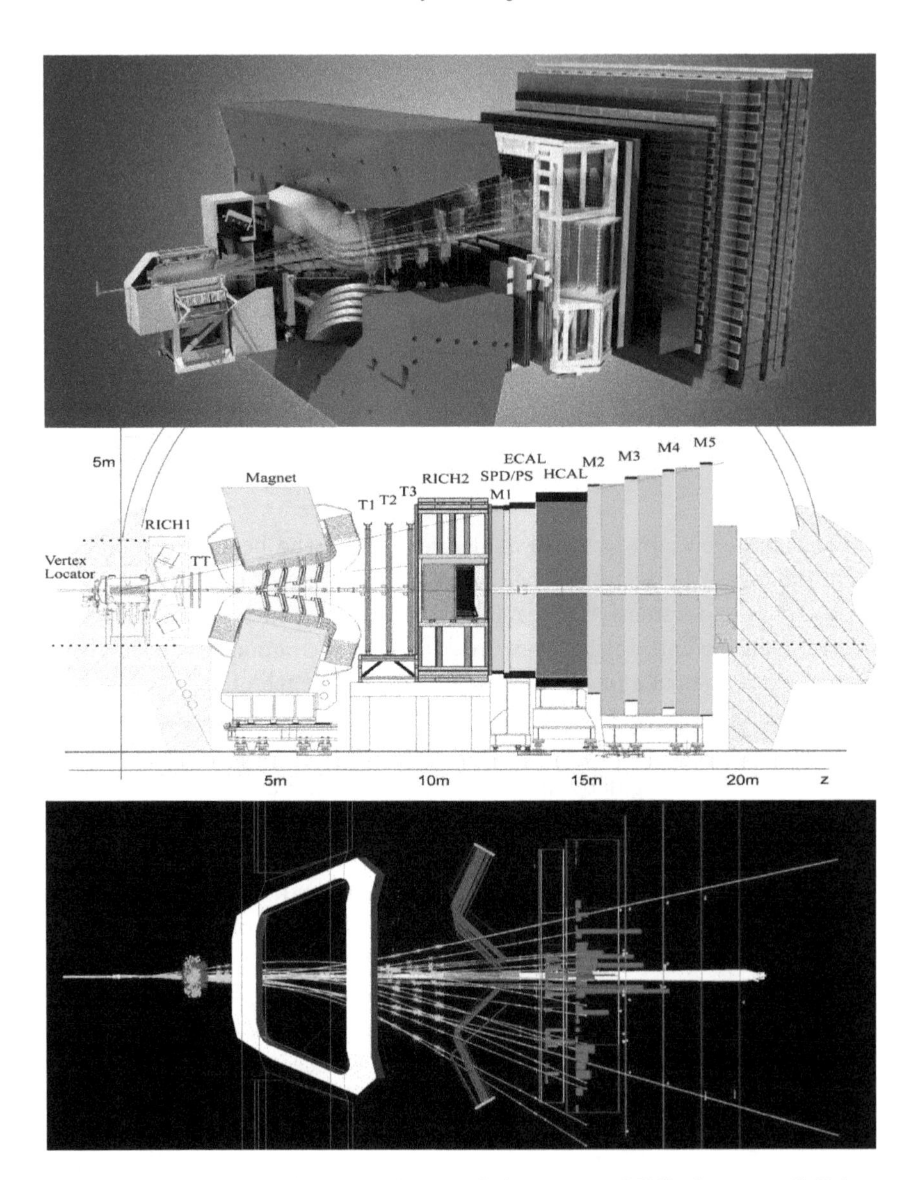

Fig. 14.2. Top: a three-dimensional view of the one-arm LHCb detector. Collisions occur on the left, and all detector elements are lined up in a configuration similar to fixed target experiments. Middle: schematic representation of the LHCb systems, from left to right, vertex detectors (VELO), two Cherenkov detectors (RICH1 and RICH2), a (warm) dipole magnet (figure 14.3), four tracking stations (TT and T1 to T3), calorimeters (SPD, PS, ECAL and HCAL), and the muon spectrometer (M1 to M5). Bottom: simulation of a $B_s^0 \to \mu^+\mu^-$ event. The two muons of opposite charge are well visible at the exit of the muon spectrometer. *Images: LHCb Collaboration.*

Fig. 14.3. LHCb's dipole magnet at the 2018 detector opening. *Image: LHCb Collaboration.*

differentiating between pions, kaons and protons. These are crucial detectors for proper identification of B-particle decay modes.

Another very delicate part of the experiment is its vertex detector, the VELO (Vertex Locator). The lifetime of a b hadron (such as a B_d^0 or B_s^0 meson, or a Λ_b baryon) is about a picosecond (10^{-12} seconds). As it is produced at the LHC with a velocity close to the speed of light, its lifetime is subject to a significant relativistic time dilatation. This allows the b hadron to travel on average about one centimetre before decaying, with most often a charmed hadron (e.g. a D meson) among its decay products. The charmed-hadron lifetime is somewhat shorter than that of the b hadron, but the decay length is nonetheless measurable. To detect these decays it was necessary to construct a silicon microstrip detector with excellent spatial resolution — of order ten micrometres — located five millimetres from the beam line. By measuring the charged particle tracks resulting from the decays of beauty and charmed hadrons, it is possible to reconstruct the exact locations where they decayed. These locations are denoted *secondary vertices* (see figure 14.4). If the secondary vertex is separate from the collision point (called *primary vertex*), it is likely due to the decay of a b or c hadron. The real *tour de force* in LHCb was to be able to include the reconstruction of these secondary vertices in the trigger system. Thanks to a computing farm of fifty thousand logical processor cores, the task can be

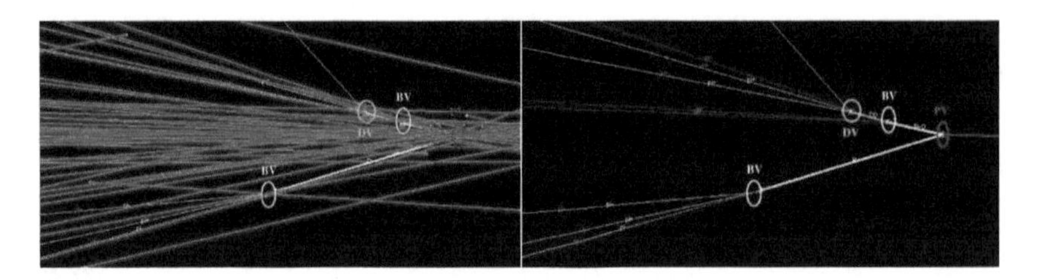

Fig. 14.4. Reconstruction of charged-particle tracks close to the interaction point in LHCb. The primary vertex where the $b\bar{b}$ pair was produced is indicated (PV), as are the sequential secondary vertices corresponding to the B-meson decay (BV) and the D-meson decay (DV). The left-hand picture shows all the tracks, whereas on the right-hand picture only tracks associated to beauty decays are displayed. *Images: LHCb Collaboration.*

performed within the available trigger latency. A specific feature of the VELO detector is that it is retractable. To avoid damaging these delicate detectors during the beam-injection phase, when the beams are still unstable, the silicon modules are retracted as two independent halves to about three centimetres from the beams. They are brought in position when the beam lines are stabilised and data taking can begin. This procedure has to be performed and monitored to very good precision together with a careful alignment to maintain the design precision of the device.

We have seen in the introduction above that LHCb has accumulated less luminosity than ATLAS and CMS during Run 1 and Run 2: about $9\,\text{fb}^{-1}$ in total. Because the LHCb flavour precision physics relies on properly resolving the vertex structure of the event, a task complicated by additional proton–proton interactions (pile-up),[4] it was decided to limit the instantaneous luminosity in the LHCb interaction point to $4 \times 10^{32}\,\text{cm}^{-2}\,\text{s}^{-1}$.[5] This is achieved by transversely separating the beams in a controlled manner.

14.1.2. *Observation of the rare decay $B_s^0 \to \mu^+\mu^-$*

On November 13th, 2012 the LHCb collaboration presented in a seminar at CERN the first measurement of the branching ratio for the decay of the rare B_s^0 meson into a pair of muons: $B_s^0 \to \mu^+\mu^-$. The observed signal had a significance of 3.5σ and the measured branching ratio was found to be

[4]Another reason for limiting the instantaneous luminosity was that pile-up increases the detector occupancy and leads to excessive reconstruction times in the high-level trigger.

[5]The forthcoming upgrade of the LHCb experiment will allow to increase the maximum instantaneous luminosity by a factor of five (see section 9.7).

Inset 14.1: Cherenkov radiation

Cherenkov radiation, named after the Russian physicist Pavel Alekseyevitch Cherenkov, is emitted by a charged particle when traversing a dielectric medium at a velocity larger than the velocity of light in that medium. The angle of emission θ_C of this Cherenkov-light shock wave relative to particle flight path is given by

$$\cos\theta_C = \frac{1}{\beta n},$$

where n is the medium's refraction index and $\beta = v/c$ is the particle velocity (particle speed v relative to the speed of light in vacuum c).

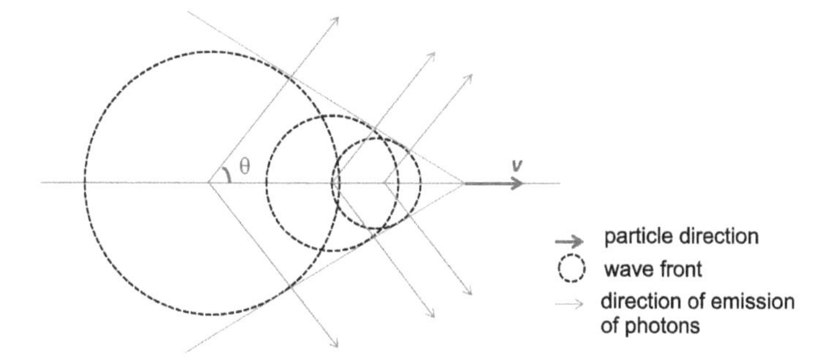

A detector measuring the Cherenkov radiation angle θ_C is thus responding only to the velocity β of the particle. Combining θ_C with a measurement of the particle momentum allows one to infer the particle mass m and thus its identity via the relation $p = \gamma mv = mv/\sqrt{1-\beta^2}$. Cherenkov radiation has a continuous spectrum enhanced at high frequency and thus appears blue to the observer (while most of it is actually in the ultraviolet spectrum). It gives the spent fuel pools of nuclear reactors their characteristic blue glow.

$3.2^{+1.5}_{-1.2} \times 10^{-9}$! Thus only three out of a billion produced B_s^0 mesons decay this way. This first evidence was found in agreement with the Standard Model prediction of $3.66 \pm 0.14 \times 10^{-9}$, but the experimental precision was yet insufficient to exclude (or detect) possible small deviations due to new physics (see figure 14.1).

The result was confirmed a year later by both LHCb and CMS, and by combining their measurements a signal significance of 6.2σ was reached,

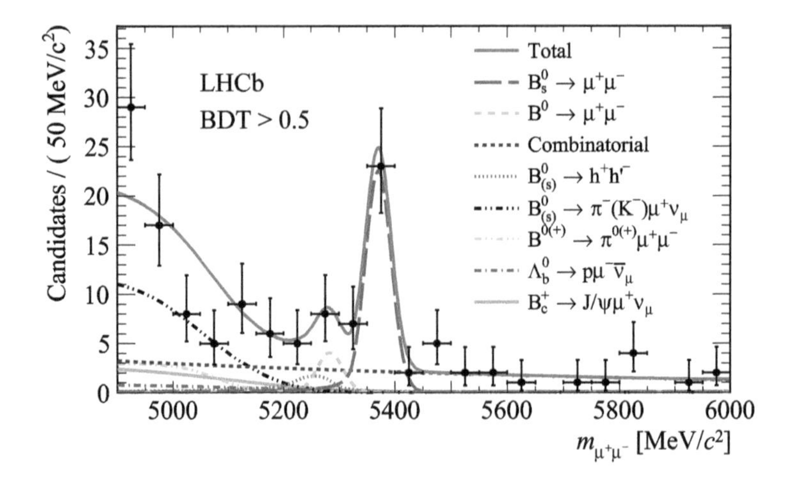

Fig. 14.5. Distribution of the two-muon invariant mass in selected events of LHCb using data recorded in Run 1 and early Run 2 (black points with statistical error bars). The dashed red and green lines represent expectations for the signals from the $B_s^0 \to \mu^+\mu^-$ and $B_d^0 \to \mu^+\mu^-$ decays, respectively. The branching fraction for a B_d^0 meson to decay into a muon pair is expected to be about thirty times smaller than for the B_s^0 meson, but its production rate is about four times larger. The signal strengths together with the expected backgrounds are all simultaneously fit to the data points. The background shape is determined in part from appropriate control data and from Monte Carlo simulations. A clear excess of events is observed at the position of the B_s^0 meson mass (red line). No significant peak is yet seen at the B_d^0 mass (green line). *Image: LHCb Collaboration, Phys. Rev. Lett. 118, 191801 (2017).*

sufficient for a firm observation of this decay. With the data recorded at the beginning of Run 2, the LHCb experiment further improved the measurement to obtain a significance of 7.8σ (figure 14.5). In 2020, by combining results from LHCb, ATLAS and CMS, the currently most precise measurement of $2.7 \pm 0.4 \times 10^{-9}$ was obtained. This is a fantastic performance, and, albeit with slight tension, yet another triumph for the experimenters and the Standard Model.

This measurement does not imply the demise of supersymmetry, but it strongly constrains it and further pushes the masses of possible supersymmetric particles to higher values, potentially out of reach for direct production at the LHC. While it is important to continue improving the measurement to reach adequacy with the theoretical uncertainty, the next major issue in this domain is the search for the $B_d^0 \to \mu^+\mu^-$ decay, which has an expected branching ratio still 30 times smaller than that of $B_s^0 \to \mu^+\mu^-$. Much larger datasets, as will become available during Run 3 and the HL-LHC, are needed to tackle this decay.

14.1.3. *Study of CP violation with B_s^0 mesons*

Among the many beautiful results released by the LHCb collaboration, the study of CP symmetry in the B_s^0 meson sector is of crucial importance in the search for new physics. In section 2.5 we introduced the unitarity triangle which allows to graphically present effects of CP violation, measured up to now essentially among B_d^0 mesons. In fact, the unitary CKM quark mixing matrix allows to construct six unitarity triangles,[6] of which one is particularly suitable to show effects related to B_s^0 mesons. According to Standard Model, effects of CP violation in the B_s^0 system should be very small and the corresponding unitarity triangle almost flat. The smallest among the angles of this triangle is usually denoted β_s. Measuring this angle is of particular importance as new heavy particles, virtually participating in the decay amplitudes of the B_s^0 system, could result in a significant increase of CP violation effects, manifesting itself through a larger magnitude of β_s (see figure 14.6).

The angle β_s can be measured through the study of the decay channel[7] $B_s^0 \to J/\psi\phi$, which gives access to the quantity $\phi_s = -2\beta_s$.[8] Measuring ϕ_s usually goes together with the measurement of another quantity characteristic of the B_s^0 system: $\Delta\Gamma_s$. Neutral B mesons exist in two so-called *proper mass states*, one slightly heavier, denoted $B_{s,H}^0$, and the other slightly lighter, $B_{s,L}^0$. They are quantum mixtures of two proper states of quark flavour, the B_s^0 (bound state of $\bar{b}s$) and the $\overline{B}_s^0$ ($b\bar{s}$).[9] The difference in the decay rates, or total widths, of these two states is given by

[6]As a major confirmation of the CKM picture, complementing the studies of the B_d^0 system at the B factories and the B_s^0 system at hadron colliders, LHCb published in 2019 the first observation of CP violation in the charm-quark system. LHCb measured a tiny (0.15%) but non-zero difference in the decay rate of D^0 and $\overline{D}^0$ mesons into $\pi^+\pi^-$ and K^+K^- pairs, using the full dataset collected between 2011 and 2018.

[7]The J/ψ particle is a meson made of a charm–anticharm quark pair. The peculiar double name stems from the fact that it was discovered almost simultaneously in 1974 on the US east cost (BNL), where it was named J, and on the US west coast (SLAC), where it was named ψ.

[8]This measurement is among the most complex analyses performed at the LHC. It requires the determination of the flavour of the B_s^0 meson upon creation (whether it was a B_s^0 or a $\overline{B}_s^0$), the precise measurement of its production and decay vertices, and a multi-dimensional parameter fit that determines the relative polarisation of the two spin-one mesons J/ψ and ϕ.

[9]The mass states are the real physical states, relevant from the point of view of weak interactions governing their decay, and that propagate in space and time. The states of flavour are relevant only at the strong-interaction production stage (with a characteristic timescale of less than 10^{-22} seconds), where the b and anti-b quarks are produced.

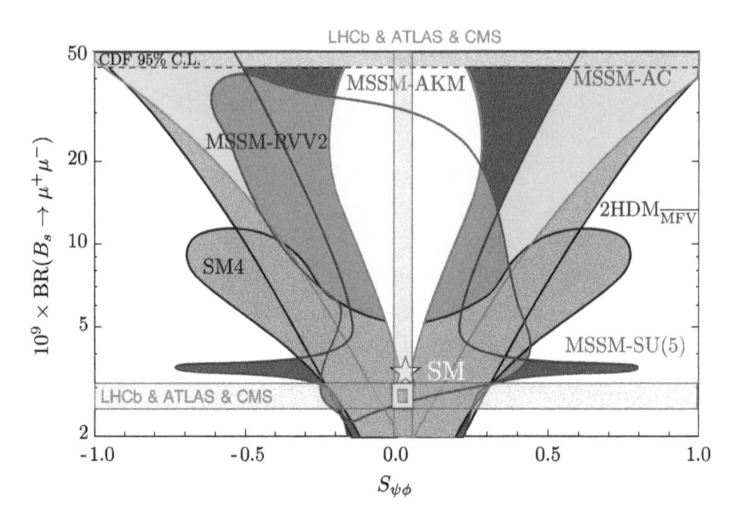

Fig. 14.6. Branching fraction of the decay $B_s^0 \to \mu^+\mu^-$ versus $S_{J/\psi\phi} \simeq -\sin\phi_s$. Coloured regions correspond to domains for these two parameters as predicted by various models for new physics: the Standard Model with a fourth generation of fermions (SM4), several variants of supersymmetric extensions of the Standard Model (MSSM), and a variant of a model with two Higgs doublets (2HDM). The star indicates the prediction of the Standard Model. The experimental results reduce this domain to a small zone, the light blue square, which is compatible with Standard Model expectation. The measurements significantly reduce the allowed parameter space of models implementing new physics. *Image: D.M. Straub, arXiv:1107.0266 (2011).*

$\Delta\Gamma_s = \Gamma_{s,L} - \Gamma_{s,H}$. It is not equal to zero, implying that the two mass states do not decay at the same rate (see inset 2.1 on page 73).

Experiments at the Tevatron were the first to attempt to measure $\Delta\Gamma_s$ and ϕ_s. However, the numbers of events recorded were not sufficient to conclude on the compatibility between experimental results and Standard Model predictions. With the advent of the LHC this picture cleared up significantly. Already in December 2011, using only $0.37\,\mathrm{fb}^{-1}$ of 7 TeV data, LHCb published a first measurement of ϕ_s proving that it was small as expected. Also the measured width difference $\Delta\Gamma_s$ agreed with the (not very precise) Standard Model prediction. With the increasing datasets, the ATLAS and CMS experiments entered the race, too, compensating their less optimal performance for this type of physics with higher integrated luminosity. Figure 14.7 gives a view of the results as of summer 2020 in the $\Delta\Gamma_s$ versus ϕ_s plane. The ellipses show the individual experimental results, and the black bar gives the Standard Model prediction, which is extremely precise for ϕ_s (driven by the suppression of CP symmetry violation in the B_s^0 meson system), and less so for $\Delta\Gamma_s$. Agreement between the experimental measurements and the Standard Model prediction is observed.

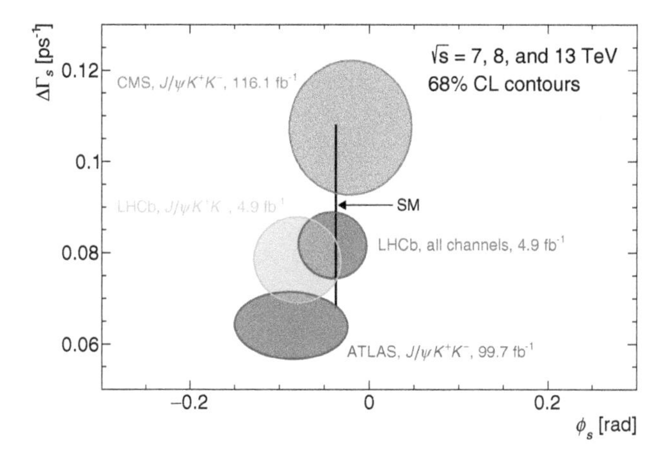

Fig. 14.7. Difference in width of the light and heavy B_s^0 mass states, $\Delta\Gamma_s$, versus the phase ϕ_s as measured by LHCb, ATLAS and CMS using $B_s^0 \to J/\psi\phi$ decays. The ellipses indicate the 68% confidence level contours. The red ellipse shows the LHCb result after combination with other B_s^0 decay modes, which are less precise than $B_s^0 \to J/\psi\phi$, but still add information. The black line shows the Standard Model prediction. Agreement between prediction and measurements is observed. *Image: T. Jakoubek (2020).*

14.1.4. *Tests of lepton flavour universality*

Lepton flavour universality assumes that lepton interactions (couplings) with the W and Z gauge bosons are the same, whatever the lepton flavour is (electron, muon, or τ lepton). In the Standard Model, this is true and changes in decay rates are solely due to differences in the lepton masses, whose effects are precisely predicted. The universality among charged leptons was tested with permil level precision at the electron–positron collider LEP operating at the Z pole in the 1990s (see section 2.4), by comparing rates of the τ-lepton decay to an electron plus two neutrinos with that of a τ-lepton decay to a muon plus two neutrinos, and by comparing these to the τ-lepton lifetime. All measurements were found in agreement with universality. However, with physicists looking for deviations from the Standard Model and hints for new physics wherever possible, a series of slightly anomalous observations related to lepton universality in the B-meson sector, none of which yet statistically significant, has sparked genuine excitement. What is this about?

In 2014, the LHCb collaboration performed a test of lepton universality by measuring the decay rates of $B^+ \to K^+\mu^+\mu^-$ and $B^+ \to K^+e^+e^-$, outside the kinematic region where the lepton pair is produced via a J/ψ resonance, using the Run-1 data. The ratio, R_K, of the two branching fractions, predicted to be 1 in the Standard Model, appeared to differ from

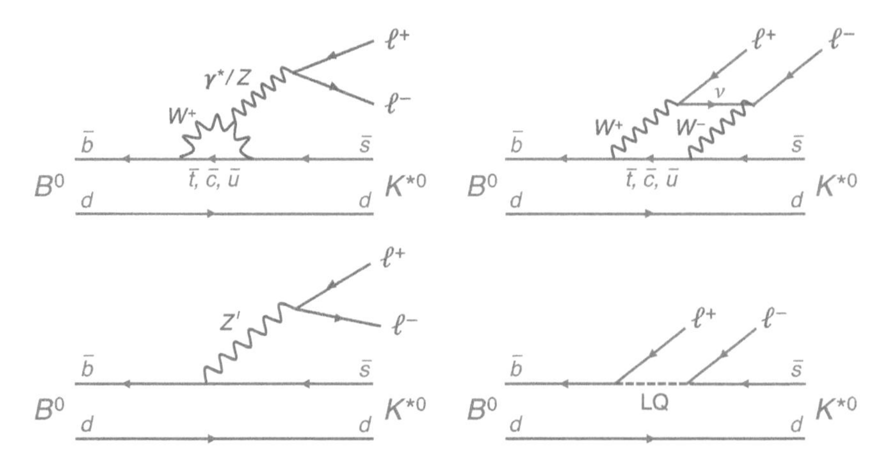

Fig. 14.8. Examples of Feynman diagrams contributing to the decay $B_s^0 \to K^{*0}\ell^+\ell^-$. The upper diagrams show the Standard Model processes occurring through exchange of a virtual W boson, which are lepton-flavour symmetric. The lower-left diagram shows a putative process involving a heavy cousin of the Z boson, denoted Z', which would interact differently with muons and electrons. The lower-right diagram shows the exchange of a hypothetical scalar leptoquark (LQ), which would interact with both quarks and leptons, but with different strength depending on the lepton flavour.

the prediction by 2.6σ.[10] A more recent LHCb measurement, including also the complete Run-2 data, found an increased deviation of 3.1σ. Such a deviation may occur in terms of statistics and would not lead to large excitement given the many searches and tests that are performed,[11] were it not for several other measurements also showing some discrepancies. For example, the ratio R_{K^*} of the branching fractions $B_d^0 \to K^{*0}\mu^+\mu^-$ and $B_d^0 \to K^{*0}e^+e^-$ (see figure 14.8), as well as the ratio $R_{D^{(*)}}$ of the branching fractions $B_d^0 \to D^{(*)-}\tau^+\nu_\tau$ and $B_d^0 \to D^{(*)-}\mu^+\nu_\mu$, differ by two to three standard deviations from their Standard Model predictions. These latter ratios also include measurements from the B factory experiments BABAR and Belle (cf. section 2.5). Finally, discrepancies have been observed in the analysis of the angular spectrum of $B \to K^*\mu^+\mu^-$ decays. However, whereas the predictions of universality tests are straightforward and exhibit

[10]While conceptually simple, and predicted with negligible theory uncertainty in the SM, the measurement of R_K poses significant challenges at a hadron collider as the trigger efficiency in the electron channel needs to be controlled at the percent level. This is achieved by normalisation to a control channel where the lepton pair is produced via a J/ψ resonance. The kinematic properties of this control channel are similar, albeit not identical to those of the signal events.

[11]The reader is referred to the discussion of the "look-elsewhere effect" in inset 13.1 on page 422.

small theoretical uncertainty, the interpretation of $B_d^0 \to K^* \mu^+ \mu^-$ decays is theoretically much more challenging and subject to doubt.

Although the latter measurement is not directly related to lepton universality, phenomenologists have made attempts to address all potential anomalies in unison by systematic extensions of the Standard Model. These extensions parameterise effects of new heavy particles entering virtually the corresponding processes (see the lower diagrams in figure 14.8). It follows a similar line of argument as the interpretation of the $B_s^0 \to \mu^+ \mu^-$ decay rate, where one searches for a deviation from the Standard Model prediction that could be due, for example, to the presence of heavy supersymmetric particles (figure 14.1). Here, the models include a heavy charged Higgs boson, a new gauge boson (Z'), or even leptoquarks.[12] A campaign of measurements of these and similar decays and ratios in the b-hadron sector has been launched and the large Run-2 dataset should allow to shed more light on the state of affair. The new e^+e^- B factory in Japan, SuperKEKB with its experiment Belle II, which started data taking in 2020, will also contribute.

14.1.5. *Exotic multi-quarks states*

In insets 1.13 on page 63 and 2.2 on page 75, we discussed only the two "white" colour configurations of quarks forming hadrons, namely quark–antiquark pairs for mesons and three-quark configurations for baryons. In recent years with much increased sensitivities achieved at the B-factories and the hadron colliders, more complex — but still "white" — configurations of quarks and gluons were discovered: *tetraquarks* ($q\bar{q}q\bar{q}$) and *pentaquarks* ($qqqq\bar{q}$) states, which can mix different quark flavours and whose precise binding structure is still under intense theoretical discussion. For example, it is yet unclear whether all quarks are tightly bound

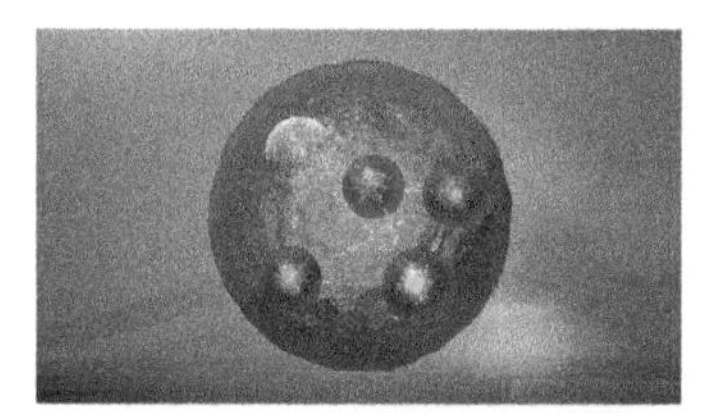

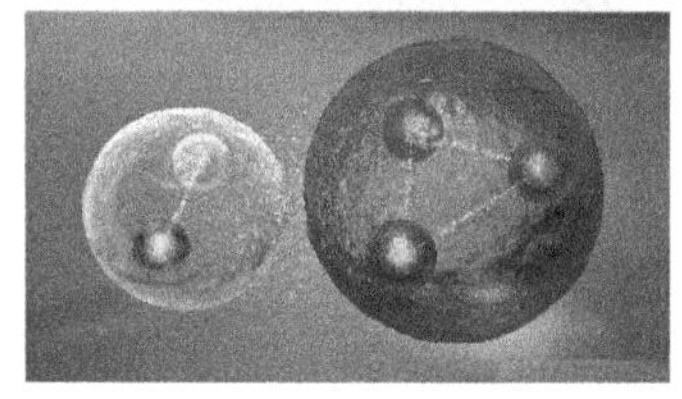

Fig. 14.9. Illustration of a pentaquark according to the hadronic bag (top) or hadro-molecule picture (bottom). *Images: CERN.*

[12]Leptoquarks (LQ) are hypothetical particles that carry both lepton and baryon numbers. They are colour-triplet bosons that allow, for instance, a quark to transform into a lepton (and vice versa) by emitting a LQ. They are encountered in various extensions of the Standard Model, such as technicolour theories, or grand unification theories based on the Pati–Salam model, SU(5) or E_6 groups. Their quantum numbers like spin, (fractional) electric charge, and weak isospin differ between the models.

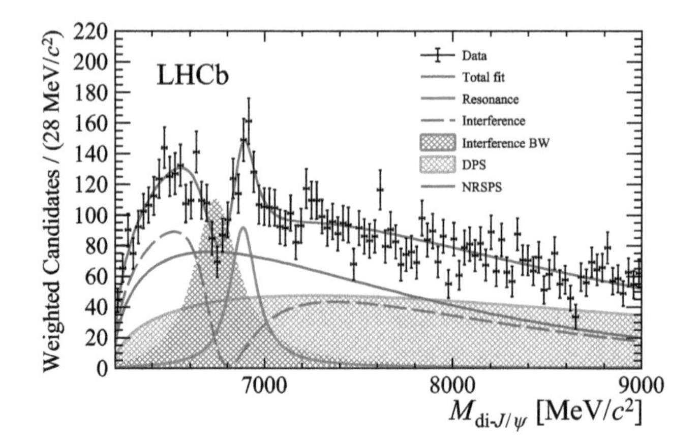

Fig. 14.10. Invariant mass spectrum of di-J/ψ candidates measured by LHCb. The data are shown by points with error bars. The curves show different background and signal contributions that are fit together to describe the data. The (hypothetical) tetraquark signal appears as a peak at 6900 MeV. Its interference with the non-resonant di-J/ψ production leads to the dip just below the peak, which is well reproduced by the fit model. The statistical significance of the signal exceeds 4σ. *LHCb Collaboration, Science Bulletin 65(23), 1983 (2020).*

in the same compact potential well or "hadronic bag", or whether they form a sort of *hadro-molecule*, a meson-baryon system with two weakly bound "bags" (figure 14.9).

By investigating the invariant mass formed by two J/ψ resonances, each decaying into a muon pair so that the final state consists of four muons, the LHCb collaboration recently discovered a resonant structure at a mass of 6.9 GeV (see figure 14.10). The new resonance may be a tetraquark composed of two charm quarks and two charm antiquarks, $c\bar{c}c\bar{c}$.[13] To help discriminate between the compact or molecule models for exotic multi-quark systems, it will be interesting to explore if even heavier tetraquarks, such as $c\bar{c}b\bar{b}$ or even $b\bar{b}b\bar{b}$, exist.

Pentaquarks have been the subject of over fifty years of exploration during which several such objects were found. None, however, stood the test of time. This was the situation until 2015, when LHCb reported the observation of a pentaquark state in the study of Λ_b^0 decays to $J/\psi pK^-$ (see figures 14.11 and 14.12). The analysis was updated with more data in 2019, confirming the pentaquark observation, but revealing a more complex structure.

[13]In the so-called di-quark model the tetraquark would be written as $[c\bar{c}][c\bar{c}]$. In these models, pairs of quarks bind together to form a single object that interacts with the third quark (in a baryon) or an anti-diquark (in a tetraquark) through gluon exchange.

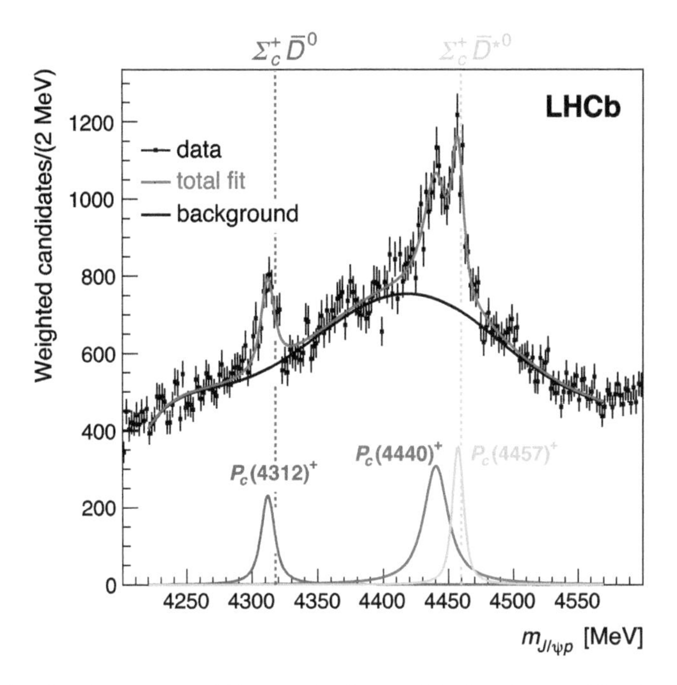

Fig. 14.11. Invariant proton-J/ψ mass in $\Lambda_b \to J/\psi p K^-$ decays. The data are represented by points with statistical error bars. The curves describe a fit of background (black) plus three pentaquark states, $P_c(4312)^+$, $P_c(4440)^+$, and $P_c(4457)^+$. The lowest (highest) mass state is close to the kinematic threshold of $\Sigma_c^+ \overline{D}^0$ ($\Sigma_c^+ \overline{D}^{*0}$) production, suggesting that they are related to the dynamics of these states. *Image: LHCb Collaboration, Phys. Rev. Lett. 122, 222001 (2019).*

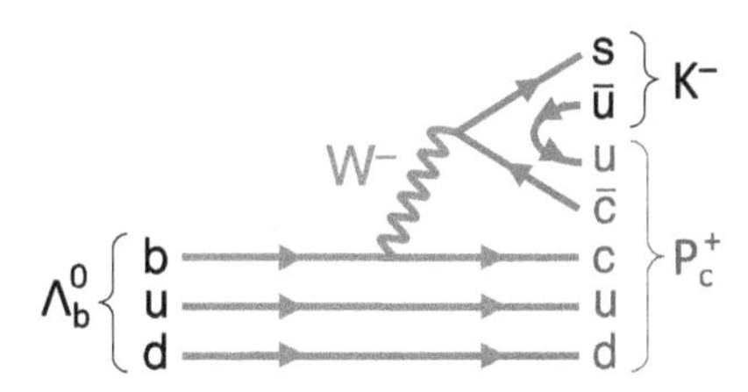

Fig. 14.12. Feynman graph for $\Lambda_b^0 \to P_c^+ K^-$, where P_c is a pentaquark state.

The proton–J/ψ invariant mass spectrum measured by LHCb is shown in figure 14.11. It exhibits three peaks, of which the higher-mass ones are close by (they could in fact not yet be resolved as two structures in the first LHCb paper), over a smooth (but a bit knotty) background shape. A multi-amplitude fit identifies with high significance the peaks with the pentaquark candidate states $P_c(4312)^+$, $P_c(4440)^+$, and $P_c(4457)^+$. The figure also indicates the mass thresholds of $\Sigma_c^+ \overline{D}^0$ and $\Sigma_c^+ \overline{D}^{*0}$ pair production (both pairs are $[duc]$, $[u\bar{c}]$ quark bound states), suggesting that they are related to the dynamics of the pentaquark states. The minimal quark content of the three pentaquark states is $duuc\bar{c}$.

14.2. Heavy-ion collision physics and the ALICE detector

From the very beginning of the LHC project it was required that the machine be able to accelerate and collide heavy ions. The ions, lead ions (Pb) in the LHC, use the same PS, SPS and LHC injector and collider chain as the protons, but upstream of the PS the injection follows a separate path. To accelerate ions, it is required to strip them off their electrons. The injection starts with a pellet of pure ^{208}Pb isotope (carrying 82 protons and electrons, and 126 neutrons), which is boiled to five hundred degrees Celsius to vaporise atoms. Some of the electrons of the lead atoms are stripped away using electrical currents. The ions are then accelerated in a linear accelerator (Linac3 on figure 6.17, tuned for $^{208}\mathrm{Pb}^{+27}$, where the right superscript indicates the charge of the ion, that is the number of electrons that have been removed), before further stripping away electrons to reach $^{208}\mathrm{Pb}^{+54}$ and injecting them into the low-energy ion ring (LEIR) for further acceleration. The ions then enter the PS after which, in the transfer line to the SPS, they are fully stripped to bare lead ions $^{208}\mathrm{Pb}^{+82}$. Their journey continues with further acceleration in the SPS and injection into the LHC for final acceleration and collision.

As the accelerating system acts only on charged particles, the protons from the nucleus are accelerated to, say, 7 TeV, but they carry with them the neutrons. In the case of lead nuclei the actual, nominal, nucleon–nucleon centre-of-mass collision energy is $(Z/A) \times 14\,\mathrm{TeV} = 5.5\,\mathrm{TeV}$. This is an increase by a factor of 25 compared to the RHIC collider at Brookhaven, USA that has dominated the heavy-ion scene in the years 2000.

14.2.1. *The quark–gluon plasma*

Heavy-ion colliders are often characterised by the collective energy of the nucleons, that is the sum of energies of all nucleons participating in the collision. This sum amounts to just a few tens of TeV in the case of a glancing collision (figure 14.13), but can reach up to $208 \times 5.5\,\mathrm{TeV} \simeq 1150\,\mathrm{TeV}$ in case of so-called *central collisions* when all nucleons in both nuclei are involved in a fully head-on

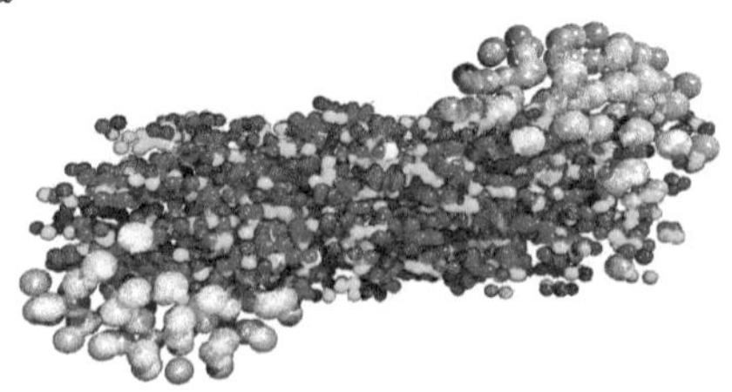

Fig. 14.13. Schematic representation of a peripheral collision between two heavy ions. In white are nucleons not participating in the collision (called spectators). In colour, the region of quark-gluon plasma formation in the collision.

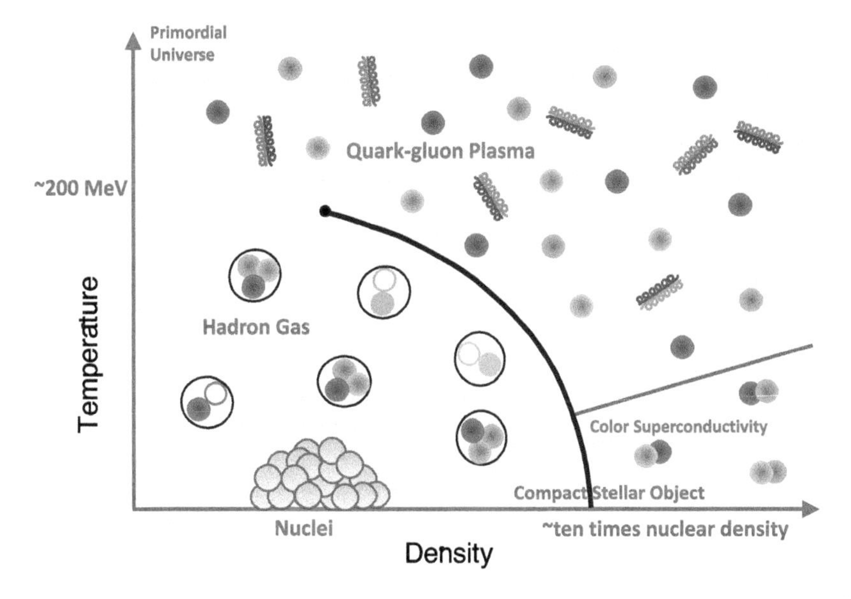

Fig. 14.14. Phase diagram for matter undergoing strong interactions as a function of temperature (ordinate) and density (abscissa). At low temperature and density, nuclear matter can exist only in the form of a hadronic gas (confined bound states of mesons and baryons). At very low temperatures and in a specific domain of densities, this gas appears as a nuclear matter (a quasi liquid) making up the atomic nuclei (elements). At higher densities, of order ten times the nuclear density, takes place a transition into a quark condensate characterised by a crystalline structure, a state probably present inside neutron stars. The curved black line is the separation line between the hadron-gas phase and the quark–gluon plasma phase, a state of matter that can be studied in high energy heavy-ion collisions.

lead–lead collision (for a nominal proton beam acceleration energy of 7 TeV).

These collisions allow the study of matter under the influence of strong interactions at *energy densities* reaching up to several tens of GeV per femtometre cube! This energy density corresponds to temperatures well beyond the critical temperature T_c of about 160 MeV (energy density of about 0.5 GeV/fm^3) where QCD matter is expected to undergo a transition from a hadronic gas to a new state of matter called *quark–gluon plasma* (figure 14.14). The quark–gluon plasma represented the state of the entire matter of the Universe up to the first microsecond after the Big Bang. At the LHC this state is formed during a heavy-ion collision and exists for only a very brief instant, roughly between 10^{-23} to 10^{-22} seconds before cooling down, evaporating, and recombining into a multitude of hadrons.

The temperature of the plasma reaches several trillion kelvin, which is about one hundred thousands times hotter than the temperature prevailing in the core of our Sun.

The primary experimental characteristics of heavy-ion collisions is the very large particle multiplicity produced, up to thirty thousand particles appearing in the detector in the case of a central lead–lead collision! Furthermore, most of these particles resulting from the evaporation of the quark–gluon plasma into a hadronic gas have transverse momenta well below a GeV, much lower than the particles produced in proton–proton collisions. The instantaneous luminosity in the heavy-ion collider mode of the LHC being a million or more times smaller (typically a few $10^{27}\,\mathrm{cm}^{-2}\,\mathrm{s}^{-1}$, the luminosity record during Run 2 was $6 \times 10^{27}\,\mathrm{cm}^{-2}\,\mathrm{s}^{-1}$) than in the proton–proton collider mode. The data-taking rate is therefore much smaller, but, despite the fact that there is no pile-up, the size of events is about twice that of proton collisions (with a standard pile-up of 25 simultaneous interactions). Even if the general-purpose detectors ATLAS and CMS can, in part, face these specific features of heavy-ion collisions and make complementary and unique contributions to heavy-ion physics as discussed below, from the very beginning of the LHC project it was decided to build a specific detector optimised for this type of studies, the ALICE detector.

14.2.2. *The ALICE detector*

ALICE[14] is a large international collaboration of more than one thousand scientific authors, 174 institutes from 39 countries, who built, run and exploit the ALICE detector (figure 14.15). The central component of the detector is a large solenoidal magnet producing an axial magnetic field of 0.5 tesla. This magnet has been recovered from the earlier L3 experiment at the LEP electron–positron collider. In the vicinity of the beam line, particle tracking is achieved through silicon micro-strip detectors. As in LHCb, the aim is to reconstruct particle trajectories, the primary vertex and the secondary decay vertices allowing identification of charmed and beauty hadrons. A large gaseous time projection chamber (TPC), about 5-metre long and with 2.5-metre outer radius, complements the track reconstruction capability of ALICE with an excellent spatial resolution, despite

[14]ALICE stands for A Large Ion Collider Experiment.

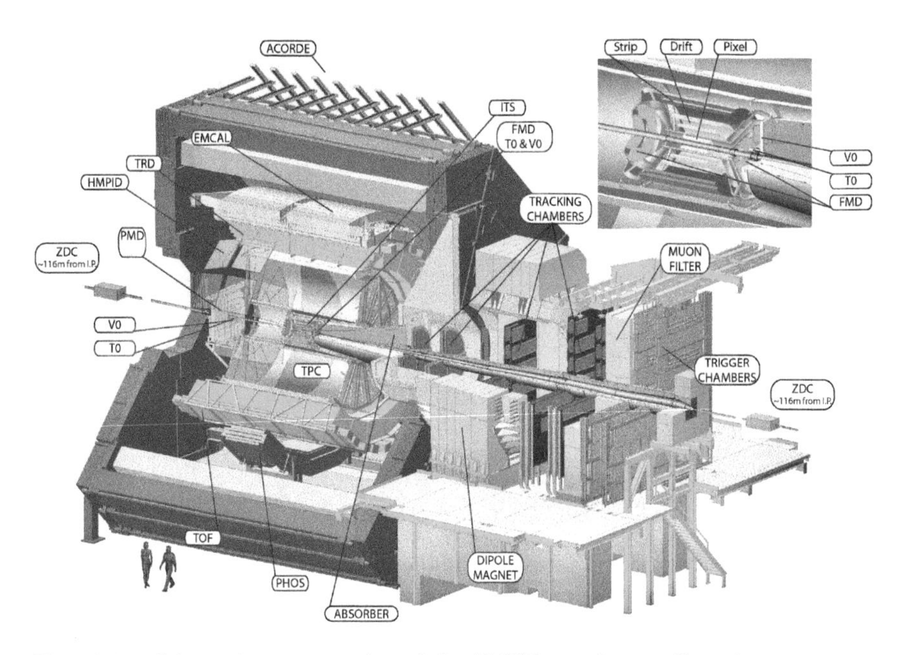

Fig. 14.15. Schematic representation of the ALICE experiment. Note the muon spectrometer arrangement on one side of the detector with a loaded plug to filter out all but muons emerging from the collision. *Image: ALICE Collaboration.*

the very large number of particles traversing it (figure 14.16). The somewhat slow data-taking rate of such a type of chamber is not detrimental, due to the relatively low luminosity and collision rate of heavy-ion collisions compared to proton–proton. Inside the TPC is the inner tracking system (ITS) made of six layers of silicon detectors exploiting three different technologies (pixel, drift and strip). It covers radii from 3.9 centimetres (innermost silicon pixel layer) to 43 centimetres for the outermost silicon strip layer.

The ALICE detector is equipped with several devices dedicated to particle identification. Making use of all known techniques, they are capable of identifying hadrons and leptons over a wide momentum range covering three orders of magnitudes, from about 100 MeV up to 100 GeV. The first device is a transition radiation detector (TRD) for the identification of electrons and positrons. A second one is based on multigap resistive plate chambers, which allows to distinguish pions, kaons and protons on the basis of their time of flight (TOF), thus their velocities, for momenta up to about 2 GeV for π/K separation and 4 GeV for K/p separation, respectively. For

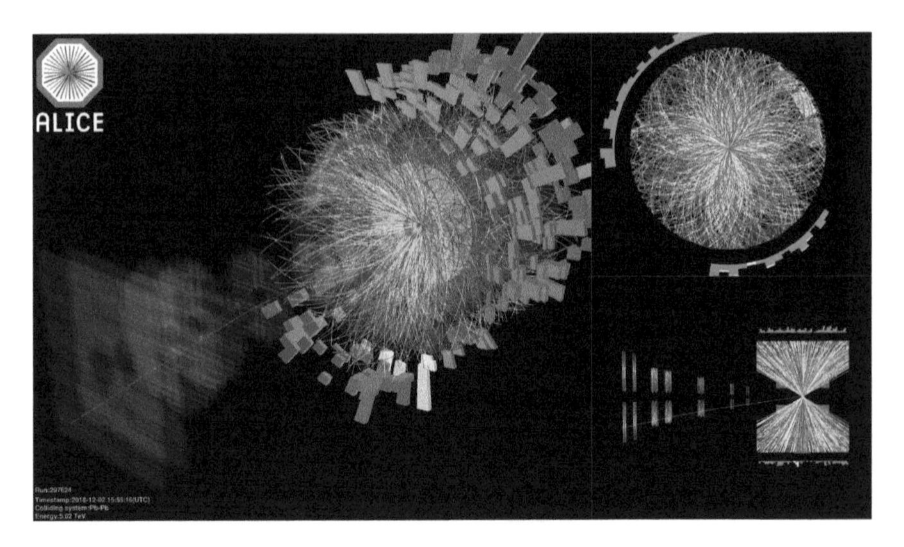

Fig. 14.16. Reconstruction of a lead–lead collision event in the ALICE detector. *Image: ALICE Collaboration.*

high-momentum particle identification (HMPID), a ring imaging Cherenkov detector (RICH) is used (see inset 14.1), which allows to discriminate pions from kaons up to a momentum of 3 GeV and from protons up to 5 GeV. A beautiful illustration of ALICE's particle identification capability is given in figure 14.17 showing the separation of particles (and antiparticles) based on their mass and charge, as obtained from the measurement of the *specific energy losses* (dE/dx, energy loss dE per unit path dx in the detector gas layer) in the TPC tracking detector, as a function of particle momentum p and charge z in lead–lead collisions. One notes in particular the populations of tracks corresponding to anti-nuclei, anti-deuteron (bound state of antiproton and antineutron), anti-triton (bound state of antiproton and two antineutrons), and anti-^{3}He (bound state of an antineutron and two antiprotons). These anti-nuclei surely cannot be produced in the fragmentation of the initial state, purely matter-made lead nuclei, but are produced in the quark–gluon plasma.

Photon energy is a powerful probe and indicator of the quark–gluon plasma temperature (figure 14.18 on the left). Electromagnetic calorimetry in ALICE is achieved through two types of detectors. A first calorimeter (PHOS) with excellent energy resolution uses lead-tungstate (PbWO4) scintillating crystals of the same type as in CMS. However, to limit the cost, the geometrical coverage of this calorimeter is limited and complemented by a lead-scintillator sandwich-type electromagnetic calorimeter (EMCAL).

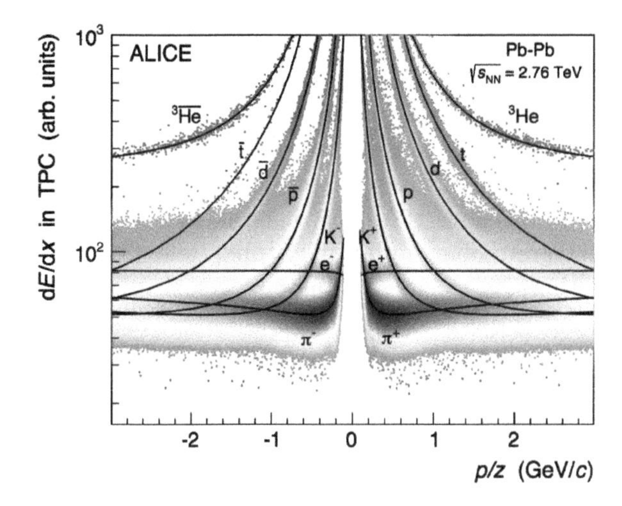

Fig. 14.17. Example of particle identification in the ALICE detector, based on the measurement of the specific energy loss in the TPC tracking detector, as a function of particle momentum (p) and charge (z) in lead–lead collision data. The various particle and antiparticle populations are well separated in the low-momentum range. Note in particular the population of tracks corresponding to anti-nuclei, anti-deuteron ($\bar{d}$), anti-triton ($\bar{t}$), and anti-^{3}He, emerging from the quark–gluon plasma. Recently, anti-^{4}He was observed, too, by ALICE. *Image: ALICE Collaboration.*

It allows, besides the measurement of electron and photon energies, also the reconstruction of hadronic jets.

In the ATLAS and CMS experiments, the muon detection systems hermetically surround the rest of the experiment. In ALICE, on the contrary, the muon spectrometer, made of flat muon-detection planes, is located on one side of the collision point only (see figure 14.15). Its front part is made of a concrete shield absorbing the hadrons emerging from the interaction region and letting through only the muons. These are then measured in detector planes placed in front and behind a dipole bending magnet to provide momentum measurement. The primary function of this spectrometer is to measure the mass of heavy *quarkonia* (bound states of $c\bar{c}$ and $b\bar{b}$) decaying to muon pairs. The mass resolution is 70–100 MeV, limited by multiple scattering in the concrete plug, but sufficient to separate the J/ψ and ψ' resonances (ground and excited $c\bar{c}$ bound states), and the $\Upsilon(1S)$, $\Upsilon(2S)$ and $\Upsilon(3S)$ states (ground and excited $b\bar{b}$ bound states), which are nearby in mass. These resonances are excellent probes of the plasma temperature allowing to test predictions from QCD. The ALICE muon spectrometer is also equipped with RPC-type chambers for the triggering of the data acquisition system.

14.2.3. *Selected heavy-ion physics results*

The first time the LHC took data in the lead–lead collision mode was for a period of one month at the end of 2010. The beam energy was about 300 TeV per nucleus, corresponding to centre-of-mass nucleon–nucleon collision energy of 2.8 TeV. During the next data-taking period at the end of 2011 the energy remained the same, but the luminosity increased significantly. An integrated luminosity of 170 μb^{-1} was delivered by the LHC, about fifteen times as much as in the previous year. Collisions were also recorded in January and February 2013, just before the two-year-long technical shutdown of the collider. During that run the data were taken in an asymmetric mode with proton–lead collisions, thus completing the variety of operation modes of the LHC. At the end of 2015, there was another one-month-long period of lead–lead data-taking, with increased nucleon–nucleon collision energy of 5.0 TeV and an integrated luminosity of 570 μb^{-1} (figures 9.11 and 9.12). At the end of the 2016 running period, there was again a month of asymmetric proton–lead collisions at 8.2 TeV nucleon–nucleon centre-of-mass energy (6.5 TeV and 2.5 TeV for the proton and lead beam energies, respectively), where the LHC delivered an integrated luminosity of 190 nb^{-1}, and now also the LHCb detector participated in the heavy-ion data collection. At the end of 2018 there was a month of final lead–lead running, again at 5.0 TeV nucleon–nucleon collision energy, where a new record integrated luminosity of 1.8 nb^{-1} was collected. In October 2017, the LHC performed a very short run colliding xenon atoms (containing 54 protons and 74 neutrons), which represent a smaller system than lead.

Already the data collected during the 2011 and 2012 lead–lead campaigns, and even more so those of 2015 at almost twice the nucleon–nucleon collision energy, provided much information on the quark–gluon plasma (QGP), complementing earlier studies at the SPS and especially RHIC. The LHC collisions form a transient QGP state for a tiny amount of matter, of order 10^{-21} grams, that cools down extremely rapidly before returning back into the hadronic phase. The properties of this state can be addressed using *soft probes*,[15] for which the ALICE detector is particularly suited: soft direct photons, low to medium transverse momentum hadrons, identification of hadrons with strange, charm, or bottom quark content, and comparing to observations with the same objects in proton–proton collisions. Another category of observables are the so-called *hard probes*:

[15]The term "soft" stands here for low-momentum particles.

high-transverse momentum ("hard") jets, heavy quarkonia, photons, Z and W bosons and even top quarks, and their behaviour in and propagation through the plasma. For these studies, the ATLAS and CMS detectors are very effective with their large geometric coverage and excellent resolution for high-energetic probes.

The understanding of heavy-ion collisions has evolved enormously since the start of the LHC. Let us look at some of the results illuminating the properties of the quark–gluon plasma.

Soft probes

One of the first ideas for a signal of the QGP was thermal radiation due to photons. Photons are penetrating and can in principle measure the properties of the QGP at early times in the collision. The challenge is, however, the huge background from resonance decays (such as $\pi^0 \to \gamma\gamma$). The left-hand panel of figure 14.18 shows the spectrum of direct photons measured in central lead–lead collisions with ALICE. There are two components in this spectrum: at larger momenta (above 4 GeV) the production of photons radiated directly from quarks is in agreement with the QCD prediction (scaled using proton–proton data); below 4 GeV, however, where the spectrum falls off exponentially, there is an additional component. A plausible interpretation is that this component is due to thermal photons radiated collectively from the hot QGP.

The study of collective behaviour in heavy-ion collisions at the LHC reveals that the QGP behaves like an ideal fluid with a shear viscosity that is ten times lower than that of any other known form of matter! It is a manifestation of deconfinement due to asymptotic freedom, i.e. the weak coupling among gluons and quarks at high energy predicted by QCD. The QGP has a collective expansion behaviour (or *flow pattern*), which in classical physics can be described with the tools of hydrodynamics, where one links in a conceptually simple way conservation laws (mass, momentum, energy) with fundamental properties of the fluid: the equation of state and transport coefficients, such as viscosity and heat conductivity. This plasma should furthermore be opaque to any strongly interacting (colour-charged) particle propagating through it. The suppression resulting from this opacity is confirmed by observation. While propagating through the QGP, strongly interacting particles suffer energy loss so that a fraction of them cannot escape, or only with much reduced energy. This phenomenon is called *quenching*.

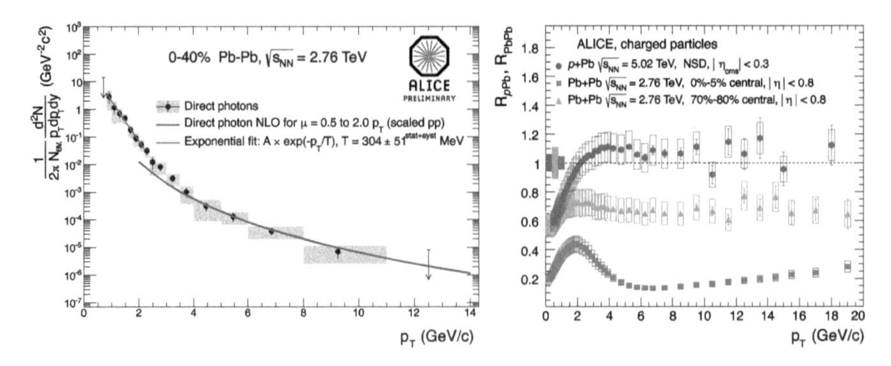

Fig. 14.18. Left: transverse momentum (p_T) spectrum of photons measured in central lead–lead collisions with ALICE. While the larger momentum spectrum is well described by QCD, the lower momentum spectrum features an exponential decrease which may be due to thermal photons generated in the quark–gluon plasma. Right: ratios of particle yields (nuclear modification factors) in central (red points) and peripheral (green points) lead–lead and proton–lead (blue points) collisions relative to proton–proton collision yields as a function of the particles' transverse momenta. Large suppression is observed which increases with the size of the system. *Images: M. Wilde (for ALICE), related published results in ALICE Collaboration, Phys. Lett. B 754, 235 (2016); ALICE Collaboration, Phys. Rev. Lett. 110, 082302 (2013).*

The right-hand panel of figure 14.18 shows the (properly normalised) ratios of the number of charged particles in lead–lead and proton–lead collisions relative to that in proton–proton collisions as a function of transverse momentum. This ratio is denoted *nuclear modification factor.*[16] It is seen that suppression (i.e. a nuclear modification factor smaller than one) is very pronounced in case of central lead–lead collisions, less pronounced in case of peripheral lead–lead collisions, and non-existent for momenta above 2 GeV in proton–lead collisions. This comparison between proton–lead and lead–lead collisions shows that the strong quenching effect at large momentum in lead–lead collisions is not due to initial-state effects, but linked to the hot dense matter (QGP) in the final state.

Collectivity has also been seen in small systems at the LHC. CMS observed as one of the first measurements at the LHC in September 2010 flow-like long-range correlations between particles in high-multiplicity proton–proton collisions. This effect, known from heavy-ion collisions,

[16]The nuclear modification factor quantifies nuclear effects, therefore the normalisation to proton collisions. The normalisation requires a correction by the average nuclear overlap (that is, how central a collision was, how many binary collisions among nucleons occurred), which is estimated with the help of a model of the nuclear collision geometry.

came as a complete surprise as such collective behaviour was not predicted by any calculation based on QCD (hydrodynamic models are not used for the description of proton collision physics). It could be revealed by applying the techniques developed for heavy-ion physics to proton–proton data.

In 2017, ALICE published a seminal paper showing for the first time collective effects also in the production of strange quarks in proton–proton collisions.[17] The observations are summarised in figure 14.19, which exhibits a rich spectrum of physics. The plot shows the measured particle yields relative to pions versus the charged particle multiplicity for different particle species with, from top to bottom, increasing strangeness[18] content. The charged particle multiplicity increases with the size of the colliding system (see sketch on top of figure): from proton–proton collisions on the left, to proton–lead, peripheral lead–lead, and central lead–lead collisions on the right. The figure shows that the hadron-to-pion ratio smoothly evolves across multiplicity reaching thermal values in central lead–lead collisions. There is a rise of particles with strangeness towards larger systems, and the rise is steeper for hadrons containing more strange quarks. No centre-of-mass dependence of these results is seen. The low multiplicity proton–proton data are described by standard QCD models, but it remains constant towards larger systems (the rise in strangeness in the data is not reproduced). The increase of the ratio could indicate thermal production of strangeness which is independent of size of system. High-multiplicity proton–proton events seem to obey the same hadro-chemistry as a fully thermalised system in lead–lead collisions. Is the same underlying physics responsible for this? This remarkable observation adds to other LHC measurements showing QGP signatures in small systems. They point towards the possible formation of QGP matter at high temperature and density also in small collisions systems such as proton–proton. If that were the case, an important question follows whether it is possible to understand the behaviour of large systems from parton (re-)scattering in small systems? Theoretical models seem to indeed allow a quantitative description along this line.

---

[17]The production of strange quarks are an interesting probe of the QGP as they are not part of the valence structure of the protons and neutrons and thus have to be created in the course of the collision. Strangeness production in the QGP proceeds mainly via gluon interactions ($gg \to s\bar{s}$), allowing strangeness to reach equilibrium within the timescale of the QGP lifetime.

[18]The *strangeness* of a particle is defined by $S = -(n_s - n_{\bar{s}})$, where n_s and $n_{\bar{s}}$ are, respectively, the numbers of strange quarks and strange antiquarks in a particle.

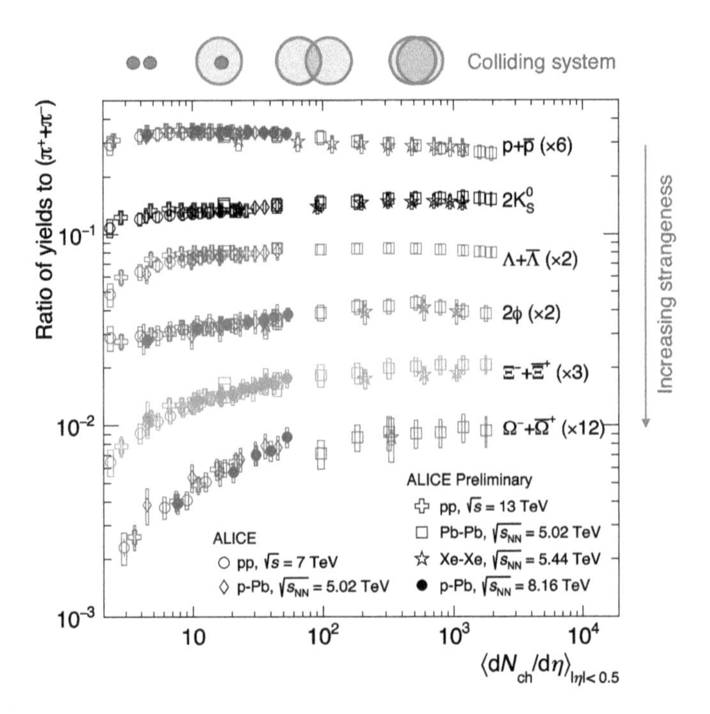

Fig. 14.19. Measured particle yields normalised to the yield of pion pairs versus the "size" of the colliding system characterised by the number of charged particles produced in an event (in the central region indicated as $|\eta| < 0.5$). Small systems like proton–proton collisions produce fewer tracks and are on the left of the abscissa. Proton–lead collisions are the next larger system, followed by lead–lead collisions with increasing centrality towards the right of the abscissa. These types of colliding systems are also illustrated on the top of the figure. The colourful data points indicate different particle species, where the amount of strangeness increases from top to bottom: K^0_S is made of $[d\bar{s}]$ valence quarks, $\Lambda^0 = [uds]$, $\phi = [s\bar{s}]$, $\Xi^- = [dss]$, $\Omega^- = [sss]$. To unclutter the data and improve the visibility, the ordinate values (ratios) are multiplied by the quoted factors. See text for a discussion of the results. *Image: ALICE Collaboration, ALI-PREL-321075 (2020).*

Hard probes

Besides the study of the QGP's collective behaviour using hydrodynamic models, one could wonder how the strongly interacting medium actually emerges from an asymptotic free theory (QCD). Is it possible to see quasi-free particles (quarks and gluons) in the QGP? To probe this, one could start with non-thermalised coloured objects and study how they interact with the thermal quark and gluon "soup". One could, for example, shoot jets through the QGP. The spectacular outcomes of this are the

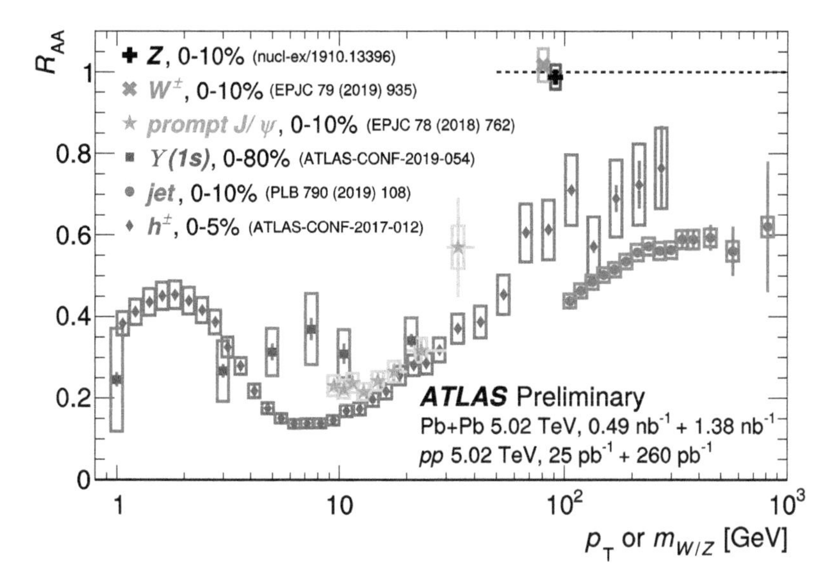

Fig. 14.20. Ratio of central lead–lead to rescaled proton–proton cross-sections (nuclear modification factor) versus transverse momentum for different hard probes. Coloured particles are suppressed, while non-coloured ones are not. *Image: ATLAS Collaboration (2020).*

famous *jet quenching* events we already encountered in section 9.4 (see the event display on figure 9.10). Indeed, the suppression of strongly interacting probes in lead–lead collisions (rising with the centrality, i.e. the size of the collision) is uniformly observed. Figure 14.20 shows the nuclear modification factor R_{AA} versus transverse momentum for strongly interacting coloured particles (hadrons, jets, heavy quarkonia) and for colourless particles (Z and W bosons), as measured by ATLAS. While the coloured particles are modified (or quenched, giving $R_{AA} < 1$), colourless particles travel through the medium without energy loss ($R_{AA} \simeq 1$). The same behaviour was also observed for photons. The recoil of colourless particles against coloured particles can therefore be used to study the quenching of coloured particles in the QGP. For example, in $pp \to \gamma + \text{jet}$ events, the imbalance in the transverse photon and jet momenta is a measure of the jet energy loss. An interesting additional feature in figure 14.20 is that jet quenching occurs even at very high jet energies up to the TeV scale. Measurements of jet properties by CMS have revealed that jet quenching in lead–lead collisions corresponds to a redistribution of energy to large angles from the jet axis: energy goes out of the jet cone into the periphery.

Heavy quarkonia

Heavy quarkonia have long been recognised as promising probes of the modifications of the fundamental QCD force in hot and dense matter.[19] At low temperature, the binding of quarks in, for example, a $c\bar{c}$ resonance (J/ψ) can be described by the simple potential $V(r) = -\alpha r^{-1} + kr$, where r is the spatial distance between the two quarks. The first term in the potential describes a Coulomb-like one gluon exchange contribution, and the second governs the confinement (it increases with r). When the $c\bar{c}$ pair is immersed in a QGP, which consists of deconfined colour charges, the binding of the pair is subject to effects of *colour screening*, reducing the effective distance i.e. attractive force between quarks. The confinement contribution in the potential disappears, and the high-colour density induces a screening of the Coulomb term of the potential, which becomes $V(r) = -\alpha r^{-1} e^{-r/\lambda_D}$. The strength of the colour screening rises with the temperature of the QGP. The length λ_D is the maximum size of a bound state. It decreases when the temperature increases, but its value depends on the binding energy of the quark–antiquark system and thus on the type of quarkonia studied. Colour screening at high-temperature dissociates ("melts") quarkonia in the QGP.[20] Their suppression pattern acts like a thermometer of the QGP, as $q\bar{q}$ pair dissociates in the plasma prior to its decay when the thermal energy in the plasma is higher than the pair's binding energy.

[19]Considerations of measuring QGP properties in heavy-ion collisions using heavy quarkonia (in particular the Υ resonance family) were of course amidst the design priorities for the dedicated experiment ALICE. They were also among the motivations influencing the design of the general purpose experiments ATLAS and CMS, already at the time of their Letter of Intents released in October 1992. The tracking detector momentum resolution had to be good enough to separate the three $\Upsilon(1S)$, $\Upsilon(2S)$, $\Upsilon(3S)$ states via the di-muon invariant mass measurement. To study the transverse momentum dependence of quarkonia production, low-momentum muons need to be measured, too. The innermost station of the CMS muon system was therefore brought as close as possible to the solenoid magnet allowing the identification of muons down to transverse momentum of 3.5 GeV, and the second station was approached as well so it can be hit by 4 GeV muons.

[20]The effect of melting is somewhat similar to the one whereby a molecule of NaCl, for example, dissociates when put into water: water molecules having a kinetic energy higher than the binding energy of NaCl provoke its dissociation in collisions.

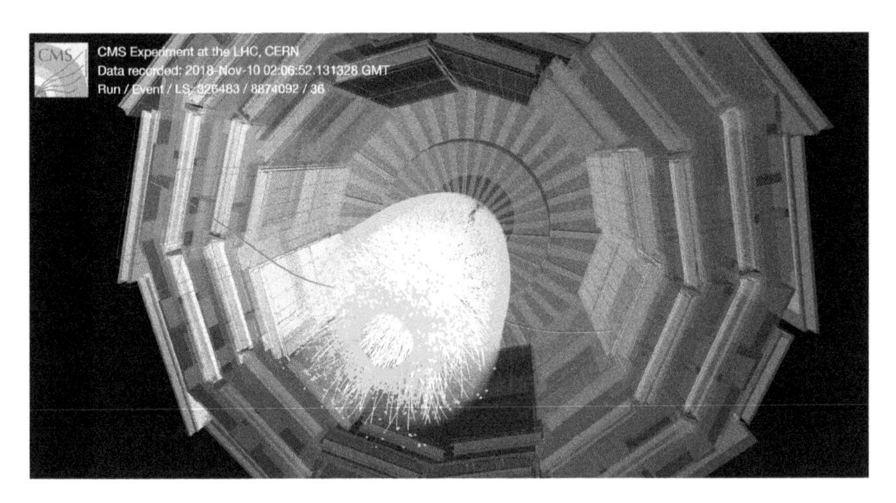

Fig. 14.21. A CMS event where a candidate Υ particle is produced in a lead–lead collision and decays to two muons (shown by the red lines), curved in the strong 3.8-tesla solenoid field (the reversed curvature from the returning field flux in the muon system is also visible). *Image: CMS Collaboration.*

For the $b\bar{b}$ pair in the $\Upsilon(2S)$ (binding energy of 550 MeV) and even more so for the $\Upsilon(3S)$ (binding energy 230 MeV), the effect of melting is much more pronounced than for the $\Upsilon(1S)$ which, as the ground state of the system, is most strongly bound and thus more compact, with a binding energy of 1050 MeV. Figures 14.21 and 14.22 show a study of the Υ system by CMS analysing decays to two muons. The plot in figure 14.22 compares the invariant mass spectrum of muon pairs as observed in central lead–lead collisions with the same mass spectrum observed in proton–proton collisions at the same nucleon–nucleon centre-of-mass energy. The melting of the $\Upsilon(2S)$ and $\Upsilon(3S)$ states relative to the $\Upsilon(1S)$ is remarkably clear.

We have seen in the above paragraph that colour screening at high-temperature dissociates quarkonia in the QGP. However, there is also a reverse effect: quarkonia can be regenerated in the QGP by recombination of heavy quark–antiquark pairs. Figure 14.23 shows the nuclear modification factor of inclusive J/ψ production versus transverse momentum as measured by ALICE. At large momentum, one notices the usual pattern of suppression in central lead–lead collisions relative to proton–proton due to colour screening dissociation. At lower momentum, however, the modification diminishes owing to soft J/ψ regeneration in the QGP. The figure

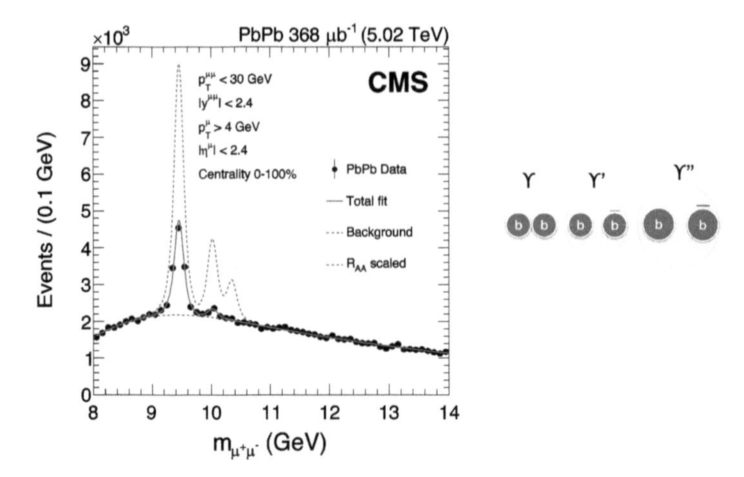

Fig. 14.22. Invariant mass spectrum for $\mu^+\mu^-$ pairs in the region of $\Upsilon(1S)$ production (big left peak), $\Upsilon(2S)$ (middle peak) and $\Upsilon(3S)$ (peak on the right), as measured by CMS. The data points are measurements for central lead–lead collisions (the blue curve is just to guide the eye). The red dashed line is the corresponding spectrum of $\Upsilon(1S, 2S, 3S)$ resonances measured in proton–proton collisions at the same nucleon–nucleon centre-of-mass energy. The deeply bound $\Upsilon(1S)$ state is less affected by colour screening and melting (see text) than the more loosely bound $\Upsilon(2S)$ and $\Upsilon(3S)$ states (see right-hand sketch). *Image: CMS Collaboration, Phys. Lett. B 790, 270 (2019).*

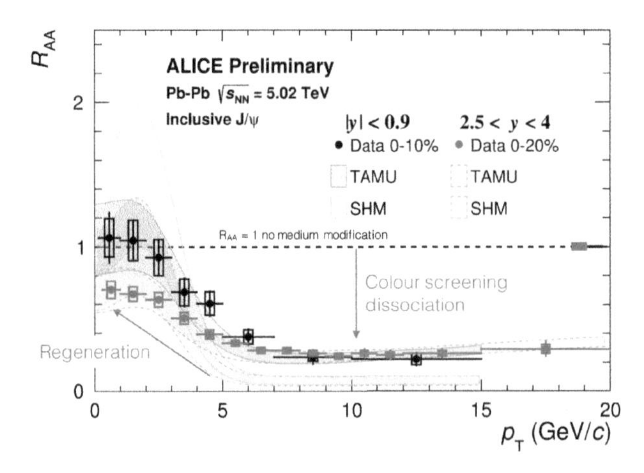

Fig. 14.23. Nuclear modification factor of inclusive J/ψ production versus transverse momentum. The black (red) markers represent the measurements in central (less central) lead–lead collision data recorded by the ALICE detector. The J/ψ suppression is reduced at low transverse momentum, especially in central collisions. The coloured bands represent predictions from phenomenological models including both J/ψ melting and regeneration. See text for a discussion of these phenomena. *Image: ALICE Collaboration ALI-DER-346483 (2020).*

also indicates that the regeneration is stronger for more central collisions than for less central ones. In fact, there is a competition between dissociation and regeneration, which approximately cancels at very low momentum and very central collisions. The effect is qualitatively reproduced by phenomenological models.

Chapter 15

Looking Ahead

It took more than twenty years to conceive and build the LHC and its sophisticated experiments. The most visible reward for this endeavour was the magnificent discovery of the Higgs boson in 2012, but this is not the whole story. A decade after the start, the LHC and its experiments have exceeded all performance promises and transformed particle physics. As we have seen in the preceding sections, many physics processes have been discovered that are used as "tools" to approach the big questions of physics: the nature of dark matter and dark energy, the hierarchy of scales and stability of the scalar sector, the matter–antimatter asymmetry in the Universe, the strong CP problem, to which direct experimental probes are yet elusive. However there is huge progress on the "*answerable questions*"[1] through measurement. Particle physics lives in data-driven times, and experiments (more not less!) are needed to guide the field to the next stage. The LHC and its detectors represent the flagship of particle physics at the energy frontier for the decade, if not decades to come.

By the end of the first phase of data taking (Run 1, 2010–2012), $5\,\mathrm{fb}^{-1}$ at $7\,\mathrm{TeV}$ centre-of-mass energy and $21\,\mathrm{fb}^{-1}$ at $8\,\mathrm{TeV}$ had been recorded by each ATLAS and CMS. To allow maintenance work and further improve the performance of accelerator and experiments, the LHC program as formulated by CERN included three major machine shutdowns lasting about two years each. The goal of the first shutdown (long shutdown 1, LS1) in 2013–2014 was to implement modifications needed to reach nominal collider energy and luminosity. The second phase of data taking (Run 2)

[1]Gavin Salam, Large Hadron Collider Physics (LHCP) conference, 2018.

lasted from 2015 to 2018 and $147\,\text{fb}^{-1}$ of integrated luminosity at 13 TeV were recorded by each ATLAS and CMS.

The successful Runs 1 and 2 were celebrated in a special 10th LHC anniversary edition of the CERN Courier in March/April 2020 (figure 15.1). If we wanted to summarise the legacy in a few headlines, we could write:

Fig. 15.1. CERN Courier edition from March/April 2020 celebrating 10 years of physics and technological prowess. *Image: CERN Courier.*

- The Higgs boson exists.
- There is — so far — no proof of physics beyond the Standard Model up to the TeV scale.
- Numerous discoveries within the Standard Model were made involving rare processes, flavour, spectroscopy, high-density strongly interacting matter.
- Accelerators, detectors, computing and analysis performed beyond expectations.
- The LHC has prompted prodigious progress in particle theory.

The second long shutdown (LS2), from 2019 to end of 2021, was needed to further increase the machine luminosity, to prepare for 14 TeV design beam energy,[2] and by the experiments to upgrade parts of their systems, which was particularly extensive for ALICE and LHCb (see section 9.7). The forthcoming Run 3 data taking, scheduled between 2022 and 2024, is expected to at least double the luminosity accumulated so far by ATLAS and CMS.

Measurements and searches will continue to focus on studies of the Higgs boson, many of which, in particular the important di-Higgs boson production to study the Higgs potential, require much larger datasets than

[2]The difference between 13 and 14 TeV proton–proton collision energy may not seem large, but it results in a noticeable increase of production cross-section for heavy final states. For example, the dominant Higgs boson production cross-section through gluon fusion increases by 13% and that of ttH and di-Higgs boson production increases by about 20%.

currently available. The study of the electroweak sector with multi-boson production requires polarisation analysis to isolate the longitudinal weak-boson scattering amplitude, which is crucial for the understanding of the Brout–Englert–Higgs mechanism at high energy. Also here, much larger datasets are needed than those currently available. In the flavour sector, more precision in CP-violation measurements is required, and a much larger dataset is needed to measure the extremely rare decay $B_d^0 \to \mu^+\mu^-$. There are many more examples of high-profile searches and measurements which are limited by the available data statistics of the LHC experiments.

15.1. Towards the High-Luminosity LHC

To evaluate whether the LHC program should be extended and upgraded beyond the baseline plan of 300 fb^{-1} integrated luminosity at 14 TeV pp collision energy, the following considerations were made: (i) what are the physics drivers and incentives; these are the most important aspects justifying additional effort and expenditure; (ii) what are the improvements that can be realised for the beam energy, luminosity and operational mode of the LHC accelerator-collider and its injectors; and (iii) what are the modifications required for the detectors to remain productive scientific instruments for an extended period and under upgraded LHC beam and operational conditions. It is also necessary to think about the positioning of the research and discovery potential of an upgraded and extended LHC program with respect to other scientific projects that could emerge on the particle physics stage world-wide in the coming decades, in particular those at the high-energy frontier (figure 15.2).

Fig. 15.2. Photo of CERN's Globe of Science and Innovation with a HL-LHC projection. On the right, the 37-metre steel sculpture "Wandering the immeasurable" by Canadian artist Gayle Hermick. The inscription describes 396 breakthroughs in the field of physics, astrophysics, and mathematics. *Image: CERN.*

These reflections were made when the *High Luminosity LHC* (HL-LHC) project was created as a design study by CERN end of 2010.[3] After the

[3]The HL-LHC is the incarnation of the *SLHC* (for Super-LHC) luminosity upgrade project, which was identified by the 2006 European Strategy of Particle Physics as the

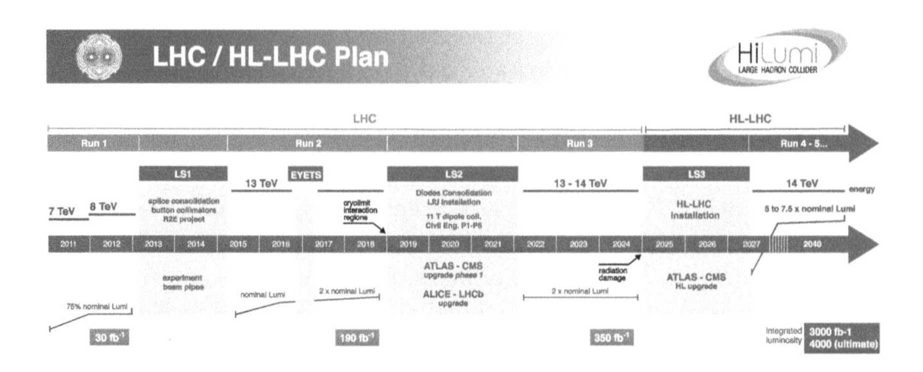

Fig. 15.3. The LHC and HL-LHC programs. *Image: HL-LHC, CERN.*

approval by the CERN Council in 2013, and the recognition of the HL-LHC budget in CERN's medium term financial planning approved by Council in 2014, the HL-LHC had become CERN's major construction project for the next decade.

Following further design reports and a detailed cost and schedule review, CERN submitted in spring 2016 the proposal for a global approval of the HL-LHC to the CERN Council, describing the physics case, upgrade goals, and the technology challenges. The document was approved in the 181th session of the CERN Council in June 2016. It became the first project formally approved by the CERN Council since the final approval of the LHC in 1996. The total material cost of the HL-LHC was estimated to 950 million Swiss francs. This does not include the upgrades of the experiments to prepare them for the harsh HL-LHC conditions.

The LHC and HL-LHC timelines with their data taking phases and technical stops, proton–proton collision energy as well as expected integrated luminosity are illustrated in figure 15.3. The main HL-LHC and detector upgrade installation occurs during the technical stop 2025–2027 (denoted long shutdown 3, LS3), and data taking is scheduled to start in the midst of 2027. LS3 is preceded by a decade of preparatory work both by the accelerator teams and the ATLAS and CMS collaborations. This includes design studies, prototyping, production, assembly and commissioning. Several parts of upgrades and civil engineering were already realised during LS2.

The planning foresees the first HL-LHC Run 4 between 2027 and 2030, followed by long shutdown 4 (LS4) in 2031, then Run 5 in 2032–2034, LS5

highest priority project after the start of the LHC. The SLHC upgrade was expected to happen in around 2015.

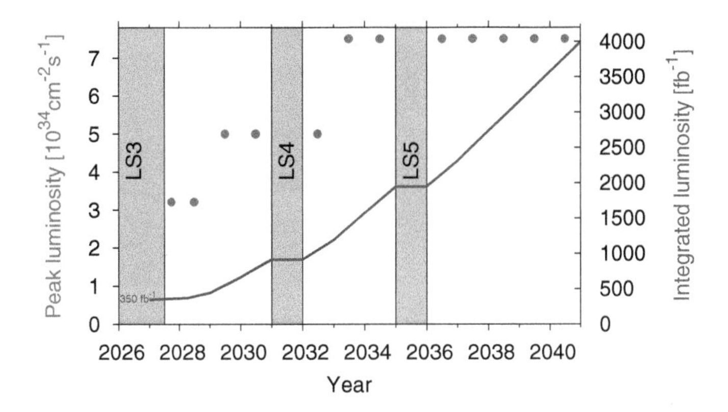

Fig. 15.4. Expected luminosity profile versus year for the HL-LHC. On the left ordinate axis and indicated by the dots the peak luminosity in multiples of $10^{34}\,\mathrm{cm}^{-2}\,\mathrm{s}^{-1}$. On the right ordinate axis and indicated by the purple line the integrated luminosity. *Image: R.T. Garcia et al., HiLumi WP2 Meeting, March 2020, and CERN-ACC-NOTE-2018-0002 (2018).*

in 2035, and the final Run 6 starting in 2036. The expected luminosity profile is illustrated in figure 15.4. The peak luminosity increases during Run 4 from $3 \times 10^{34}\,\mathrm{cm}^{-2}\,\mathrm{s}^{-1}$ to $5 \times 10^{34}\,\mathrm{cm}^{-2}\,\mathrm{s}^{-1}$, with pile-up levels rising from about 80 to 140 interactions per proton–proton bunch crossing. The *virtual peak luminosity*, which is the peak luminosity the LHC could reach if not levelled, reaches $10^{35}\,\mathrm{cm}^{-2}\,\mathrm{s}^{-1}$ during Run 4. Levelling, that is, artificially reducing the peak luminosity by slightly separating the beams, reduces the amount of pile-up in the detectors. Since the virtual luminosity is significantly larger than the actual one, the LHC can compensate proton *burn-off* (loss of protons due to collisions) by reducing the amount of levelling and effectively stay with a constant peak luminosity for many hours, which benefits the integrated luminosity. During Run 5 the LHC may reach the ultimate peak luminosity of $7.5 \times 10^{34}\,\mathrm{cm}^{-2}\,\mathrm{s}^{-1}$ corresponding to a breathtaking number of 200 simultaneous pile-up interactions at each proton bunch crossing. The intensity will reach 220 billion protons per bunch, which is 1.8 times the peak value reached during Run 2. The projected yearly integrated luminosity will rise from $120\,\mathrm{fb}^{-1}$ to $260\,\mathrm{fb}^{-1}$ during Run 4, and attain ultimately $400\,\mathrm{fb}^{-1}$. The total integrated luminosity for each ATLAS and CMS will reach $3000\,\mathrm{fb}^{-1}$ towards the end of 2037, or $4000\,\mathrm{fb}^{-1}$ if operation continues until 2040. Such a data sample would contain 250 million Higgs boson events and 160 thousand rare di-Higgs boson events according to the Standard Model.

Fig. 15.5. Left: two of the four 6.3-metre long 11-tesla dipole magnets using Nb_3Sn superconducting cables that will be installed during LS3 to make space for additional beam collimators. Right: wiring machine producing a long length Rutherford cable, made from 40 Nb_3Sn strands, for use in a 11-tesla HL-LHC dipole magnet. The wiring machine is the only one left in Europe able to perform such a job. *Images: CERN.*

15.1.1. *Accelerator-collider complex*

Reaching such ambitious goals requires improving the performance of the entire collider and injector chain (figure 6.17). We have mentioned above the increase of the beam intensity by almost a factor of two above the nominal value. Achieving this is the goal of the LHC Injector Upgrade (LIU) project, completed during LS2 (2019–2021). It includes the new more powerful Linac4 (replacing Linac2, see chapter 6), upgrades of PS booster and SPS for improved beam brightness and reliability for proton physics, and also improvements of the heavy-ion injection chain. New high-field dipole magnets (11-tesla, niobium-three-tin, Nb_3Sn, superconductors, see figure 15.5) were originally planned to be installed during LS2 to gain space for additional beam cleaning collimators that absorb off-momentum particles. However, due to reliability risks requiring more tests the installation has been postponed to LS3. In addition to more intense beams, the luminosity can be increased by reducing the cross-section of the luminous region of the colliding protons (see inset 6.2 on page 178). This will be achieved by a stronger optical focusing of the beams in the interaction points through reducing the focusing length β^*. It requires larger aperture high-field (11–12 tesla field strength) quadrupole magnets based on highly delicate Nb_3Sn superconducting wires and a special shielding system (figure 15.6).[4] The larger aperture

[4]The manufacturing process is as follows. A magnet coil is wound with multi-filament wire cables made of chemical Nb and Sn precursors in a copper matrix. The wires are insulated with thin (few millimetres) glass fibre. This set-up allows to deform the cables without breaking them. The coils are then heated to about 650°C to form Nb_3Sn out of the precursor elements. The resulting coil is brittle and must be handled with

entails a larger crossing angle of the beams. To avoid loss of luminosity due to the reduced luminous overlap, another improvement considered are special magnetic cavities providing *crab crossings.* These allow a small rotation of the proton bunches just before and just after bunch crossing to achieve full overlap of the bunches at crossing time in presence of a crossing angle, for maximal bunch collision efficiency and thus luminosity. Finally, the increased beam intensity and luminosity lead to increased radiation along the LHC, which requires the development of new radiation-hard electronics and, in particular, power converters supplying the currents for the superconducting magnets.

Fig. 15.6. View of the new 11-metre long octagonal beam screen for the HL-LHC magnets that perform the final focusing of the proton beams before collision. The beam screens shield the magnet coils and cryogenic system from the heat loads and other damage by the highly penetrating collision debris. *Image: CERN.*

15.1.2. *Detector upgrades*

Both collaborations ATLAS and CMS have put in place a detailed program of research and developments to improve in several stages during LS2 and, mostly, LS3 their detector systems to be able to operate under HL-LHC pile-up conditions (see figure 15.7) with similar performance and efficiency as during Run 2, and to withstand the tenfold increased radiation levels. In first place, the silicon pixel and microstrip inner tracking detectors, strongly exposed to radiation and a large track occupancy due to their proximity to the beam line, must be replaced. Both ATLAS and CMS benefit from the opportunity to increase the forward acceptance of their trackers to $|\eta| \sim 4$, corresponding to a polar angle relative to the bean line of only two degrees. This extension will improve missing energy reconstruction, the separation of signal forward jets from pile-up, and extend the b-quark tagging capability of the detectors. The granularity of the devices is increased by more than a factor of ten to over a billion channels to avoid overlapping tracks in the

greatest care. It is reinforced by polymer injection filling the space between cables. These manufacturing steps, although conceptually simple, are very delicate rendering the mass production of large Nb_3Sn-based superconducting magnets very challenging and expensive.

Fig. 15.7. Simulation of vector-boson fusion Higgs boson production with decay to a pair of τ-leptons in CMS. The Higgs boson is produced together with 200 proton–proton pile-up interactions as is expected during the HL-LHC phase. Most of the reconstructed tracks belong to pile-up vertices. *Image: CMS Collaboration.*

same silicon pixel or strip. A digital model of the ATLAS Inner Tracking detector with silicon pixel and silicon strip layers is shown in figure 15.8.

During LS3 the CMS experiment will replace the end-cap (pseudorapidity range from 1.5 to 3, see inset 10.5 on page 307) electromagnetic crystal calorimeter due to insufficient radiation hardness (and also part of the hadronic one) with a high-granularity calorimeter made of silicon pad detectors as active elements and iron (or tungsten) as absorber material — figure 15.9. This technique will allow to measure shower developments in three dimensions (as currently possible through the longitudinal segmentation in the ATLAS liquid-argon calorimeter) through the depth of the calorimeters thereby greatly improving individual particle tracking-showering, needed for hadron versus π^0 separation and for pile-up rejection through better spatial and angular resolution.

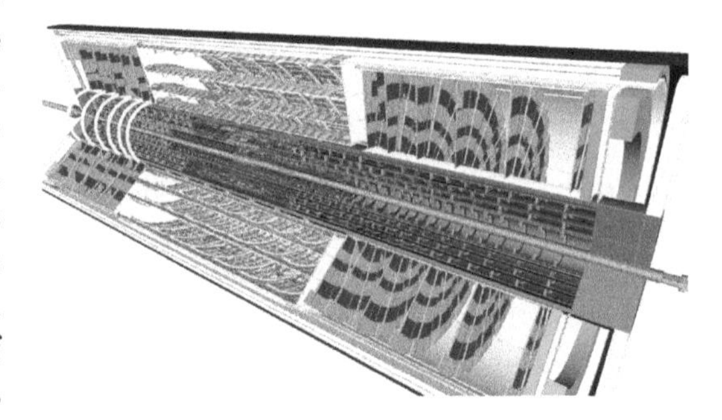

Fig. 15.8. Schematic model of the ATLAS Inner Tracker (ITk) upgrade for the HL-LHC. Close to the beam pipe are arranged five silicon pixel layers followed by four silicon strip layers in the barrel. The end-caps have an extended acceptance up to $|\eta| = 4$, a complex structure of six strip discs and a number of pixel rings arranged for an efficient recognition of the track patterns. *Image: ATLAS Collaboration, CERN-LHCC-2017-005 (2017).*

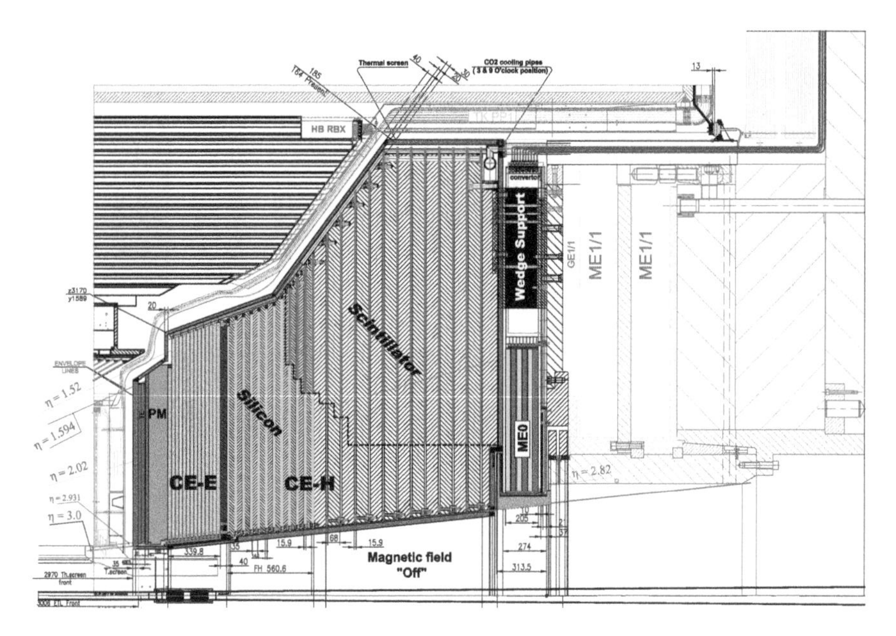

Fig. 15.9. Schematic drawing of upper half of one CMS end-cap calorimeter for the HL-LHC. It consists of a 34-centimetre thick electromagnetic compartment (CE-E) followed by a hadronic compartment (CE-H). The green region to the lower left is instrumented with silicon detectors and the blue region to the upper right with scintillator tiles. The CE-E is instrumented with 28 sampling layers of silicon detectors and the CE-H with 12 planes of stainless steel absorbers between which sit silicon modules and scintillator tileboards. *Image: CMS Collaboration, CERN-LHCC-2017-023 (2017).*

For further redundancy against pile-up, both collaborations develop timing devices based on novel silicon-based low-gain avalanche detectors (LGAD), whose short signal rise time of less than 0.5 nanoseconds allows an excellent time resolution of better than 30 picoseconds per hit.[5] This is to be compared with the expected 50-millimetre longitudinal Gaussian spread of the luminous region, corresponding to a time spread of 175 picoseconds. This allows to separate the interaction vertex of the *hard scattering* event of interest from vertices of pile-up collisions. While the ATLAS timing detector covers the end-caps only ($2.4 < |\eta| < 4.0$), where the pile-up rates are highest and the vertex resolution capability of the tracking detector weakest, CMS extends the timing device into the barrel with the advantage of a more precise timing reference of the hard-scattering event from which the

[5]This is the timing resolution before any irradiation. The resolution will deteriorate by about 60% at maximum expected irradiation during HL-LHC operation.

pile-up is to be separated. Overall, the timing detectors are expected to improve the tracking-based pile-up jet suppression by up to 50%.

The muon detection systems also require improved rejection of fake muons. An additional layer of GEM-type muon chambers (see footnote 3 on page 471) in the rapidity range 2.0–2.5 is foreseen for CMS to improve the selectivity of forward muon triggering. ATLAS plans to replace all muon chambers in the first end-cap wheels during LS2 with new wheels equipped with higher precision small TGC and MicroMegas gas chambers (see inset 7.6 on page 215).

To face the increasing instantaneous luminosity and pile-up rates, both experiments substantially reinforce their hardware (first level) and software (second level) trigger systems. The allowed latency of the first-level trigger, which is the available time interval during which the trigger system has to take the decision whether or not to retain an event for further study by the software trigger, will be increased by a factor of four from currently 2.5/3.2 microseconds (ATLAS/CMS) to about 10 microseconds (which includes some contingency). This allows significantly more processing and hence the application of more sophisticated algorithms to distinguish signal from background events. This increase of latency comes with a price: while the first-level trigger evaluates its decision, all detector systems must continue to accept data from the bunch crossings occurring every 25 nanoseconds (modulo empty bunches) and store the information in their local pipeline memories. In ATLAS, the latency will be mainly limited by the 512-bunch-crossing buffer of the strip detector readout chip. If the buffers overflow data are lost, the increase from the current 100-bunch-crossing pipeline memories to the deeper HL-LHC pipelines requires the exchange and upgrade of all readout systems and associated electronics. The ATLAS trigger and data acquisition system allows first level rates of 1 megahertz and software trigger rates of up to 10 kilohertz, which is the rate of events moved to permanent storage. CMS targets a maximum first-level trigger rate of 750 kilohertz, which is possible without loss of physics events thanks to the use of tracking information to reduce contamination from pile-up.

Finally, all LHC experiments invest into improving the computational efficiency of their event reconstruction. The increase of computing time with pile-up is nonlinear due to the rise of combinatorial options to form tracks from the measured hits. The collaborations explore the use of modern computation acceleration technologies, such as highly parallel graphics processing units (GPU) and fast field-programmable gate arrays (FPGA), in addition to standard central processing unit (CPU) based computers.

The material cost of the ATLAS and CMS upgrades amounts to about 270 million Swiss francs for each experiments. The entire upgrade program is closely followed by the LHCC (cf. section 8.5) from the scientific justification and instrumentation point of view[6] and by specially set-up upgrade cost groups scrutinising the cost aspects, ultimately reporting to CERN management and the RRB. The approvals of the various detector upgrade projects were achieved between 2017 and 2020 (see, for example, figure 15.10).

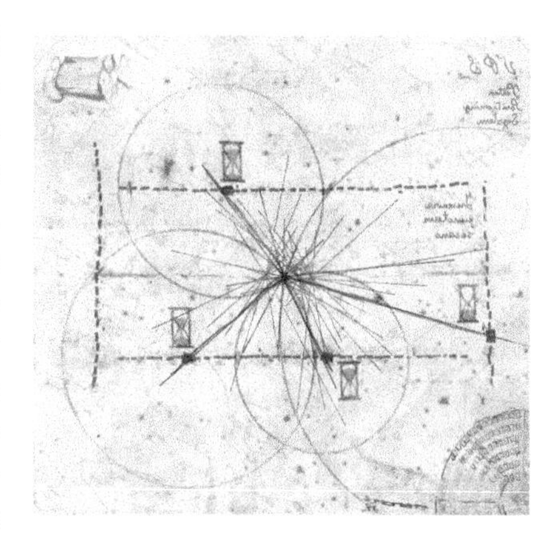

Fig. 15.10. Illustration in Leonardo da Vinci style on the cover of the Technical Design Report (TDR) of the timing detector (MTD) for the HL-LHC upgrade of CMS. All CMS TDR covers feature a similar drawing by CERN scientist Sergio Cittolin. *Image: Sergio Cittolin and CMS Collaboration, CMS-PHO-OREACH-2009-001.*

The ALICE and LHCb detectors underwent substantial and ambitious upgrades during LS2, which we have described in section 9.7. LHCb's goal is to collect an integrated luminosity of $50\,\mathrm{fb}^{-1}$ until the end of Run 4. Both collaborations are currently preparing the case of another upgrade for installation during LS4. With this upgrade, LHCb expects to operate with an instantaneous luminosity of $2 \times 10^{34}\,\mathrm{cm}^{-2}\,\mathrm{s}^{-1}$, ten times that of the Run-3 detector, allowing the experiment to collect up to $300\,\mathrm{fb}^{-1}$ by the end of the HL-LHC program.

15.1.3. *Physics goals*

The first and foremost justification for all these efforts and expenses must be the expected returns in physics. What can and must be better understood or clarified within the Standard Model, where could it possibly fail and provide indications for future developments, or even provide answers to some of most outstanding questions in physics such as dark matter?

Before we discuss some of the physics aspirations in more detail, we should emphasise that the analysis of the data collected during the LHC's Run 1 and Run 2 showed that the extrapolations of the LHC's physics potential from the early times were, without exception, on the conservative

[6]Special upgrade review groups were created under the LHCC providing detailed monitoring of the progress of each collaboration.

side. As we had stated at the beginning of this chapter, accelerators, detectors, computing and analysis performed beyond expectations. The reasons for the better than expected physics sensitivity are multi-fold. The faithfulness of the detector and physics modelling allowed precise signal and background modelling; data-driven background determination methods and *in-situ* constraints allowed to reduce systematic uncertainties; sophisticated analysis methods allowed physicists to reconstruct leptons to very low momenta, and hadronically decaying particles with very large Lorentz boost using substructure techniques; the widespread use of multivariate methods and machine learning for classification and regression improved background suppression and calibration. Previous experiments at the Tevatron and LEP made a similar experience. When faced with real data rather than simulation, physicists have more imagination, daring and creativity. We can therefore expect that the estimates discussed in the following pages concerning LHC results over the next fifteen and more years will also likely turn out to be on the conservative side.

15.1.4. *The Higgs sector*

The study of the Higgs boson will continue to be key for the HL-LHC physics program. None of the characteristics and features of this most curious force particle — the only known spin-zero elementary particle — can be taken for granted and must thus be measured. The Higgs boson studies may be categorised into: property measurements such as mass and a CP admixture; the search for rare Higgs boson processes such as $H \to \mu\mu$, $H \to Z\gamma$, di-Higgs boson production and constraints on the Higgs boson self-coupling; the search for very rare and forbidden decays; the precise study of the Higgs boson couplings through the study of Higgs boson production and anomalous couplings by differential cross-section measurements as well as global and partially global Higgs boson coupling fits; and finally the search for new physics in Higgs production or other scalar states.

The Higgs boson mass is not predicted by theory and is an essential parameter for any theoretical calculation involving the Higgs boson. As of 2020, it has already been measured with a one permil precision, which will be further improved once the final results with Run-2 data have been released and combined. Further improvement in precision by a factor of about four is expected with the HL-LHC. The question of the Higgs boson's spin-parity property is essentially settled to scalar (0^+). However a possible pseudoscalar (0^-) admixture, giving rise to CP violation, is still possible within the current experimental precision and is among the priority studies.

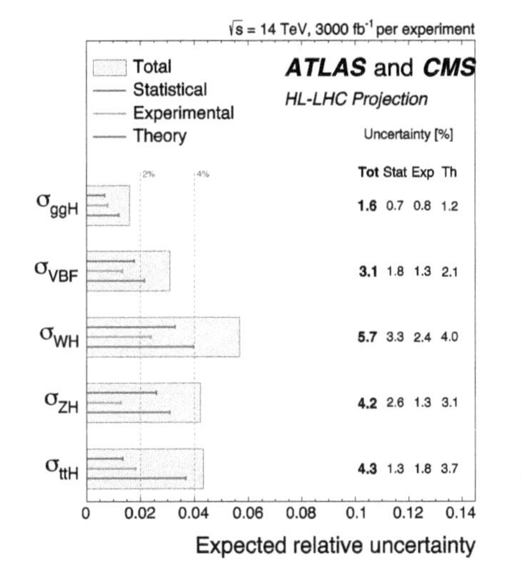

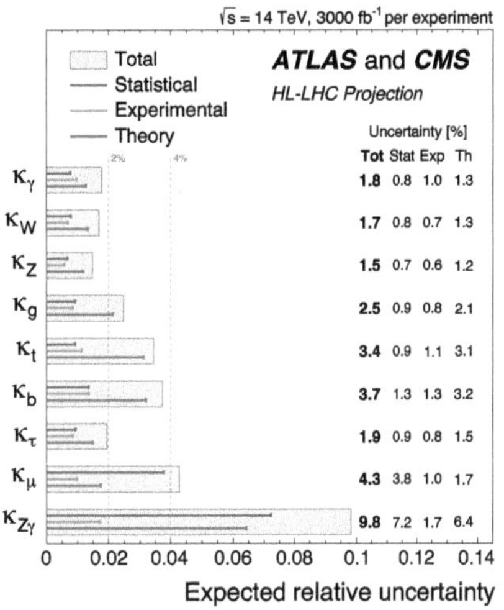

Fig. 15.11. Expected relative precision in percent on Higgs coupling measurements for 3000 fb^{-1} at the HL-LHC, combining ATLAS and CMS projections. Couplings to photons, vector bosons W and Z, $Z\gamma$, gluons, top and bottom quarks, and to τ-leptons and muons can be measured. Couplings to photons, gluons and $Z\gamma$ are effective only, as they appear through loop diagrams responsible for the decays to two photons (figure 11.5) and $Z\gamma$, or to Higgs production through gluon fusion (figure 11.4). The left panel shows the expected precision for the measurement of the inclusive cross-sections per production mode and the right panel shows that for the linear coupling strength modifiers. *Images: CERN Yellow Reports: Monographs, CERN-2019-007 (2019).*

The HL-LHC dataset will allow the measurement of the Higgs boson couplings to photons, vector bosons (W and Z) and to fermions (t, b, τ, and μ) to unprecedented detail and precision. Figure 15.11 gives a flavour of the constraints on the various production cross-sections and coupling strength modifiers κ_i (ratio of measured to predicted coupling strength) for the combined ATLAS and CMS projections. Percent-level precision is achieved for gluon fusion production cross-section and the Higgs boson couplings to photon, W, Z, and τ-lepton. The couplings to the heaviest and lightest accessible fermions, top quark, bottom quark, and muon, will be measured with about 4% precision. Each experiment should be able to observe the rare Higgs boson decay to $Z\gamma$. The Higgs boson decay to a pair of charm quarks is expected to be out of reach for a direct observation (via associated production WH and ZH), but limits of a few times the Standard Model prediction should be achievable.

In addition to the inclusive coupling studies, it will be crucial to perform detailed differential cross-section measurements with the HL-LHC dataset. A projection by CMS for various measurements of the Higgs boson transverse momentum is shown in figure 15.12. By combining ATLAS and CMS a per-bin precision of better than 5% can be achieved below $p_T^H \sim 200\,\text{GeV}$. The different momenta are sensitive to different underlying physics phenomena. Large momenta are expected to have sensitivity to heavy particles affecting the Higgs boson production. While most of the spectrum is dominated by the di-photon and Z-boson pair measurements, $H \to bb$ decays become relevant at high momentum where the other decays are lacking statistics due to their low-branching fractions. The lower panel shows the expected relative uncertainties of measurement and theoretical prediction, which are of similar size.

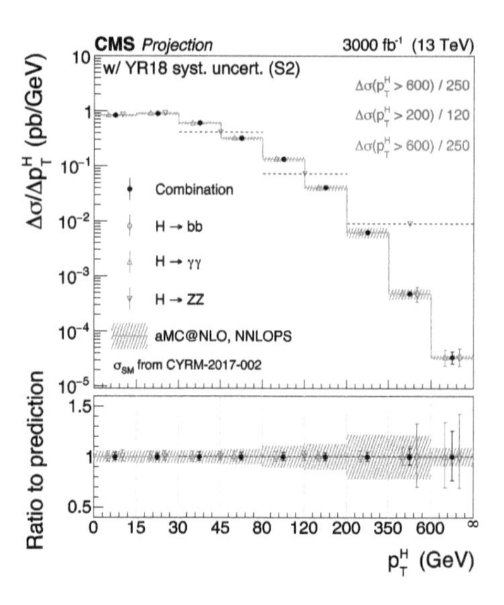

Fig. 15.12. HL-LHC projections by CMS of differential cross-section measurements of the Higgs boson transverse momentum using decays to, respectively, photon, Z-boson, and b-quark pairs, and their combination. The central values are assumed to follow the theoretical prediction shown by the hatched area. The lower panel shows the expected relative uncertainties of measurement and theoretical prediction. *CERN Yellow Reports: Monographs, CERN-2019-007 (2019).*

The measurement of the Higgs boson width Γ_H through the analysis of contributions far off the mass shell (cf. section 11.7) will benefit from the HL-LHC dataset. It is expected that Γ_H can be determined with a precision between 20% and 40%, depending on the theoretical progress in the calculation of the interfering amplitudes.

Among the most important scientific objectives of the HL-LHC is to verify the existence of triple-Higgs boson coupling (or Higgs self-coupling), $\lambda_{HHH}^{\text{SM}} = m_H^2/(2v^2) \simeq 0.13$, where v is the vacuum expectation value of the Higgs field. The existence of such a coupling is related to the *non-abelian* nature[7] of the electroweak theory.

[7]In practice this means that carriers of an interaction couple to the *charge* generating the interaction. They are thus themselves carriers of this charge and can interact among themselves. For example, quantum electrodynamics is an abelian interaction, as photons do not carry electric charge, thus there is no direct photon–photon coupling. On the other hand, the electroweak interaction with weak charges g and g' is non-abelian so that there are $WW\gamma$ or WWZ couplings, for example. The term *abelian* is derived from

Inset 15.1: New physics effects on Higgs boson couplings

How large may deviations in the Higgs boson couplings relative to the Standard Model expectation turn out? Denoting with $g_X^{\rm SM}$ the Higgs coupling to particle X in the Standard Model and by g_X this same coupling as expected in a theory introducing new physics at an energy scale $\Lambda_{\rm NP}$, it is possible — in a generic and simplified way — to parameterise the expected deviation as follows:

$$\frac{g_X}{g_X^{\rm SM}} \approx 1 + \alpha \left(\frac{1\,{\rm TeV}}{\Lambda_{\rm NP}}\right)^2$$

here $|\alpha|$ can vary from few percent to about 10% depending on particle X and the specific theory considered. With the HL-LHC it should thus be possible to probe a number of theories. For example, in the minimal supersymmetric Standard Model the deviation for the coupling to τ-leptons would be of order:

$$\frac{g_\tau}{g_\tau^{\rm SM}} \approx 1 + 10\% \cdot \left(\frac{0.4\,{\rm TeV}}{m_A}\right)^2$$

where m_A is the mass of the additional (pseudoscalar) Higgs boson A of supersymmetry. This means that if m_A were of order 1 TeV, the deviation would be just about 2%, at the limit of what can be seen with HL-LHC.

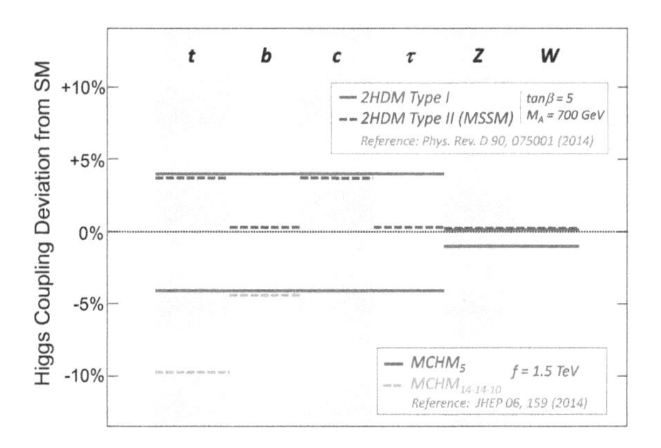

the abelian group, which comprises mathematical operations that are invariant under a change of the order with which they are applied. It is named after the Norwegian mathematician Niels Henrik Abel (1802–1829) who died prematurely of tuberculosis.

The above figure shows examples of potential deviations from Standard Model expectations of the Higgs boson couplings to Standard Model particles (top, bottom, charm quarks, τ-lepton, Z and W bosons) in four new physics scenarios that are not yet excluded by LHC searches. The 2HDM scenarios stand for models featuring two Higgs doublets (recall the Standard Model has one Higgs doublet). The Type-II scenario corresponds to the minimal supersymmetric extension of the Standard Model (MSSM). The two parameters m_A (mass of the pseudoscalar neutral Higgs boson) and $\tan\beta$ (ratio of vacuum expectation values of the two Higgs doublets) fully define the model when neglecting quantum corrections. The MCHM scenarios refer to minimal composite Higgs models where the Higgs boson is the manifestation of a broken symmetry at a higher mass scale f. The pattern of the deviations would be indicative of the specific physics scenario at work. They do not exceed 10%.

Given the wealth of scenarios beyond the Standard Model, each with different unknown parameters, a more systematic approach to pin down the presence and the structure of new physics, allowing also to consistently interpret different searches, would be in order. This can be achieved through the formalism of *effective field theories*, which were introduced in inset 11.3 on page 358.

A deviation from the Standard Model expectation could indicate that the observed Higgs boson is not an elementary particle, or it could point to models with extended Higgs sectors, or simply that the Higgs potential is more complicated than currently posited. The problem is that processes giving direct access to the triple-Higgs coupling (two Higgs boson final states, see the two right-hand diagrams in figure 15.13) have exceedingly small cross-sections (36 fb at 14 TeV for inclusive HH production), more than a thousand times smaller than that of single Higgs boson production. Di-Higgs boson production is thus at the limit of observability even at the HL-LHC. The irreducible contribution from di-Higgs final states not due to triple-Higgs coupling, is given, for example, by the Feynman diagram on the left in figure 15.13. The interference between this and the triple-Higgs coupling diagrams is destructive, reducing the di-Higgs boson production cross-section in the Standard Model.

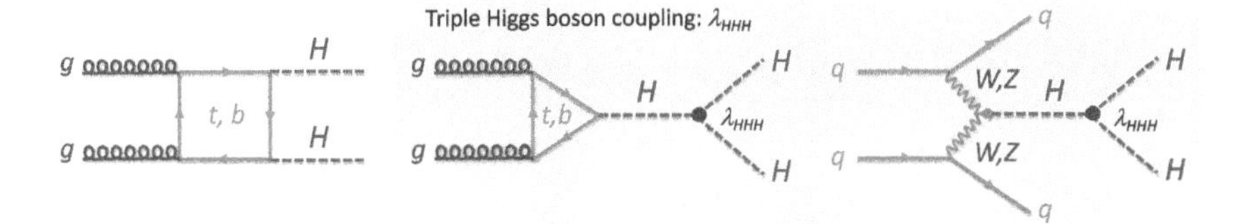

Fig. 15.13. Feynman diagrams for the production of two Higgs bosons in proton–proton collisions. Only the two diagrams on the right imply triple-Higgs boson couplings; the left-most diagram also leads to two Higgs bosons in the final state. The measurement of the di-Higgs boson final state gives access to the modulus of the sum of the three quantum mechanically interfering amplitudes, from which the triple coupling must then be extracted.

A deviation from $\lambda_{HHH}^{\text{SM}}$ will show up as a difference from the Standard Model di-Higgs boson cross-section but also in an alteration of the di-Higgs boson invariant mass spectrum. For example, the spectrum for $\kappa_\lambda = \lambda_{HHH}/\lambda_{HHH}^{\text{SM}} = 10$ would be softer than that predicted for the Standard Model (cf. left panel of figure 15.14). It is thus important to not only observe di-Higgs boson production but to also measure its kinematic properties. The most suitable channels for this study are $HH \to bb\gamma\gamma$, $HH \to bb\tau\tau$, and $HH \to bbbb$, where the former two channels have similar sensitivity and provide stronger constraints than the latter channel. These complex final states are well adapted for the application of machine learning methods to suppress reducible backgrounds, and one may expect significant sensitivity improvements in the coming years. Figure 15.14 shows HL-LHC projections for $3000\,\text{fb}^{-1}$ integrated luminosity. The left-hand plot shows distributions of the di-Higgs boson invariant mass for varying values of κ_λ. Both, integral and shape of the distributions are exploited to measure triple Higgs boson coupling. The expected constraints on κ_λ from the individual di-Higgs boson decay channels and their ATLAS and CMS combination are shown on the right panel. In case of a Standard Model outcome, $\kappa_\lambda = 1$ can be measured with an uncertainty of 50%. At this stage, we should point out that the triple Higgs boson interaction vertex also affects single-Higgs boson production through Higgs boson quantum corrections, allowing the experimenters to derive complementary information on κ_λ from differential cross-section measurements.

The Higgs physics studies at the HL-LHC will enhance the sensitivity to new physics, exploiting, in addition to the indirect probes via precision and rare decay measurements discussed above, a host of direct search targets. These range from exotic decays of the Higgs boson such as decays to light

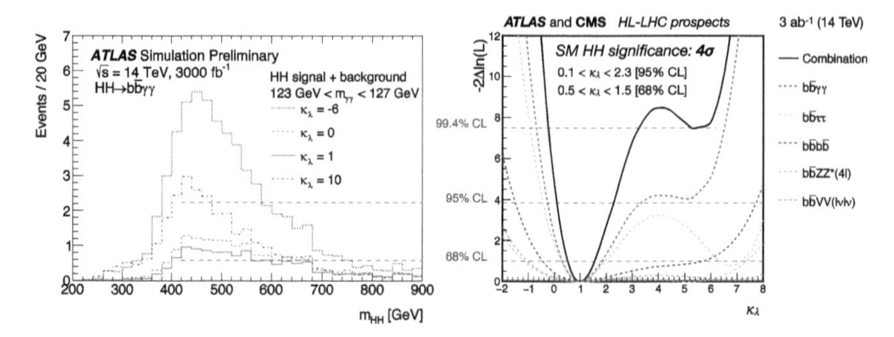

Fig. 15.14. Monte Carlo simulations of di-Higgs boson production at the HL-LHC for an integrated luminosity of $3000\,\mathrm{fb}^{-1}$. The left-hand panel shows ATLAS distributions of m_{HH} for $HH \to bb\gamma\gamma$ signal and background events passing a tight analysis selection for different values of $\kappa_\lambda = \lambda_{HHH}/\lambda_{HHH}^{\mathrm{SM}}$. The differing height and form of the distributions reflect the varying di-Higgs boson interference patterns with respect to the diagrams in figure 15.13 as a function of κ_λ. The right-hand panel shows combined ATLAS and CMS constraints on κ_λ expressed in units of a test statistic denoted by $-2\Delta\ln(L)$, which is similar to a χ^2 function. The deviation from the minimum expresses less likely outcomes. The values at the 68% crossing give the 1σ uncertainty interval for κ_λ, while the 95% crossings determine excluded values. The dotted curve shows the constraints from the individual HH decay channels and the solid black curve their combination. The form of the curve reflects the interference pattern. *Images: ATLAS Collaboration, ATL-PHYS-PUB-2018-053 (2018), CERN Yellow Reports: Monographs, CERN-2019-007 (2019).*

scalars, light dark photons or axion-like particles, and decays to long-lived particles, to the production of new Higgs bosons, neutral and charged, at masses above or below the Higgs boson. The HL-LHC will have access to new Higgs bosons as heavy as 2.5 TeV.

15.1.5. *Triple and quartic vector boson couplings*

The HL-LHC will allow ATLAS and CMS to explore and probe to significant detail another prediction of electroweak theory related to its non-abelian nature: triple and quartic vector boson couplings as occur in di-boson and tri-boson production, respectively, and in vector boson scattering processes (see example diagrams in figure 15.15). The large available data sample will give access to measuring the vector boson polarisation (recall that the scattering of longitudinally polarised W bosons is moderated by Higgs boson exchange diagrams), and will allow to study these processes at large momentum transfer and final state mass, offering enhanced sensitivity to high-scale new physics. The HL-LHC will also enable the experimenters to quantitatively study quartic vector boson couplings.

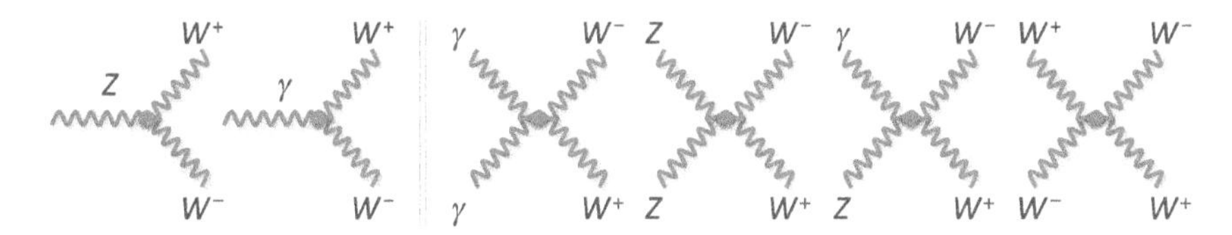

Fig. 15.15. Diagrams corresponding to triple vector boson couplings (on the left) and of quadruple (quartic) vector boson couplings (right).

15.1.6. *Direct searches for new physics*

Although the HL-LHC operates at the same 14 TeV proton–proton collision energy as that expected for Run 3, the tenfold increased data sample will allow to produce events of rare occurrences where the colliding partons carry very large fractions of the proton momenta. Therefore, higher data statistics effectively allows to probe larger parton collision energies. The growth is, however, not as significant as if the LHC beam energies had been increased. The large data samples primarily allows to study rare new physics processes at accessible masses, such as via electroweak production, with compressed mass spectra, or similar.

Searches for supersymmetry will continue to explore the strong production of supersymmetric squarks and gluinos for which the sensitivity in mass can be improved by about 700 GeV compared to the full Run-2 dataset results. The sensitivity to top squarks can be improved by about 500 GeV as their production cross-section is smaller than that of first and second generation squarks and gluinos so that the benefit from the data increase is larger. The data will allow to set 95% confidence level limits up to about 1.7 TeV per experiment. The largest effect will be on electroweak supersymmetry production of gauginos and sleptons. For example, the sensitivity to the process depicted in figure 13.16 for equal $\tilde{\chi}_1^\pm$ and $\tilde{\chi}_2^0$ masses will increase by up to 500 GeV to a limit of more than 1.1 TeV (see figure 15.16). Compressed spectra for degenerate gaugino masses will be addressed by using events where high transverse momentum jets were radiated off the incoming protons, thus giving the final state particles a Lorentz boost which allows to better separate them from background events. This should allow to significantly improve the sensitivity of so-called higgsino and wino scenarios.

By comparing the 5σ discovery reach for supersymmetric particles with the 95% confidence level exclusion limits from the full Run-2 analyses, one finds similar mass values. This means that it will be challenging to reach

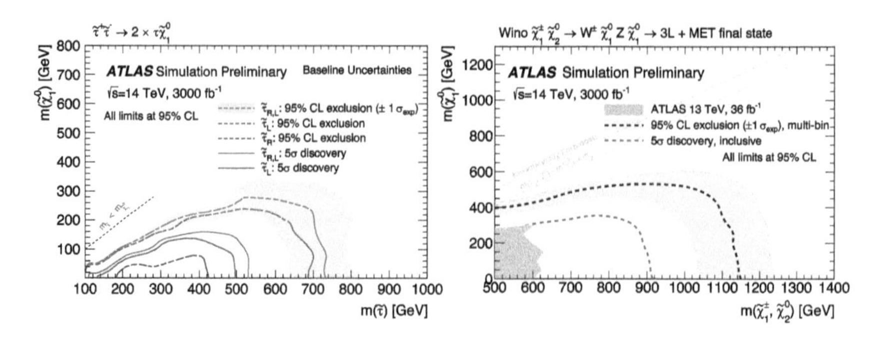

Fig. 15.16. Examples for the extension of the search domain in stau pair production (left) and chargino-neutralino production and decay through a WZ-boson pair (right) as projected by ATLAS for the HL-LHC. The plots show the expected domain where the HL-LHC has sensitivity to achieve discovery significance of 5σ, as well as the expected 95% confidence level limits in absence of a signal. *Images: ATLAS Collaboration, ATL-PHYS-PUB-2018-048 (2018).*

sufficiently large significance for a firm discovery should supersymmetry lie in the domain only accessible to the HL-LHC. However, it may of course provide strong evidence, which would allow the physicists to set directions for future research. Should supersymmetry finally show up at the LHC, its discovery would much exceed in significance the discovery of the Higgs boson. The study of supersymmetric particles, determining the sparticle spectrum, understanding the mechanism of supersymmetry breaking, as well as illuminating the many complex facets of this theory, will require a long-term effort over several decades and higher energy colliders.

Searches for possible new heavy gauge bosons W' and Z' as those appearing in a number of grand unified theories, in left–right symmetric models, or in theories with additional space dimensions (sections 4.2 and 4.3), are expected to extend their sensitivity by about 1.5 TeV for sequential Standard Model W' and Z' bosons over that achieved with the Run-2 dataset. A similar sensitivity gain is expected for a right-handed charged gauge boson $W_R \rightarrow tb \rightarrow \ell\nu b$ with Standard Model couplings, which may be excluded for masses up to 4.9 TeV. Other resonance searches such as for Kaluza–Klein states due to extra spatial dimensions or top–antitop resonances expect all improved mass sensitivity between 1.5 and 2.0 TeV at the HL-LHC compared to the Run-2 analyses.

Searches in the flavour sector will continue and benefit from the improved trigger capabilities of ATLAS, CMS, and LHCb. The $B_s^0 \rightarrow \mu^+\mu^-$ branching fraction is expected to be measured with a precision of a few percent by CMS and LHCb (see figure 15.17 for CMS).

The HL-LHC should also allow CMS and LHCb to observe the ultra-rare decay $B_d^0 \to \mu^+\mu^-$ and verify the Standard Model prediction, which is about thirty times smaller than that for the corresponding B_s^0 meson decay (cf. section 14.1).

The particularly interesting as theoretically clean double ratio of $B_s^0 \to \mu^+\mu^-$ branching fraction to mass difference between heavy and light B_s^0 mass eigenstates divided by the ratio of $B_d^0 \to \mu^+\mu^-$ branching fraction to mass difference between heavy and light B_d^0 mass eigenstates could be measured to 10% (which is, however, insufficient for a stringent test).

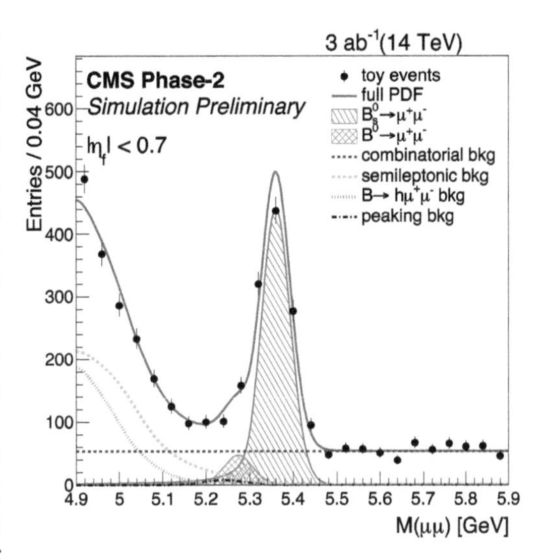

Fig. 15.17. Projected di-muon invariant mass distribution with overlaid fit results for the $B_{d,s}^0 \to \mu^+\mu^-$ analysis by CMS, for an integrated luminosity of 3000 fb^{-1}. *Image: CMS Collaboration, CMS-PAS-FTR-18-013 (2018).*

A huge amount of other searches such as for long-lived particles, heavy resonances, or new physics in exchange diagrams contributing only to modifications in tails of invariant mass distributions, searches for vector-like quarks, and many other theoretical scenarios developed to moderate the hierarchy problem will be improved at the HL-LHC.

Searches for the production of dark matter particles will continue at the HL-LHC through the various $X + E_T^{\text{miss}}$ signatures, where X is a visible particle, such as a jet (figure 15.18) or a photon, and E_T^{miss} the recoiling missing transverse energy. The search for invisible Higgs boson decays and the overall coupling combination is expected to reach a precision to probe a branching fraction as low as 2%, offering strong constraints on low-mass dark matter. Monojet signatures will allow to search for WIMP pair production, but also for electroweak supersymmetric wino or higgsino production with compressed spectra, as discussed above. Mono-photon and vector boson fusion signatures will also help for these when the data samples increase. It is expected that $\tilde{\chi}_1^0$ masses up to 300 GeV or more can be probed by the latter two signatures. The sensitivity of the $X + E_T^{\text{miss}}$

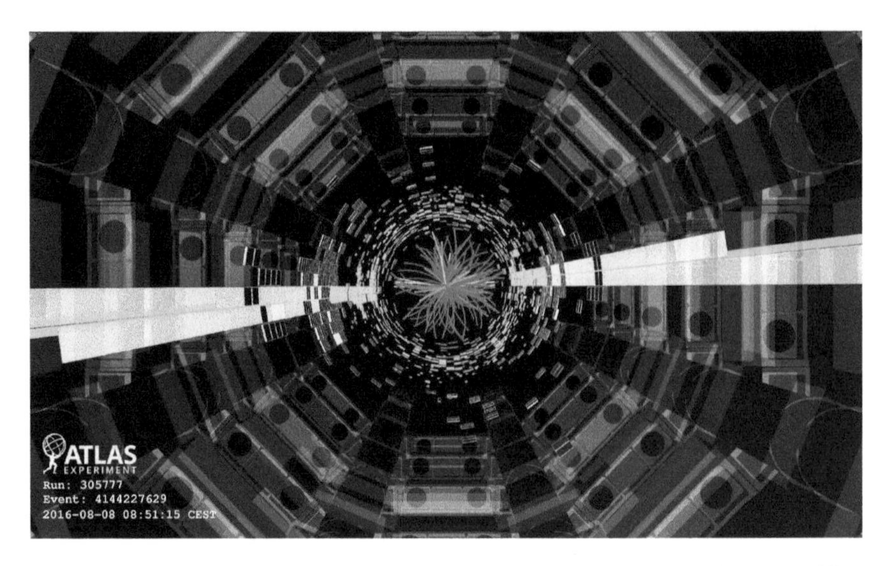

Fig. 15.18. A central di-jet event with 8.0 TeV invariant mass measured by ATLAS in 2016. *Image: ATLAS Collaboration, Phys. Rev. D 96, 052004 (2017).*

analyses to mediators is expected to be improved by factors of three to six in mass scale at the HL-LHC over that using the full Run-2 dataset.

15.2. The next big thing

Particle physics in the coming decade will be dominated by the LHC and its high-luminosity upgrade the HL-LHC discussed above. But, given the preparation time large-scale accelerator projects need to take off grounds, it is important that the world physics community discusses the next-after projects and moves ahead with studying their physics case, technological requirements, site, size, and cost. This process has been ongoing since decades already for some projects, such as the International Linear electron–positron collider and a muon collider discussed below.

Other projects, such as a new circular electron–positron collider or a new energy frontier proton–proton collider with high-field magnets are more recent. The different world regions organise strategic planning and deliberation processes during which the scientific community, leaders of the large laboratories contributing to the field, and funding agencies of the various countries are consulted to discuss the scientific and economic impact of projects.

In Europe, for example, the European Strategy for Particle Physics is a process initiated in 2006 that *"forms the cornerstone of Europe's decision-making for the future of our field"*.[8] It was updated in 2013 and again in 2020. We will discuss the recommendations from that process, which are submitted to and approved by CERN Council, below. Other world regions and countries have similar proceedings.

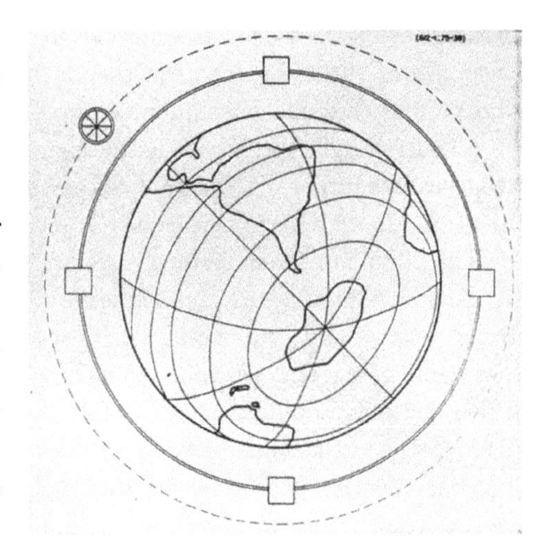

Fig. 15.19. In a talk given in 1954 Enrico Fermi envisioned an accelerator that would encircle the Earth and achieve energies never probed before. The "Globatron" (which was to be built in 1994) with current LHC magnet technology and as a proton–proton collider would reach 20 000 TeV (20 PeV) centre-of-mass energy. *Image: Fermilab Today (2007).*

In the US, such a process is ongoing in 2020–2022, in order to update the priorities stated in 2014 by the Particle Physics Project Prioritization Panel (P5). We would like to cite from the preamble of the 2014 P5 report.[9]

> *Wondrous projects that address profound questions inspire and invigorate far beyond their specific fields, and they lay the foundations for next-century technologies we can only begin to imagine. Particle physics is an excellent candidate for such investments. Historic opportunities await us, enabled by decades of hard work and support. Our field is ready to move forward.*

---

[8]Extract from the summary of the "2020 Update of the European Strategy for Particle Physics", CERN-ESU-015.

[9]The 2014 US P5 panel report "Building for Discovery, Strategic Plan for U.S. Particle Physics in the Global Context", distilled the inputs issued from the discussions within the particle physics community to five essential science drivers for particle and astroparticle physics: (1) use the Higgs boson as a new tool for discovery; (2) pursue the physics associated with neutrino mass; (3) identify the new physics of dark matter; (4) understand cosmic acceleration: dark energy and inflation; (5) explore the unknown: new particles, interactions, and physical principles. The LHC directly addresses three of these drivers (1, 3, 5), and indirectly the other two.

The highest priority for the immediate future of particle physics in both Europe and the US is the success of the large-scale facilities HL-LHC in Europe and the long-baseline neutrino experiment DUNE[10] at Fermilab in the US. In Japan, the Hyper-Kamiokande long-baseline neutrino project, a twenty times larger replacement of the successful Super-Kamiokande experiment, is among the priorities of the particle physics community.

The 2020 European strategy update has identified an electron–positron Higgs boson factory as the highest-priority next collider. This choice is driven by two considerations: (i) the potential of very high-precision measurements of the Higgs boson and electroweak sector of the Standard Model, which the clean environment of electron–positron collisions can offer, and (ii) to bridge the time needed for research and development to achieve the cost-effective production of high-field magnets needed to realise a new energy-frontier hadron collider that significantly surpasses the LHC in beam energy. We shall discuss these projects in the following.

We conclude this introduction by recalling that fundamental discoveries may occur at any time at the LHC or elsewhere. They may lead to a change in strategy, in particular if the discovery indicates the energy scale where new physics is to be expected.

15.2.1. *Future electron–positron colliders*

The mass of the Higgs boson being known and relatively light, a most suitable tool for high-precision Higgs boson studies would be an electron–positron collider with centre-of-mass energy of 240–300 GeV. Similar to the searches performed at LEP (section 2.4), the Higgs boson would be produced in association with a Z boson through the Higgsstrahlung process $e^+e^- \to ZH$ (see figure 15.20).[11] This clean final state gives access to all Higgs boson decay modes with branching fractions larger than a few 10^{-4}. In particular, it would be possible to measure with percent-level precision the $H \to c\bar{c}$ decay, and to constrain possible Higgs boson decays into invisible particles (see discussion in section 15.1.4) with a precision up to ten times better than at the HL-LHC.

[10]The acronym DUNE stands for Deep Underground Neutrino Experiment.

[11]Because the electron mass, $m_e \simeq 0.51$ MeV, is so small, the electron-Higgs boson coupling is extremely weak ($g_{Hee} = \sqrt{2}m_e/v \simeq 2.9 \cdot 10^{-6}$, with $v = 246$ GeV the well-known Higgs vacuum expectation value), so that the direct production of a Higgs boson via $e^+e^- \to H$ (like for the Z boson at LEP) has a very small cross-section. Such a direct production is only possible at a muon collider via $\mu^+\mu^- \to H$ with a $(m_\mu/m_e)^2 \simeq 42\,000$ enhanced cross-section.

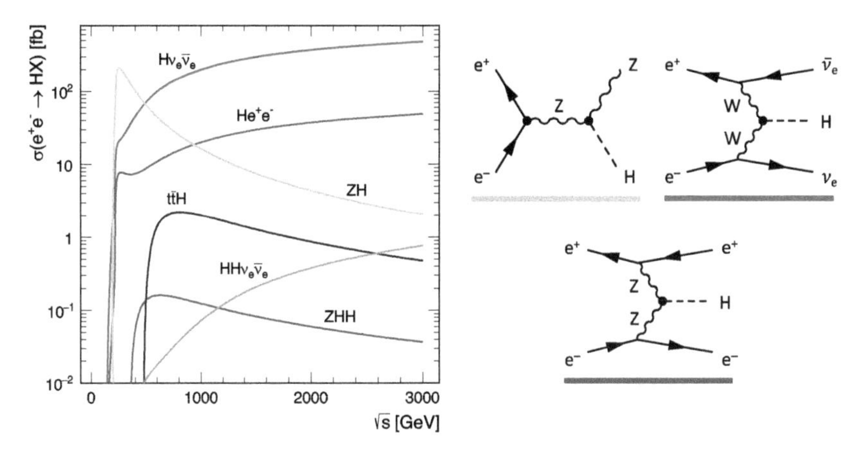

Fig. 15.20. Higgs production cross-section in an e^+e^- collider as a function of the centre-of-mass energy $\sqrt{s}$. On the right, Feynman diagrams of the three dominant processes ZH Higgsstrahlung (yellow, upper left diagram), W-boson fusion (red, upper right diagram), and Z-boson fusion (blue, lower diagram). In the first stage of the machine ($\sqrt{s} \simeq 250\,\text{GeV}$), only the ZH final state contributes with significant rate. *Image: CLICdp.*

With increasing centre-of-mass energy $\sqrt{s}$, the rate of the Higgsstrahlung process declines with $1/s$. Another interesting channel is $e^+e^- \to H\nu_e\bar{\nu}_e$ through WW fusion whose cross-section increases logarithmically as $\log(s/m_H^2)$ with the centre-of-mass energy: it reaches about 15% of the ZH cross-section at 300 GeV (figure 15.20) and becomes the most abundant channel beyond 450 GeV centre-of-mass energy. The corresponding ZZ fusion channel $e^+e^- \to He^+e^-$ also rises logarithmically but is suppressed compared to WW fusion. Access to triple-Higgs coupling through the mechanism $e^+e^- \to ZH^* \to ZHH$ (purple line in figure 15.20) requires a centre-of-mass energy of at least 400 GeV, but the cross-section remains two orders of magnitude smaller than that of ZH production, which limits the reachable precision. More promising at much larger $\sqrt{s}$ is the vector boson fusion process $e^+e^- \to H^*\nu_e\bar{\nu}_e \to HH\nu_e\bar{\nu}_e$, which overtakes ZHH above about 1.2 TeV centre-of-mass energy.

As at the HL-LHC, due to destructive interference with other processes leading to di-Higgs boson final states but not involving triple Higgs boson coupling,[12] the actual cross-section depends strongly on the value of that

[12]For example, in the Higgsstrahlung process, producing ZH, the Z could be virtual (off its mass shell) and decay into ZH giving a ZHH final state. Similarly, vector boson fusion could have an additional internal vector boson line allowing two Higgs bosons to be produced at two different VVH vertices.

Fig. 15.21. Simulated $e^+e^- \to t\bar{t}$ event as reconstructed in one of the possible detectors of a future electron–positron collider. One notes in particular the high-granular silicon-based calorimeter allowing to reconstruct particle showers with unprecedented detail. The event shows the individually reconstructed neutral and charged particles. The colour code indicates the particle flow algorithm reconstruction without explicit reference to the Monte Carlo generator information. *Image: The ILC, A Global Project, arXiv:1901.09829 (2019).*

coupling. While the top-Higgs boson coupling via ttH production will be quite precisely studied at the HL-LHC, an electron–positron collider can also produce this final state if its collision energy achieves 500 GeV at least (black line in figure 15.20).

Beyond Higgs-boson physics, electron–positron colliders also allow high-precision measurements of fundamental Standard Model parameters such as the W-boson and top-quark masses (figure 15.21) or the weak mixing angle (or Weinberg angle, $\sin^2\theta_W$, introduced in inset 1.10 on page 45) with tenfold or more improved precision over that reached at the HL-LHC. These measurements allow very powerful tests of the Standard Model. The huge samples of Z bosons as well as heavy quarks and leptons allow to probe rare or suppressed phenomena such as charged-lepton flavour violation with increased sensitivity. A proposed schedule of operations is presented in figure 15.22.

The conceptual challenge with circular e^+e^- colliders is the synchrotron radiation beam energy loss varying as E^4/R, where E is the electron and positron beam energy and R the radius of the collider. Assuming we wanted to build a 500 GeV circular e^+e^- collider with two 250 GeV beams: limiting the energy loss per electron or positron per turn to the LEP-2 value of about 3 GeV, a gigantic ring with circumference of almost 900 kilometres would be required! This is the reason why over past years, in fact past decades, e^+e^- linear acceleration and collider techniques have been developed. Two projects of this type are under active study. While linear colliders have negligible synchrotron radiation (except for electromagnetic beam–beam radiation effects), the challenge of this technology is that the (majority of) electrons and positrons who did not collide are not recovered, and their energy is lost after one passage through the accelerator. We will in the following briefly introduce the three main future electron–positron collider concepts.

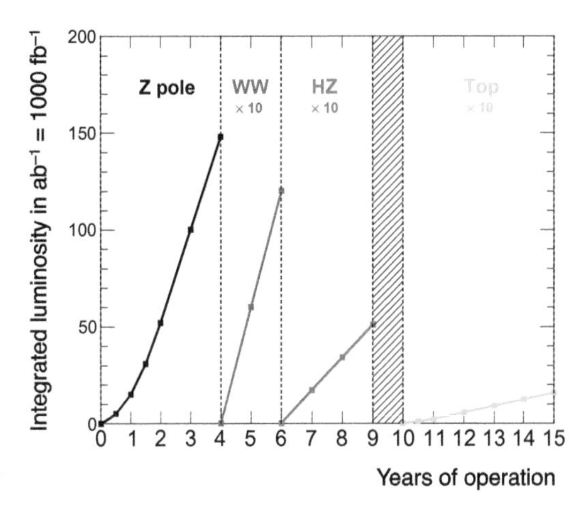

Fig. 15.22. Expected integrated luminosity versus number of years of operation for one of the future electron–positron collider projects, a circular collider based at CERN. For illustration, $150\,\text{ab}^{-1}$ ($1\,\text{ab}^{-1} = 1000\,\text{fb}^{-1}$) operation at the Z boson mass pole ($\sqrt{s} = m_Z \simeq 92\,\text{GeV}$) correspond to 3×10^{12} visible Z boson decays (for comparison, this is ten thousand times the sample available at LEP); $5\,\text{ab}^{-1}$ ZH correspond to one million ZH events. *Image: FCC-ee: The Lepton Collider, EPJ Special Topics 228, 261 (2019).*

International Linear Collider

The first and most senior project is the International Linear Collider (ILC), an e^+e^- collider with a centre-of-mass collision energy of 250 GeV in a first phase of operation, that would be increased to about 500 GeV in a second phase, which could be ultimately increased to 800 GeV. The ILC, first outlined in 2003, builds upon the concept of the TESLA consortium developed during the 1990s, as well as developments from concurrent

Fig. 15.23. Artistic view of the ILC tunnel in Japan. *Image: Rey Hori, KEK.*

projects such as the Japanese Linear Collider, the Global Linear Collider and the Next Linear Collider. A global design effort under the mandate of the International Committee for Future Accelerators (ICFA) produced the ILC reference design report in 2007, culminating in a technical design report in 2013. Since that time, the luminosity goals of the ILC have been revised to keep up with the new competition from circular electron–positron colliders.

The ILC is based on a 1.3 GHz superconducting radio-frequency accelerating technology providing an average accelerating gradient of 32 mega-volt (MV) per metre. It has a length of about 30 kilometres (figures 15.23 and 15.24) in its 500 GeV configuration and could be built with techniques that are mastered[13] today.

Figure 15.25 shows a superconducting radio-frequency cavity for the ILC under test. The ILC allows to polarise the positrons which provides an advantage for some physics measurements (such as the weak mixing angle). The critical aspects are the forming and welding techniques of bulk niobium as the full ILC project (800 GeV) requires 16 000 cavities of pure niobium.

A possible ILC operating schedule over more than twenty years would be to start at the Z-mass pole for one up to three years collecting about $0.1\,\mathrm{ab}^{-1}$ ($100\,\mathrm{fb}^{-1}$), then move to the Higgs boson factory mode at 250 GeV where an integrated luminosity up to $2.0\,\mathrm{ab}^{-1}$ could be collected in twelve years of operation, corresponding to the production of almost half

[13]The final-focusing technique, required to bring the beam size at the collision point down to about 500×6 square nanometres in order to achieve the luminosity goals remains a challenge, a factor of five has still to be gained.

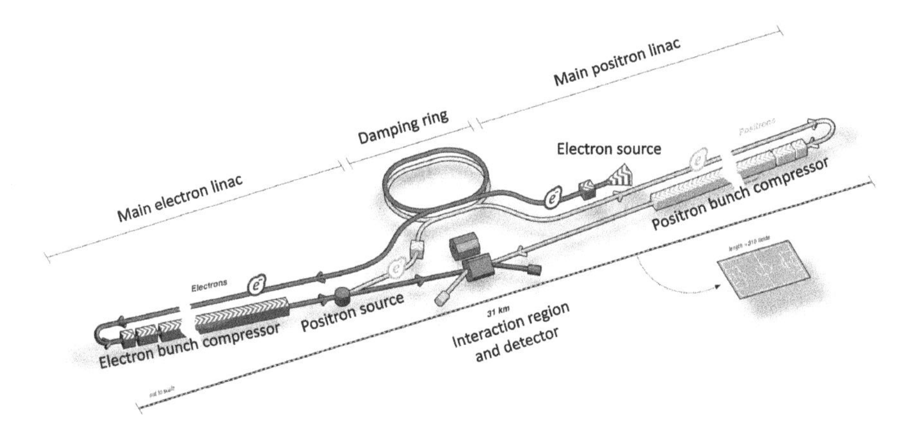

Fig. 15.24. Schematic view of the ILC, with its two linear accelerators (linacs) for electrons and positrons. A part of the accelerated electrons are used to produce positrons by passing the high-energy electrons through a long helix-shaped undulator to create an intense circularly polarised photon beam. The photon beam hits a thin conversion target to produce e^-e^+ pairs. The electrons and positrons are injected at 5 GeV into the centrally placed 3.2-kilometre long damping rings to reduce the injected beam emittance (angular and momentum spread) by six orders of magnitude in the short interval between machine pulses. *International Linear Collider, Technical Design Report, arXiv:1306.6327 (2013).*

Fig. 15.25. Superconducting radio-frequency cavity made of pure niobium for the ILC under test. It operates at 1.3 GHz frequency with an accelerating gradient of about 32 MV/m. The ILC will need 16 000 of these one metre long cavities in its $\sqrt{s} = 500$ GeV configuration. *Image: DESY.*

a million Higgs bosons, followed by an intermediate extension of the project to the top–antitop production threshold at 350 GeV for one year collecting 0.1 ab^{-1}, before reaching the final 500 GeV centre-of-mass energy where in nine years about 4.0 ab^{-1} could be collected.

Japan may be willing to cover about half of the cost of the ILC, provided it is built in Japan. The construction (material) cost amounts to about 8 billion US dollars for the 500 GeV phase (of which 1.5 billion US dollars for tunnel and buildings), while the 250 GeV phase is estimated to cost about 5 billion US dollars. The timeline of the project assumes four years of preparatory phase and nine years of construction. If the decision could be taken rapidly, the ILC could start operating in the 250 GeV phase in the early 2030s.

Compact Linear Collider

The other, quite different, e^+e^- linear collider concept considered is the Compact Linear Collider (CLIC). About 50-kilometre long, this is an even more ambitious project, aiming at reaching collision energies as high as 3 TeV. Besides the study of the Higgs boson and electroweak physics (although not as well as the other e^+e^- colliders on the Z pole), CLIC would allow a more precise measurement of di-Higgs production than the other e^+e^- collider projects, and has an increased search sensitivity for new heavy particles.

The attractiveness and relevance of this machine is thus closely related to the results of the new physics searches obtained at the LHC at a scale not exceeding about 2.5 TeV. Figure 15.26 shows the possible layout of CLIC in the Geneva area. Three stages are considered for this project with centre-of-mass collision energies at 380 GeV, 1.5 TeV, and 3 TeV, and lengths of, respectively, 11, 29 and 50 kilometres.

The CLIC accelerator scheme (figure 15.27) is based on a novel system of room-temperature (that is, normal-conducting) cavities with gradients of 100 MV/m,[14] fed with radio-frequency power through a second electron beam of only a few GeV but extremely high intensity, and parallel to the main beam. The needed radio-frequency power would be generated by the passage of the secondary beam through resonant cavities in the immediate vicinity of the accelerating cavities of the main beam. This system would

[14]Gradients at CLIC are three times that of the ILC.

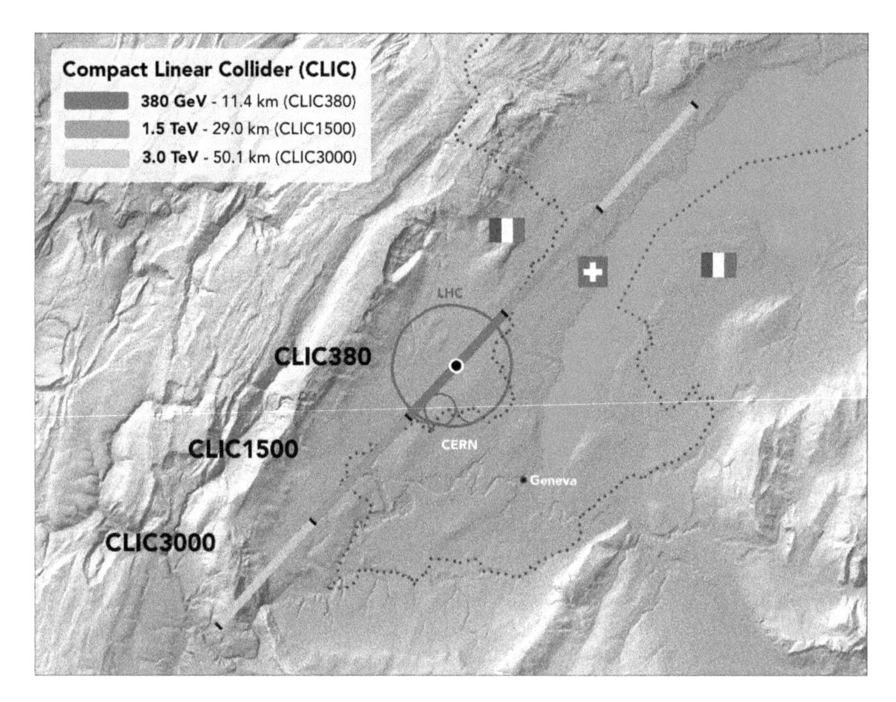

Fig. 15.26. Possible layout of CLIC in the CERN area close to Lake Geneva. Three stages are considered for this project with centre-of-mass collision energies at 380 GeV, 1.5 TeV and 3 TeV, and lengths of, respectively, 11, 29 and 50 kilometres. *Image: CLIC accelerator footprint, CERN CDS OPEN-PHO-ACCEL-2017-027 (2017).*

replace the klystrons traditionally used for feeding accelerators with radio-frequency power. The two-beam concept underlying the CLIC design was demonstrated in a dedicated test facility at CERN (CTF3) that operated until 2016 (see figure 15.28). The limiting factors to achieve the required accelerating gradient are potential radio-frequency breakdowns, the need for very high quality copper, micrometre-level precision turning and milling, and very rigorous brazing techniques. The final focusing of the beams at the 1–3 nanometre level in one transverse dimension is also very challenging. The operation of CLIC at 3 TeV collision energy would require 590 MW of electric power (170 MW for the first 380 GeV stage), which is an important consideration. It is similar to that of the FCC-hh project discussed below, and significantly exceeds the 100 MW foreseen for the HL-LHC, and the 160 MW of the 500 GeV ILC (at $2 \times 10^{34}\,\mathrm{cm}^{-2}\,\mathrm{s}^{-1}$ peak luminosity). The construction and installation of the first 380 GeV stage is expected to take about five years, two further years would be required for the commissioning,

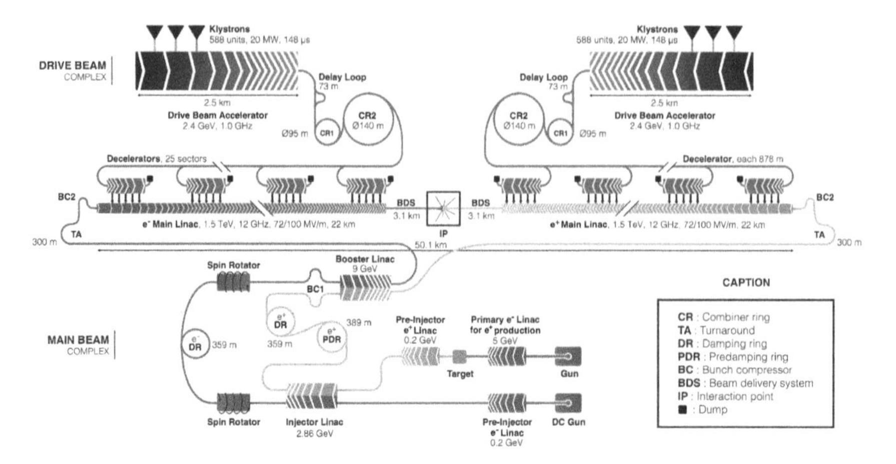

Fig. 15.27. Layout of the 3 TeV CLIC configuration. The accelerating radio-frequency power is produced by the passage of two high intensity 2.4 GeV, 2.5-kilometre long electron beams (drive beams) in 25 decelerating sectors (each 878-metre long) per accelerated beam (electron and positron linacs). The first CLIC stage of 380 GeV centre-of-mass energy requires only four decelerating sectors per beam and a single drive-beam complex. The location of the detector at the interaction point is indicated by IP. The design includes the polarisation of electrons. *Image: CLIC accelerator schematic diagrams, CERN CDS OPEN-PHO-ACCEL-2019-002 (2019).*

followed by about eight years of physics operation that would allow CLIC to produce 1.0 ab^{-1} of integrated luminosity. The next two extension stages would increase the produced integrated luminosity by factors of, respectively, 2.5 and 4 per year of operation.

Fig. 15.28. Test facility at CERN used to study the two-beam concept of CLIC. *Image: CTF3, CERN.*

The cost of the project is estimated to almost 6 billion Swiss francs for the first stage, with additional 5 and 7 billion Swiss francs for the extensions to 1.5 TeV and 3 TeV, respectively. For comparison, the cost of the LEP project (including both stages, LEP-1 and LEP-2) would today amount to about 2.6 billion Swiss francs (estimate corrected for inflation).

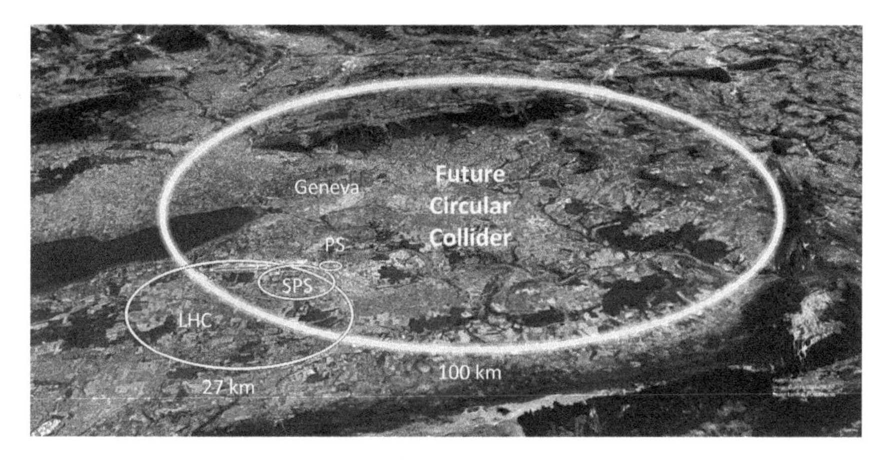

Fig. 15.29. Possible location for the FCC-ee and FCC-hh projects. With its 100-kilometre (dotted line) circumference, the tunnel would encompass the entire Geneva area. The existing PS, SPS and LHC rings are shown to the left in white. *Image: Future Circular Collider, CERN CDS OPEN-PHO-ACCEL-2019-001 (2019).*

Future circular electron–positron collider

Despite the difficulty due to synchrotron radiation circular electron–positron colliders have to face, several projects are being proposed too. The LEP-3 project, a comparatively cheap option, was discussed for some time during the early 2010s. It consisted of an e^+e^- collider ring inside the existing LEP-LHC tunnel with very high luminosity and 240 GeV collision energy. The scientific potential of the LEP-3 project, limited to the detailed study of Higgs boson physics and of very rare Z decay modes, as Z bosons would be produced by billions, did not reach that of the other e^+e^- projects. However, its advantage was that it could be operational much earlier, although it would require to dislodge the LHC, an option that became impossible with the approval of the HL-LHC project that will operate until the mid to late 2030s.

The most recent — and most ambitious — among the proposed CERN projects is called FCC (Future Circular Collider). It would be located in a 100-kilometre long circular tunnel facility surrounding Geneva (figures 15.29 and 15.30), which would host several accelerator-colliders over many decades of scientific exploitation. An e^+e^- collider, FCC-ee (initially called TLEP, for Tera-LEP), with collision energy up to 350 GeV

Fig. 15.30. Artistic view of the FCC tunnel with access shafts and the LHC in the Franco-Geneva region. *Image: Polar Media and Future Circular Collider, CERN CDS OPEN-PHO-ACCEL-2019-001 (2019).*

could be the first stage of the FCC.[15] Such a collider would deliver substantially higher luminosity and event yields than the ILC and CLIC (cf. figure 15.22). In units of $10^{34}\,\text{cm}^{-2}\,\text{s}^{-1}$, peak luminosities per interaction point of, respectively, 230, 28, 8.5, 1.6 are targeted at centre-of-mass energies of 92, 160, 240, 365 GeV. An integrated luminosity of $150\,\text{ab}^{-1}$ collected at the Z-boson pole would correspond to 3×10^{12} visible Z boson decays, which is about 10 000 times the full sample available at LEP, including its four experiments. In Higgs boson factory mode at 240 GeV, the FCC-ee could collect $5\,\text{ab}^{-1}$, corresponding to about one million ZH events.[16]

An important asset of circular colliders is the possibility to have several experiments simultaneously taking data, whilst the ILC or CLIC can accommodate only one running experiment at a time. With such a high luminosity, the main challenge for the FCC-ee is the very short lifetime of the beams, few minutes only, due to the extreme focusing required at the interaction points, which provokes a blow-up of the particle bunches. This would be remedied by constructing an additional booster synchrotron ring in the collider tunnel whose function is to regularly replenish the main ring when the luminosity has decreased too much (*top-up injection*). The

[15]This energy limit is set by the requirement of consuming at most 50 MW electrical power per beam.
[16]The energy loss by synchrotron radiation per turn amounts to: 0.036 GeV at the Z pole, 0.34 GeV at $\sqrt{s} = 160$ GeV (WW threshold), 1.72 GeV at 240 GeV (ZH factory), and 9.21 GeV at 365 GeV ($t\bar{t}$ threshold), according to the E^4 scaling law.

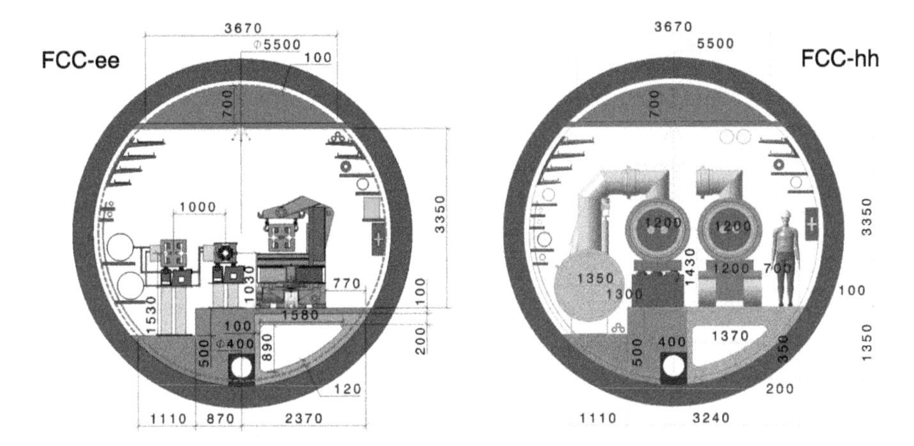

Fig. 15.31. Conceptual design (as of mid-2020) of the 5.5-metre inner diameter FCC tunnel in the e^+e^- collider configuration (left-hand drawing) and proton–proton collider configuration (right-hand side). In the FCC-ee version, the left-most magnet shows the separate e^+ and e^- beam tubes and the magnet at the centre corresponds to the replenishing accelerator (booster). The right-most magnet just shows its temporary position at installation time. Also shown are several service elements. For the FCC-hh configuration (right-hand drawing), the left-most structure is the cryogenic system and the central one to the left corresponds to the two-in-one high-field (16 tesla) bending magnet surrounding the beam pipes. The right structure illustrates magnet transport. *Image: FCC, CERN.*

FCC-ee will use a 6 GeV linear accelerator to pre-accelerate electrons and produce positrons, whose emittance will be reduced in a small damping ring. The existing SPS could be used to further accelerate the electrons and positrons to 20 GeV before injection into the top-up booster. The feasibility of the FCC-ee (and CEPC, see below) concepts is underpinned by the experience from the LEP (CERN), DAPHNE (Frascati, Italy), PEP II (SLAC, Stanford, USA), KEKB and SuperKEKB (Japan) colliders it is based on. Figure 15.31 illustrates the conceptual design of the FCC tunnel in the e^+e^- collider configuration (left-hand drawing). Its inner diameter of 5.5 metres is 1.7-metre larger than the LHC tunnel.

The cost of the project exceeds that of the first stages of ILC and CLIC and can only be justified in the framework of the complete FCC program. The collider and injector complex are estimated to about 3 billion Swiss francs, the civil engineering including the tunnel boring to 5.4 billion Swiss francs, and another 2 billion Swiss francs are needed for technical infrastructure, giving a total of 10.5 billion Swiss francs. Following the 2020 update of the European Strategy for Particle Physics, the next years foresee the completion of a Technical Design Report for the FCC projects, with a decision to be taken by the next strategy update in the midst 2020s. This could allow the FCC-ee to start operation before 2040.

China also proposes a 100-kilometre circular e^+e^- machine called CEPC (Circular Electron Positron Collider), with a design very similar to the FCC-ee. This project designed to host two detectors was developed soon after the discovery of the Higgs boson in 2012 as a dedicated Higgs boson factory operating at 240 GeV centre-of-mass energy. As for the FCC-ee, the tunnel for such a machine could also host a super proton–proton collider (SPPC) to reach a new high-energy frontier. Permanent top-up injection will be used to maintain constant luminosity. A 10 GeV linear accelerator built at ground level, accelerates both electrons and positrons. The emittance of the positrons is reduced by passing them through a low-energy damping ring. The overall power consumption of the CEPC is estimated to 270 MW, dominated to 40% by the accelerating radio-frequency cavities (60 MW per beam) and to 20% by the magnet power supplies. According to the 2018 CEPC Conceptual Design Report, the project could start construction in 2022 with full completion and begin of physics exploitation by 2030. This timeline is significantly shorter than that of the FCC-ee and similar to the ILC. The total CEPC project cost is estimated to about 5 billion US dollars.

Summary

Figure 15.32 compares the expected peak luminosities of the linear and circular e^+e^- colliders discussed above, as a function of the centre-of-mass energy. For ILC and CLIC, several design upgrades are also shown. The luminosity of the circular machines decreases with energy as it is limited by the cost of electricity needed to compensate for synchrotron radiation losses, whereas linear colliders become more cost effective at energies above about 350 GeV. The luminosity of the FCC-ee at the Z pole would be so large that it could produce almost half a trillion Z bosons per year and per experiment, to be compared to the 80 million Z bosons produced during the entire LEP lifetime for the sum of the four experiments.

Together with runs at the WW threshold, at the maximum of the ZH production rate (see figure 15.22), and at the $t\bar{t}$ threshold, all colliders would allow to scrutinise the Standard Model with a precision orders of magnitude better than the present one (cf. figure 15.33). Projections for the expected precision on the Higgs boson coupling modifiers (see section 15.1.4) obtained

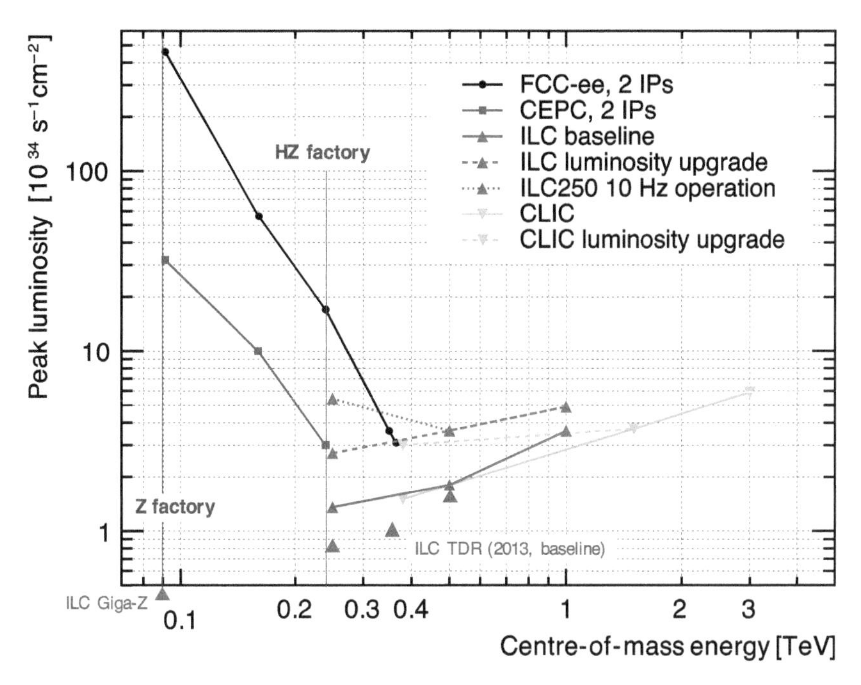

Fig. 15.32. Expected peak luminosity versus centre-of-mass energy for the various e^+e^- collider projects. The figure indicates several luminosity upgrades in the designs of ILC and CLIC. The lower CEPC luminosity compared to FCC-ee is in part due to different electric power bounds assumed. *Image: J. List (2019); figure modified by authors.*

by the various collider options and upgrade stages are given in figure 15.34. While there are no large differences between the e^+e^- colliders, the integrated FCC measurements including the FCC-hh project discussed below obtain the best precision. All e^+e^- colliders, but most prominently the FCC-ee, are electroweak boson and flavour factories, allowing more than five times improved indirect constraints on new physics in the electroweak precision sector compared to the HL-LHC, three to four orders of magnitude improved sensitivity on charged-lepton flavour violating decays such as $Z \to e\mu$. It is expected that the FCC-ee produces 200 billion τ-leptons, and decays such as $B_s^0 \to \tau^+\tau^-$ and $B_d^0 \to K^*\tau^+\tau^-$ can be cleanly measured using a sample of two trillion $b\bar{b}$ events. They allow powerful searches for weakly coupled particles or new weak forces.

The avalanche of data must be paired with theoretical calculations that significantly improve the precision of the currently available Standard Model predictions. While there may be no conceptual obstacles to this, the

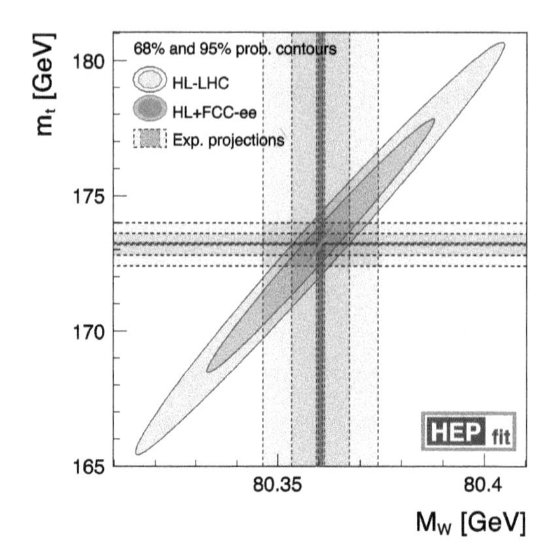

Fig. 15.33. Projection of constraints in the top-quark mass versus W-boson mass plane after the HL-LHC (grey) and the FCC-ee (blue). The vertical and horizontal bands indicate the precision of the direct measurements within their 68% and 95% confidence levels. The ellipses show the indirect constraint from the fit of electroweak precision observables. These projections do not include theoretical uncertainties, which are expected to limit the precision of the fit after the FCC-ee. *Image: J. de Blas (2019).*

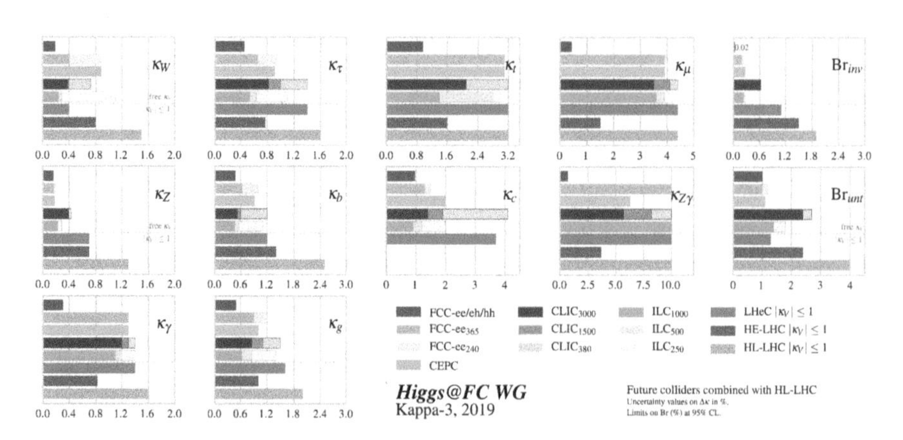

Fig. 15.34. Expected precision (in percent) on the Higgs boson couplings (modifiers κ) for the various future collider options and upgrade stages (cf. also figure 15.11). *Image: J. de Blas et al., JHEP 01, 139 (2020).*

program has been estimated to an equivalent of 500 person-years over the next 20 years.

In addition to new physics searches, the higher collision energy of the ILC and, in particular, CLIC provide interesting opportunities for measurements of top-quark to Z boson couplings and the Higgs boson self-coupling. The parameter κ_λ of trilinear Higgs boson coupling is expected to be measured to 50%, 30%, and 10% at the HL-LHC, ILC, and CLIC, respectively. These measurements as well as the searches for new particles would however be much outperformed by the FCC-hh collider project which we will discussed below.

15.2.2. *Future hadron collider*

The 100-kilometre circumference FCC tunnel is a facility that could be built to directly host a very large hadron collider capable of accelerating and colliding protons as well as heavy ions, or as an e^+e^- Higgs boson factory followed by a hadron collider (figures 15.31 and 15.37). The future hadron collider project at CERN is denoted by the acronym FCC-hh, where "hh" stands for hadron–hadron. If equipped with technologically fully mastered 8.3-tesla NbTi LHC-type dipole magnets, it would provide 55 TeV proton–proton collisions (see figure 15.35). With 16-tesla Nb_3Sn magnets (figure 15.36), however, a collision energy of 100 TeV could be reached. Developments are also being pursued on high-temperature superconductors, which could lead to 20-tesla magnets, allowing to reach more than 130 TeV proton–proton collision energy in the FCC tunnel.

A 100 TeV FCC-hh collider positioned adjacent to CERN's LHC and SPS would use these accelerators as injectors, either via the LHC at 3.3 TeV proton beam energy, or via the SPS using new superconducting magnets at a beam energy of 1.3 TeV. Larger even than the technical challenge of achieving 16 tesla dipole fields, is to produce them at affordable cost and industrial quantity. A cost-effectiveness per TeV of a factor three to five has to be gained. To improve the performance of the Nb_3Sn conductor and reduce its cost, a global research and development program involving laboratories and industry from the USA, Russia, Europe, Japan and Korea has been initiated in 2018.

Other challenges are the beam dump to safely discard the order of magnitude higher stored energy in the beam than at the HL-LHC (the FCC-hh will carry 10 400 proton bunches compared to less than 2800

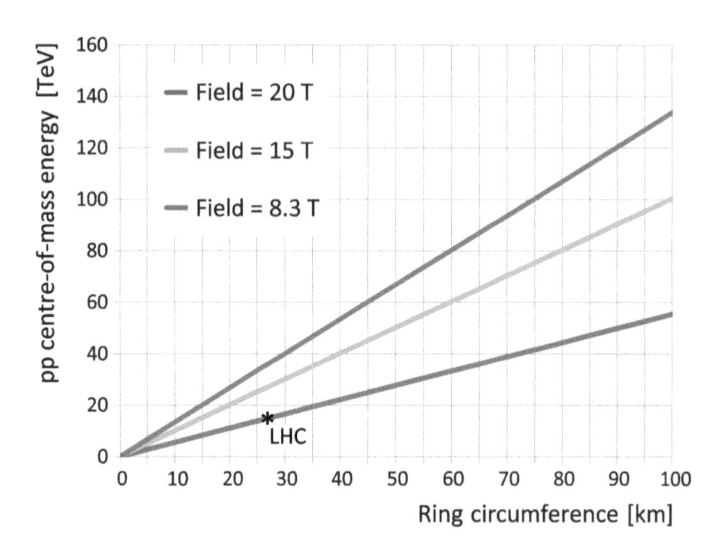

Fig. 15.35. Collision energy in the centre-of-mass ($E_{\rm CM}$) of a proton–proton collider as a function of circumference (C) and for three different values of the dipole magnetic field (B) needed to keep the protons on a circular trajectory. The curves are obtained using the relation $E_{\rm CM} \simeq 0.3 \cdot B \cdot C \cdot \delta/\pi$, assuming about two-third of the circumference can be instrumented with dipole magnets ($\delta \approx 2/3$). The smallest value, 8.3 tesla, corresponds to the present LHC dipoles at 7 TeV beam energy (14 TeV centre-of-mass). The next one, 15 tesla, is for niobium-three-tin (Nb_3Sn) magnets presently under study, but requiring at least one or two decades of research and development to achieve cost-effective production at industrial scales. The highest value, 20 tesla, would require the use of high-temperature superconductors and will need even more extended research.

Fig. 15.36. Model coil for a 16-tesla FCC-hh dipole magnet with Nb_3Sn cable. *Image: N. Caraban/CERN.*

at the LHC), as well as the synchrotron radiation. The FCC-hh will emit about 5 MW at 100 TeV centre-of-mass energy, which requires the development of a new beam screen (cf. figure 15.6) that protects the cryogenic system against a beam-induced temperature of 50 kelvin, compared to 5–20 kelvin at the LHC. About 100 MW cooling power will be needed to dissipate the 5 MW synchrotron radiation heat.

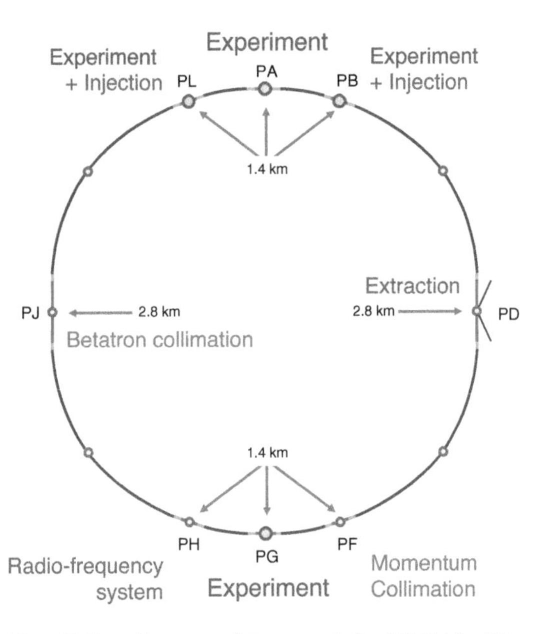

Fig. 15.37. Conceptual layout of the FCC-hh. The two general-purpose experiments are located at the interactions points PA and PG (in red). Additional experiments may be located at PL and PB together with the injection of the counter-rotating beams. Momentum and betatron collimation systems are located at PF and PJ (blue). The beam extraction (dump) is located at PD, and the accelerating radio-frequency system are at PH. *Image: FCC Collaboration, EPJ Special Topics, 228, 755 (2019).*

Similar to the LHC, the FCC-hh is designed to host two general-purpose detectors, and possible additional dedicated-purpose experiments (figure 15.37). The general-purpose detectors would run at the highest achievable peak luminosity of almost $30 \times 10^{34}\,\text{cm}^{-2}\,\text{s}^{-1}$, a factor of four above the HL-LHC maximum, generating the daunting number of one thousand simultaneous proton–proton collisions (pile-up) at each bunch crossing. A total integrated luminosity of $30\,\text{ab}^{-1}$ is targeted per experiment over about 25 years of operation.

Figures 15.38 and 15.39 (for the cross-section) show the conceptual general-purpose design of a detector for the FCC-hh, largely inspired by the best features of ATLAS and CMS, but extended in many respects. The large centre-of-mass energy has the consequence that the known particles with mass of 170 GeV or less will be produced at lower proton momentum fraction (x), leading to more pronounced forward production than is the case at the LHC. The FCC-hh detector must therefore provide precision tracking, calorimetry, and muon detection and measurement beyond

Fig. 15.38. Conceptual design as it stands in 2020 of a FCC-hh general-purpose detector. The overall length and diameter are 50 and 20 metres, respectively. The central tracker is shown in grey, surrounded in the central part of the detector by the electromagnetic (dark blue) and hadronic (light blue and green) calorimeters and a 4-tesla solenoid. To maximise the rapidity coverage, tracking devices (grey) surrounded with additional 4-tesla solenoids are added in the forward parts, completed again by electromagnetic and hadronic calorimeters. The outside layer consists of the muon spectrometer (orange). Note that there is no iron return yoke outside the central solenoid to shield the magnetic field of the central solenoid, as this would be too heavy and costly for such a large magnet. Equipment in the cavern needs thus to be protected from the large stray fields. The cost of such a detector would be on the scale of one billion Swiss francs, twice that of ATLAS or CMS. *Image: FCC Collaboration, EPJ Special Topics, 228, 755 (2019).*

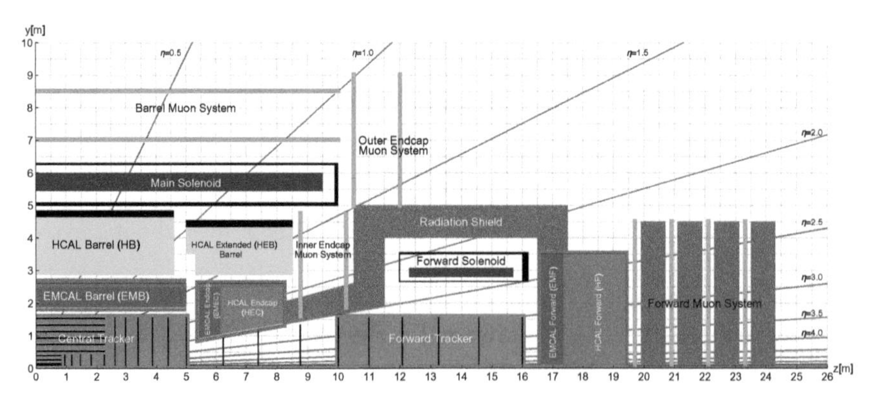

Fig. 15.39. Longitudinal cross-section of the FCC-hh detector design shown in figure 15.38. The installation and opening of the detector requires a cavern length of 66 metres (compared to the 53-metre long ATLAS cavern). *Image: FCC Collaboration, EPJ Special Topics, 228, 755 (2019).*

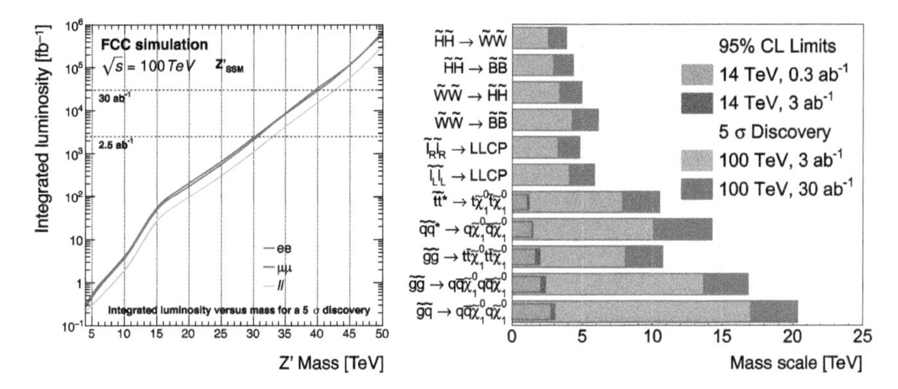

Fig. 15.40. Left: luminosity required at the FCC-hh for the discovery of a Standard Model like Z' decaying to an electron or muon pair versus its mass. Right: expected LHC and HL-LHC limits on various searches for supersymmetric processes and particles (in red) and corresponding 5σ discovery potential at the FCC-hh (blue) for different integrated luminosity value. *Image: FCC Collaboration, EPJ C 79, 474 (2019).*

pseudo-rapidity values of four. In addition, the detector needs to be capable of measuring multi-TeV jets, leptons and photons from heavy resonances with masses up to 50 TeV, while at the same time measuring Standard Model processes with moderate and low momenta. Finally, very radiation resistant detector systems and micro-electronics need to be developed for the FCC-hh.

The large leaps of a factor of seven in collision energy and a factor of ten in integrated luminosity offers huge physics potential compared to the HL-LHC, similar to that of the LHC after the Tevatron. New phenomena may be discovered up to masses of 40 TeV for excited quarks, 43 TeV for Standard Model like Z' bosons (see left panel of figure 15.40), 16 TeV for supersymmetric gluinos, 10 TeV for top squarks (right panel of figure 15.40), WIMP dark matter can be discovered between 1 and 3 TeV, to name a few prominent examples. That reach would allow to discover or definitely exclude TeV-scale supersymmetry or dark matter, and strongly constrain the paradigm of naturalness. Studies of Standard Model processes such as high-mass longitudinal vector boson scattering can be measured to 3% precision, Drell–Yan processes can be measured with 10% precision up to 15 TeV di-lepton mass. The FCC-hh also offers a very attractive heavy-ion physics program with nucleon–nucleon centre-of-mass energies of 39 TeV for lead–lead collisions, 63 TeV for proton–lead, and a luminosity exceeding the HL-LHC by a factor between ten and thirty. With a quark–gluon plasma temperature of 1 GeV abundant in-medium (thermal) production of charm

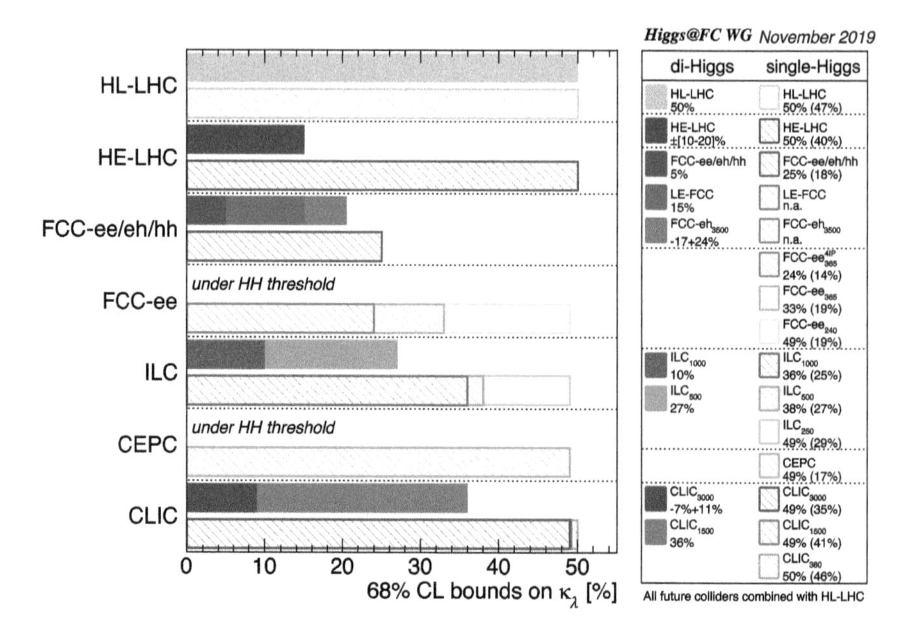

Fig. 15.41. Expected sensitivity for the various electron–positron and hadron collider projects for the Higgs boson trilinear self-coupling parameter κ_λ. All results are combined with the expected 50% sensitivity of the HL-LHC. The single-Higgs boson constraints on κ_λ are obtained indirectly through precise measurements of production and decay spectra sensitive to quantum corrections involving the Higgs boson self-coupling. They depend on the assumed theoretical and experimental precision. The FCC-ee and CEPC colliders do not have sufficient energy for di-Higgs production in the *ZHH* channel, and therefore only contribute through indirect constraints. *Image: J. de Blas et al., JHEP 01, 139 (2020).*

quarks would occur with the perspective of new insight into the temperature evolution of the plasma.

The FCC-hh would be the ultimate Higgs boson factory. It would produce 200 times more inclusive Higgs bosons than the HL-LHC, 530 times more ttH events, and almost 400 times more di-Higgs boson events. The latter events would allow the FCC-hh experiments to determine the κ_λ parameter governing Higgs boson self-coupling to 5% precision (cf. figure 15.41). Invisible Higgs decays to dark matter candidates could be seen down to a branching fraction of less than one permil. Table 15.1 gives in the upper row the number of single and double Higgs boson events expected after completion of the FCC-hh program, that is, 25 years of operation and 30 ab^{-1} integrated luminosity. One billion Higgs bosons would be produced per year, and the di-Higgs production cross-section would reach

Table 15.1. Number of Higgs boson events expected at the FCC-hh and comparison with the HL-LHC for the main single Higgs boson production modes and double Higgs boson production. The first row gives the number of events times one million assuming 25 years of running at a collision energy of 100 TeV giving an integrated luminosity of 30 ab^{-1}. The second row gives the ratios of the event yields at the FCC-hh and the HL-LHC, the latter for an integrated luminosity of 3 ab^{-1} taken at 14 TeV collision energy.

	$gg \to H$	VBF	WH	ZH	$t\bar{t}H$	HH
$N_{\text{FCC-hh}}$ [million events]	24 000	2100	460	330	960	36
$N_{\text{FCC-hh}}/N_{\text{HL-LHC}}$	180	170	100	110	530	390

1.5 pb. The lower row compares the FCC-hh to the HL-LHC after 10 years of operation and 3 ab^{-1} integrated luminosity collected.

The cost of the project depends on whether or not it is preceded by the FCC-ee. In case of preexistence of the tunnel infrastructure, the collider and injector are estimated at 13.6 billion Swiss francs with additional 3.4 billion Swiss francs for technical infrastructure and civil engineering. In case the FCC-hh is built standalone the technical infrastructure and civil engineering cost increases to 10.4 billion Swiss francs. The integrated FCC-ee and FCC-hh projects has a total cost of 28.6 billion Swiss francs. The overall project duration including implementation and operation of the integrated FCC spans seven decades: a preparatory phase of 8 years, followed by 10 years of construction, 15 years of operation and exploitation of the FCC-ee, then 10 years of FCC-ee removal, infrastructure work and installation of the hadron machine and detectors, followed by 25 years of operation. Assuming a project decision may be taken in 2026, the start of the hadron collider operation for physics could thus be around 2045 for the standalone FCC-hh project, and the later 2060s for the integrated FCC.

In this context we should also mention the proposal put forward by Chinese physicists for adapting the 100-kilometre circumference circular tunnel of the CEPC for proton–proton collisions with 50 TeV or more collision energy, depending on the magnet technology available. The main advantage of this proposal is the advantageous cost for the tunnel excavation and a likely earlier project realisation. A challenge will be the availability of the human and technical expertise and infrastructure which has grown over decades at particle physics centres such as CERN.

Another possibility beyond 2035 for CERN could consist in replacing all present 8.3-tesla NbTi dipole magnets in the LHC tunnel with new 16-tesla Nb_3Sn ones. This project, called the High-Energy LHC (HE-LHC), would provide proton–proton collisions at a centre-of-mass energy of 27 TeV. Over 20 years of operation the HE-LHC experiments could each collect about

$15\,\mathrm{ab}^{-1}$, which would provide twenty times the statistics of di-Higgs boson production of the HL-LHC and should allow a measurement of κ_λ with 15% precision. New phenomena of excited quarks may be discovered up to a mass of 12 TeV, Standard Model like Z' up to 13 TeV, supersymmetric gluinos up to 6 TeV, and top squarks up to 3.2 TeV mass. The cost of the project would amount to 5 billion Swiss francs for the new collider, and 2.2 billion Swiss francs for injector, technical infrastructure and civil engineering, giving a total of 7.2 billion Swiss francs. The timeline foresees 8 years of preparation and about the same time for the construction. Physics operation could be expected in the mid-2040s.

A comparison of the expected sensitivities of the e^+e^- and hadron collider projects discussed above for the Higgs boson self-coupling parameter κ_λ is provided in figure 15.41. Shown are results from direct constraints via the measurement of di-Higgs boson production as well as indirect constraints through precise Higgs boson coupling measurements.

While a scenario with an e^+e^- Higgs boson factory followed by a very high energy hadron collider in a 100-kilometre tunnel is favoured at CERN for the medium and long-term future (provided that political and societal support can be secured within the CERN member states, and funds are made available in due time, in particular for the large tunnel infrastructure), it is not excluded that, before a decision is made by the CERN Council, Japan or China decide to build an e^+e^- collider (ILC or CEPC) for exploitation in the years 2030–2040. In such a case, another project for the medium-term future (with a construction in parallel with the HL-LHC running) could be considered at CERN: an electron–hadron collider, the LHeC, presented in the next section.

Inset 15.2: Superconducting magnets and society

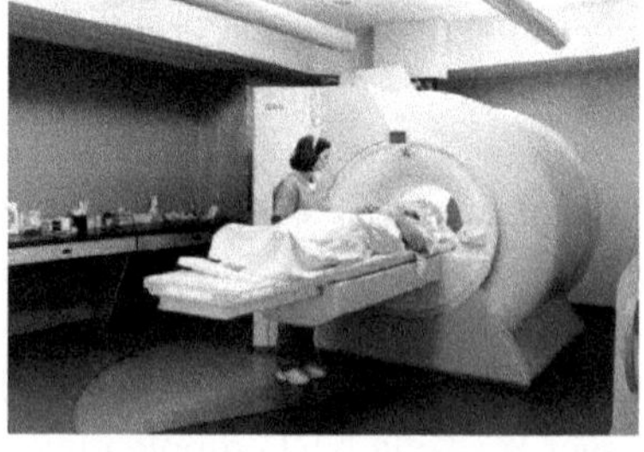

High-field superconducting magnets are an invaluable tool for scientific research, in particular in high-energy physics to achieve powerful accelerator, collider and detector magnets, as discussed in this book. Their developments for fundamental research have led to applications in nuclear magnetic resonance spectrometers (NMR) with spin-offs in medicine in form of high resolution magnetic resonance imaging (MRI) scanners (the image above shows one of the 36 000 MRI scanner in use in the world as of 2020) and for ion-beam cancer therapy.

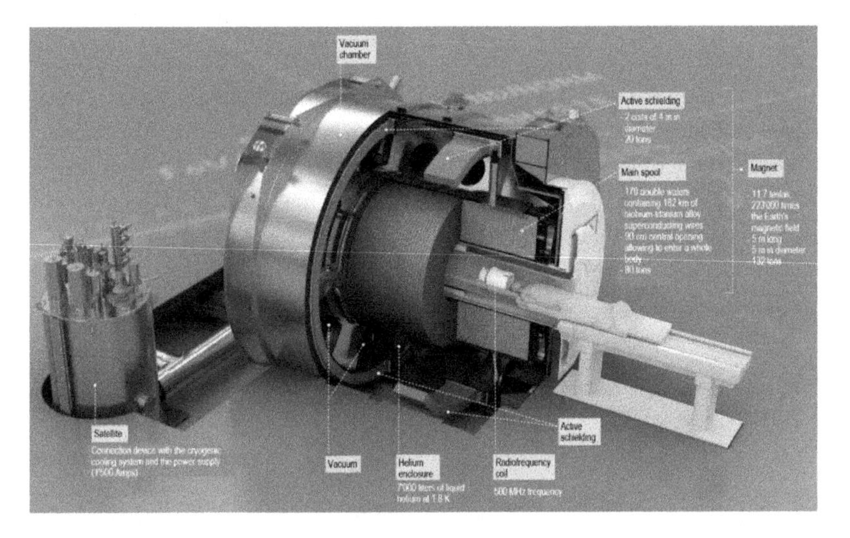

The image above (credit: CEA, France) shows a sketch of the Iseult magnet at Neurospin (CEA) which has reached a nominal magnetic field strength of 11.7 tesla on July 18th, 2019. This is a world record for a whole-body magnetic resonance imaging scanner.

Another notable application is fusion energy reactors (superconducting tokamaks, a toroidal apparatus to confine the hot fusion plasma in the shape of a torus).

The image on the left (credit: ITER Organization) shows an illustration of the International Thermonuclear Experimental Reactor (ITER) tokamak in construction in south of France by a large international collaboration to study nuclear fusion. Fields of up to 12 tesla are produced in a large toroidal magnet.

The production of a first plasma is expected for 2025 and full operation with fusion energy production by 2035.

Other applications are electric power distribution lines and power systems, high speed transport (magnetic levitation train, Maglev, a system that suspends, guides and propels trains using a large number of magnets for lifting and propulsion without use of wheels, axles, transmissions, and overhead wires). The figure below (credit: Wikipedia) shows a train of the L0 Series on SCMaglev test track in the Yamanashi Prefecture, Japan. It is powered by the magnetic fields induced either side of the vehicle by the passage of the vehicle's superconducting magnets. It reached a speed of 603 km/h in 2013.

Until to date, virtually all superconducting magnets have been made from Nb-based low-temperature superconductors: NbTi with a superconducting transition temperature T_c of 9.2 kelvin, and Nb_3Sn with a T_c of 18.3 kelvin. The key parameter in developing high field magnets is the critical current density J_c in units of ampere per square millimetre that a strand of superconductor in its long-length form can sustain under high magnetic field conditions.

The figure below (credit: T. Shen and L. Garcia Fajardo, Instruments 2020, 4(2), 17, modified by the authors) shows the critical density that can be achieved at 4.2 kelvin (unless otherwise specified) as a function of the applied magnetic field for strands of different types of superconducting material. The grey band shows the range of current densities used for the LHC dipole magnets (at 1.8 kelvin). Most large-scale applications make use of NbTi superconductors for which the manufacturing process of long strands is well mastered.

The more challenging Nb_3Sn magnets (cf. the manufacturing discussion in footnote 4) are key to the ITER tokamak, to the upgrade of the LHC with its 11-tesla quadrupole magnets (see section 15.1), and even more for the FCC-hh or SPPC projects, where dipole magnets with field strength of 16 tesla are foreseen (section 15.2), which seems to be the upper limit that can be reached with a Nb_3Sn collider magnet.

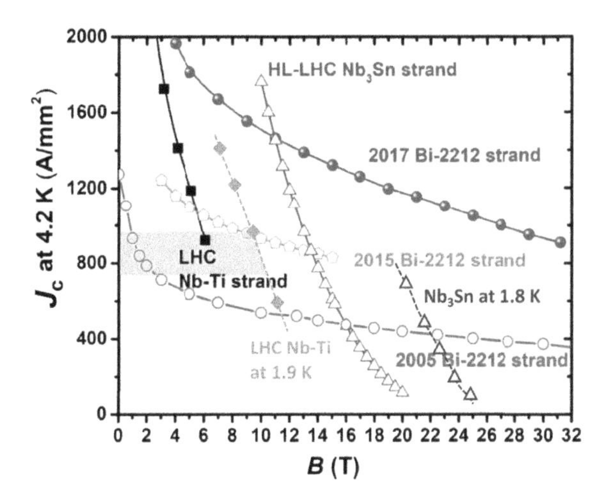

The technical capability of manufacturing very high-field magnets could be greatly expanded by high-T_c superconducting materials with an upper critical magnetic field exceeding 50 tesla at 4 kelvin, much above that of NbTi (about 14 tesla at 1.8 kelvin) and Nb_3Sn (26–27 tesla at 1.8 kelvin). The figure above shows the progress made between 2005 and 2017 in the development of a class of high-temperature superconductors, $Bi_2Sr_2CaCu_2O_x$, called Bi-2212 and discovered in 1988, which does not contain a rare-earth element and is a promising candidate for manufacturing cables in various geometries. Once mass fabrication is understood and costs are controlled, Bi-2212 could be first employed in small very high-field NMR devices (beyond the 23.5 tesla limit presently achieved with Nb_3Sn), and later (in the 2040s?) in large collider dipole magnets with magnetic fields of 20 tesla or even beyond.

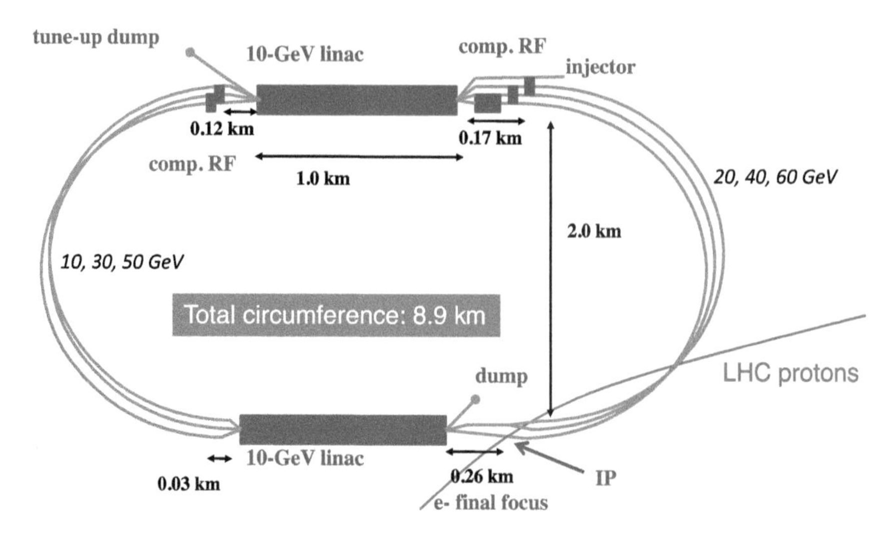

Fig. 15.42. Layout of the 60 GeV energy recovery linac for the LHeC project. A 500 MeV electron bunch coming from the injector is accelerated in each of the two 10 GeV superconducting linacs during three revolutions through respective arcs with two-kilometre diameter, after which an energy of 60 GeV is reached. The beam is then focused and brought into collision with the proton beam from the LHC in the interaction point (IP), where an asymmetric particle detector is positioned (because of the larger proton energy the collision debris are boosted along the proton beam direction). The remaining electrons in the bunch are then sent back to the first linac at a decelerating radio-frequency phase, and after three revolutions with deceleration, converting their energy back into the radio-frequency system, the bunch is dumped. The total circumference of the racetrack-shaped accelerator-collider is 8.9 kilometres, which is one-third of the LHC circumference. *Image: LHeC Collaboration, J. Phys. G, 39, 075001 (2012).*

15.2.3. *Electron–hadron collider*

By adding to the existing 7 TeV LHC proton accelerator ring a new electron accelerator with an energy of about 60 GeV, one can obtain a Large Hadron–Electron Collider (LHeC) with 1.3 TeV centre-of-mass energy.[17] Figure 15.42 gives a sketch of the baseline LHeC design as a so-called energy recovery linac (linear accelerator). It consists of two superconducting linacs each about one-kilometre long, arranged in a "racetrack" configuration tangential to the LHC. Each linac provides about 10 GeV acceleration. The electrons thus make three turns in the racetrack reaching an energy of

[17]The centre-of-mass energy of a head-on collision of two particles with different energy and negligible mass compared to their kinetic energy is given by $E_{\mathrm{CM}} \simeq 2\sqrt{E_1 E_2}$, so for a 60 GeV electron colliding with a 7 TeV proton one finds $E_{\mathrm{CM}} = 1.3$ TeV.

60 GeV before colliding with the 7 TeV protons or 2 TeV lead ions. The present design is optimised for a low-power consumption by decelerating the electron beam after the collision and recovering nearly all its energy into the radio-frequency cavities of the linacs, a principle called *energy recovery*. The linacs would be operated in so-called continuous wave operation mode, as opposed to the usual pulsed mode, which is possible for the moderate accelerating gradient needed to achieve 60 GeV electron acceleration. A design study for an energy recovery linac test platform is being pursued at CERN and at the IJClab in Orsay (France).

The LHeC with a design instantaneous luminosity of $10^{33}\,\text{cm}^{-2}\,\text{s}^{-1}$ would allow physicists to investigate electron–proton and electron–ion collisions at unprecedented energy and rate, much beyond those achieved at the electron–proton collider HERA at DESY in Hamburg, which ended operation in 2007.[18] Measurements at the LHeC would lead to a very precise determination of the parton (quark and gluon) density functions of the proton and nucleons in kinematic regimes, in term of the variables x and Q^2, much closer to those of the LHC collisions.[19] A reanalysis of the LHC and HL-HLC data with thus improved parton density functions would allow to reduce the systematic uncertainties in many analyses, in particular those aiming at precision tests of the Standard Model. In addition, with a 60 GeV electron beam, Higgs bosons can be produced rather abundantly through the charged (dominant) and neutral current vector boson fusion mechanism, which involves the electron and a quark in the initial state, leading to a much cleaner final states compared to the dominant gluon fusion production process at the HL-LHC. The dominant Higgs boson decay to a b-quark pair would have a signal to background ratio of larger than one. For an integrated luminosity of $1\,\text{ab}^{-1}$ collected after about 10 years of operation, the Higgs boson to b-quark coupling (κ_b) could be measured with a precision at the 1–2% level. Also κ_c would be accessible and could be measured with 4% precision, which would be far beyond the precision obtained at the HL-LHC.

[18]HERA achieved a peak luminosity of $7.5 \times 10^{31}\,\text{cm}^{-2}\,\text{s}^{-1}$ and delivered an integrated luminosity of $0.8\,\text{fb}^{-1}$ to each of its two detectors, H1 and ZEUS, between 1992 and 2007.

[19]The variable x is the fraction of the proton momentum carried by the parton, as defined in inset 10.4 on page 306, and Q^2 is the momentum transfer between the probe particle and the target parton inside the proton.

The cost of the LHeC in its energy recovery linac version is estimated to about 1.8 billion Swiss francs, of which almost half is due to the superconducting radio-frequency electron acceleration system. It is thus significantly cheaper than the FCC project, but it also does not offer an equivalent physics potential.

An electron–hadron collider version is also considered for the integrated FCC project, following closely the design of the LHeC. Here, 50 TeV protons would collide with 60 GeV electrons providing 3.5 TeV centre-of-mass energy. The peak luminosity of this FCC-eh collider would reach 15 times that of the LHeC.

We should mention also another electron–hadron collider project, the 3.9-kilometre circumference Electron–Ion Collider (EIC), which has been officially launched at Brookhaven National Lab (BNL), USA, in September 2020. The EIC will be constructed and operated in partnership with the Thomas Jefferson National Accelerator Facility (JLAB). Part of the existing RHIC setup will be re-used after RHIC completes operation in 2024 (cf. figure 15.43 and chapter 14). The EIC is expected to start taking data by 2030. It will operate in a very different energy regime compared to the LHeC, colliding with high luminosity ($10^{33-34}\,\text{cm}^{-2}\,\text{s}^{-1}$) polarised electrons in the 5–10 GeV energy range (upgradable to 20 GeV) with polarised protons and light nuclei with energies up to 60 GeV, spanning centre-of-mass collision energies in the 20–100 GeV range (upgradable to 140 GeV).

Fig. 15.43. Areal picture of the current RHIC heavy-ion collider at BNL, Long Island, USA, which will be replaced by the 1.6–2.6 billion US dollar Electron–Ion Collider (EIC). *Image: Brookhaven National Lab, USA.*

The EIC will study the nuclear structure through deep inelastic scattering in a new kinematic regime of very low proton and ion momentum fraction x and with high sensitivity owing to the polarised beams. Detailed nucleon tomography should help understand how sea quarks and gluons, as well as their spins, are distributed in space and momentum inside the nucleon, and how the nucleon properties emerge from the partons and their interactions. Further, how do the confined hadronic states emerge from these quarks and gluons, and what happens to the gluon density in nuclei? Does it saturate at high energy (low x), giving rise to gluonic matter, the so-called colour glass condensate[20] with universal properties in all nuclei?

15.2.4. *Muon colliders*

Hadron colliders offer the possibility to reach the highest centre-of-mass energies in the collisions between elementary constituents (quarks and gluons), but at the cost of a suppression of high parton momentum fractions x due to the parton density functions, very complex collision debris, and not knowing the initial constituents' momenta. Electron–positron colliders, on the other hand, lead to very clean events with a well-controlled initial state, but they are limited to particle production via the electroweak force and their centre-of-mass energy is strongly limited by the energy loss due to synchrotron radiation. As the radiated power is proportional to $1/m^4$ and $1/R^2$, where m is the mass of the accelerated particle and R the radius of the circular collider, the choice is either to increase R (as for instance in the 100-kilometre FCC-ee project) or to use a heavier particle. The muon which is 200 times heavier than the electron, leading to a radiation power reduced by a factor 10^9, would be the obvious (and only) candidate.

As in the electron–positron case, a muon–antimuon collider would have a well-defined initial state and produce clean collision events. A problem is, however, the short muon lifetime of 2.2 microseconds at rest, although it increases with the Lorentz factor $\gamma = E_\mu/m_\mu$ in the laboratory (relativistic time dilation). In practice, for a 14 TeV centre-of-mass energy collider, the average available time laps between muon production at low energy (a few GeV), acceleration, and collision at high energy inside a detector is of the order of one-tenth of a second or less if too much time is spent at low energy.

---

[20]The colour glass condensate (CGC) is a hypothetical dense gluonic state in hadrons, which universally appears in the high-energy limit of scattering processes. The logarithmically increasing cross-section of proton–lead collisions with energy, as if the proton's size expands beyond the geometric cross-section $\pi r_c^2 \simeq 30\,\text{mb}$ (where r_c is the proton radius, of about one femtometre), is seen as an experimental manifestation of the CGC. At high energy (low x), due to Heisenberg's uncertainty relation, quantum fluctuations may live long. Gluon self-interaction has thus the time to create multiple daughter gluons which further interact among each others, creating more fluctuations until the transverse area of the proton (πr_c^2) is fully filled, i.e. saturated. In this picture, it is assumed that the CGC is the initial state of the quark–gluon plasma generated in heavy-ion collisions.

It is thus important to bring the produced muons and antimuons as fast as possible to the required collision energy and into collision.[21]

Because the elementary components of the proton follow parton density distributions (cf. inset 10.4 on page 306) and individually carry in average less than 10% of the proton's momentum, a 14 TeV muon collider could be considered effectively equivalent to a 100 TeV proton collider in terms of new physics reach (figure 15.44).[22] Such a collider could be built with 16-tesla dipole bending magnets, similar to the ones of the FCC-hh or HE-LHC using Nb_3Sn superconducting cables, inside the existing LHC tunnel. If enough luminosity can be delivered, it could also be a useful machine to study the Higgs boson. For example, with an integrated luminosity of 20 ab^{-1} accumulated at 14 TeV collision energy possibly after twenty years of operation at peak luminosity of 10^{35} cm^{-2} s^{-1}, 20 million Higgs bosons would be produced dominantly via vector boson fusion, and 90 thousand di-Higgs boson events. These abundances outperform any of the electron–positron colliders.

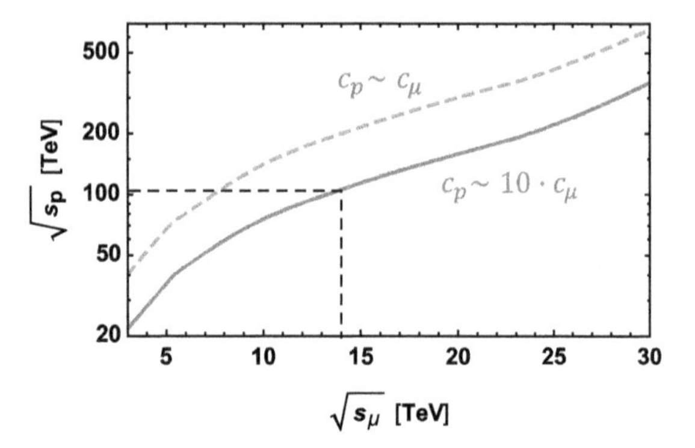

Fig. 15.44. Centre-of-mass energy at which the cross-section of a scattering process at a muon–antimuon collider equals that of a proton–proton collider. The dashed line indicates equivalent couplings between the produced particle(s) to muon and proton, and in blue for a ten times enhanced proton coupling as could be the case, for instance, in strong interaction production. *Image: Muon Collider Working Group, Input to the European Particle Physics Strategy Update (2018).*

[21]Because of the exponential decay law there will always be a fraction of muons that decay on the way between production and collision. The shorter the time spent, the smaller that fraction. For example, if the time spent before collision corresponds to the dilated muon lifetime, about 63% of the muons decay on the way, while only 39% decay if the time laps can be halved, or 86% decay if the time laps corresponds to twice the muon lifetime.

[22]In both, muon and proton colliders, the production cross-section of a particle with mass m scales with the inverse of the centre-of-mass energy squared, $1/s$. In the case of the proton collider, the cross-section is further reduced by an $m/\sqrt{s}$ dependent factor taking

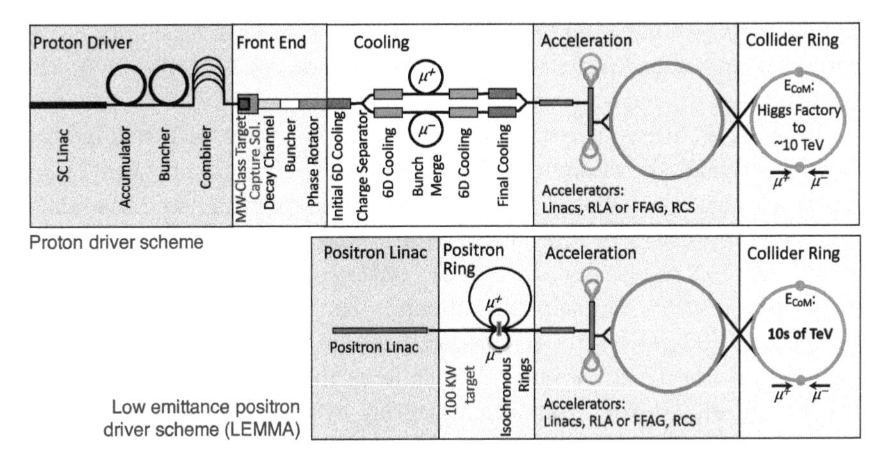

Fig. 15.45. Schematic designs of two versions of a high-energy muon collider facility. In the top one, bunched protons, accelerated in a high flux linac and hitting a target, produce muons from the decay of pions and kaons (front-end section, light blue). The following section (cooling, in yellow), the most critical part, aims at reducing by a factor 10^5 the size of the beam phase space (a six-dimensional space: three space, two angular and one momentum dimension) before reaching the acceleration and collision phases (red and green, respectively). In the bottom design, $\mu^+\mu^-$ pairs are produced by a high-flux positron beam. No cooling is needed before the acceleration and collision sections. In both schemes, energy efficient acceleration requires to set up recirculation, similar to that of the LHeC discussed above, which allows multiple passages through high-gradient radio-frequency acceleration regions. The high-energy muons are injected in a continuing top-up scheme into the collider ring and brought to collision at interaction points and detectors that must be robust enough to cope with large beam-induced backgrounds. *Image: M. Boscolo et al., Rev. Accel. Sci. Tech. 10, 189 (2019).*

The first concept of a muon collider (top row in figure 15.45) is based on the very large production rate of muons and antimuons in fixed-target proton–nucleus collisions via pion and kaon decays. The initial 7 GeV proton beam is required to deliver a power between one and four megawatts. The main difficulty of this scheme, besides manufacturing a target that withstands such a beam power, is that the momenta and angles of the produced muons are broadly distributed, which renders difficult the formation of intense and coherent muon beams at well defined energy and direction. A procedure to reduce the spread with an acceptable efficiency, the so-called

into account the steeply falling parton luminosities. The actual suppression depends on the coupling of the particle to the constituents of the proton.

dE/dx cooling,[23] has been experimentally tested, but the feasibility of the entire beam stacking and cooling scheme has not yet been proven at the required level of efficiency.

In 2017, an alternative way of producing $\mu^+\mu^-$ pairs with well-defined momentum and direction was proposed via the use of positron–target collisions, thus evading the tricky cooling steps. It is denoted low emittance muon accelerator scheme (LEMMA) and illustrated in the bottom row of figure 15.45 (see also inset 15.3). Other schemes based on the photo-production of $\mu^+\mu^-$ pairs are also investigated. They would make use of a high-energy (60 GeV) and very intense (few mA) pulsed electron beam as described in the LHeC project. A very large flux of high-energy photons could be produced from Compton scattering by firing a high-power pulsed laser beam onto the electron beam. Muon pairs of energies in the range 15–20 GeV could be abundantly photo-produced on a heavy nuclei target (gold or tungsten), but still requiring a moderate cooling stage.

In both schemes depicted in figure 15.45, multi-pass rings for muon acceleration and luminosity production can be used. It results in muon colliders that are more compact and more energy efficient than electron–positron machines (linear or circular) at high energy. For example, for the proton driver scheme at 3 TeV centre-of-mass energy, four times more luminosity for the same electrical power can be obtained with the muon collider than with CLIC. This relative advantage of the muon collider further increases with rising beam energy.

In parallel to the energy-frontier muon collider option, a low-energy circular $\mu^+\mu^-$ Higgs boson factory at a centre-of-mass collision energy which can be varied around 125 GeV so as to measure directly the Higgs boson width (expected to be 4.1 MeV in the Standard Model), has been proposed by the former CERN Director-General Carlo Rubbia. This collider could be extremely compact, 300-metre circumference only, so that it would fit, for example, in the existing CERN ISR tunnel, and could be realised at

[23]The principle is based on a succession of steps, each composed of a (small) energy loss dE in a thin liquid-hydrogen slab of length dx, which slightly reduces the muon momentum (and hence the transverse component p_T), followed by an acceleration with radio-frequency cavities which increases the longitudinal component p_L. After this reduction of the transverse emittance, a bending dipolar magnetic field followed by a wedge-shaped slab ensures that muons with higher momentum traverse more material in order to reduce the global momentum spread. All these operations have to take place inside a strong magnetic field oriented along the mean muon trajectory to keep the spiralling particles inside a confined volume.

a much lower cost (probably less than one billion Swiss francs) compared to the ILC, CLIC, or FCC-ee projects. However, it would deliver a peak luminosity of less than $10^{32}\,\mathrm{cm}^{-2}\,\mathrm{s}^{-1}$, and thus only about 10 000 Higgs bosons per year via the direct annihilation production $\mu^+\mu^- \to H$. In terms of Higgs boson couplings, it would not significantly improve over the HL-LHC, not even for the total Higgs boson width, and could not compete with any of the future e^+e^- colliders.

Although rather advanced designs of a muon collider facility exist on paper and in simulation, much more experimental research and development programs are needed to show that a machine with a large enough luminosity can be built within a reasonable timescale as a more cost effective

Inset 15.3: A new scheme for a muon collider

A way to produce $\mu^+\mu^-$ pairs at a well-defined momentum, thus evading the tricky and time-consuming (in terms of muon lifetime) cooling steps, has been proposed by the LEMMA project in 2017. Muons are produced directly in the reaction $e^+e^- \to \mu^+\mu^-$ by shooting a high-flux positron beam on a target containing electrons at rest. With a 45 GeV positron beam, the centre-of-mass energy is just above the $2m_\mu$ threshold and the μ^+ and μ^- are produced at 22.5 GeV in the forward direction in the laboratory frame. The figure below shows a simplified layout of such a muon collider facility. The acceleration stage in linacs could make use of one of the various plasma wakefield acceleration schemes which are presently studied in several laboratories worldwide and which may offer accelerating gradients well above 1 GeV/m, thus maximising the available lifetime in the collider ring.

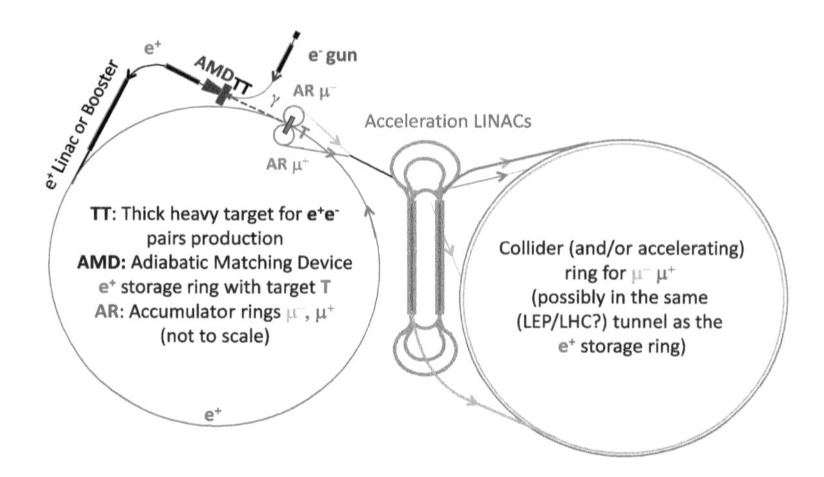

Among the various technical challenges, because of the small cross-section of muon pair production, a positron production rate of at least 10^{16} positrons per second would be required to obtain an acceptable collider luminosity, which is about two orders of magnitude above what is foreseen for ILC or CLIC.

alternative to the integrated FCC.[24] Another potentially serious difficulty we did not mention yet may come from the neutrino radiation localised in the plane of the collider ring and along the extensions of the accelerating linacs. Although the neutrinos interact very weakly with matter, the high-energy neutrino flux coming from the muon decays would be so huge, that in some places at the Earth' surface, the radiation hazard could become close to the limit presently tolerated for human settlement. Several schemes to

[24]Another proposal put forward in 2012 is that of a photon–photon annihilation Higgs boson factory, also operating at 125 GeV centre-of-mass energy and possibly producing up to 10 000 Higgs bosons per year. The photon beams are produced by firing a very high-power pulsed laser beam onto a high-intensity polarised electron beam just in front of the collision point. The electron beam energy would not exceed 80 GeV, thus the ring could be limited to 9-kilometre circumference, reducing the cost compared to ILC or FCC-ee. The technical feasibility of this scheme remains to be proven as of 2020.

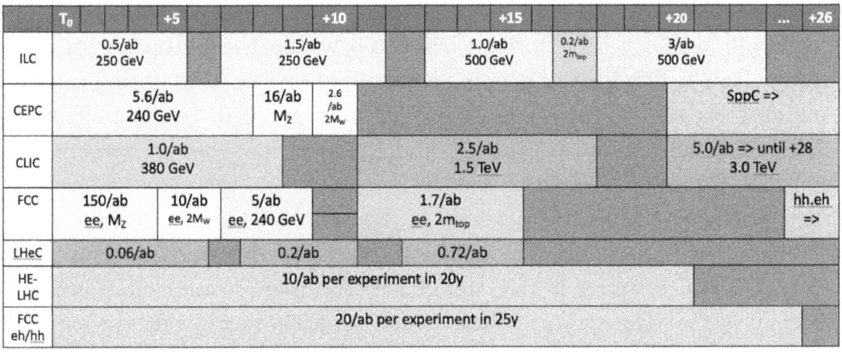

	T_0 … +5 … +10 … +15 … +20 … … +26
ILC	0.5/ab 250 GeV; 1.5/ab 250 GeV; 1.0/ab 500 GeV; 0.2/ab $2m_{top}$; 3/ab 500 GeV
CEPC	5.6/ab 240 GeV; 16/ab M_Z; 2.6 /ab $2M_W$; SppC =>
CLIC	1.0/ab 380 GeV; 2.5/ab 1.5 TeV; 5.0/ab => until +28 3.0 TeV
FCC	150/ab ee, M_Z; 10/ab ee, $2M_W$; 5/ab ee, 240 GeV; 1.7/ab ee, $2m_{top}$; hh,eh =>
LHeC	0.06/ab; 0.2/ab; 0.72/ab
HE-LHC	10/ab per experiment in 20y
FCC eh/hh	20/ab per experiment in 25y

Fig. 15.46. Timeline and stages of various collider projects starting at time T_0 as submitted to the 2020 Update of the European Strategy of Particle Physics. *Image: J. de Blas et al., JHEP 01, 139 (2020).*

mitigate the problem, such as introducing wiggles along the muon trajectories in order to increase the spatial dispersion of the neutrino flux, are under study.

15.3. Courage

There is thus no lack of original and technologically as well as scientifically ambitious collider projects at CERN and elsewhere in the world for the coming decades. Several of these projects, such as the integrated FCC, cover very long timescales, beyond the professional life of a generation of physicists (and policy makers), as summarised in figure 15.46 for the projects discussed in the preceding sections (not including the muon collider as more fundamental technical challenges are yet to be addressed). These timescales are the scientific and societal price to pay for progress in this science that takes us into the deepest secrets of space, time, matter and force. It would be a fundamental mistake to assume progress can be made faster and cheaper, on the top of a table, by just thinking or trying harder. The knowledge to conceive and build these machines is hard to acquire, but easy to lose if not preserved over generations of physicists and engineers through ambitious and compelling projects that attract the best scientific minds. It is thus of utmost importance to educate societies and policy makers about the value of pursuing this type of research in view of advancing technology and our understanding of Nature, despite the financial restrictions, particularly in Europe.

The various options we have presented, ILC, CLIC, FCC, CEPC, and muon collider, have been actively discussed within the European Strategy Group (ESG). The ESG converged on its final deliberation during a week-long drafting session held in Germany in January 2020. In June 2020, the CERN Council announced that it had updated its strategy, intended to guide the future of particle physics in Europe within the global landscape. The document highlights the need to pursue an *electron–positron Higgs factory* as the highest-priority facility to start shortly after the completion of the HL-LHC program in the later 2030s. Another recommendation of the strategy is that Europe, in collaboration with the worldwide community, should undertake a feasibility study for a next-generation hadron collider at the highest achievable energy (FCC-hh if implemented at CERN), in preparation for the longer-term scientific goal of exploring the high-energy frontier, with an electron–positron collider as a possible first stage (FCC-ee or CEPC). If the ILC moves forward in Japan during the coming years, it is expressed that CERN and the European particle physics community would wish to collaborate.

The technological and economic benefits and spin-offs that this type of research necessarily generates, in addition to the scientific, educational and cultural feats, are crucial assets for our societies to value the large investments required to realise future collider projects. At the beginning of the 20th century, research about relativity or quantum mechanics was in the domain of pure science, with no obvious perspectives for applications in the near or even longer term. Nonetheless, it is thanks to the discoveries in these two domains that present-day technical revolutions, based on electronics, lasers and information technologies could occur, leading us into the present "silicon age". Even if it is difficult to perceive any practical utility or value in the discovery of the Higgs boson, it represents a very major step in our understanding of the physical laws that govern the Universe. Spontaneous symmetry breaking takes place in many areas of Nature, now it is proven to also occur in its most fundamental settings. It could lead to developments we are incapable of predicting today.

Chapter 16

Conclusions

Which lesson can we learn from CERN's history, from its inception in 1954 up to the start of the LHC and the discovery of the Higgs boson, the culminating moment of this endeavour thus far? Europe, destroyed and ruined at the exit of Second World War, has been able to set up a laboratory unique in the world. The CERN founders put in place an administrative, financial, and, above all, scientific framework that allowed Europe to gradually, but surely, catch up and regain its place in the orchestra of world physics. CERN grew to become the largest particle physics research laboratory in the world and a model for other research institutions such as ESO (European Southern Laboratory), ESA (European Space Agency), or the EMBL (European Molecular Biology Laboratory). Today with the multitude of laboratories and universities around the world taking part in the LHC programme, CERN has effectively become a world laboratory.

During the 1970s CERN contributed to particle physics and the making of the Standard Model thanks to beautiful and successful experiments such as Gargamelle and CDHS for neutrino physics or BCDMS and EMC for muons. In the 1980s, with the discovery of the W and Z gauge bosons and of jets in hadronic collisions thanks to the Sp$\bar{\text{p}}$S collider and the UA1 and UA2 experiments, CERN came to the forefront of the field. The LEP era then followed, from 1990 till 2000, during which the Standard Model was consolidated and tested to unprecedented detail, making it the most complete and thoroughly tested physics theory to date. In the meantime, the LHC and its experiments were conceived, designed and went through various stages of prototyping and testing, followed by their construction and commissioning. During the decade between LEP dismantling in 2000 and the beginning of LHC high-energy operation in 2010, progress in particle physics continued thanks to several large colliders in the world: HERA in Hamburg shedding light on the proton structure, the Tevatron at Fermilab with the discovery and study of the top quark — just to mention the most important contribution, and B factories in California and Japan with the discovery and precise measurement of matter–antimatter symmetry violation in the B-meson system. As for non-accelerator related experiments, the most outstanding results during this period and the 1990s were the discovery of neutrino oscillation and thus of non-zero neutrino masses, the (unexpected) observation of an accelerating universal expansion at work since about six billion years, and the confirmation by newer and more powerful satellite experiments (WMAP and Planck) of the enigmatic phenomena of dark matter and dark energy in the Universe. In 2010, the LHC started operation and the increase in luminosity was so swift that within few months CERN became again a world focal point of particle physics research. The discovery of the Higgs boson in 2012, followed by the award of the Nobel Prize in Physics in October 2013 to François Englert and Peter Higgs who proposed in 1964 the mass-generation mechanism, was monumental scientific success. Since the discovery of the W and Z bosons in 1982–1983, confirming gauge theory and electroweak unification, the discovery of the Higgs boson is undoubtedly the most significant result, possibly even since the early 1960s and the evidence of quarks forming the composite of all hadrons, including the protons and neutrons.

The LHC is also a tremendous technological success. This achievement has been possible thanks to a stable long-term financing policy guaranteed by the European governments. Through the LHC, particle physics has been a driver of innovation. We saw in this book how the advances in superconducting magnet technology required for the LHC have been transferred to industry. Particle physics instrumentation and acceleration technologies provide new advanced applications in medicine, and even led to discoveries in so remote fields like Egyptology and volcanology. It cannot be overemphasised how much the invention of the World Wide Web at CERN in 1989, to allow communication among physicists around the world, has transformed our societies. Today, the usage of the world wide LHC computing grid, introduced to analyse the colossal amount of data produced by the LHC experiments, shows the way for other large-scale data-producing activities in physics, medicine, biology and other fields.

CERN is also a unique meeting and discussion place for physicists from all over the world. With the fully globalised LHC experiments, CERN allows to foster contact and build bridges with world regions where tensions are often breaking the news. In former times, for example during the cold war, CERN and particle physics played already a similar role allowing scientists from East and West to talk to each other, at a time when this was neither easy nor common. Later in the 1990s, during the Balkanic wars, a number of scientific conferences and schools were held in the region to maintain relations between local scientists at the initiative of physicists supported by CERN and UNESCO. With the return of peace, almost all newly formed states in that area wished to contribute to CERN projects or become associated to the Laboratory. CERN was seen as a waiting room for association to, or membership in the European Union. In other world regions we are witnessing the rapid rise of particle physics research in China, with already notable results in neutrino physics, charm-quark physics, and the active involvement of the Chinese particle physics community in future large colliders projects we discussed in this book.

The discovery of the Higgs boson at the LHC provided part of the answer to one of the fundamental questions in physics: the origin of electroweak symmetry breaking and mass of elementary particles. In the decades ahead of us, the LHC and future collider projects at the energy frontier are expected to shed light on other still unresolved questions related to the Higgs boson and beyond. For example, which form did Nature choose for the Higgs potential — a question intimately related to the fate of our Universe. Is the discovered Higgs boson unique, or does it have siblings or distant cousins? Is it an elementary scalar or a composite particle emerging from new symmetries at high energy? Is the hierarchy problem indeed evidence for a non-natural universe or does new physics provide a solution to it? Do electroweak and strong interactions unify at high energy (small distance)? Is the Higgs boson a portal to a yet unknown dark sector, a mirror world whose experimental hint is the all-pervading dark matter in the Universe? Will future studies of *CP* violation in the quark and lepton sectors help us understand why antimatter disappeared from the Universe, while a small fraction of matter particles survived

Dark matter, dark energy, gravity... ? Can they fit in the Big Puzzle? *Image: Lison Bernet.*

the cataclysm of primordial annihilation? Will the study of heavy-ion collisions teach us more about the temporal evolution and intrinsic properties of the quark–gluon plasma state of matter?

The Standard Model has resisted until now all attempts by particle physicists to break it, but we know that the model is incomplete. With the High-Luminosity LHC coming into operation in 2027, conceived to provide a data sample exceeding in size by an order of magnitude the LHC's initial design, we expect percent level precision for Higgs boson couplings to bosons and fermions, the observation of rare Higgs boson decays, and evidence for Higgs boson pair production allowing us to glean information about the Higgs potential. Alternative studies of electroweak symmetry breaking through multi-gauge-boson interactions will become possible. Searches for rare new physics processes, including dark-matter production will significantly improve their sensitivity. In a joint and complementary effort between accelerator and non-accelerator-based experiments, dark-matter particle candidates might surface during the coming one or two decades. Experiments at the LHC and Super-KEKB in Japan should clarify any of the anomalies currently seen in charged weak interactions and lepton universality tests.

Smaller scale, but yet highly sophisticated experiments in Switzerland, at Fermilab, USA, and in Japan will measure the magnetic moment and lepton quantum number of the muon to unprecedented precision, thereby resolving or confirming a longstanding tension with the Standard Model. *CP*-violating electric dipole moments of leptons, nucleons and molecules, indicators of new physics, are probed with ever increasing precision by several elaborate experiments worldwide. Feebly interacting axion fields, possibly emerging from a new symmetry responsible for the suppression of *CP* violation in strong interactions and a candidate for dark matter, are a hot topic in contemporary particle physics. Ingenious axion detection techniques are being developed and investments are made in the Asia, Europe and the US to realise them in new experiments.

Significant progress is also expected in neutrino physics during the next ten to fifteen years with the KATRIN experiment probing the electron-neutrino mass at the sub-eV level, improved neutrinoless double-beta decay searches to explore the nature of neutrinos, and finally with the huge long-baseline DUNE and Hyper-Kamiokande experiments under construction in the US and Japan. These experiments will be able to conclude whether there is *CP* violation in the lepton sector and how neutrino masses are ordered. They will also provide much improved sensitivity to proton decay, a messenger of grand unification, and watch out for neutrino bursts from supernovae explosions. Continuing and improved short-baseline experiments will study whether there are non-interacting sterile neutrinos. Altogether this is an amazing physics programme on the most mysterious particles of the particle zoo.

In 2020, the new underground cryogenic gravitational wave telescope KAGRA in Japan has joined LIGO and Virgo for concerted studies of gravitational waves, a consortium that will later be expanded by LIGO-India. Elements of the next generation triangular underground gravitational waves observatory, the Einstein Telescope, are being tested. The observatory should allow to study very early cosmology, as early as the dark age, and provide sufficient sensitivity to allow measurements of neutron star inspirals. Space-born interferometers such as LISA, an ESA-led space antenna, are also being developed with the goal of extending the detection range to lower frequency, allowing in the 2030s to significantly extend our knowledge of black-hole physics, possibly even finding hints for quantum gravitational effects. More speculative signals from cosmic strings or primordial gravitational waves generated during cosmological inflation may be detected with the upcoming generation of cosmic microwave background experiments. Among these are the large ground-based CMB-S4 experiment, operating at the South Pole and the Chilean Atacama plateau, and space-born missions such as LiteBIRD (JAXA, NASA) and CORE (ESA) for launch in the late 2020s and 2030s. Progress on the construction and exploitation of very large ground-based optical telescopes will provide new insights into the secrets of dark matter and dark energy. The Vera C. Rubin Observatory, located on the Cerro Pachón ridge in north-central Chile, with its very large aperture telescope will be able to make a complete survey of the southern sky within a few days. With a first light foreseen in 2022, it will perform the Legacy Survey of Space and Time (LSST) and provide unprecedented sensitivity to transient phenomena like supernovae and γ-ray bursts. It will moreover compile astronomical catalogues thousands of times larger than any existing ones, and include a wide and detailed survey of gravitational lensing effects on galaxies induced by dark matter. Breakthrough science on the formation of the first objects such as primordial stars, galaxies and black holes is expected from the Extremely Large Telescope (ELT), a European project featuring a 39-metre diameter primary mirror and currently under construction on top of Cerro Armazones in the Atacama Desert of northern Chile. There is thus no lack of ambitious projects in fundamental astrophysics and cosmology and thus groundbreaking scientific news are to be expected.

We can think of particle colliders as machines that allow us to travel backwards in time. The collisions among heavy ions at the LHC allow us to investigate the period occurring up to about one microsecond after the Big Bang, when matter transitioned from the quark–gluon plasma into the hadronic phase of matter. The LHC's proton–proton collisions, on the other hand, allow us to explore the earlier electroweak transition phase occurring between 10^{-15} and 10^{-11} seconds after the Big Bang (figure 16.1). Future generations of physicists will push forward the power of colliders and explore yet earlier phases in the evolution of the Universe. New particles emerging from yet unknown symmetries that governed the early history of our Universe may unfold from these collisions, giving us clues about the fundamental questions we cannot answer today, and help us extrapolate ever closer towards the Big Bang and ever deeper into the structure of space and time. Outcomes between evolution and revolution of our knowledge of fundamental physics are to be expected. Will matter and force be federated in an overarching super-symmetry? Will we be able to resolve extra space dimensions? Can we break fermions or bosons into yet more fundamental particles? Physics is an experimental science and only the continued investment into the empirical study of Nature can guide us among a profusion of ideas. Even if according to some antique believes *ad initio erat verbum* (St. John, I,1) — at the beginning there was the Word — the beginning, if there was, were symmetries and elementary particles, who knows, maybe in form of tiny strings.

In troublesome times where powerful, democratically elected institutions propagate the existence of "alternative facts", where social isolation allows groups to grow their own uncritical "truth", and where the enlightenment of the scientific revolution has become volatile, we need to remind ourselves of the scientific principles. Science admits ignorance, doubt and error. The architects of the Brout–Englert–Higgs mechanism were awarded the Nobel Prize not because they had a beautiful idea. Scientific history is full of beautiful (and revolutionary) ideas, which did not lead to a Nobel Prize or are forgotten. It is the very fact that their idea could be empirically and universally proven, through the discovery of the Higgs boson at the LHC, which led to the award. The creators of supersymmmetry have not (yet) been awarded, and may well never be, if no proof of existence of supersymmetry is found. Science is a story of trial and error, of mistakes and detours (and, alas, also of belief and prejudice, of power and

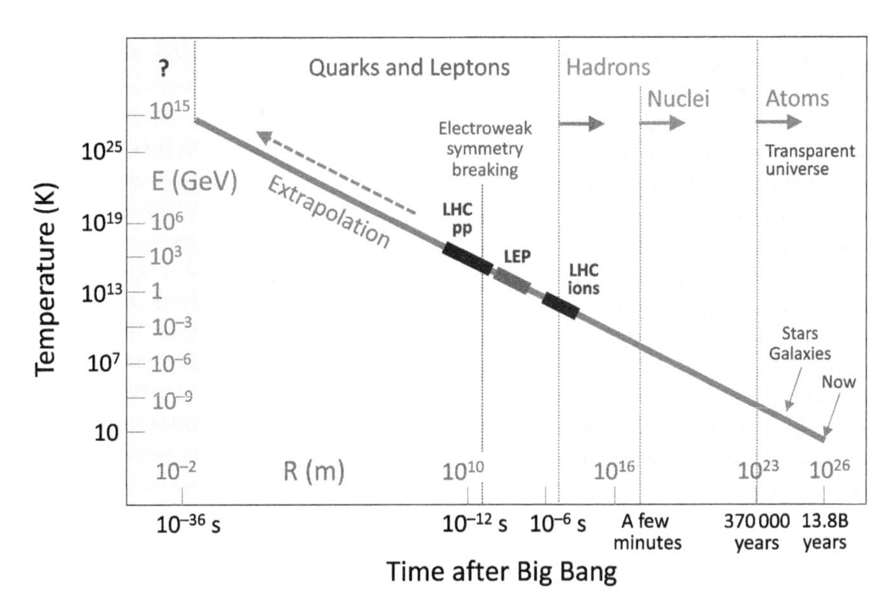

Fig. 16.1. Phases of the evolution of the Universe expressed as temperature (kinetic energy) versus time since the Big Bang (both axes are in logarithmic scale). Also shown is the corresponding radius R of the observable Universe. Observational astronomy (telescopes, CMB experiments) is limited to the period after an age of about 370 000 years when the Universe became transparent to photons. In a near future, new probes such as gravitational waves, may access earlier ages. Nuclear and particle physics experiments, at LEP or LHC, reproduce the energy per particle E (from Boltzmann's relation $E = 3k_BT/2$) that prevailed at much earlier times. The extrapolation to even earlier times, and higher temperatures and energies will require new instruments and possibly a change of theoretical paradigm.

influence). Similar to the course of natural selection, science erects marvels of thought and technology without a preconceived goal. Its mutations are ideas founded on fertile grounds of curiosity, scientific culture and education, its selection mechanisms are predictive power and the authority of experiment. Steven Weinberg wrote "*The Standard Model would have seemed unsatisfying to many natural philosophers [...] It is impersonal; there is no hint in it of human concerns [...] there is no element of purpose in the Standard Model*". Science is fundamentally built upon a critical mind in an utterly complex universe. There is no simple solution, and we do not even know whether the human brain has the capacity to penetrate the complexity and depth of the physical world, despite our skills to invent machines that help us with this task. We can only feel our way and move forward along the universal principles of science but with humankind's own imagination and ambition.

In view of the scientific and societal challenges we face, we shall close with CERN physicist Paris Sphicas, who concluded the 2020 International Conference of High-Energy Physics reminding us why Homo sapiens so successfully made its way through prehistory. Sphicas cited from a Scientific American article[1] as follows: *"Compared with other human species it turns out we were the friendliest. What allowed us to thrive was a kind of cognitive superpower: a particular type of affability called cooperative communication. We are experts at working together with other people, even strangers. We can communicate with someone we have never met about a shared goal and work together to accomplish it. [...] It allows us to plug our minds into the minds of others and inherit the knowledge of generations."* We may not have always been friendly, but we survived because we cooperated.

[1] "Humans Evolved to Be Friendly" by Brian Hare and Vanessa Wood, Scientific American, August 2020.

Units of Length, Time, Mass–Energy, and Some Typical Physical Scales

10^{-4} cm	1 micrometre (μm)	Wavelength of infrared and visible light
10^{-5} cm	0.1 micrometre	Size of cells and viruses
10^{-8} cm	1 angström (Å)	Atomic radia
10^{-13} cm	1 fermi (fm)	Radius of a proton, range of the strong force
10^{-16} cm		Range of the weak interaction (weak force)
10^{-18} cm		Upper limit on the size of a quark or lepton
10^{-33} cm		Planck's length
10^{9} cm		Earth radius
10^{13} cm		Radius of Earth's orbit
10^{18} cm		One light year
10^{23} cm		Radius of our galaxy — the Milky Way
10^{28} cm		Radius of the observable Universe
10^{-24} cm^2	1 barn (b)	Nuclear cross-section
10^{-36} cm^2	1 picobarn (pb)	Cross section for production of a Higgs boson
10^{-44} cm^2		Interaction cross-section for a solar neutrino

10^{17} s	~13.8 gigayears	Time since the Big Bang
	~4.6 gigayears	Time since the formation of the solar system
10^{13} s	a few million years	Rise of humankind
10^{2} s	100 seconds	Time after the Big Bang for primordial nucleosynthesis
10^{-6} s	1 microsecond (μs)	Lifetime of a muon, a charged pion
		Time for the transition from the quark–gluon plasma phase to the hadronic phase after the Big Bang
10^{-9} s	1 nanosecond (ns)	Lifetime of hyperons
		Crossing time of proton bunches in the LHC (25 nanoseconds)
10^{-12} s	1 picosecond (ps)	Lifetime of charmed and beauty hadrons
		End of the electroweak transition in the Big-Bang model
10^{-23} s		Lifetime of the W, Z, H, top quark
10^{-33} s		Time after the Big Bang for the end of the grand-unification phase and likely appearance of the matter–antimatter asymmetry in the Universe
10^{-44} s	Planck time	Quantum gravity, limit of applicability of known physics, of separation between space and time
1 eV	1 electron-volt	Binding energy of an electron in the outer shells of an atom (1.6×10^{-19} joules)
10^{6} eV	1 mega-electron-volt (MeV)	Binding energy of a nucleon in a nucleus (typically 10 MeV)
		Electron mass ($m_e = 0.5\,\text{MeV} = 9.1\times10^{-28}\,\text{g}$)

10^9 eV	1 giga-electron-volt (GeV)	Neutron, proton mass ($m_p = 0.938\,\text{GeV} = 1.67 \times 10^{-24}\,\text{g}$)
10^{11} eV	100 GeV	Electroweak scale
		Mass of W, Z, H, top quark
10^{12} eV	1 tera-electron-volt (TeV)	Energy of the LHC proton beams (7 TeV)
10^{15} GeV		Grand-unification scale (hypothetical)
10^{19} GeV		Planck mass or energy
10^{28} g		Mass of the Earth (6×10^{27} g)
10^{33} g		Mass of the Sun (2×10^{33} g)

Bibliography

Lectures in Physics

- *The Feynman Lectures on Physics*, R.P. Feynman, R.B. Leighton and M. Sands, New Millennium Edition, Basic Books, 2011.

The Standard Model

Popularisation

- *The Particle Hunters*, Y. Neeman and Y. Kirsh, Cambridge University Press, 2nd edition, 1996.
- *Collider: The Search for the World's Smallest Particles*, P. Halpern, John Wiley & Sons, 2010.
- *Quantum Field Theory for the Gifted Amateur*, T. Lancaster and S.J. Blundell, Oxford University Press, illustrated edition, 2014.

More advanced

- *Introduction to High Energy Physics*, D.H. Perkins, Cambridge University Press, 4th edition, 2000.
- *Leptons and Quarks*, L.B. Okun, World Scientific, 2014.
- *Introduction to Elementary Particles*, D. Griffiths, Wiley-VCH, 2nd edition, 2008.
- *Particle Physics*, B.R. Martin and G. Shaw, John Wiley & Sons, 3rd edition, 2008.

- *Gauge Theories in Particle Physics*, I.J.R. Aitchinson and A.J.G. Hey, CRC Press, 4th edition, 2012.
- *Gauge Theories of the Strong, Weak, and Electromagnetic Interactions*, C. Quigg, Frontiers in Physics Series, Benjamin/Cummings, 1983.
- *Quarks and Leptones: An Introductory Course in Modern Particle Physics*, F. Halzen and A.D. Martin, John Wiley & Sons, 1984.
- *The Standard Theory of Particle Physics*, L. Maiani and L. Rolandi (editors), World Scientific, 2016.
- *Modern Particle Physics*, M. Thomson, Cambridge University Press, 2013.
- *Collider Physics within the Standard Model*, G. Altarelli, Springer International Publishing, Lecture Notes in Physics, 2017.
- *Concepts of Elementary Particle Physics*, M.E. Peskin, Oxford University Press, 2019.

Discoveries of the W and Z bosons, results from LEP/SLC and the Tevatron

Popularisation

- *The God Particle: If the Universe is the Answer, What is the Question*, L. Lederman and D. Teresi, Dell Publishing, New York, 1993.
- *Prestigious Discoveries at CERN*, R. Cashmore, L. Maiani, J.P. Revol (editors), Springer-Verlag, Berlin, 2003.
- *Anomaly*, T. Dorigo, World Scientific, 2017.

Professional

- UA1 Collaboration, *Phys. Lett. B* 122 (1983) 103, *Phys. Lett. B* 126 (1983) 398.
- UA2 Collaboration, *Phys. Lett. B* 122 (1983) 476, *Phys. Lett. B* 129 (1983) 130.
- CERN — the second 25 years, J. Ellis *et al.* (editors), *Phys. Rep.* 403–404, 2004.

– *60 Years of CERN, Experiments and Discoveries*, H. Schopper and L. Di Lella (editors), World Scientific, 2015.

Big Bang, Cosmology

Popularisation

– *A Brief History of Time*, S. Hawking, Bantam Books, 10th edition, 1998.

– *The Nature of Space and Time*, Stephen Hawking and Roger Penrose, Princeton University Press, Revised edition, 2015.

– *Enquête sur l'Univers*, J. Audouze and Jean-Pierre Chièze, Nathan, 1991.

– *The First Three Minutes: A Modern View of the Origin of the Universe*, S. Weinberg, Basic Books, 2nd edition, 1993.

More advanced

– *Cosmology*, E. Harrison, Cambridge University Press, 2000.

– *Fundamentals of Cosmology*, J. Rich, Springer-Verlag, Berlin, 2001.

– *The Early Universe* (Frontiers in Physics), E.W. Kolb and M.S. Turner, CRC Press, 1994.

– Cosmic microwave background anisotropies, W. Hu and S. Dodelson, *Annu. Rev. Astron. Astrophys.* 40 (2002) 171.

– N. Suzuki *et al.*, *Astrophys. J.*, 746 (2012) 85.

– T. Delubac *et al.*, *Astron. Astrophys.* 574 (2015) A59.

– Cosmological constraints from baryonic acoustic oscillation measurements, J.M. Le Goff and V. Ruhlmann-Kleider (2015), *Scholarpedia* 10 (2015) no. 9, 32149.

– Planck collaboration, *Astron. Astrophys.*, 571 (2014) A16, *Astron. Astrophys.* 594 (2016) A13.

– Reheating in inflationary cosmology: theory and applications, R. Allahverdi, R. Brandenberger, F. Cyr-Racine, A. Mazumdar, *Ann. Rev. Nucl. Part. Sci.* 60 (2010) 27.

Accelerators, the LHC

Popularisation

- *The Large Hadron Collider*, D. Lincoln, The Johns Hopkins University Press, Baltimore 2009.
- *LHC: Large Hadron Collider*, P. Ginter *et al.*, Lammerhuber, edition, Lammerhuber, 2013.
- *ATLAS: A 25-Year Insider Story of the LHC Experiment*, The ATLAS Collaboration, World Scientific, 2019.

More advanced/professional

- *An Introduction to Particle Accelerators*, E. Wilson, Oxford University Press 2001.
- *Design Study of the Large Hadron Collider*, CERN 91-03, May 1991.
- *LHC, Conceptual Design*, CERN/AC/95-05, October 1995.
- The Large Hadron Collider project: historical account, G. Brianti, *Phys. Rep.* 403–404 (2004) 349.
- *The Large Hadron Collider: a Marvel of Technology*, L. Evans (editor), EPFL Press, 2014.
- *The High Luminosity Large Hadron Collider*, O. Bruning and L. Rossi (editors), World Scientific, 2015.
- *Particle Accelerators: From Big Bang Physics to Hadron Therapy*, U. Amaldi, Springer International Publishing, Switzerland, 2015.
- *Challenges and Goals for Accelerators in the XXIst Century*, O. Bruning and S. Myers (editors), World Scientific, 2016.

Instrumentation and methods

- *Techniques for Nuclear and Particle Physics Experiments: A How-to Approach*, W.R. Leo, Springer-Verlag, 2nd edition, 1994.
- *The Experimental Foundations of Particle Physics*, R. Cahn and G. Goldhaber, Cambridge University Press, 2009.

- *Technology Meets Research, 60 Years of CERN Technology: Selected Highlights*, Advanced Series on Directions in High Energy Physics: Volume 27, C. Fabjan, T. Taylor, D. Treille and H. Wenninger (editors), World Scientific, 2017.
- *Statistical Methods for Data Analysis in Particle Physics*, L. Lista, Springer-Verlag, 2nd edition, 2017.

LHC experiments

- ALICE, Technical Proposal, CERN-LHCC-95-71, 1995.
- ATLAS, Letter of Intent, CERN-LHCC-92-04, 1992.
- ATLAS, Technical Proposal, CERN-LHCC-94-43, 1994.
- CMS, Letter of Intent, CERN-LHCC-92-03, 1992.
- CMS, Technical Proposal, CERN-LHCC-94-38, 1994.
- LHCb, Technical Proposal, CERN-LHCC-98-004, 1998.
- LHCf, Technical Proposal, CERN-LHCC-2005-032, 2005.
- MoEDAL, Technical Design Report, CERN-LHCC-2009-006, 2009.
- TOTEM, Technical Proposal, CERN-LHCC-99-007, 1999.
- CMS-TOTEM Precision Proton Spectrometer, Technical Design Report, CERN-LHCC 2014-021, 2014.

Physics at LHC

- Large Hadron Collider Workshop, Aachen 4–9 October 1990, Volumes I, II and III, CERN-90-10, ECFA-90-133, 1990.
- ATLAS detector and physics performance, Technical Design Report, CERN-LHCC-99-014, CERN-LHCC-99-015, 1999.
- CMS Physics, Technical Design Report, Volumes I and II, CERN-LHCC-2006-001, CERN-LHCC-2006-021, 2006.
- *Perspectives on LHC Physics*, G. Kane and A. Pierce (editors), World Scientific, 2009.

- *A Zeptospace Odyssey: A Journey into the Physics of the LHC*, G. Giudice, Oxford University Press, 2010.
- Introduction to the physics of the total cross-section at the LHC, G. Pancheri and Y. Shrivastava, *Eur. Phys. J. C* 77 (2017) 3.

Dark Matter, Black Holes

- Particle Dark Matter: evidence, candidates and constraints, G. Bertone, D. Hooper, J. Silk, *Phys. Rep.* 405 (2005) 279.
- Study of potentially dangerous events during heavy-ion collisions at the LHC: report of the LHC safety study group, J.-P. Blaizot *et al.*, CERN 2003-001, 2003.
- Review of the Safety of LHC Collisions, J. Ellis *et al.*, *J. Phys. G* 35 (2008) 115004.
- *Particle Dark Matter — Observations, Models and Searches*, G. Bertone (editor), Cambridge University Press, 2010.

Search and discovery of the Higgs boson

Popularisation

- *Le LHC détectera le boson de Higgs . . . s'il existe*, F. Englert, La Recherche, page 58, 2008.
- *Higgs — the Invention and Discovery of the "God Particle"*, J. Baggott, Oxford University Press, 2012.
- *The Discovery of the Higgs Boson at the LHC*, P. Jenni and T.S. Virdee, chapter 1 in 60 Years of CERN Experiments and Discoveries, Advanced Series on Directions in High Energy Physics: Volume 23, H. Schopper, L. Di Lella (editors), World Scientific, 2015.
- *A la recherche du boson de Higgs*, C. Grojean, L. Vacavant, Librio, 2013.
- *A caccia del bosone di Higgs*, Luciano Maiani, Romeo Bassoli, edizione Mondadori Universita, Milano, 2013.
- *Higgs Discovery: The Power of Empty Space*, L. Randall, Ecco, 2013.

– *Le boson et le chapeau mexicain*, G. Cohen-Tannoudji, M. Spiro, Gallimard, 2013.

– *The Particle at the End of the Universe: How the Hunt for the Higgs Boson Leads Us to the Edge of a New World*, S. Carroll, Dutton, 2012.

Professional

– F. Englert and R. Brout, *Phys. Rev. Lett.* 9 (1964) 321.

– P. Higgs, *Phys. Rev. Lett.* 16 (1964) 508.

– Observation of a new particle in the search for the Standard Model Higgs boson with the ATLAS detector at the LHC, ATLAS Collaboration, *Phys. Lett. B* 716 (2012) 1.

– Observation of a new boson at a mass of 125 GeV with the CMS experiment at the LHC, CMS Collaboration, *Phys. Lett. B* 716 (2012) 30.

– *Discovery of the Higgs Boson*, A. Nisati and V. Sharma (editors), World Scientific, 2016.

– Handbook of LHC Higgs cross sections: Deciphering the nature of the Higgs sector, report of the LHC Higgs Cross Section Working Group, CERN Yellow Reports, CERN-2017-002-M, 2017.

HL-LHC and HE-LHC

– Letter of Intent for the Phase-II Upgrade of the ATLAS Experiment, CERN-LHCC-2012-022, 2012.

– ATLAS Phase-II Upgrade Scoping Document, CERN-LHCC-2015-020, 2015.

– Technical Proposal for the Phase-II Upgrade of the CMS Detector, CERN-LHCC-2015-010, 2015.

– Report on the Physics at the HL-LHC, and Perspectives for the HE-LHC, CERN Yellow Reports, CERN-2019-007, arXiv:1902.10229, 2019.

Projects beyond the LHC

- 2020 Update of the European Strategy for Particle Physics, CERN-ESU-015-2020, 2020.

ILC and CLIC

- The International Linear Collider Technical Design Report (in 4 volumes), T. Behnke *et al.*, ILC-REPORT-2013-040, 2013.
- The Compact Linear Collider (CLIC) — 2018 Summary Report, CERN-2018-005-M, 2018.
- The Compact Linear Collider (CLIC) — Project Implementation Plan, CERN-2018-010-M, 2018.
- The Compact Linear e^+e^- Collider (CLIC): Physics Potential, CLICdp-Note-2018-010, 2018.

FCC-ee and CEPC

- First Look at the Physics Case of TLEP, M. Bicer *et al.*, *JHEP* 1401 (2014) 164.
- FCC Physics Opportunities, Future Circular Collider Conceptual Design Report Volume 1, *Eur. Phys. J. C* 79 (1019) 474.
- FCC-ee: The Lepton Collider, Future Circular Collider Conceptual Design Report Volume 2, *Eur. Phys. J. Spec. Top.* 228 (2019) 261.
- CEPC Conceptual Design Report: Volume 1 — Accelerator, IHEP-CEPC-DR-2018-01, 2018.
- CEPC Conceptual Design Report: Volume 2 — Physics & Detector, IHEP-CEPC-DR-2018-02, 2018.

FCC-hh

- Physics at the FCC-hh, a 100 TeV pp collider, CERN Yellow Report, CERN-2017-003-M, 2017.
- FCC-hh: The Hadron Collider, Future Circular Collider Conceptual Design Report Volume 3, *Eur. Phys. J. Spec. Top.* 228 (2019) 755.

LHeC

- A Large Hadron Electron Collider at CERN: Report on the Physics and Design Concepts for Machine and Detector, J.L. Abelleira Fernandez *et al.*, *J. Phys. G* 39 (2012) 075001.
- The Large Hadron-Electron Collider at the HL-LHC, P. Agostini *et al.*, CERN-ACC-Note-2020-0002, 2020.

Muon colliders

- Muon Colliders, J.P. Delahaye *et al.*, arXiv:1901.06150, 2019.
- Further searches of the Higgs scalar sector at the ESS, C. Rubbia, arXiv:1908.05664, 2019.
- On The Feasibility of a Pulsed 14 TeV c.m.e. Muon Collider in the LHC Tunnel, D. Neuffer and V. Shiltsev, FERMILAB-PUB-18-508-APC, arXiv:1811.1069, 2018.

Index

Lightning Source UK Ltd.
Milton Keynes UK
UKHW020616191121
394242UK00002B/34

9 789813 236080